Spiro Compounds

Spiro Compounds

Synthesis and Applications

Edited by

Ramon Rios Torres

University of Southampton
Southampton, United Kingdom

WILEY

This edition first published 2022

Registered Office
John Wiley & Sons, Inc., 111 River Street, Hoboken, NJ 07030, USA

Editorial Office
111 River Street, Hoboken, NJ 07030, USA

For details of our global editorial offices, customer services, and more information about Wiley products visit us at www.wiley.com.

Wiley also publishes its books in a variety of electronic formats and by print-on-demand. Some content that appears in standard print versions of this book may not be available in other formats.

Library of Congress Cataloging-in-Publication Data

Names: Riós Torres, Ramón, author.
Title: Spiro compounds : synthesis and applications / Ramon Rios Torres, University of Southampton, Southhampton, United Kingdom.
Description: First edition. | Hoboken, NJ : Wiley, 2022. | Includes index.
Identifiers: LCCN 2021031988 (print) | LCCN 2021031989 (ebook) | ISBN 9781119567639 (hardback) | ISBN 9781119567622 (adobe pdf) | ISBN 9781119567653 (epub)
Subjects: LCSH: Organic compounds–Structure.
Classification: LCC QD255.4 .R56 2022 (print) | LCC QD255.4 (ebook) | DDC 547/.5–dc23
LC record available at https://lccn.loc.gov/2021031988
LC ebook record available at https://lccn.loc.gov/2021031989

Cover Design: Wiley
Cover Image: © Connect world/Shutterstock

Set in 9.5/12.5pt STIXTwoText by Straive, Pondicherry, India
10 9 8 7 6 5 4 3 2 1

MIX
Paper from responsible sources
FSC www.fsc.org FSC® C013604

Contents

List of Contributors

Vikas R. Aswar
Department of Chemistry, Indian Institute of Technology Bombay, Mumbai, India

Lu Bai
Key Laboratory of Synthetic and Natural Functional Molecule of the Ministry of Education, College of Chemistry & Materials Science, Northwest University, Xi'an, China

Matthias Baud
School of Chemistry and Institute for Life Sciences, University of Southampton, Southampton, UK

Hak-Fun Chow
Department of Chemistry, Institute of Molecular Functional Materials and The Center of Novel Functional Molecules, The Chinese University of Hong Kong, Shatin, Hong Kong

Xavier Companyó
Department of Chemical Sciences, University of Padova, Padova, PD, Italy

Xiao-Yuan Cui
School of Chemistry and Molecular Engineering, East China Normal University, Shanghai, China

Luca Dell'Amico
Department of Chemical Sciences, University of Padova, Padova, PD, Italy

Sambasivarao Kotha
Department of Chemistry, Indian Institute of Technology Bombay, Mumbai, India

Dietmar Kuck
Department of Chemistry and Center for Molecular Materials (CM2), Bielefeld University, Bielefeld, Germany

Jingjing Liu
Key Laboratory of Synthetic and Natural Functional Molecule of the Ministry of Education, College of Chemistry & Materials Science, Northwest University, Xi'an, China

Xinjun Luan
Key Laboratory of Synthetic and Natural Functional Molecule of the Ministry of Education, College of Chemistry & Materials Science, Northwest University, Xi'an, China

Marta Meazza
School of Chemistry, University of Southampton, Southampton, UK

Albert Moyano
Secció de Química Orgànica, Departament de Química Inorgànica i Orgànica, Facultat de Química, Universitat de Barcelona, Barcelona, Catalonia, Spain

Suva Paria
Department of Chemical Sciences, University of Padova, Padova, PD, Italy

Ramon Rios
Department of Organic Chemistry, University of Southampton, Southampton, UK

Yellaiah Tangella
Department of Chemistry, Indian Institute of Technology Bombay, Mumbai, India

Alberto Vega-Peñaloza
Department of Chemical Sciences, University of Padova, Padova, PD, Italy

Jan Veselý
Department of Organic Chemistry, Faculty of Science, Charles University, Prague, Czech Republic

Michal Urban
Department of Organic Chemistry, Faculty of Science, Charles University, Prague, Czech Republic

Peng-Wei Xu
School of Chemistry and Molecular Engineering, East China Normal University, Shanghai, China

Jin-Sheng Yu
School of Chemistry and Molecular Engineering, East China Normal University, Shanghai, China

Jian Zhou
School of Chemistry and Molecular Engineering, East China Normal University, Shanghai, China

Preface

When I was a kid, I dreamt to be an engineer and build planes, cars, etc. However, in high school, I felt in love with Organic Chemistry. I changed my dreams of building things for how to assemble C, H, N, and other atoms to form molecules. I was always astonished of the inherent difficulty of **organic** structures, not only for their diversity and ordered assembly but also for their 3D shape. With time, I enjoyed learning Organic Chemistry: starting from my bachelor in organic chemistry, followed by a PhD in organometallic chemistry, and several postdocs in organometallic chemistry and organocatalysis. Later on, as academic, I have always tried to develop new reactions that can allow to obtain difficult structures in an enantiopure form. However, one of the greatest challenges that I faced has been the synthesis of spiro compounds. My own way to be interested in the synthesis of spiro compounds first started in the synthesis of quaternary carbons in an enantiopure fashion. Precisely, while studying how to synthesize quaternary carbons, I developed an interest in spiro compounds, how difficult it will be to join two cycles by a single atom, and do it enantioselectively. In my research group, we have been lucky enough to develop several reactions that led to spiro compounds in an enantiopure form (or almost enantiopure), but I still remember the first time that we got a spiro compound in an enantiopure form as an incredible feeling of achievement. Now, several years later, I decided to honor this type of compounds by editing a book, summarizing the achievements that Organic Chemists have been done in the last decades.

I also want to dedicate this book to Professor Dieter Enders who left us during the writing of this book. He was always a source of inspiration, since his early success with RAMP and SAMP chemistry, until the development of highly complex domino reactions, showing a commitment and brilliance to organic chemistry that inspired me in my career. I still remember his kind hospitality in The Domino cat symposium in Aachen. Professor Enders, you left a huge footprint in organic and synthetic chemistry.

Finally, I want to thanks all the authors for their work and commitment in those difficult COVID times.

1

Spiro Compounds: A Brief History

Marta Meazza

School of Chemistry, University of Southampton, Southampton, UK

Policyclic molecules containing at least two rings joined together by a single atom, mostly a carbon atom, previously named spiranes, are called spiro compounds or spirocycles, and the single central atom is referred to as the spiro atom [1]. We should mention that apart from carbon, other elements such as nitrogen, phosphorus, and arsenic may represent the spiro atom.

The term was coined by the German chemist Nobel laureate Adolf von Baeyer who created the first spirane in 1900 [2].

This peculiar structural feature is present in natural products and has long been the subject of methodological studies and synthetic efforts [3].

Several synthetic procedures for spiro compounds have been developed and will be extensively discussed in the next chapters. However, the asymmetric synthesis of spirocycles that allow the creation of stereogenic quaternary centers represent a demanding task for organic chemists. Even the concepts of spiro aromaticity and spiro antiaromaticity can be applied when spiroconjugation is possible [4].

The search for the key term "spiro" in SciFinder^{n} database, at the end of October 2019, resulted in more than 40 700 references with an exponential growth starting from the middle of the last century and an increasing attention to this subject is expected in the future (Figure 1.1).

These massive research efforts cover a wide range of fields from organic and medicinal chemistry to material sciences and engineering, to name a few.

The enormous interest in spiro compounds rely on their distinctive properties often associated with the three-dimensional stereochemical features, reflecting on their pharmacological properties that include, among others, bactericidal, fungicidal, anticancer, cytotoxic, antidepressant, antihypertensive, insecticidal, herbicidal, and plant growth regulatory effects [5]. These properties are due to the tetrahedral nature of the spiro carbon and consequent asymmetric features associated with it.

Spiro Compounds: Synthesis and Applications, First Edition. Edited by Ramon Rios Torres.

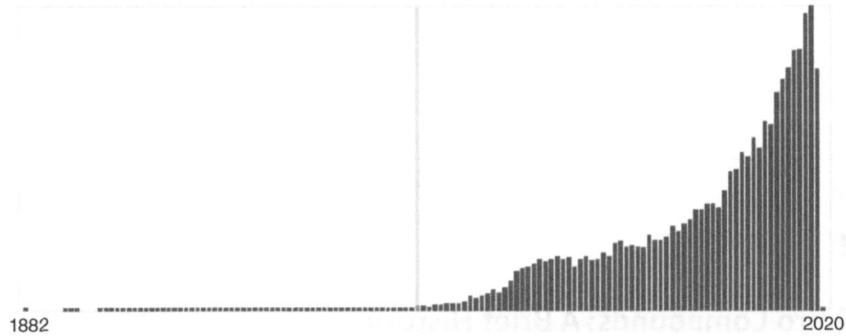

Figure 1.1 Growing interest in spiro compounds in chemical literature.

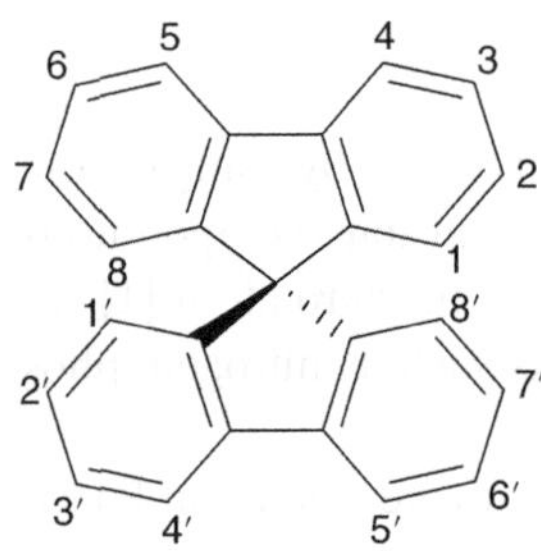

1

Figure 1.2 Dye sensitizer 9,9-spirobifluorene. *Source:* Lupo et al. [7].

In addition, many other practical utilizations include optoelectronic devices, ophthalmic lenses, and solar cells [6]. Compounds like 9,9-spirobifluorene **1** (Figure 1.2) have application in dye-sensitized solar cells (DSCs) and represent the most efficient alternative to the current solar cell technologies [7].

Spirocyclic compounds find technological application as efficient charge-transfer molecules due to their intramolecular donor–acceptor structural feature amplified by spiroconjugation. The desired optical properties can be achieved by careful design of the spiro donor–acceptor characteristic as illustrated in Figure 1.3 [8]. When structural characteristics make it possible, spiro compounds can equilibrate with their non-spiro analogues exhibiting photochemical phenomena like photochemical memory.

We report here some examples of carbocyclic and heterocyclic naturally occurring compounds containing the spiro moiety (Figure 1.4). One of the simplest compounds is the pheromone of the olive fly *Dacus oleae* **5**. Phelligridin G **6** from the fungus *Phellinus igniarius* has been long used in Traditional Chinese Medicine for the treatment of gonorrhea [9]. The antimycotic drug griseofulvin **7**, isolated from a penicillium mold in 1939, found application in the treatment of fungal skin

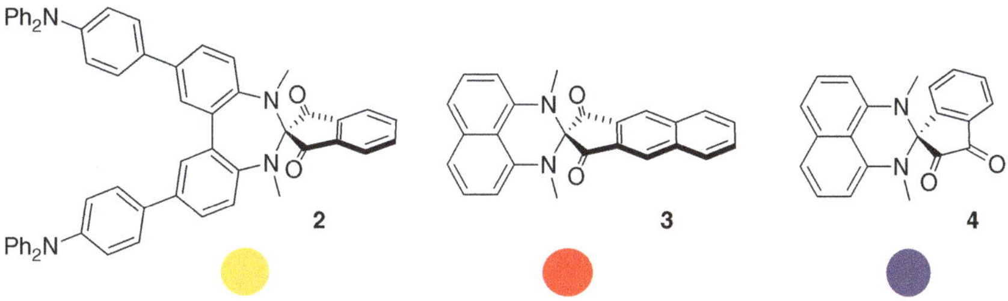

Figure 1.3 Donor–acceptor spiro compounds and colors displayed by them. *Source:* Wössner et al. [8].

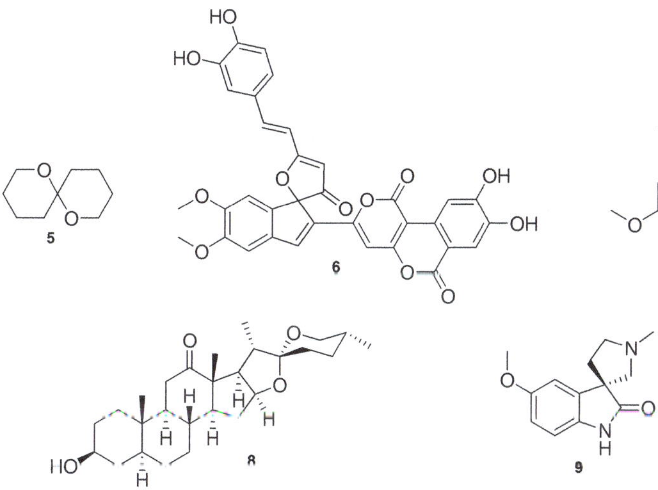

Figure 1.4 Examples of naturally occurring compounds containing the spiro moiety.

infections since 1957. Hecogenin **8**, the aglycone part of a steroid saponin found in the plant Agave sisalana, is responsible for many therapeutic effects and is also used as a starting material in the synthesis of corticosteroids [10]. Horsfiline **9** is an oxindole alkaloid having analgesic effect, isolated from the plant *Horsfieldia superba* [11].

A classic example of the importance of the presence of a spiro functionality is the retention of the biological activity of perhydrohistrionicotoxin **10**, the completely reduced analogue of the potent nicotinic receptor antagonist alkaloid (−)-histrionicotoxin **11**, isolated from "dart-poison" frogs, that clearly suggests the fundamental role of the spiropiperidine moiety in determining a strong receptor binding. The massive synthetic efforts on this topic are collected in a book chapter [12] (Figure 1.5).

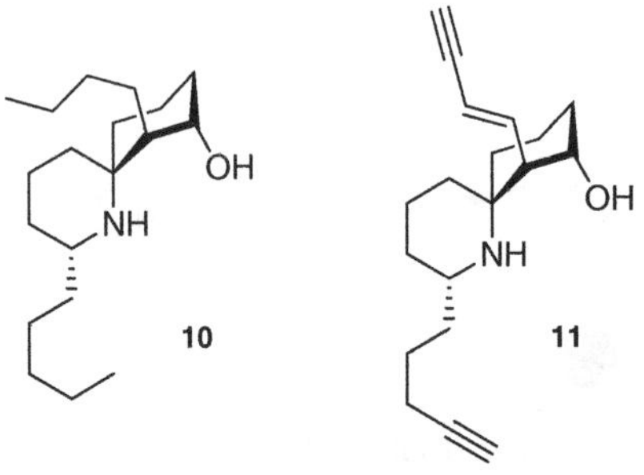

Figure 1.5 Spiro functionality in nicotinic receptor antagonists. *Source:* Hart [12].

As stated before, spirocycles are present in successfully developed medications and represent attractive synthetic targets included in chemical libraries for diversity-oriented synthesis within drug discovery projects. In this context, the spiro moiety has been and can be employed both as core structure and as an activity modulator, appended to decorate the peripheral part of the molecule [13].

The major advantage of spirocycles in biological applications as core structure or pharmacophores originates from their 3-D nature and the associated conformational features that allow for a better ability to interact with the target protein enzyme. The tetrahedral feature of the spiro atom renders the two ring planes nearly perpendicular to each other with a limited number of potential conformations. When added in the periphery of the molecule, the spirocycle acts as a modulator of physicochemical properties such as log P and water solubility, as well as affecting the metabolic stability of the molecule. Not least, from an intellectual property perspective, the introduction of spirocycles offers the possibility of obtaining a free patent space in a me-too research approach.

Prominent examples of marketed spirocompounds, illustrating these concepts, include fluspirilene **12**, spiraprilat **13,** and cevimeline **14**, while experimental compounds in different stages of clinical development are ETX0914 **15,** a DNA gyrase inhibitor; tofoglifozin CSG452 **16,** an inhibitor of hSGLT2 for the treatment of Type 2 diabetes; AZD1979 **17,** an antagonist of melanin-concentrating hormone receptor; and rolapitant **18,** a neurokinine 1 receptor antagonist [13, 14] (Figure 1.6).

We wish once more to draw the attention of the readers on the potential usefulness and uniqueness of the spiro motif in the interaction with a specific biological target spanning from drugs to agrochemicals.

The enzyme Acetyl-coenzyme A carboxylases (ACCs) have crucial roles in fatty acid metabolism in most living organisms, among which include humans, insects, and plants. The experimental ACC inhibitor compounds for the treatment of human metabolic disease contain a spirocyclic moiety as in Takeda compound **19** [15] and in Pfizer PF-05221304 **20**. The last one is currently in phase II clinical trials for the treatment of Non-Alcoholic Steatohepatitis (NASH) [16] (Figure 1.7).

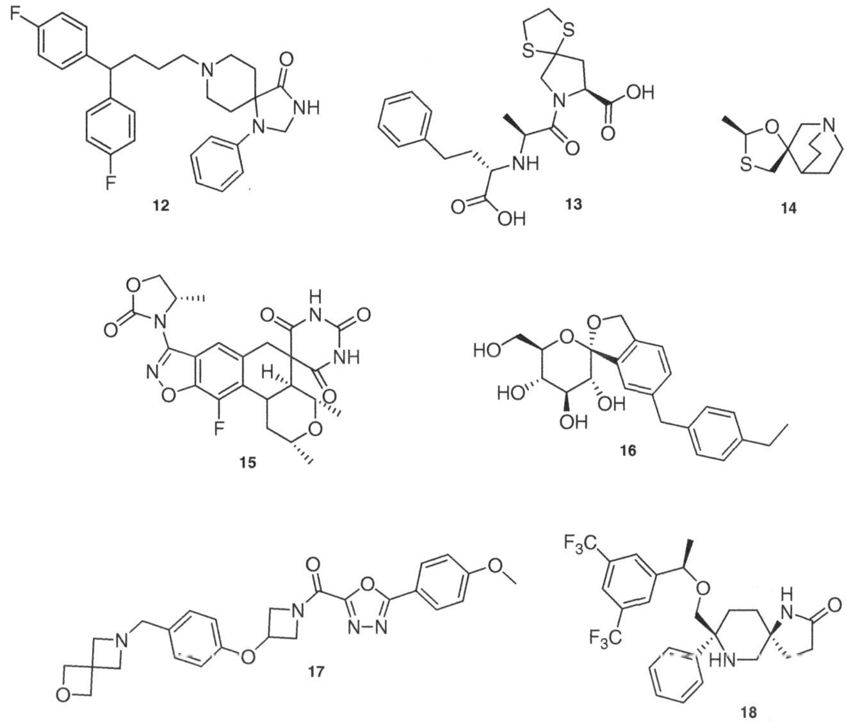

Figure 1.6 Examples of marketed spiro compound drugs. *Sources:* Based on Zheng and Tice [13]; Zheng et al. [14].

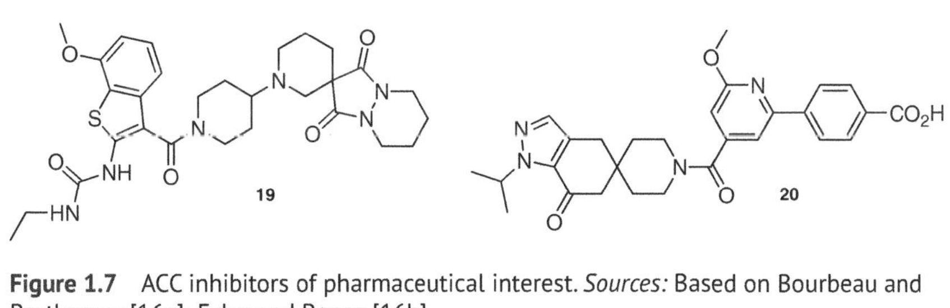

Figure 1.7 ACC inhibitors of pharmaceutical interest. *Sources:* Based on Bourbeau and Bartberger [16a]; Esler and Bence [16b].

The commercial insecticide/acaricide products spirotetramat **21**, spiromesifen **22**, and spirodiclofen **23** from Bayer CS and spiropidion **24** from Syngenta, acting as insect ACC inhibitors, all have spirocyclic structures [17] (Figure 1.8).

New spirocyclic herbicide compounds with the representative formula **25** have been recently patented [18]. It is noteworthy that compounds **21** and **24**, sharing similar molecular features with **25**, do not show any phytotoxic effect (Figure 1.9).

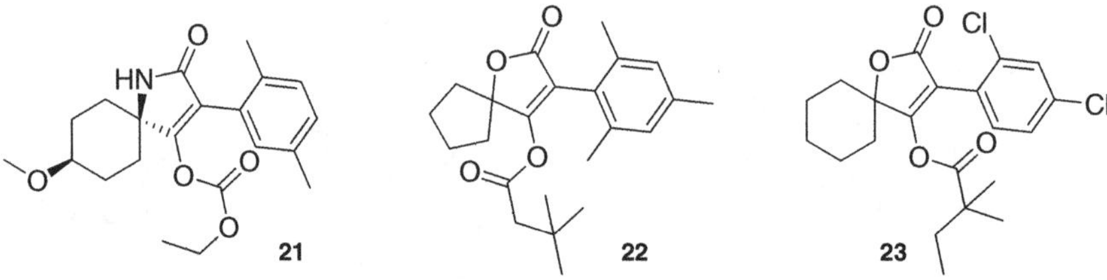

21 22 23

Figure 1.8 Commercial spirocyclic insecticide/acaricide products. *Source:* Jeschke et al. [17].

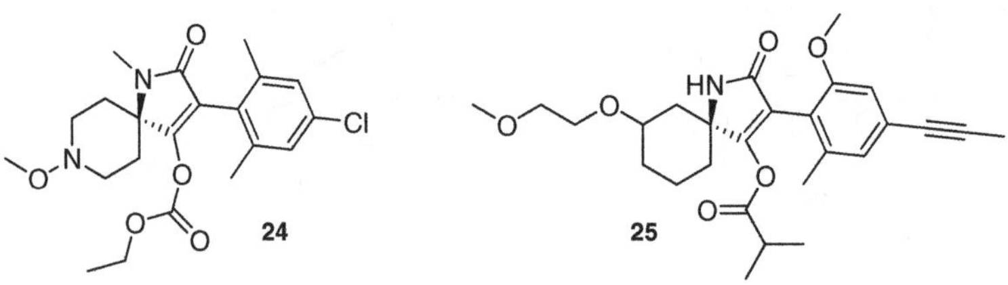

24 25

Figure 1.9 Recently patented spiro compound of agrochemical interest.

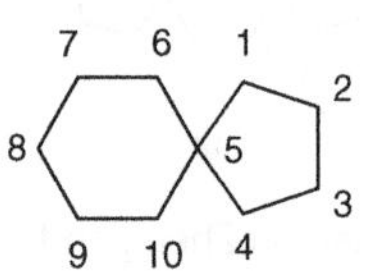

26

Figure 1.10 Example of numbering of spirocyclic compounds.

As presented in this chapter, spirocyclic scaffolds find application in a large number of sectors for their own peculiar architecture characteristics, displaying valuable application properties, or simply because of the introduction of structural novelty that guarantee patentability and intellectual property rights.

1.1 Notes on IUPAC Rules for Spiro Compounds

Naming spirocycles could be quite complex. The accepted rules are collected in the IUPAC blue book [1, 19].

Simplifying with two examples, the structure **26** is numbered starting from the smallest cycle (Figure 1.10). The name comes from the prefix spiro followed by square brackets containing the number of atoms of the two cycles starting from the smallest and excluding the spirocenter. In this case, the functional group is an alkane so that the name became spiro[4,5]decane.

(*R*)-1,7-dioxaspiro[5.5]undecane   (*S*)-1,7-dioxaspiro[5.5]undecane

Figure 1.11 Example of naming chiral spiro compounds.

When the compound is chiral because it contains a chiral center, the CIP rules are followed. In the case in which the substituents on the spirocenter are the same, but the structures display an axial chirality as in Figure 1.11, we assign arbitrarily the priority to one of the cycles and then, within each cycle the order follows the CIP rules: a>a′>b>b′.

References

1 Moss, G. (**1999**). *Pure Appl. Chem.* 71: 531–558.
2 Baeyer, A.V. (**1900**). *Ber. Dtsch. Chem. Ges.* 33: 3771–3775.
3 (a) Singh, G.S. and Desta, Z.Y. (**2012**). *Chem. Rev.* 112: 6104–6155. (b) Smith, L.K. and Baxendale, I.R. (**2015**). *Org. Biomol. Chem.* 13: 9907–9933.
4 Fleming, I. (**2010**). Molecular Orbitals and Organic Chemical Reactions, Reference Edition. Chichester, UK: Wiley & Sons, Ltd.
5 Molvi, K.I., Haque, N., Awen, B.Z.S., and Zameeruddin, M. (**2014**). *World J. Pharm. Sci.* 3: 536–563.
6 Saragi, T.P.I., Spehr, T., Siebert, A. et al. (**2007**). *Chem. Rev.* 107: 1011–1065.
7 Lupo, D., Salbeck, J., Schenk, H. et al. (**1998**). Spiro compounds and their use as electroluminescence materials. Patent: US5840217, 24 November 1998.
8 Wössner, J.S., Grenz, D.C., Kratzert, D., and Esser, B. (**2019**). *Org. Chem. Front.* 6: 3649–3656.
9 Xu, J.-P. (**2017**). Cancer Inhibitors from Chinese Natural Medicines. Boca Raton, FL: CRC Press.
10 Cruz, M.S., Barroso, S.C., Navoni, J.A. et al. (**2016**). *Toxicol. Rep.* 3: 539–543.
11 Jossang, A., Jossang, P., Hadi, H.A. et al. (**1991**). *J. Org. Chem.* 56: 6527–6530.
12 Hart, D.J. (**2011**). Organic Synthesis via Examination of Selected Natural Products. Singapore: World Scientific Publishing Co Pte Ltd.
13 Zheng, Y. and Tice, C.M. (**2016**). *Expert Opin. Drug Discovery* 11: 831–834.
14 Zheng, Y., Tice, C.M., and Singh, S.B. (**2014**). *Bioorg. Med.Chem. Lett.* 24: 3673–3682.
15 Kamata, M., Yamashit, T., Kina, A. et al. (**2012**). *Bioorg. Med. Chem. Lett.* 22: 4769–4772.

16 (a) Bourbeau, M.P. and Bartberger, M.D. (**2015**). *J. Med. Chem.* 58: 525–536. (b) Esler, W.P. and Bence, K.K. (**2019**). *Cell. Mol. Gastroenterol. Hepatol.* 8: 247–267.
17 Jeschke, P., Witschel, M., Krämer, W., and Schirmer, U. (eds.) (**2019**). Modern Crop Protection Compounds, 3e. Weinheim, Germany: Wiley-VCH Verlag GmbH.
18 Angermann, A., Bojack, G., Buscato Arsequell, E., et al. (**2019**). Specifically substituted 2-alkyl-6-alkoxyphenyl-3-pyrrolin-2-ones and their use as herbicides. Patent: WO2019219588, 13 May 2019.
19 Favre, H.A. and Powell, W.H. (**2013**). Nomenclature of Organic Chemistry: IUPAC Recommendations and Preferred Names. RSC.

2

Selected Applications of Spirocycles in Medicinal Chemistry

Matthias Baud

School of Chemistry and Institute for Life Sciences, University of Southampton, Southampton, UK

2.1 Introduction

Spiro compounds contain two rings, connected by a single sp^3 hybridized quaternary center, the "spiroatom" [1]. The latter is often a carbon, although a number of quaternary *N*-spiro ammoniums have also been reported. Trospium chloride (**1**) (Table 2.1) is a good example, and its spiro ammonium motif can be readily prepared by double *N*-alkylation of endo-nortropine [3]. Spirocyclic systems are found in a wide range of natural products [4], including spiro-ketals [5, 6], lactones [7], lactams [8, 9], and oxindoles [10–12]. An early and illustrative example of spirocyclic natural product which has attracted the attention of medicinal chemists is the antibiotic platensimycin (**2**). It is a metabolite from *Streptomyces platensis* which represents a structurally unusual example of bioactive molecule containing a carbaspirocyclic scaffold. Its antibiotic activity was reported by Merck in 2006, as part of a screening campaign to identify inhibitors of beta-ketoacyl synthases I/II (FabF/B) enzymes [13]. Inhibition of FAB enzymes by platensimycin leads to impaired biosynthesis of key fatty acids required bacterial cell membrane integrity [14]. Platensimycin displays activity against a range of Gram-positive bacteria, including strains showing resistance to other potent antibiotics such as methicillin, vancomycin, linezolid, or macrolide. Structural studies on an *Escherichia coli* FabF(C163Q) in complex with platensimycin highlighted important interactions underlying complex formation. The shape complementarity and conformational restriction provided by the spiro motif are important contributors to the potency of platensimycin, allowing polar interactions and hydrophobic contacts at the binding site entrance (Figure 2.1) [13]. The first total synthesis of racemic platensimycin was reported by Nicolaou on the same year (Scheme 2.1) [15], involving a key ruthenium-catalyzed enyne cycloisomerization [16]. Since then, stereoselective syntheses of platensimycin spirocyclic core based on rhodium-catalyzed asymmetric cycloisomerization and hypervalent iodine-mediated de-aromatizing cyclization [17], decarboxylative allylation [18], and intramolecular Diels–Alder [19] have been reported.

Spiro Compounds: Synthesis and Applications, First Edition. Edited by Ramon Rios Torres.

Table 2.1 Selected examples of FDA- approved drugs containing spirocyclic motifs.

Structure	ID	Name (Trade name)	Indication
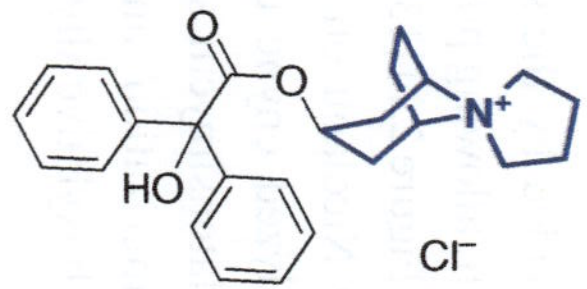	1	Trospium chloride (Flotros)	Overactive bladder
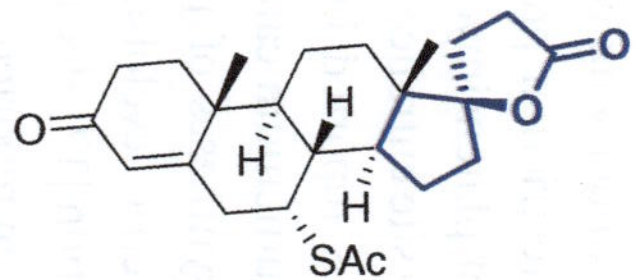	3	Spironolactone (Aldactone)	Heart failure, edema, hypertension
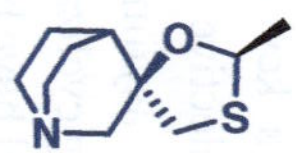	4	Cevimeline (Evoxac)	Dry mouth (Sjögren's Syndrome)
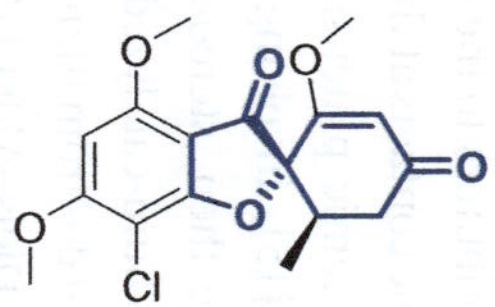	5	Griseofulvin (Crivicin)	Antifungal antibiotic for ringworm infections
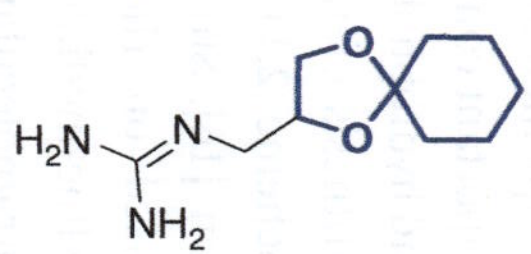	6	Guanadrel (Hylorel)	Hypertension

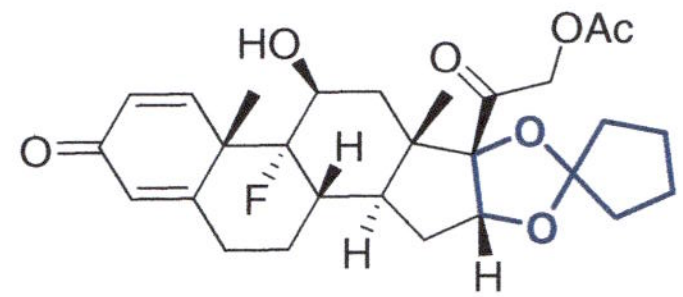	7	Amcinonide (Cyclocort)	Inflammatory and pruritic manifestations
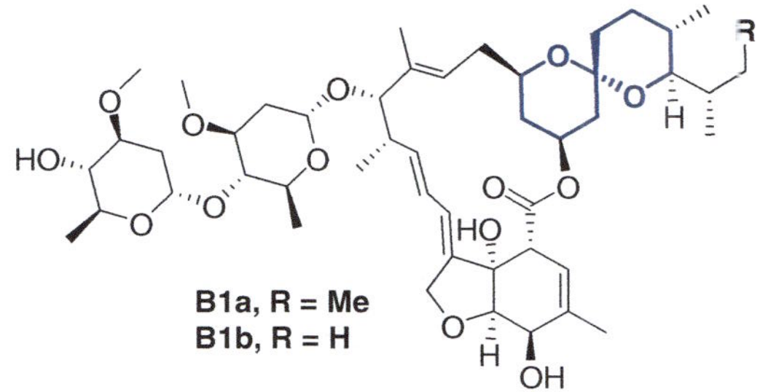	8	Ivermectin (Ascapil)	Anti-parasitic
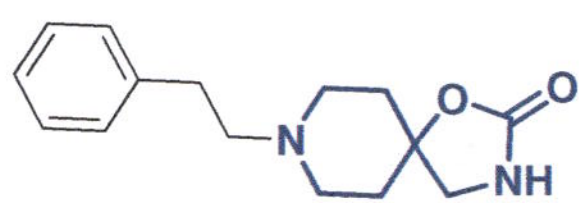	9	Fenspiride (Eurespal)	Antitussive
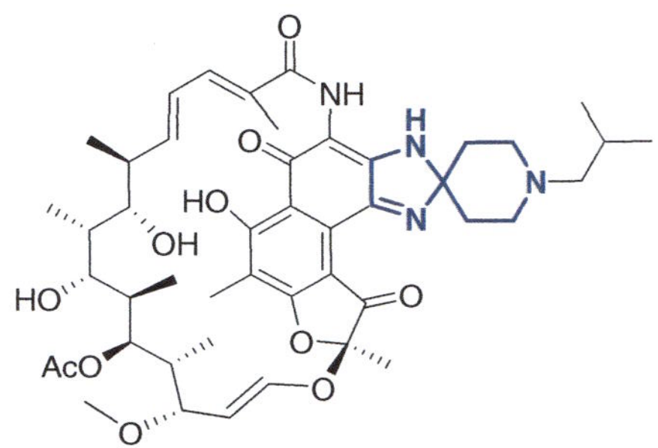	10	Rifabutin (Ansatipin)	Antibiotic, tuberculosis

(*Continued*)

Table 2.1 (Continued)

Structure	ID	Name (Trade name)	Indication
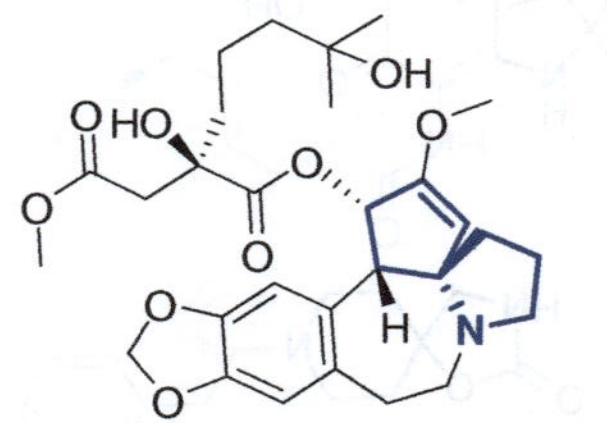	**11**	Homo-harringtonine (Ceflatonin)	Chronic myeloid leukemia
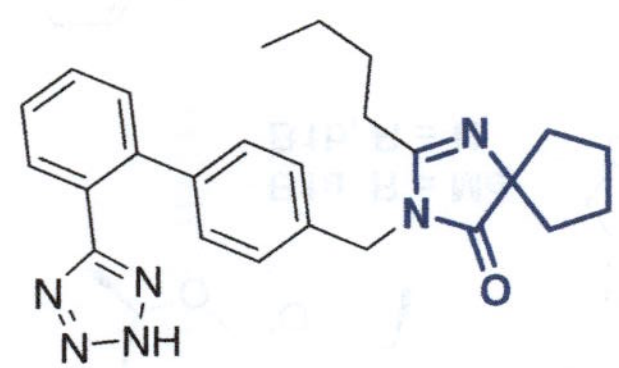	**12**	Irbesartan (Avapro)	Hypertension
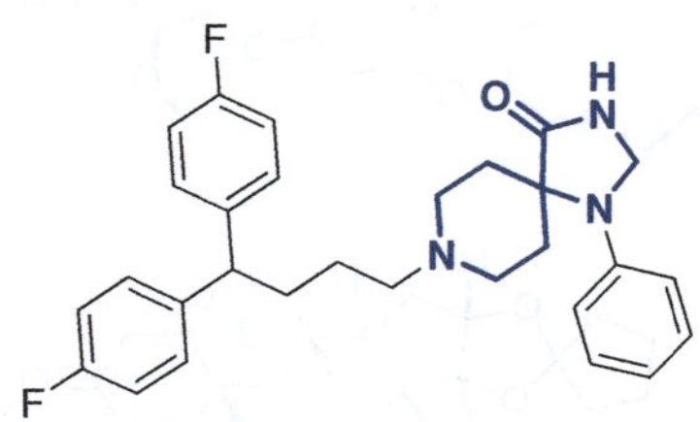	**13**	Fluspirilene (Imap)	Schizophrenia

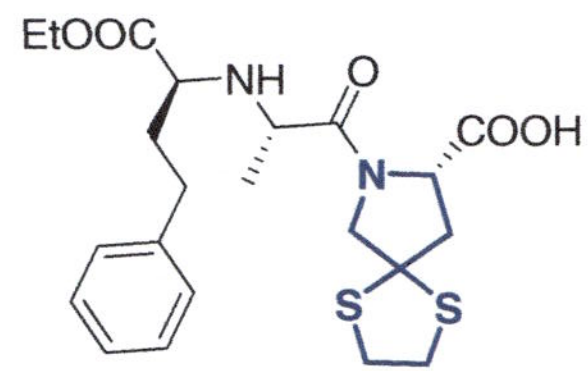	**14**	Spirapril (Rencrmax)	Hypertension
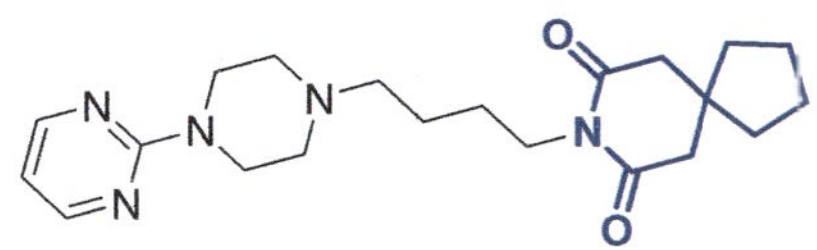	**15**	Buspirone (Buspar)	Anxiety disorders

Source: Adapted from Knox et al. [2].

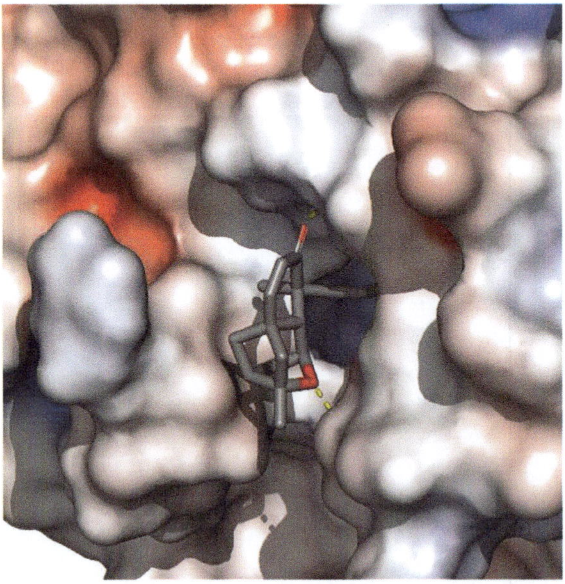

Figure 2.1 X-ray crystal structure (pdb 2gfx[12]) of Platensimycin (**2**, sticks representation) bound to the active site of FabF (surface representation) highlights hydrophobic contacts and hydrogen bonds in the complex.

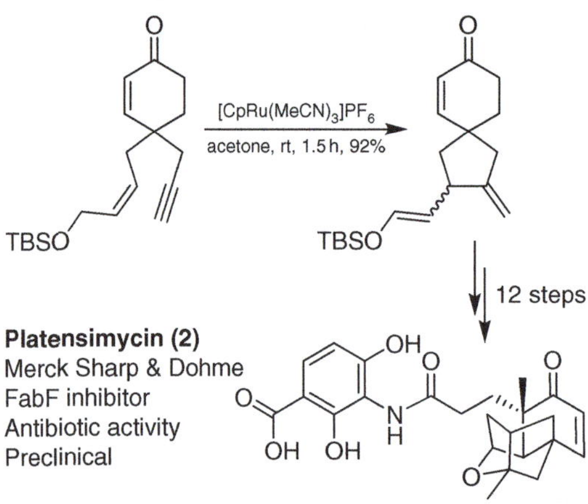

Scheme 2.1 Key enyne cycloisomerization step in Nicolaou's total synthesis of platensimycin. *Source:* Adapted from Nicolaou et al. [15].

In contrast, a comparatively small number of spirocycle containing drugs have been investigated in the last decades, and spirocyclic molecules are still under-represented in marketed drugs [2]. Selected examples of spirocycle containing drugs are shown in Table 2.1, including spironolactone **3** which has been known for over 50 years [20]. It seems fair to state that historically drug design strategies in medicinal chemistry have been heavily inspired (or biased?) by the advances in synthetic chemical methods toward new molecular scaffolds. This raises the question of chemical diversity and unconscious bias toward traditional/ubiquitous building blocks and reliable reactions/protocols for assembling increasingly complex molecules. As a compelling example, the phenyl ring and its related (hetero)aromatic analogues are over-represented and virtually all best-selling, over-the-counter drugs contain at least one such unit in their core scaffold or as a substituent. It is not particularly surprising as a plethora of synthetic methods for aromatic functionalization have been conceptually and experimentally well established for decades [21]. Such methods include, for example, a wide range of additions and substitutions, cross-couplings, and aromatization strategies. While this is obviously a significant advantage for the rapid access to a wide range of analogues from a core structure, this bias is an important factor slowing down the discovery of new bioactive scaffolds. Historically, spirocycles have been considered difficult motifs to synthesize, mainly due to the presence of a quaternary carbon center. While chirality is often found in evolutionary optimized, complex natural product-derived drugs (e.g. Griseofulvin **5**, Ivermectin **8**), early examples of synthetic spirocyclic drugs (e.g. fluspirilene **13**) are achiral. The introduction of stereogenic centers allows higher structural complexity and diversity across the three dimensions, and it has been suggested to be advantageous for the discovery of ligands targeting biological receptors, such as an enzyme active site or a protein–protein interaction (PPI) interface [22]. Equally important is the control of their configuration for achieving optimal interaction with the chiral target receptor. It is well illustrated by the contrasted potencies and/or pharmacologies of various stereoisomers of a wide range of chiral drugs [23, 24], including the infamous thalidomide [25]. Controlling the stereochemical outcome, hence chirality of spirocycles, has been a major focus of synthetic organic chemists in the last decade, and this is reflected by the exponentially growing number of articles in organic chemistry journals reporting the syntheses of increasingly complex spirocyclic systems. Overall, spirocycles have gained increasing popularity in modern drug discovery, and the recent development of robust synthetic methodologies have made them an integral part of the medicinal chemists' arsenal of pharmacophores for the development of small synthetic molecules targeting therapeutically important biological pathways. This brief chapter captures several important literature examples that illustrate some key applications of spirocycles for the development of bioactive molecules of relevance to the pharmaceutical sector. The breadth of synthetic methods available for the preparation of a wide range of spirocycles will be discussed in the subsequent chapters.

2.2 General Features

The sp^3-rich character of spirocycles and resulting extended range of exit vectors spanning across all three spatial dimensions represent an exciting opportunity for expanding molecular complexity while maintaining structural rigidity. These two parameters are also of prime importance for optimizing the thermodynamic fingerprint of chemical probes and drugs, and achieve optimal potency and selectivity profiles. At the same time, structural novelty opens valuable avenues for exploring new chemical spaces and IP protection of novel bioactive scaffolds. Spirocycles give access to more rigid scaffolds and better-defined directionality of exit vectors, particularly for spirocycles containing small rings (e.g. cyclobutanes, oxetanes, azetidines, and thietanes, Figures 2.2 and 2.3), which are conformationally highly restricted. Studies by Lovering showed a positive correlation between the simple indicator of saturation Fsp^3 = (number of sp^3 hybridized carbons/total

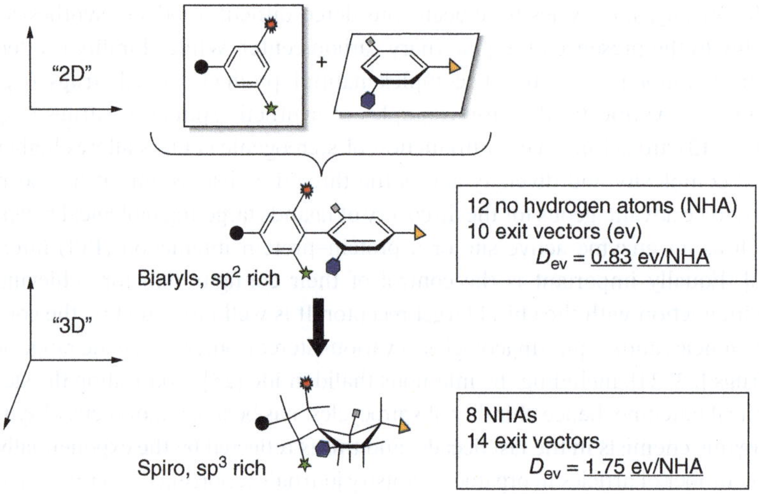

Figure 2.2 Schematic comparison of substituted biaryl and spiro[3,4]octane scaffolds: the overall conservation of the relative orientation and high density of exit vectors can be exploited for bioisosteric replacement strategies.

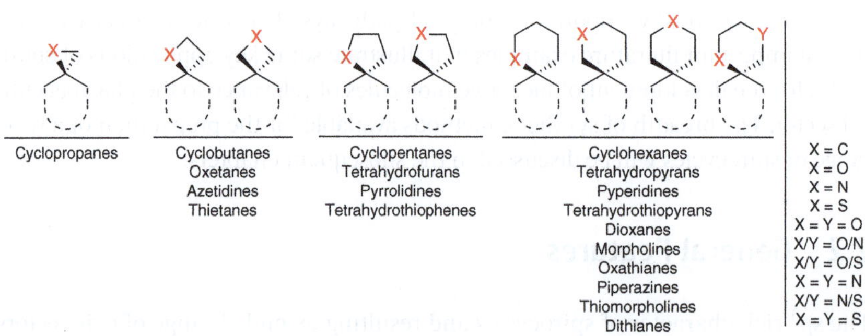

Figure 2.3 Examples of saturated spirocyclic ring systems frequently encountered in drug discovery.

carbon count) and the number of stereogenic centers as compounds progress through clinical trials, suggesting an enrichment in these features along clinical development [26]. Another direct consequence of their sp^3 richness is their higher density of exit vectors (D_{ev}, Figure 2.2) compared with sp^2-rich systems. D_{ev} can be a useful primary indicator to rapidly estimate the three-dimensionality of a scaffold and its potential for molecular complexity and diversity generation through synthesis. We use this indicator regularly in our lab as a criterion to prioritize fragment hit-to-lead optimization. This contrasts with some scaffolds that have been at the center of drug discovery over the last decades. Those include the typical poly(hetero)aromatic structures which result from linking of multiple two-dimensional units, and provide limited scope for targeting the three dimensions. As a result, the 3D character of spirocycles and greatly extended range of accessible vectorizations are regarded as powerful ways to "escape from flatland" [26, 27] through the bioisosteric replacement of "flat", sp^2-rich scaffolds found in historical fragment and lead screening libraries.

The constrained and directional exit vectors projected from "spiro" scaffolds allow fine-tuning the relative orientations of functional groups within bioactive molecules in order to achieve optimal molecular interaction with a target receptor. Such rigidity can also be exploited for the bioisosteric replacement and conformational restriction of flexible moieties in biomolecules, hence reducing detrimental entropic penalties for binding [28]. Spirocycles have also been used to modulate physicochemical properties and pharmacokinetics of biomolecules, including aqueous solubility and metabolic stability [26, 27, 29]. A number of reports have also highlighted that a high sp^2/aromatic content is generally detrimental to the physicochemical properties of drug candidates [26, 27, 30, 31].

2.3 Property Optimization in Bioactive Compounds

Spirocyclic structures displaying a wide range of biological activities have been reported [32, 33] (Figure 2.4). For example, the representative SAR405838 (**16**) belongs to a family of potent spirooxindole derivatives potently inhibiting MDM2 [34, 35]. MDM2 inhibition has attracted high attention for the development of anticancer therapeutics acting *via* reactivation of the tumor suppressor protein p53. The oxindole ring mimics the Trp23 side chain of p53, and makes key contacts with Leu54 backbone carbonyl in MDM2 *via* its NH group (protein data bank codes 1YCR and 3LBL). At the same time, the spirooxindole motif provides *conformational restriction* to the scaffold, "locking" the molecule in the optimal conformation for binding MDM2 with sub-nanomolar affinity. **16** was recently examined in phase 1 clinical trials in patients with advanced solid

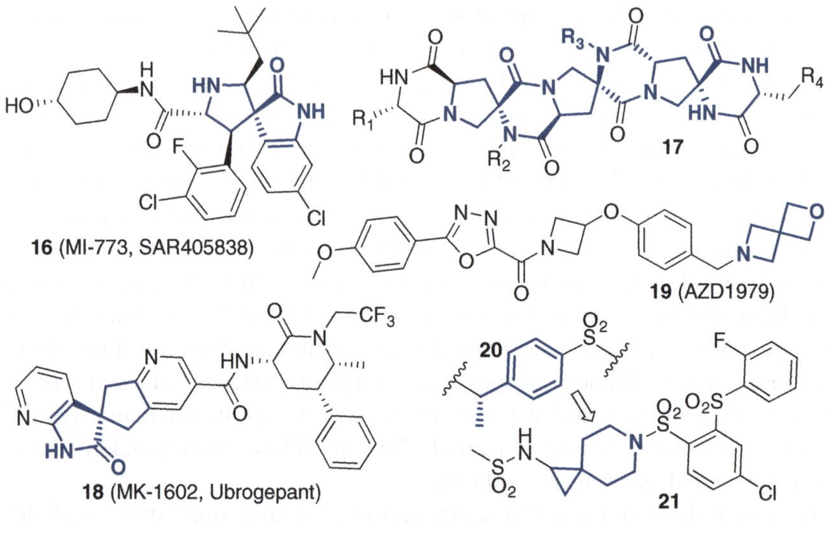

Figure 2.4 Examples of lead optimization employing spirocycles. *Sources:* Based on Zheng and Tice [32]; Zheng et al. [33].

tumors [36, 37]. Conformational restriction has also previously been employed by Schafmeister and coworkers for the development of covalently constrained helix mimetics targeting MDM2 [38]. Solid-phase synthesis of the peptidic backbone followed by formation of the diketopiperazine linkages led to the rigid spiro-oligomer **17** (K_d = 400 nM) [38–40], whose hydrophobic R-groups project in a (i, i+4, i+8) fashion. Those mimic the side chains of residues Phe19, Trp23, and Leu26 of p53 and make key contacts at the MDM2 surface. **17** displayed cell permeability and activity, and induced p53 upregulation in Huh7 cells. It is worth noting that the backbone of these constrained helix mimetics have very low conformational freedom, making their *de novo* design challenging and as a result, only few examples have been reported. Another good example of spirocycle functionalized compound is the Calcitonin Gene-Related Peptide (CGRP) Receptor antagonist **18**, which has progressed to Phase 3 clinical trials and is expected to reach the market for the treatment of migraine in the early 2020s [41]. Introduction of the spirocycle in this molecule led to enhanced *metabolic stability* compared with non-spirocyclic analogues, for example, bearing labile and oxidation prone dihydroquinazolinone units [42]. Lead molecule **19** was discovered as part of a lead optimization effort at AstraZeneca aimed at improving the properties of molecules targeting the central nervous system (CNS), more precisely the melanin-concentrating hormone receptor 1 (MCHr1), a target protein for the treatment of obesity and its downstream effects [43]. An important focus has been to identify a chemical series with attractive physicochemical properties, including reduced interaction with the ether-a-go-go-related gene (hERG) channel [44] while retaining *fine-tuned lipophilicity* profiles and sufficient CNS exposure [45]. The appended spirocyclic moiety provided improved physicochemical properties required for CNS exposure while *lowering off-target interactions* toward hERG, compared with morpholine and open ring analogues. Another early example is the triaryl bis-sulfone derivative **20**, which was identified by Merck as a potent type 2 canabinoid receptor (CB2) antagonist [46]. Modulation of CB2 has been proposed as a strategy for the development of immunomodulators. Replacement of the central phenyl ring of **20** with a spirocyclopropyl piperidine, saturated unit led to sulphonamide analogue **21**. The latter displayed a similar binding mode, potency, and selectivity profile against CB2 as parent **20**. Reduction in the number of aromatic rings is known to affect the physicochemical properties, hence developability of leads [31, 47]. Pleasingly, **21** showed reduced off-target effects as exemplified by a c. 10-fold reduction in affinity against the rat calcium channel and the cytochrome P450 2C9, while also presenting a beneficial clogP drop from 2.81 to 1.48.

2.4 Four-Membered Rings: Synthesis, Applications, and New Design Principles

A broad class of spirocycles which has attracted significant interest in drug discovery are those containing a four-membered ring [48]. This is mainly due to their structural novelty and historically poor exploration of the associated chemical and intellectual property space, hence presenting important opportunities for the development and patenting of new drugs without overlapping with other traditionally densely populated IP spaces [48]. Among them, substituted spirocyclobutanes/oxetanes/thietanes/azetidines are among the most represented due to their chemical robustness. A recent survey of the literature by Carreira and Fessard suggested that the incidence of drug-like spirocyclic structures containing at least one four-membered ring in research articles and patents remained relatively low and steady from the early 1960s to c. 2000, while it has followed an exponential growth and approximately quintupled between 2000 and 2013. Unsurprisingly, this correlated with major advances in the synthesis of such motifs during the same period. Those will be discussed in the following chapters of this book. This illustrative example is another testament that the general interest for new molecular scaffolds in medicinal chemistry usually evolves hand in hand with the development of new synthetic methodologies to access them, and is fuelled by the downstream associated benefits in research and development. Such benefits include, for example, structural novelty, improved physicochemical and pharmacokinetic properties of drug candidates, and the potential for exploiting new intellectual property. Of note, epoxides, aziridines, and episulfides are generally avoided due to their intrinsic reactivity. For the same reason, functional groups bearing heteroatoms in a 1,2-relationship (dioxolanes, pyrazolidines, disulfides, isoxazolidines, oxathiolanes, and isothiazolidines) and 1,3-relationship (ketals, aminals, 1,2-dithianes, oxazolidines, oxathiolanes, and thiazolidines) are generally avoided due to their chemical susceptibility to reducing agents and nucleophiles [49]. They are generally used as protecting groups in synthesis and sometimes for prodrug approaches, rather than as critical structural drug motifs.

A survey of the literature reveals several key strategies for the synthesis of oxetanes, thietanes, and azetidines, some of which are shown in Scheme 2.2. Such

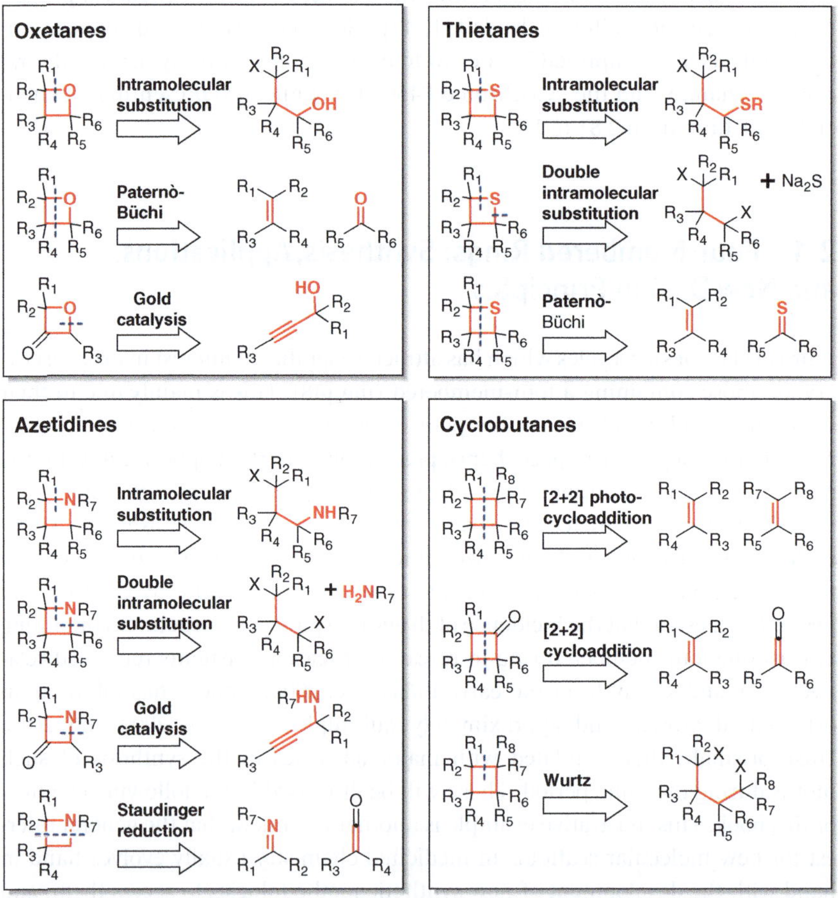

Scheme 2.2 Summary of common strategies for the synthesis of common four-membered rings encountered in spirocyclic systems of medicinal chemistry interest.

strategies include: (i) intramolecular displacement of a leaving group (e.g. halogen or activated alcohol nucleofuge) by the heteroatom; (ii) ring formation *via* the reaction of a nucleophile with doubly activated substrates bearing LGs in a 1,3 relationship; (iii) [2+2]-cycloaddition reactions, including Paternò–Büchi or reduction of lactams obtained by Staudinger ketene cycloaddition. Controlling the regioselectivity of cycloadditions is difficult, and often limits the scope and utility of such approaches; (iv) Gold-catalyzed cyclization of propargylic alcohols and amines toward their 3-oxo-substituted products. The wide range of methods available to prepare optically active propargyl alcohols and amines [9] make this fourth approach particularly attractive, greatly expanding the scope of substitution patterns that can be introduced onto the azetidine and oxetane scaffolds.

Oxetanes have been known for 140 years [50]; however, historically they have received little attention in medicinal chemistry [51] compared with epoxides, tetrahydro-furans, and pyrans. This is mainly due to their perceived reactivity: due to ring strain, they are generally seen as too reactive for their incorporation in drug candidates, despite the harsh conditions needed for their opening. Early reports have suggested that despite displaying similar ring strain to epoxide [52], their opening with nucleophiles is actually much slower [53]. Only in the last decade have oxetanes and their spiro analogues found applications in medicinal chemistry [48, 51, 54]. Due to its geometry and electronic properties, the oxetane ring has been used as a bioisosteric replacement for *gem*-dimethyl and carbonyl units (Figure 2.5). Despite useful to increase the affinity of a ligand to a target receptor, the introduction of a bulky hydrophobic *gem*-dimethyl motif generally increases the lipophilicity and decreases the solubility of a lead compound, which is often unwanted. The oxetane geometry mimics that of a *gem*-dimethyl, while being more hydrophilic due to its oxygen atom.

Carreira's lab and collaborators from Hoffman-La Roche examined the physicochemical and biological properties of spiro-oxetanes **22–30**, along with their

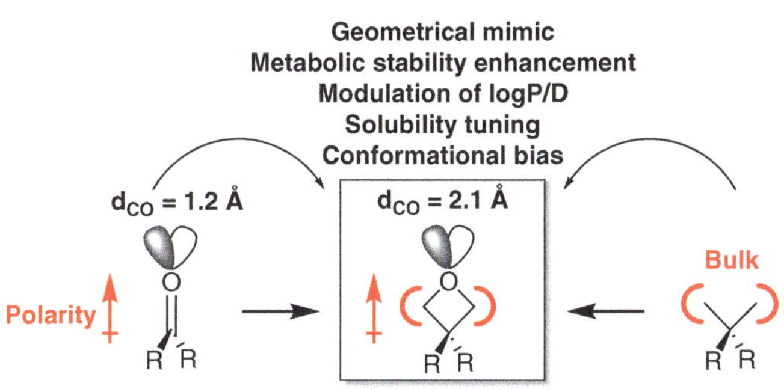

Figure 2.5 Comparison between carbonyl, gem-dimethyl, and oxetane groups, highlighting similar spatial arrangement of lone pairs and hydrophobic bulk.

corresponding carbonyls and *gem*-dimethyl analogues (Table 2.2) [55]. The oxetane ring generally provides higher metabolic stability, lower log P/D, and higher aqueous solubility than its *gem*-dimethyl counterparts across all series scrutinized. Unsurprisingly, this trend also correlated with a lower pK_a (1.2–1.6 units) for the amine, resulting from the inductive effect of the oxygen. While being less water soluble than carbonylated analogues, spiro-oxetanes displayed much improved metabolic stability toward oxidation in the piperidine and pyrrolidine series. Properties of the azetidin-3-one derivative could not be determined due to its low stability.

Four-membered ring containing spirocycles have also been scrutinized for their potential as bioisosteres of morpholines (Y = O), piperidines (Y = CR_2), piperazines (Y = NR), and thiomorpholines (Y = S, SO, SO_2), for the development of lead molecules with improved physicochemical and pharmacokinetic properties (Figure 2.6a) [48, 56]. While both classes display a similar relative positioning of polar and apolar groups, the comparison has its limitations. For example, the N–Y distance in these six-membered heterocycles is approximately 3 Å, while it is significantly higher and is slightly above 4 Å in their spirocyclic counterparts (Figure 2.6a). Similarly, the C–O distance in an oxetane is around 2 Å, while the C–O distance in the parent carbonyl group is approximately 1.2 Å. Another important difference is the introduction of exit vectors "out-of-plan" in these isosteres compared with the parent groups (Figure 2.6b). Nevertheless, such differences also contribute to the unique properties of the spiro analogues and make them interesting structural features for bioisosteric replacement (Figure 2.6c). For example, incorporation of morpholine rings into drug scaffolds is an established strategy for improving aqueous solubility. However, oxidative metabolism is a known inactivation pathway of morpholines, most often resulting from ring opening. Spiro-oxetanes **26** and **29** (Table 2.3) were proposed as replacements for morpholine **31**, due to the similar relative spatial disposition of their hydrophobic and polar features. Interestingly, **29** displays higher solubility and lower logP than the parent morpholine, while also remaining stable to oxidative metabolism. A number of other potential spirocyclic morpholine mimics based on [3.3], [3.4], and [3.5] motifs have been reported by Carreira, displaying a range of unique dipoles and exit vectors for tunable properties [55].

Another illustrative successful example is the bioisosteric replacement of the metabolically labile morpholine unit of BTK inhibitor **32** (Figure 2.7). This led to the development of 2-oxa-6-azaspiro[3.3]heptane functionalized analogue **33**, which displays enhanced stability and solubility while retaining similar potency to its target compared with the parent molecule [57].

Thalidomide **34** (Figure 2.8) was introduced on the market in the late 1950s–early 1960s as an antiemetic and sedative, and banned a few years later due to its shattering side-effects. Overall, thalidomide is thought to have caused

Table 2.2 Physicochemical and biochemical properties of selected oxetanes, and their carbonylated and gem-dimethyl counterparts.

R = 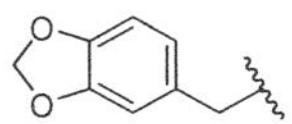	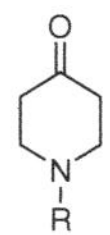 22	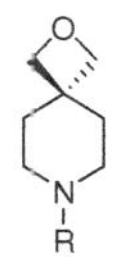 23	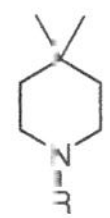 24	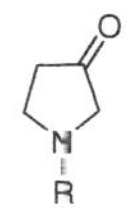 25	 26	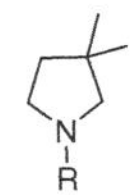 27	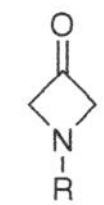 28	 29	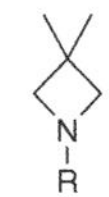 30
Solubility (μg ml^{-1})	4000	**1400**	220	4100	**730**	40	n.d.	**24 000**	290
Lipophilicity logD (logP)	1.2 (1.6)	**1.0 (2.0)**	2.3 (4.4)	−0.1 (−0.1)	**0.7 (1.5)**	1.4 (3.7)	n.d.	**0.5 (1.2)**	0.8 (3.1)
hCL_{int} (min^{-1} mg^{-1} μl)	120	**6**	23	100	**2**	10	n.d.	**3**	0
mCL_{int} (min^{-1} mg^{-1} μl)	88	**22**	31	580	**27**	39	n.d.	**7**	16
pK_a amine	7.5	**8.3**	9.5	6.1	**8.1**	9.7	n.d.	**8.0**	9.6

Source: Based on Wuitschik et al. [55].

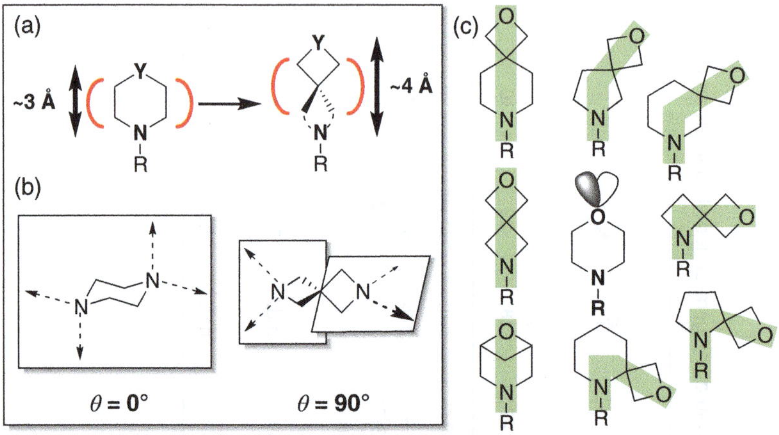

Figure 2.6 (a) Spiro-oxetane as a morpholine bioisostere. *Sources:* Based on Carreira and Fessard [48]; Burkhard et al. [56]; (b) comparison of exit vectors in the piperidine motif and its spiro counterpart; (c) spirocyclic oxetane based on mimics of morpholine.

Table 2.3 Comparison of the properties of spirocyclic oxetanes with their morpholine parent.

R = 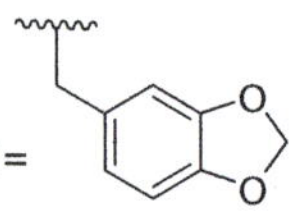	 **26**	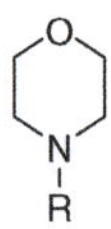 **31**	 **29**
Solubility (μg ml^{-1})	730	8000	**24 000**
Lipophilicity logD (logP)	0.7 (1.5)	1.5 (1.6)	**0.5 (1.2)**
hCL_{int} (min^{-1} mg^{-1} μl)	2	9	**3**
mCL_{int} (min^{-1} mg^{-1} μl)	27	8	**7**
pK_a amine	8.1	7.0	**8.0**

Source: Adapted from Carreira Wuitschik et al. [55].

Figure 2.7 Replacement of morpholine by 2-oxa-6-azaspiro[3.3]heptane in BTK inhibitors.

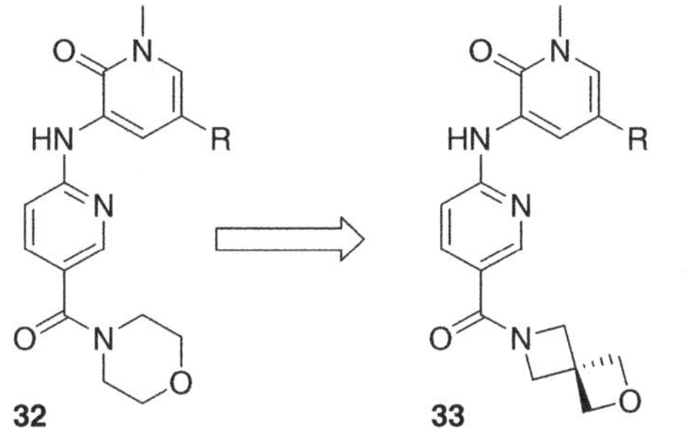

32
BTK binding EC_{50} = 0.2 nM
hCL = 30 ml/min/mg
Sol. (LYSA) < 1 mg/l

33
EC_{50} = 0.6 nM
hCL = 19 ml/min/mg
Sol (LYSA) 3 mg/l

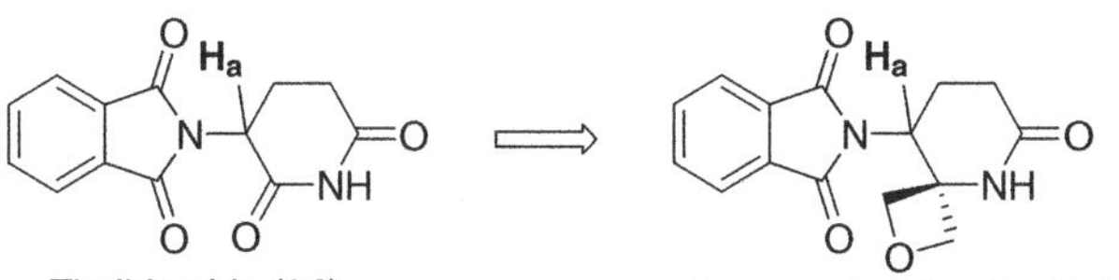

Thalidomide (**34**)

Oxetano-thalidomide (**35**)

	logD[a]	pK_a[b]	Aq. Sol.[c]	Human hepatocyte clearance[d]	Plasma stability, 5 h[e]	
34	0.24	10.7	29	3.7	40	Unstable
35	0.0	>11	78	2.3	74	Stable
(+)-35				2.6	72	Stable
(−)-35				2.5	68	Stable

Figure 2.8 Top: Structure of thalidomide **34** and its oxetano analogue **35**. Bottom: comparison of the physicochemical properties of thalidomide and oxetano-thalidomide. *Source:* Modified from Burkhard et al. [56]. Notes: *a*Logarithmic n-octanol/water distribution coefficient at pH 7.4. *b*Ionization constants determined at 23 °C by spectrophotometry in water. *c*Intrinsic solubility measured at pH 6.5 by a Lyophilization Solubility Assay [μg ml^{-1}]. *d*Intrinsic clearance rates [μl min^{-1} 10^{-6} cells] measured in human hepatocytes. *e*Stability in human plasma after five hours incubation time, expressed as a percentage of the initial concentration. The reported values represent the average of three runs (n = 3).

numerous and severe birth defects in more than 10 000 children during that time [25]. While thalidomide was initially administered as a racemic mixture, it has since been shown that the (*R*) enantiomer has sedative effects and the (*S*) enantiomer is teratogenic. It is also known that racemization spontaneously takes place *in vivo* due to the acidity of H_a (Figure 2.8), although as of today the exact mechanism underlying the teratogenicity of thalidomide is still under investigation. In 2013, Carreira and coworkers reported derivative **35** where one of the imide carbonyls is replaced by an oxetane [56], with the goal of limiting *in vivo* metabolism and racemization, and potentially altering teratogenicity (Figure 2.8). Not unexpectedly, **35** displayed higher pK_a, lower lipophilicity, and increased solubility. It displayed similar intrinsic clearance rates in human microsomes compared with **34**; however, showed much improved plasma stability after five hours. Pleasingly, **35** was configurationally stable to racemization in human blood plasma after five hours, thereby showing that a carbonyl to oxetane switch can be a viable approach to mitigate epimerization at adjacent stereocenters.

2.5 Further Examples

The unique range of properties provided by spirocycles has resulted in the development of a large number of bioactive lead compounds, and would easily deserve the attention of an entire book. These agents are targeting a wide range of protein/enzymes involved in diverse disorders, such as skin conditions, nausea, neurodegeneration, chronic pain, or even malaria. A few representative examples are presented in Figures 2.9 and 2.10. For example, the carba-spirocyclic natural product Ingenol mebutate (**36**) acts as a protein kinase C activator, and was approved by the FDA (tradename Picato®) in 2012 for the treatment of actinic keratosis. It can be prepared on a large scale in a protection/acylation/deprotection three step sequence from the naturally abundant ingenol [58]. The aza-spirocyclic drug

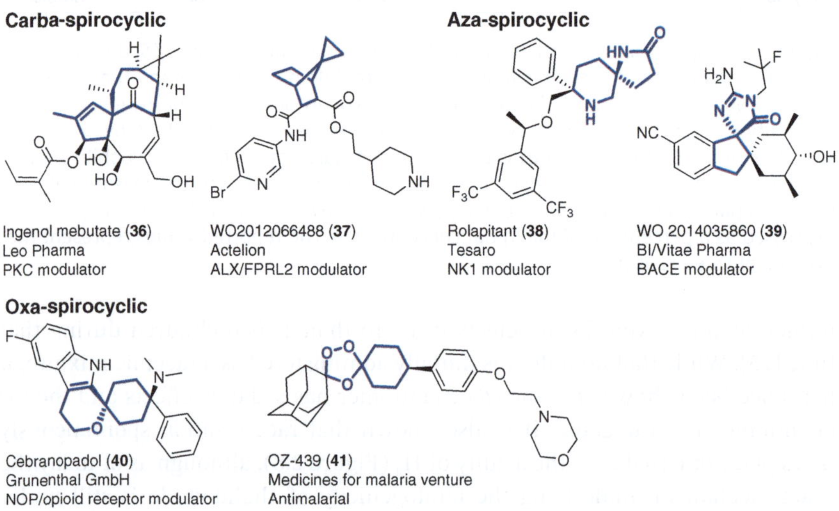

Figure 2.9 Representative examples of spirocyclic bioactive lead molecules.

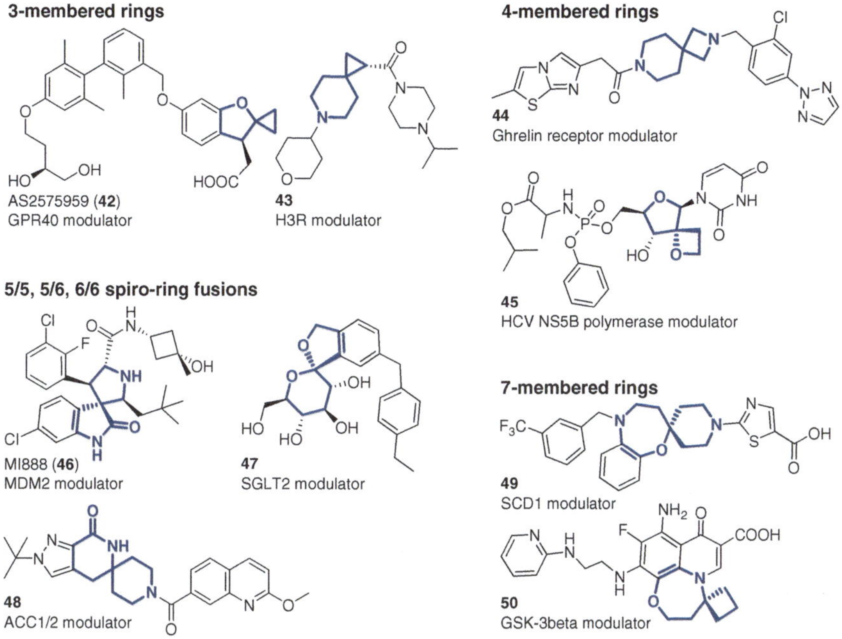

Figure 2.10 Additional examples further illustrating the breadth of ring types and sizes found in lead molecules of medicinal chemistry interest.

Rolapitant (**38**) is a potcnt and sclcctivc Ncurokinin-1 (NK-1) rcccptor antagonist which was recently approved for the prevention of delayed chemotherapy-induced nausea and vomiting (CINV) in adults, in combination with other antiemetics. Its spirocyclic scaffold can be accessed *via* the Michael addition of the corresponding nitro-piperidine onto methyl acrylate, followed by reduction and subsequent lactamization. The oxa-spirocyclic, investigational molecule Cebranopadol **40** is a potent opioid receptor agonist, and is currently under development for the treatment of a range of acute and chronic pain conditions. Key to the synthesis of **40** is a Lewis acid-mediated oxa-Pictet–Spengler reaction of the substituted parent cyclohexanone and indole fragments [59]. Of note, a review of the recent literature shows that five- and six-membered rings still dominate the pool of spiro compounds. As discussed earlier in this chapter, this trend is not surprising. This is mainly due to their restricted conformation compared with larger ring systems, and the historical synthetic challenges associated with smaller ones. It is expected that the developments of synthetic methods to prepare four-membered heterocyclic spiro systems in the last decade will naturally lead to a better representation of these features in the literature [60].

2.6 Conclusions

In this chapter, a number of useful applications of spirocycles in medicinal chemistry and their impacts in drug discovery were highlighted. The introduction of spirocycles has proven valuable for optimizing key parameters such as conformational freedom/restriction, lipophilicity, solubility, basicity, and metabolic stability, in addition to opportunities in exploring novel chemical and IP space. As mentioned, the synthetic challenge associated with spirocyclic compounds along with the wider range of chemical transformations of historical building blocks such as reactive arenes have been major contributors to their lower occurrence in the drug discovery patent literature and in commercial small-molecule screening libraries compared with non-spirocyclic compounds. Controlling the stereochemistry of the spirocycle also adds to the challenge, and will be discussed in the subsequent chapters of this book. This last point also perhaps provides an explanation why the majority of synthetic spirocyclic compounds reported in the literature contain only one spirocycle. It is also worth mentioning that sp^2-rich systems also present some indubitable advantages when it comes to screening technologies (e.g. fragments), such as the ability to monitor their behavior (e.g. binding) in fluorescence-based assays and ligand-observed NMR experiments under water suppression conditions. The lack of suitable fluorophore or ^1H chemical shifts in sub-aromatic regions can make spirocyclic systems unsuitable in a number of biophysical assays if sp^2 functionalities are not present elsewhere in the molecule. Additionally, despite the mounting evidence highlighting the potential of spirocycles for enhancing the metabolic stability of a wide range of compounds, examples of studies on their metabolites [61] are less common and much remains to be done to evaluate the wider stability landscape of individual spirocyclic classes *in vitro* and in full organisms. A systematic assessment of the molecular mechanisms and pathways underlying their metabolic activation in a context-dependent manner will be paramount for the design of next-generation spirocyclic structures with optimal stability. Nevertheless, the advances in chemical transformations and diversity-oriented synthesis approaches [62–67] in the last two decades have changed the landscape. As a result, spirocycles have now made their way toward screening compound libraries and an ever-increasing number of novel spirocyclic building blocks are now commercially available (Figure 2.11). Many institutions and commercial vendors have incorporated spirocyclic lead-like compounds and fragments in their historical libraries, expanding the scope of available scaffolds for structure-based drug design and novel IP generation. Four, five, and six-membered rings containing spirocycles have historically received most attention for structure-based drug design. This is in great part due to their inherent chemical stability compared with three-membered rings, low conformational freedom compared with larger ring systems, hence greater rigidity providing highly

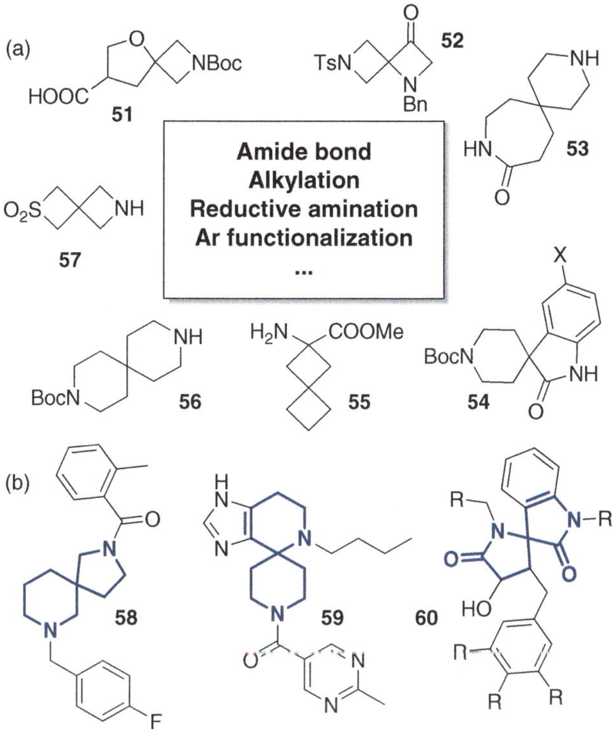

Figure 2.11 Representative examples of (a) multifunctional spiro building blocks and (b) lead compounds for screening that are currently available from commercial vendors.

directional vectors for derivatization and efficient receptor targeting. A recent review by Carreira and Fessard highlighted the potential of spirocycles for generating new IP in medicinal chemistry and beyond [48]. Overall, while the potential of spirocycles in medicinal chemistry is still to be fully exploited, the steadily expanding scope of synthetic methods, bioactive leads, and associated biological data promises to unlock a wide range of novel molecular entities and pharmacologies, with much needed benefits to patients lurking around the corner.

References

1 Moss, G. (**1999**). Extension and revision of the nomenclature for spiro compounds. *Pure Appl. Chem.* 71 (3): 531–558.

2 Knox, C., Law, V., Jewison, T. et al. (**2011**). DrugBank 3.0: a comprehensive resource for "omics" research on drugs. *Nucleic Acids Res.* 39: D1035–D1041.

3 Krebs, A., Schafer, B., and Kochner, A. (**2010**). Method for producing azoniaspironortropine esters and nortropan-3-one compounds, US2010048903.

4 Smith, L.K. and Baxendale, I.R. (**2015**). Total syntheses of natural products containing spirocarbocycles. *Org. Biomol. Chem.* 13 (39): 9907–9933.

5 Zhang, F.-M., Zhang, S.-Y., and Tu, Y.-Q. (**2018**). Recent progress in the isolation, bioactivity, biosynthesis, and total synthesis of natural spiroketals. *Nat. Prod. Rep.* 35 (1): 75–104.

6 Perron, F. and Albizati, K.F. (**1989**). Chemistry of spiroketals. *Chem. Rev.* 89 (7): 1617–1661.

7 Quintavalla, A. (**2018**). Spirolactones: recent advances in natural products, bioactive compounds and synthetic strategies. *Curr. Med. Chem.* 25 (8): 917–962.

8 Anthoni, U., Chevolot, L., Larsen, C. et al. (**1987**). Marine alkaloids. 12. chartellines, halogenated beta-lactam alkaloids from the marine bryozoan *Chartella papyracea*. *J. Org. Chem.* 52 (21): 4709–4712.

9 Manam, R.R., Teisan, S., White, D.J. et al. (**2005**). Lajollamycin, a nitro-tetraene spiro-β-lactone-γ-lactam antibiotic from the marine actinomycete *Streptomyces nodosus*. *J. Nat. Prod.* 68 (2): 240–243.

10 Arulananda Babu, S., Padmavathi, R., Ahmad Aslam, N., and Rajkumar, V. (**2015**). Chapter 8 - recent developments on the synthesis and applications of natural products-inspired spirooxindole frameworks. In: Studies in Natural Products Chemistry, vol. 46 (ed. R. Attaur), 227–339. Elsevier.

11 Hong, L. and Wang, R. (**2013**). Recent advances in asymmetric organocatalytic construction of 3,3′-spirocyclic oxindoles. *Adv. Synth. Catal.* 355 (6): 1023–1052.

12 Singh, G.S. and Desta, Z.Y. (**2012**). Isatins as privileged molecules in design and synthesis of spiro-fused cyclic frameworks. *Chem. Rev.* 112 (11): 6104–6155.

13 Wang, J., Soisson, S.M., Young, K. et al. (**2006**). Platensimycin is a selective FabF inhibitor with potent antibiotic properties. *Nature* 441 (7091): 358–361.

14 Häbich, D. and von Nussbaum, F. (**2006**). Platensimycin, a new antibiotic and "superbug challenger" from nature. *ChemMedChem* 1 (9): 951–954.

15 Nicolaou, K.C., Li, A., and Edmonds, D.J. (**2006**). Total synthesis of platensimycin. *Angew. Chem. Int. Ed.* 45 (42): 7086–7090.

16 Trost, B.M., Surivet, J.-P., and Toste, F.D. (**2004**). Ruthenium-catalyzed enyne cycloisomerizations. Effect of allylic silyl ether on regioselectivity. *J. Am. Chem. Soc.* 126 (47): 15592–15602.

17 Nicolaou, K.C., Li, A., Edmonds, D.J. et al. (**2009**). Total synthesis of platensimycin and related natural products. *J. Am. Chem. Soc.* 131 (46): 16905–16918.

18 Trajkovic, M., Ferjancic, Z., Saicic, R.N., and Bihelovic, F. (**2019**). Enantioselective synthesis of the platensimycin core by silver(I)-promoted cyclization of Δ6-α-iodoketone. *Chem. Eur. J.* 25 (17): 4340–4344.

19 Ghosh, A.K. and Xi, K. (**2007**). Enantioselective synthesis of (−)-platensimycin oxatetracyclic core by using an intramolecular Diels–Alder reaction. *Org. Lett.* 9 (20): 4013–4016.

20 Taylor, F.F. and Faloon, W.W. (**1959**). The role of potassium in the natriuretic response to a steroidal lactone (SC-9420). *J. Clin. Endocrinol. Metab.* 19 (12): 1683–1687.

21 Roughley, S.D. and Jordan, A.M. (**2011**). The medicinal chemist's toolbox: an analysis of reactions used in the pursuit of drug candidates. *J. Med. Chem.* 54 (10): 3451–3479.

22 Clemons, P.A., Bodycombe, N.E., Carrinski, H.A. et al. (**2010**). Small molecules of different origins have distinct distributions of structural complexity that correlate with protein-binding profiles. *Proc. Natl. Acad. Sci. U. S. A.* 107 (44): 18787–18792.

23 Brooks, W.H., Guida, W.C., and Daniel, K.G. (**2011**). The significance of chirality in drug design and development. *Curr. Top. Med. Chem.* 11 (7): 760–770.

24 Nguyen, L.A., He, H., and Pham-Huy, C. (**2006**). Chiral drugs: an overview. *Int. J. Biomed. Sci.* 2 (2): 85–100.

25 Matthews, S.J. and McCoy, C. (**2003**). Thalidomide: a review of approved and investigational uses. *Clin. Ther.* 25 (2): 342–395.

26 Lovering, F., Bikker, J., and Humblet, C. (**2009**). Escape from flatland: increasing saturation as an approach to improving clinical success. *J. Med. Chem.* 52 (21): 6752–6756.

27 Lovering, F. (**2013**). Escape from flatland 2: complexity and promiscuity. *Med. Chem. Commun.* 4 (3): 515–519.

28 Wipf, P., Skoda, E.M., and Mann, A. (**2015**). Chapter 11 – conformational restriction and steric hindrance in medicinal chemistry. In: The Practice of Medicinal Chemistry, 4e (eds. C.G. Wermuth, D. Aldous, P. Raboisson and D. Rognan), 279–299. San Diego: Academic Press.

29 Leeson, P.D. and Springthorpe, B. (**2007**). The influence of drug-like concepts on decision-making in medicinal chemistry. *Nat. Rev. Drug Discov.* 6 (11): 881.

30 Aldeghi, M., Malhotra, S., Selwood, D.L., and Chan, A.W.E. (**2014**). Two- and three-dimensional rings in drugs. *Chem. Biol. Drug Des.* 83 (4): 450–461.

31 Ritchie, T.J. and Macdonald, S.J.F. (**2009**). The impact of aromatic ring count on compound developability – are too many aromatic rings a liability in drug design? *Drug Discov. Today* 14 (21): 1011–1020.

32 Zheng, Y.-J. and Tice, C.M. (**2016**). The utilization of spirocyclic scaffolds in novel drug discovery. *Expert Opin. Drug Discov.* 11 (9): 831–834.

33 Zheng, Y., Tice, C.M., and Singh, S.B. (**2014**). The use of spirocyclic scaffolds in drug discovery. *Bioorg. Med. Chem. Lett.* 24 (16): 3673–3682.

34 Zhao, Y., Aguilar, A., Bernard, D., and Wang, S. (**2015**). Small-molecule inhibitors of the MDM2-p53 protein-protein interaction (MDM2 inhibitors) in clinical trials for cancer treatment. *J. Med. Chem.* 58 (3): 1038–1052.

35 Skalniak, L., Surmiak, E., and Holak, T.A. (**2019**). A therapeutic patent overview of MDM2/X-targeted therapies (2014–2018). *Expert Opin. Ther. Pat.* 29 (3): 151–170.

36 de Jonge, M., de Weger, V.A., Dickson, M.A. et al. (**2017**). A phase I study of SAR405838, a novel human double minute 2 (HDM2) antagonist, in patients with solid tumours. *Eur. J. Cancer* 76: 144–151.

37 de Weger, V.A., de Jonge, M., Langenberg, M.H.G. et al. (**2019**). A phase I study of the HDM2 antagonist SAR405838 combined with the MEK inhibitor pimasertib in patients with advanced solid tumours. *Br. J. Cancer* 120 (3): 286–293.

38 Brown, Z.Z., Akula, K., Arzumanyan, A. et al. (**2012**). A spiroligomer α-helix mimic that binds HDM2, penetrates human cells and stabilizes HDM2 in cell culture. *PLoS One* 7 (10): e45948.

39 Brown, Z.Z. and Schafmeister, C.E. (**2010**). Synthesis of hexa- and pentasubstituted diketopiperazines from sterically hindered amino acids. *Org. Lett.* 12 (7): 1436–1439.

40 Schafmeister, C.E., Brown, Z.Z., and Gupta, S. (**2008**). Shape-programmable macromolecules. *Acc. Chem. Res.* 41 (10): 1387–1398.

41 MK-1602 Clinical trials. https://clinicaltrials.gov/ct2/results?term=MK-1602&age_v=&gndr=&type=&rslt=&phase=2&Search=Apply (accessed September 2019).

42 Bell, I.M. (**2014**). Calcitonin gene-related peptide receptor antagonists: new therapeutic agents for migraine. *J. Med. Chem.* 57 (19): 7838–7858.

43 Johansson, A., Löfberg, C., Antonsson, M. et al. (**2016**). Discovery of (3-(4-(2-Oxa-6-azaspiro[3.3]heptan-6-ylmethyl)phenoxy)azetidin-1-yl)(5-(4-methoxyphenyl)-1,3,4-oxadiazol-2-yl)methanone (AZD1979), a melanin concentrating hormone receptor 1 (MCHr1) antagonist with favorable physicochemical properties. *J. Med. Chem.* 59 (6): 2497–2511.

44 Högberg, T., Frimurer, T.M., and Sasmal, P.K. (**2012**). Melanin concentrating hormone receptor 1 (MCHR1) antagonists — still a viable approach for obesity treatment? *Bioorg. Med. Chem. Lett.* 22 (19): 6039–6047.

45 Bingham, M. and Rankovic, Z. (**2012**). Chapter 18 medicinal chemistry challenges in CNS drug discovery. In: Drug Discovery for Psychiatric Disorders (eds. Z. Rankovic, R. Hargreaves and M. Bingham), 465–509. The Royal Society of Chemistry.

46 Lavey, B.J., Kozlowski, J.A., Shankar, B.B. et al. (**2007**). Optimization of triaryl bis-sulfones as cannabinoid-2 receptor ligands. *Bioorg. Med. Chem. Lett.* 17 (13): 3760–3764.

47 Ishikawa, M. and Hashimoto, Y. (**2011**). Improvement in aqueous solubility in small molecule drug discovery programs by disruption of molecular planarity and symmetry. *J. Med. Chem.* 54 (6): 1539–1554.

48 Carreira, E.M. and Fessard, T.C. (**2014**). Four-membered ring-containing spirocycles: synthetic strategies and opportunities. *Chem. Rev.* 114 (16): 8257–8322.

49 Baell, J.B. and Holloway, G.A. (**2010**). New substructure filters for removal of pan assay interference compounds (PAINS) from screening libraries and for their exclusion in bioassays. *J. Med. Chem.* 53 (7): 2718–2740.

50 Reboul, M. (**1878**). Oxyde de Propylene Normal et Polyoxypropylenes. *Ann. Chim. (Paris)* 14: 495–497.

51 Bull, J.A., Croft, R.A., Davis, O.A. et al. (**2016**). Oxetanes: recent advances in synthesis, reactivity, and medicinal chemistry. *Chem. Rev.* 116 (19): 12150–12233.

52 Ringnér, B., Sunner, S., and Watanabe, H. (**1971**). The enthalpies of combustion and formation of some 3,3-disubstituted oxetanes. *Acta Chem. Scand.* 25: 141–146.

53 Pritchard, J.G. and Long, F.A. (**1958**). The kinetics of the hydrolysis of trimethylene oxide in water, deuterium oxide and 40% aqueous dioxane[1]. *J. Am. Chem. Soc.* 80 (16): 4162–4165.

54 Rogers-Evans, M., Knust, H., Plancher, J.-M. et al. (**2014**). Adventures in drug-like chemistry space: from oxetanes to spiroazetidines and beyond! *CHIMIA Int. J. Chem.* 68 (7-8): 492–499.

55 Wuitschik, G., Rogers-Evans, M., Buckl, A. et al. (**2008**). Spirocyclic oxetanes: synthesis and properties. *Angew. Chem. Int. Ed.* 47 (24): 4512–4515.

56 Burkhard, J.A., Wuitschik, G., Plancher, J.-M. et al. (**2013**). Synthesis and stability of oxetane analogs of thalidomide and lenalidomide. *Org. Lett.* 15 (17): 4312–4315.

57 Dewdney, N.J., Kennedy-Smith, J., Kondru, R.K., et al. (**2009**). Inhibitors of Bruton's tyrosine kinase - Google patents. US Patent 20090306041A1.

58 Grue-Sørensena, G., Liang, X., Högberga, T., et al. (**2012**). Ingenol-3-acylates iii and ingenol-3-carbamates - Google patents. WO Patent 2012083953A1.

59 GmbH, G., Gruss, M., Kluge, S., and Pruehs, S. (**2013**). Crystalline (1R,4R)-6'-Fluoro-*N*,*N*-dimethyl-4-phenyl-4',9'-dihydro-3'H-spiro[cyclohexane-1,1'-pyrano[3,4,b]indolo]-4-amine - Google patents. WO Patent 2013007361A8.

60 Mergelsberg, I., Scherer, D.H., Huttenloch, M.E., et al. (**2013**). Process and intermediates for the synthesis of 8-[-methyl]-8-phenyl-1,7-diazaspiro[4.5]decan-2-one compounds - Google patents. US Patent 20140024834A1.

61 Li, X.-Q., Hayes, M.A., Grönberg, G. et al. (**2016**). Discovery of a novel microsomal epoxide hydrolase–catalyzed hydration of a spiro oxetane. *Drug Metab. Dispos.* 44 (8): 1341–1348.

62 Kidd, S.L., Osberger, T.J., Mateu, N. et al. (**2018**). Recent applications of diversity-oriented synthesis toward novel, 3-dimensional fragment collections. *Front. Chem* 6 (460).

63 Sveiczer, A., North, A.J.P., Mateu, N. et al. (**2019**). Spirocycles as rigidified sp^3-rich scaffolds for a fragment collection. *Org. Lett.* 21 (12): 4600–4604.

64 King, T.A., Stewart, H.L., Mortensen, K.T. et al. (**2019**). Cycloaddition strategies for the synthesis of diverse heterocyclic spirocycles for fragment-based drug discovery. *Eur. J. Org. Chem.* 2019 (31–32): 5219–5229.

65 Hung, A.W., Ramek, A., Wang, Y. et al. (**2011**). Route to three-dimensional fragments using diversity-oriented synthesis. *Proc. Natl. Acad. Sci.* 108 (17): 6799–6804.

66 Tan, W., Zhu, X.-T., Zhang, S. et al. (**2013**). Diversity-oriented synthesis of spiro-oxindole-based 2,5-dihydropyrroles via three-component cycloadditions and evaluation on their cytotoxicity. *RSC Adv.* 3 (27): 10875–10886.

67 Liu, Y., Wang, H., and Wan, J. (**2013**). Recent advances in diversity oriented synthesis through isatin-based multicomponent reactions. *Asian J. Org. Chem.* 2 (5): 374–386.

3

Recent Advances in the Asymmetric Synthesis of Spiro Compounds Through Cycloadditions

Alberto Vega-Peñaloza[1], *Suva Paria*[1], *Luca Dell'Amico*[1], *and Xavier Companyó*[1, 2]

[1] *Department of Chemical Sciences, University of Padova, Padova, Italy*
[2] *Section of Organic Chemistry, Department of Inorganic and Organic Chemistry, University of Barcelona, Barcelona, Spain*

3.1 Introduction

Spiro compounds are organic molecules that present a twisted structure where two perpendicular cycles are linked together by a single tetrasubstituted stereogenic center, known as the spiro center. Since its definition in early 1900 [1], spiro compounds have attracted the attention of the scientific community owing to their unique features. The spiranic motif is present in several natural and non-natural compounds possessing a wide range of biological activities [2]. As a result, the introduction of spiranic scaffolds in drug candidates is gaining interest in modern drug discovery programs [3]. Their well-defined 3D structure along with the enhanced conformational rigidity offer distinctive features for new drug candidates, such as enhanced drug–target recognition or improved physicochemical properties [3]. Hence, the increasing demand of new compounds derived from privileged scaffolds in chemistry, biology, and medicinal research has fueled the development of novel methodologies toward the synthesis of spiro compounds [4]. In addition, the construction of spiro centers in a catalytic, stereocontrolled manner represents by itself a formidable synthetic challenge. In this light, methodologies based on the catalysis by organometallic complexes offer a robust approach for the discovery of novel cycloaddition reactions [5]. The versatility of the diverse transition metals, such as Cu, Rh, or Ni, together with the development of new chiral ligands make the use of transition-metal catalysts a very powerful approach. Recently, the advent of organocatalysis has paved the way for unprecedented activation modes and reaction pathways [6], including cyclization, cycloaddition, and cascade transformations [7], representing a complementary and straightforward approach to construct spiro compounds in asymmetric fashion.

The present chapter assesses the most relevant synthetic methodologies toward the asymmetric catalytic construction of spirocyclic compounds by means of either cycloaddition or formal cycloaddition processes published between 2015

Spiro Compounds: Synthesis and Applications, First Edition. Edited by Ramon Rios Torres.

and 2019. Therefore, the most significant transformations have been selected regardless of the concerted or stepwise nature of the reaction mechanism. For each transformation, the substrate scope, the generality, and limitations as well as the key reaction mechanism are discussed. The chapter is divided between organometallic and organocatalytic strategies, and within each section, the different methodologies are classified according to the number of atoms involved in the formation of the spiro cycle.

3.2 Organometallic Methodologies

Over the past decades, transition-metal catalysis has emerged as a powerful tool for the construction of complex molecules. Also, rigid and strained chiral scaffolds have been synthesized in a straightforward manner under catalytic routines. Across these molecules, spirocompounds represent an important structural family with diverse and specific biological activity. In the following pages, it will be demonstrated how transition metals have overcome the synthetic issues connected to the assembly of the rigid all-substituted quaternary spirocenter. Drawbacks of some transition-metal-based catalytic systems remain the relatively low stability under aerobic conditions, while the most recent transformations have found important applications in concert with organocatalytic approaches. The section has been divided into three subcategories according to the involved reactions: [3+2] cycloaddition, [4+2] cycloaddition, and miscellaneous reactions.

3.2.1 Organometallic [3+2] Cycloaddition Strategies to Construct Spiro Compounds

In 2014, Franz and coworkers reported the asymmetric annulations of alkylidene oxindole **1** and allyl silane **2** for the synthesis of spirocyclic oxindole **3** (Scheme 3.1) [8]. The reaction was conducted in the presence of $ScCl_3$ and the chiral (*R*,*S*)-indapybox ligand **4** and was proposed to proceed by the formation of β-silyl carbocation **5** through a Michael/1,2-silyl shift/cyclization sequence.

Mechanistically, NaBArF played a dual role in the process, primarily is responsible to generating a cationic scandium complex and secondly is stabilizing the β-silyl carbocation **5**. Remarkably, this transformation represents the first example of a [3+2] annulation process involving allylsilanes and unsaturated carbonyl compounds to furnish chiral cyclopentanes in high yields (73–99%) and good to excellent stereocontrol (6 : 1–>20 : 1 dr and 64–99% ee).

Arai, Yamanaka, and coworkers presented in 2015 a [3+2] cycloaddition mediated by $Cu(OTf)_2$ and the chiral PyBidine ligand **13** for the synthesis of a variety of functionalized spiro[pyrrolidin-3,3′-oxindole]s **14** (Scheme 3.2) [9]. The

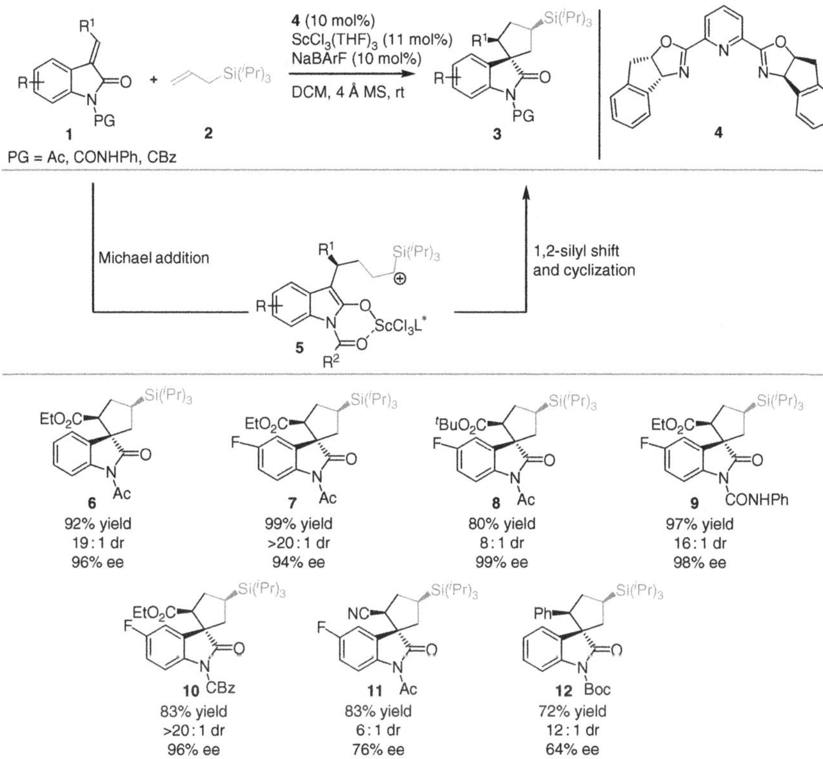

Scheme 3.1 Scandium-catalyzed enantioselective carboannulation between alkylidene oxindole and allylsilanes. *Source:* Modified from Ball-Jones et al. [8].

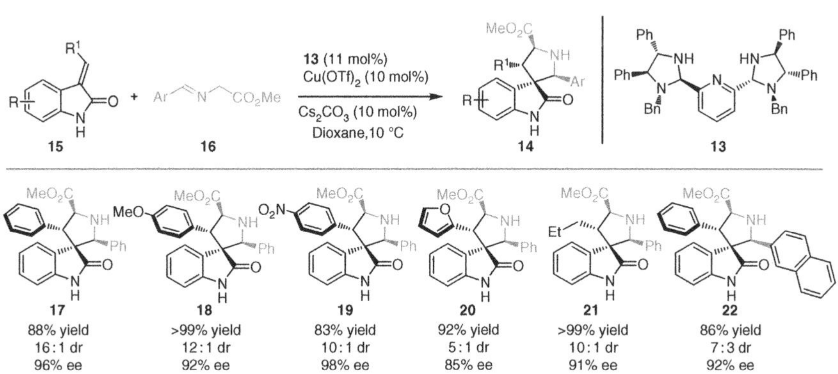

Scheme 3.2 Copper-catalyzed asymmetric [3+2] cycloaddition between alkylidene oxindole and imino esters. *Source:* Modified from Arai et al. [9].

catalytic system, which was previously used for the endo-selective [3+2] cycloaddition of iminoesters with nitroalkenes [10], furnished the spiranic products in excellent yields (83%–>99%) and high stereoselectivities (5 : 1–>20 : 1 and 91%–98% ee).

In the same year, the group of Yu and Deng reported the first asymmetric 1,3-dipolar cycloaddition of azomethine ylides **23** with α-alkylidene succinimides **24** to obtain dispiropyrrolidine derivatives **25** in good yields (up to 90%) and high stereoselectivities (>20 : 1 dr and 87–96% ee) (Scheme 3.3) [11]. The methodology uses catalytic amounts of chiral *N,O*-ligand **26** in the presence of $Cu(OAc)_2$ to impart high stereocontrol over the [3+2] cyclization process. The catalytic system was also applied for the synthesis of pyrrolidinyl-spirooxindole skeletons with excellent diastereo- and enantiocontrol.

Later on, the authors extended the application of their catalytic system (using a different ligand) to the 1,3-dipolar cycloaddition of azomethine ylides to 5-alkylidene thiazolidine-2,4-diones for the synthesis of spirocyclic pyrrolidine-thiazolidinediones [12].

Shortly after, Feng, Liu, and coworkers reported that a combination of $Mg(OTf)_2$ and the chiral *N,N′*-dioxide ligand **34** is a highly effective catalytic system for the

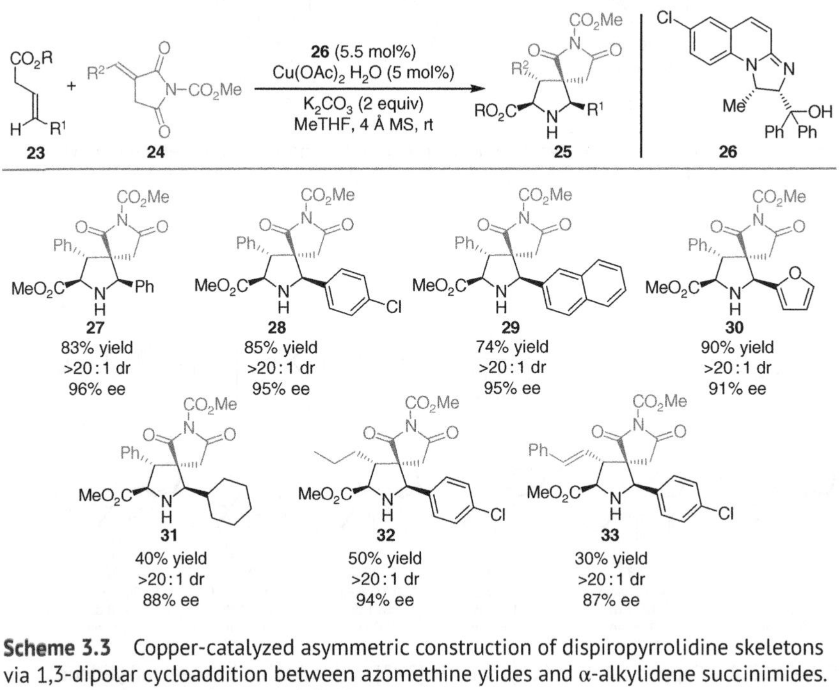

Scheme 3.3 Copper-catalyzed asymmetric construction of dispiropyrrolidine skeletons via 1,3-dipolar cycloaddition between azomethine ylides and α-alkylidene succinimides. *Source:* Modified from Yang et al. [11].

asymmetric [3+2] cycloaddition of methyleneindolinones **1** with *N,N′*-cyclic azomethine imines **35** to produce pyrazolidine-based spirooxindoles **36** [13]. This strategy allows the construction of valuable contiguous tertiary and quaternary stereocenters (Scheme 3.4).

A weak positive nonlinear effect was observed between the enantiomeric excess of the product **36** and the chiral ligand **34**. Hence, the authors suggest the presence of both monomeric and oligomeric catalytic species in the reaction medium, with the monomeric complex being more active. Subsequently, an asymmetric [3+2] cycloaddition of methyleneindolinones with nitrones was also reported by the same authors using the combination of chiral *N,N′*-dioxide ligand and $Co(BF_4)_2{\cdot}H_2O$ to synthesize spirooxindole derivatives in good yield and excellent diastereo- and enantiocontrol [14].

In 2016, Cramer and coworkers reported the [3+2] annulation between alkynes **47** and cyclic sulfonimines **46** catalyzed by a chiral Rh complex for the synthesis of spirocyclic sultams **44** (Scheme 3.5) [15]. The methodology involves the use of a chiral cyclopentadienyl ligand **45**. Under the optimized reaction conditions, the enantioenriched sultams are obtained in good to excellent yields (50–99%) and

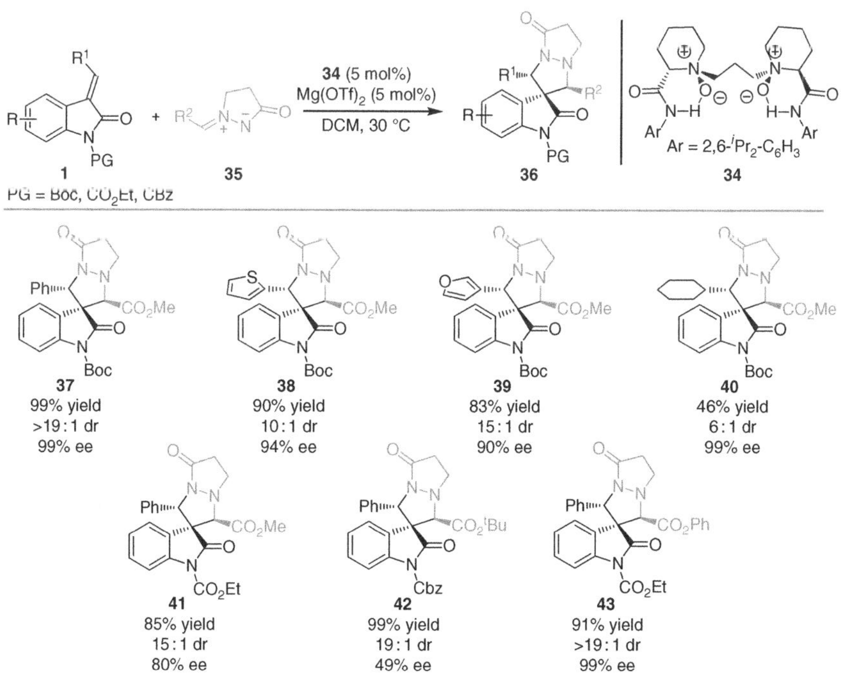

Scheme 3.4 Magnesium-catalyzed asymmetric [3+2] cycloaddition between methyleneindolinones and *N,N′*-cyclic azomethine imines.

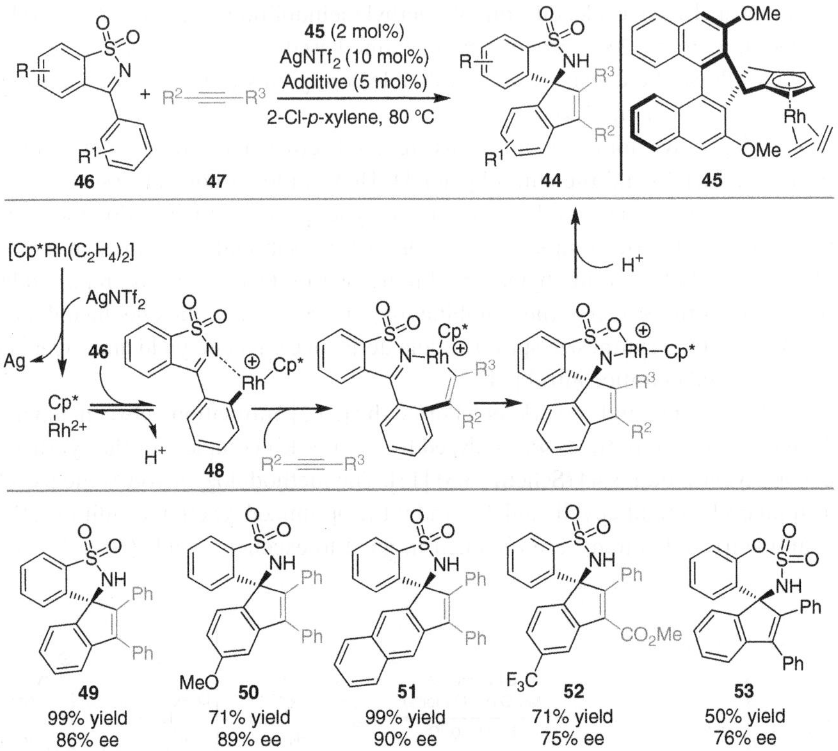

Scheme 3.5 Rhodium-catalyzed enantioselective [3+2] annulation to form spirocyclic sultams. *Source:* Modified from Pham and Cramer [15].

enantioselectivites (75–90%). The proposed catalytic cycle is initiated by the C–H activation of the sulfonimine **46**, which is directed by the *N*-sulfonyl imino group, to generate the rhodacycle **48**. Alkyne migratory insertion of intermediate **48** followed by the addition to the imine bond generates the spiro compound **44**. Low regioselectivities were observed in the case of unsymmetrical aryl alkynes **47**. This is the major limitation of this process.

In the same year, the group of Zhao reported the asymmetric [3+2] cycloaddition of *p*-quinone methides **54** with vinyl cyclopronanes **55** and epoxides **56** (Scheme 3.6) [16]. The reaction proceeds through a 1,6-conjugate addition of Pd-activated vinyl cyclopronanes **55** and epoxides **56** to *p*-quinone methides **54** followed by the annulation event. Ferrocene-based chiral phosphine **57** was the ligand of choice for the annulations of epoxides, while chiral spiro phosphene **58** was found to be optimal in the case of vinyl cyclopropanes, obtaining in both the cases the products **61-65** in good to high yields (71–98%) and excellent stereocontrol (3 : 1–>20 : 1 dr and 51–99% ee).

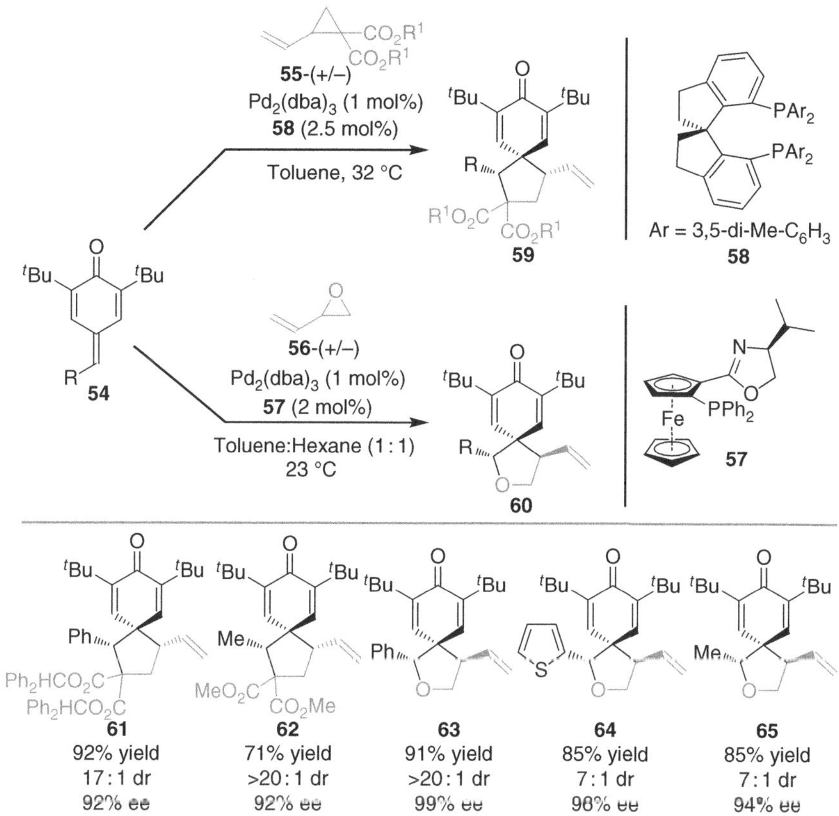

Scheme 3.6 Palladium-catalyzed stereoselective 1,6-conjugate addition/annulation of vinyl epoxides/cyclopropanes to *p*-quinone methides. *Source:* Modified from Ma et al. [16].

In 2018, the group of Enders established the first Cu-catalyzed enantioselective Kinugasa/Michael domino reaction for the synthesis of spirocyclic β-lactams **66** by desymmetrization of prochiral cyclohexadienone **67** with nitrones **68** (Scheme 3.7) [17]. $Cu(OTf)_2$ and indane-BOX ligand **69** was found to be the optimal catalytic system for the reaction. Mechanistically, the reaction starts with the formation of Cu-acetylide intermediate **70** which underwent [3+2] cycloaddition with nitrones **68** to form Cu-bound isoxazoline intermediate **71**. A rearrangement of the intermediate **71** affords the tethered four-membered copper enolate intermediate **72**, which subsequently participates in desymmetric Michael addition to afford the spirocyclic β-lactams **66** in a highly enantio- and diastereoselective fashion (>20 : 1 dr and 82–96% ee).

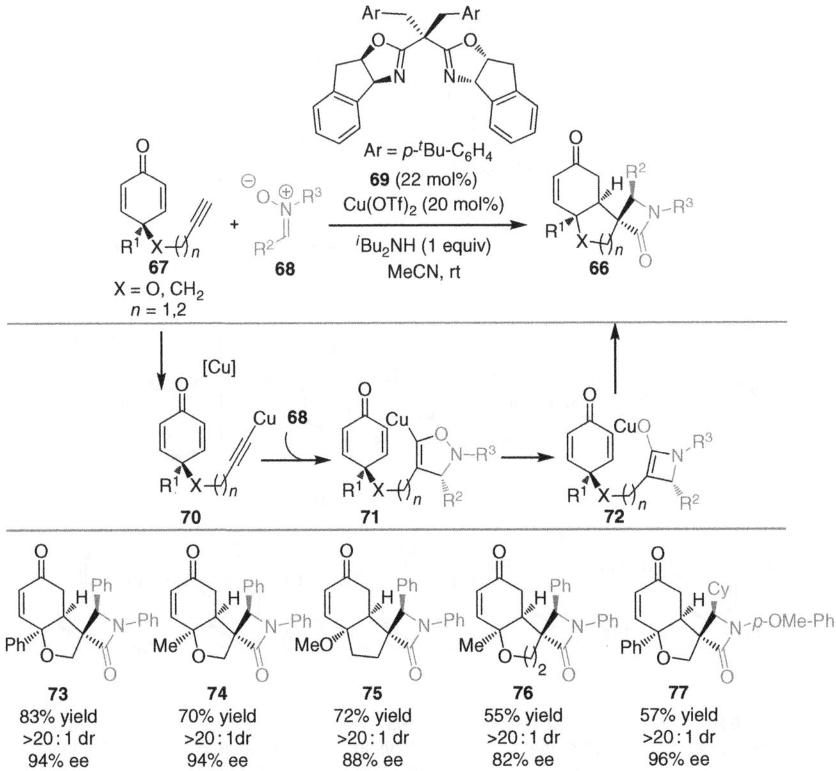

Scheme 3.7 Copper-catalyzed asymmetric synthesis of spirocyclic β-lactams through Kinugasa/Michael domino sequence. *Source:* Modified from Shu et al. [17].

In 2018, Waldmann and Antonchick reported the enantioselective synthesis of spirotropanyl oxindoles **78** by a bimetallic relay strategy (Scheme 3.8) [18]. Transient azomethine ylides **79** were generated from *E*-oximino α-diazo ketones **80** in the presence of an achiral Rh(II) complex, which subsequently underwent intermolecular [3+2] cycloaddition with 3-alkenyloxindole **81** catalyzed by the chiral *N,N*-dioxide Nd(III) Lewis acid complex. The products are obtained in high yields (54–97%) with high enantio- and diastereoselectivity (3.2 : 1 dr and 77%–>99% ee).

3.2.2 Organometallic [4+2] Cycloaddition Strategies to Construct Spiro Compounds

In 2015, Liu, Feng, and coworkers reported a regio- and enantioselective aza-Diels–Alder reaction of 3-vinylindoles **90** with isatin-derived ketimines **91**

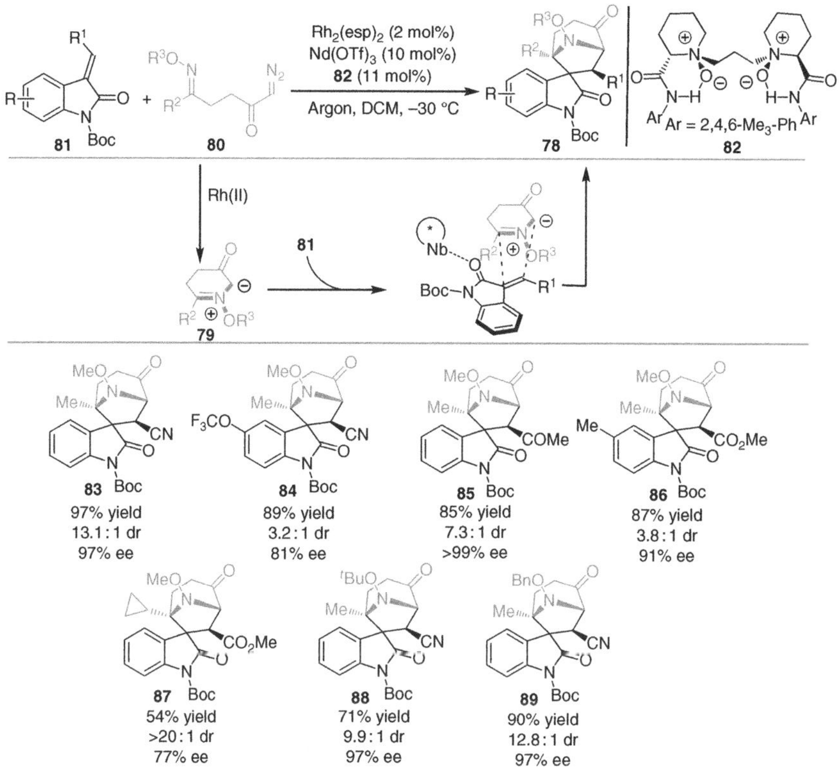

Scheme 3.8 Bimetallic relay catalysis for the enantioselective synthesis of the spirotropanyl oxindole. *Source:* Modified from Jia et al. [18].

(Scheme 3.9) [19]. The reaction uses Ni(II) complex together with the chiral *N,N′*-dioxide ligand **92**, providing a wide range of spiroindolones **93** in good to high yields (75–96%), high enantioselectivities (62–94% ee), and complete diastereocontrol. Mechanistically, the reaction commenced by the activation of isatin-derived ketimines **91** by the chiral L-PiPr$_2$/Ni(II) complex through coordination with the two carbonyl groups, via a lowering LUMO strategy. The 2,4,6-triisopropylaniline moieties in the chiral *N,N′*-dioxide ligand **92** ensured the *Si*-face approach of the 3-vinylindoles **90**, providing the product exclusively in *exo* fashion.

Recently, the same group established the asymmetric Diels–Alder reaction/[3,3]sigmatropic rearrangement cascade of methyleneindolinones **1** with 1-thiocyanatobutadienes **101** for the synthesis of spiranic cyclohexenyl isothiocyanates **102** (Scheme 3.10) [20]. The reaction uses Ni(II) complex and the chiral *N,N′*-dioxide ligand **103**, providing a range of spiro cyclohexenyl isothiocyanates

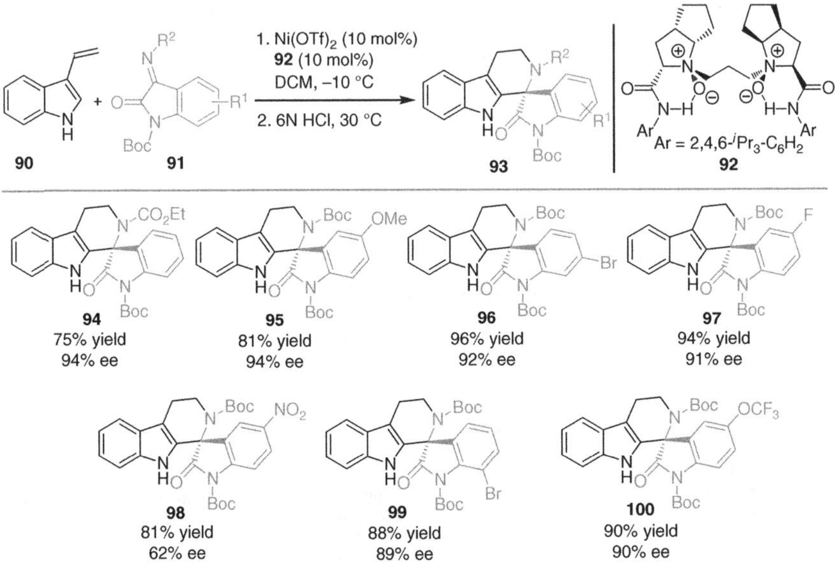

Scheme 3.9 Regio- and enantioselective aza-Diels–Alder reactions of 3-vinylindoles. *Source:* Modified from Zheng et al. [19].

in high yields (64–97%) with excellent diastereo (>15 : 1 in all the cases) and enantioselectivities (65–94% ee). Mechanistically, the reaction involves the initial activation of methyleneindolinones **1** by the chiral L-PiPr$_2$/Ni(II) complex through coordination with the two carbonyl groups, generating intermediate **104**. Subsequently, the 1-thiocyanatobutadienes **101** participates in a stereoselective quasi-concerted Diels–Alder cycloaddition from the *Re* face of **104** to afford intermediate **105**, which further undergoes a suprafacial [3,3]sigmatropic rearrangement to furnish the final product **102**. Interestingly, the authors applied the developed method to the gram-scale synthesis of NITD609, an antimalarial active spirocyclic indole. Shortly after, the same group also developed a catalytic asymmetric Diels–Alder reaction for the synthesis of optically pure spiro[cyclohexane-oxindoline] derivatives using Zn(OTf)$_2$ and a chiral *N,N′*-dioxide ligand [21].

3.2.3 Organometallic Miscellaneous Strategies to Construct Spiro Compounds

The following reports are based on the construction of strained small ring spirocyclic scaffolds and on the utilization of merged metal- and organocatalytic strategies. The first example deals with an enantioselective [2+2] cycloaddition of disubstituted ketenes **114** and isatins **115** for the synthesis of spirocyclic β-lactones

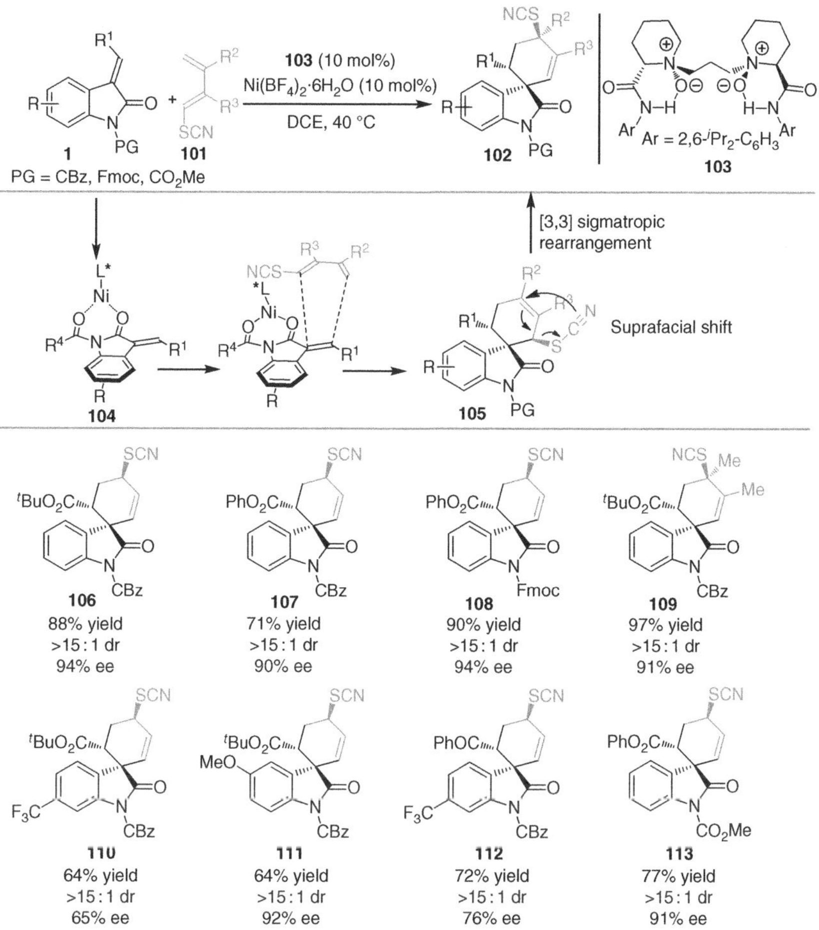

Scheme 3.10 Ni-catalyzed asymmetric Diels–Alder/[3,3]sigmatropic rearrangement cascade between methyleneindolinones with 1-thiocyanatobutadienes. *Source:* Modified from Zhou et al. [20].

116, reported by the group of Feng in 2014 (Scheme 3.11) [22]. The method utilizes $Sc(OTf)_3$ as Lewis acid catalyst in the presence of the chiral *N,N′*-dioxide ligand **117**, responsible for controlling the stereochemistry of the product **116**. A variety of chiral β-lactones were obtained in excellent yield (80–96%) with very high diastereo- and enantioselectivity (>20 : 1 dr and 88–97% ee).

In 2015, Tanaka and coworkers outlined an enantioselective [2+2+2] cycloaddition approach to construct spiro-cyclohexadienes **123**, starting from terminal alkynes **124**, acetylenedicarboxylates **125**, and cyclopropylideneacetamides **126** (Scheme 3.12) [23].

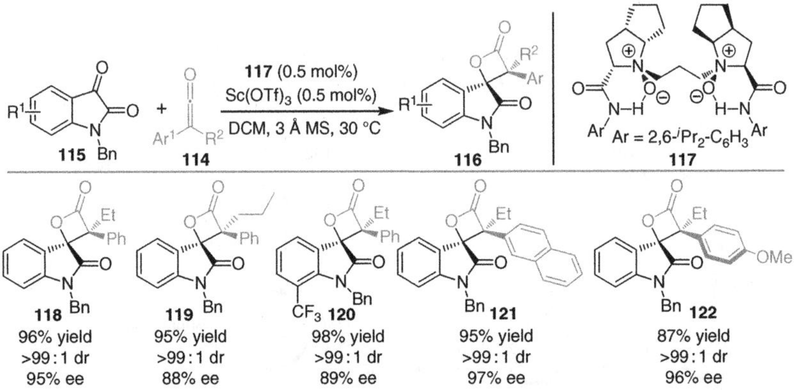

Scheme 3.11 Scandium-catalyzed asymmetric cycloaddition between ketenes and isatins. *Source:* Modified from Hao et al. [22].

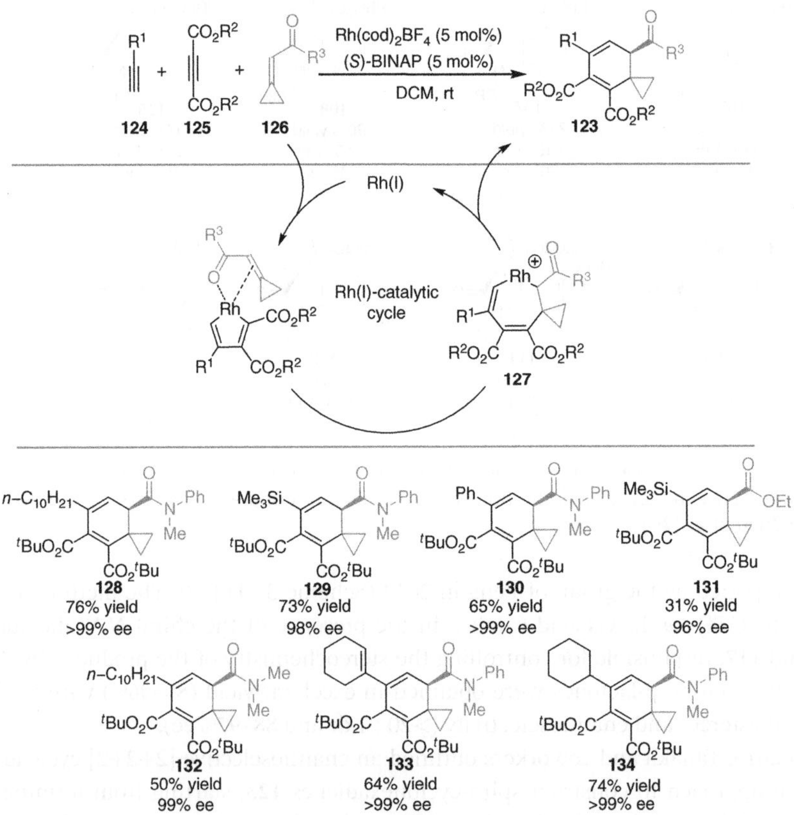

Scheme 3.12 Rhodium-catalyzed [2+2+2] cycloadditions between alkynes and cyclopropylideneacetamides. *Source:* Based on Yoshida et al. [23].

In the reported reaction mechanism, a rhodium(I)/(*S*)-binap cationic complex is responsible for the excellent stereocontrol (96%–>99% ee), affording the spiro cyclohexadiene products **123** in moderate to good yield (31–76%). Mechanistically, the reaction proceeds through the formation of intermediate **127** by coupling the two terminal alkynes **124** and **125** with cyclopropylideneacetamides **126** in the presence of Rh(I) complex. Regioselective insertion of **126** produces the intermediate **127** which undergoes direct reductive elimination to furnish the final spiranic product **123**. In a subsequent report, the authors extended the scope of the [2+2+2] cycloaddition to 1,6-enynes instead of alkynes [24].

In 2018, the group of Rios reported an enantioselective synthesis of spiropyrazolones **135** by a formal [6+2] cycloaddition reaction (Scheme 3.13) [25]. The

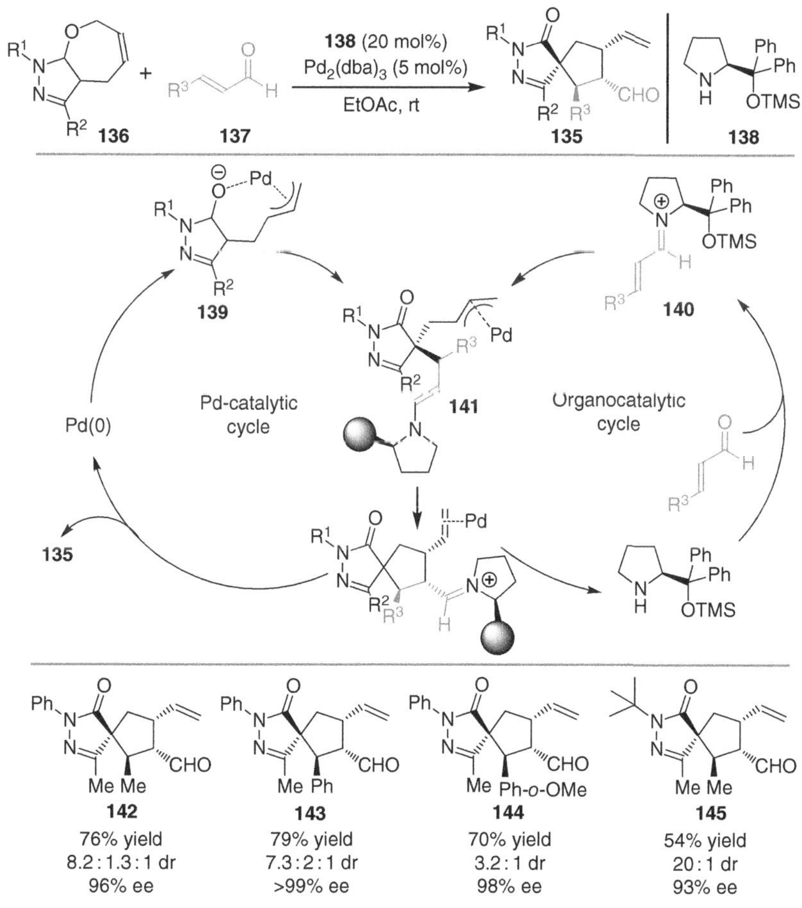

Scheme 3.13 Synergistic palladium/chiral secondary amine-catalyzed formal ring contraction for the enantioselective synthesis of spiropyrazolones. *Source:* Modified from Meazza et al. [25].

method relies on the catalytic synergistic activation of the seven-membered pyrazolone derivatives **136** by Pd along with the activation of enals **137** by the chiral secondary amine **138**. Specifically, the reaction is initiated by the simultaneous formation of highly active Pd-complex **139** and the chiral iminium ion intermediate **140**. Stereoselective interception of the iminium ion by the nucleophilic Pd-complex **139** provides the enamine intermediate **141** bearing the Pd-coordinated allyl cation. Intramolecular ring closing between the enamine moiety and allyl cation in **141** furnishes the final spiranic products **135**. The [5,5]spiropyrazolone products are obtained in good yields (54–79%), moderate to good diastereoselectivities (3.2 : 1 to 20 : 1), and excellent enantiocontrol (93%–>99%).

3.3 Organocatalytic Methodologies

The versatility and robustness of the different organocatalytic activation modes have been implemented in a number of stereoselective cascade processes, addressing the requests of atom and step economy of modern synthetic chemistry. Remarkably, organocatalysts are generally robust, readily available, and their easy scaffold modifications allow the straightforward generation of different structural variants for the development of highly diversified asymmetric transformations. In this scenario, organocatalytic methodologies represent a powerful strategy for the preparation of complex spiro compounds with high optical purity. The section has been divided into four subsections according to the involved reactions: [3+2] cycloadditions, [4+2] cycloadditions, [4+3]-, [2+2] cycloadditions and switchable strategies, and miscellaneous reactions.

3.3.1 Organocatalytic [3+2] Cycloaddition Strategies to Construct Spiro Compounds

The group of Glorius envisaged an a^3–d^3 umpolung reactivity strategy of β,β-disubstituted enals to construct spirocyclic oxindoles **146** bearing two highly congested contiguous quaternary carbon centers (Scheme 3.14) [26]. In this manner, the diastereoselective annulations of a variety of isatins **147** with β,β-disubstituted enals **148** were successfully developed under *N*-heterocyclic carbene catalysis **149**. The use of a pivalic acid as Brønsted acid cocatalyst revealed crucial in order to ensure high yields and diastereoselectivities (8 : 1–>20 : 1 dr and 68–98% yields). Moreover, the authors carried out an enantioselective variant using a chiral NHC catalyst where the acid cocatalyst also had a beneficial effect on the reaction enantioselectivity (one example, 84% ee). The authors proposed that after the formation of Breslow intermediate, the acid cocatalyst stabilizes the intermediate **150** *via* two hydrogen bonds, promoting the diastereoselective formation of the adduct

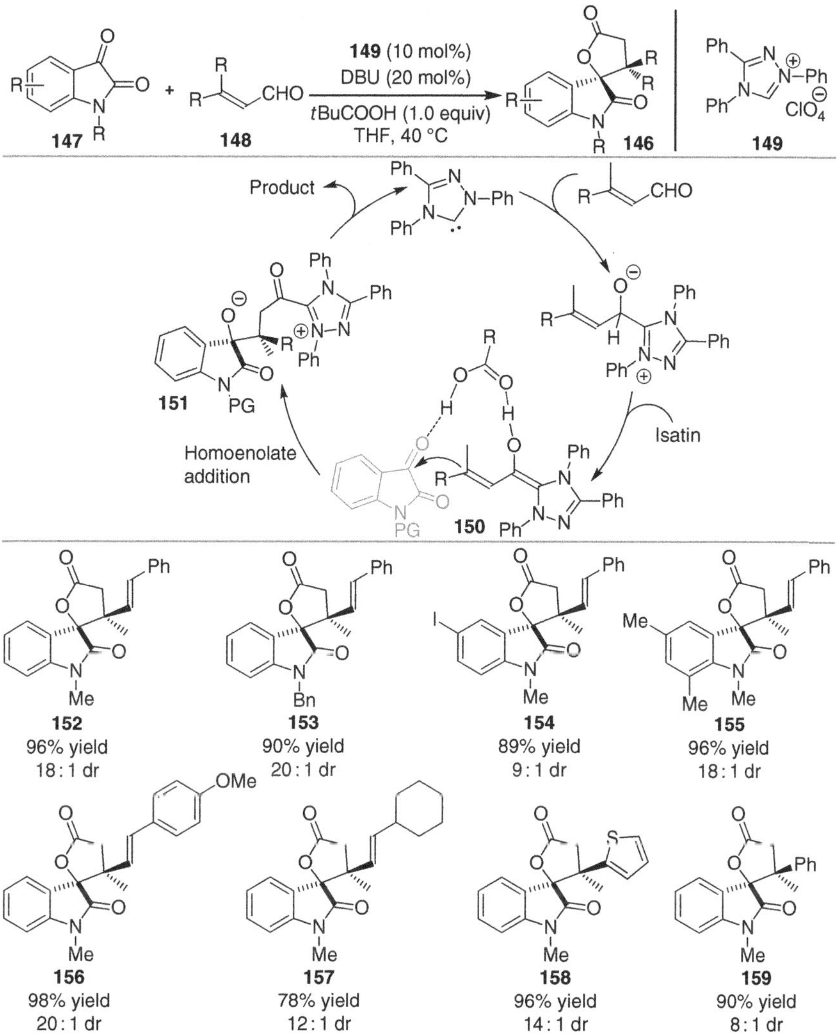

Scheme 3.14 Conjugate umpolung of β,β-disubstituted enals and isatins promoted by *N*-heterocyclic carbene/Brønsted acid dual catalysis. *Source:* Modified from Li et al. [26].

151. Subsequently, intramolecular attack of the alkoxide moiety at the carbonyl group in **151** affords the desired spirooxindole **146** and regenerates the NHC catalyst.

Substrates that bear multiple nucleophilic and electrophilic sites are highly useful syntons to rapidly generate architecturally complex molecules and bioactive compound libraries. In this context, Sun and coworkers reported a method for the construction of five-membered spirocyclic oxindoles **160** based on a Michael–Mannich cascade reaction of a ketimine intermediates, characterized by a dichotomous reactivity profile, with methyleneindolinones **1** (Scheme 3.15) [27]. The one-pot process is catalyzed by the bifunctional quinine-derived squaramide **161** in an enatioselective fashion. Several β-dicarbonyl substrates **162** bearing different alkyl chains and both electron-withdrawing and electron-donating groups on the

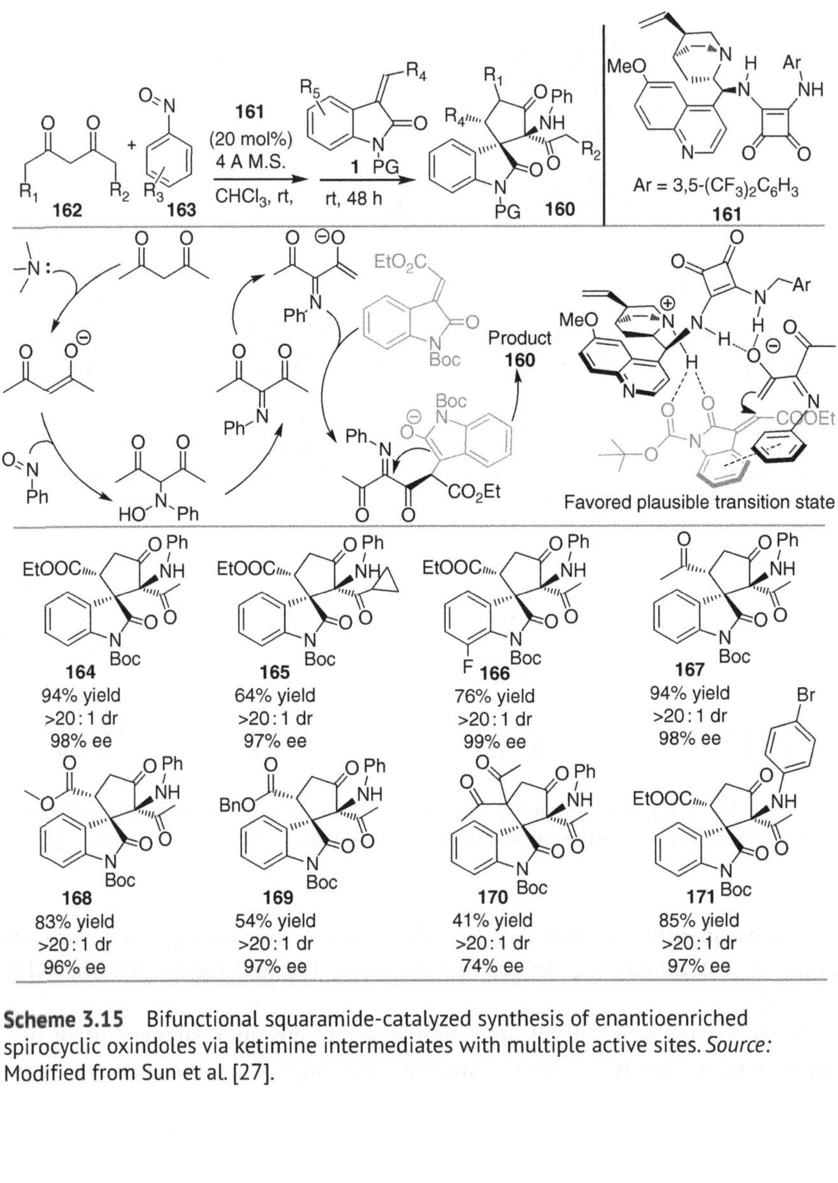

Scheme 3.15 Bifunctional squaramide-catalyzed synthesis of enantioenriched spirocyclic oxindoles via ketimine intermediates with multiple active sites. *Source:* Modified from Sun et al. [27].

aryl ring of the methyleneindolinone **1** and the nitrosoarene **163** are well tolerated (31–94% yields, >20 : 1 dr and 95–99% ee). Also, various electron-withdrawing groups (R^4) at the methylene position afforded the desired products with high stereocontrol. Control experiments pointed out the importance of the Boc group as a hydrogen-bond acceptor for activating the methyleneindolinones. The proposed mechanism is depicted in Scheme 3.15; enolization of acetylacetone promoted by the tertiary amine group of the bifunctional catalyst is followed by *N*-selective addition to nitrosobenzene under basic conditions. The resulting ketimine is deprotonated and the methyleneindolinone is activated by the same catalyst to promote the intermolecular Michael addition. The final irreversible Mannich-type cyclization proceeds smoothly to afford the enantioenriched spiranic products **160**.

Wang, Hong, and coworkers reported the first stereoselective 1,3-dipolar cycloaddition reaction between azlactones **172** and methyleneindolinones **1** catalyzed by chiral phosphoric acids (Scheme 3.16) [28]. The methodology gives

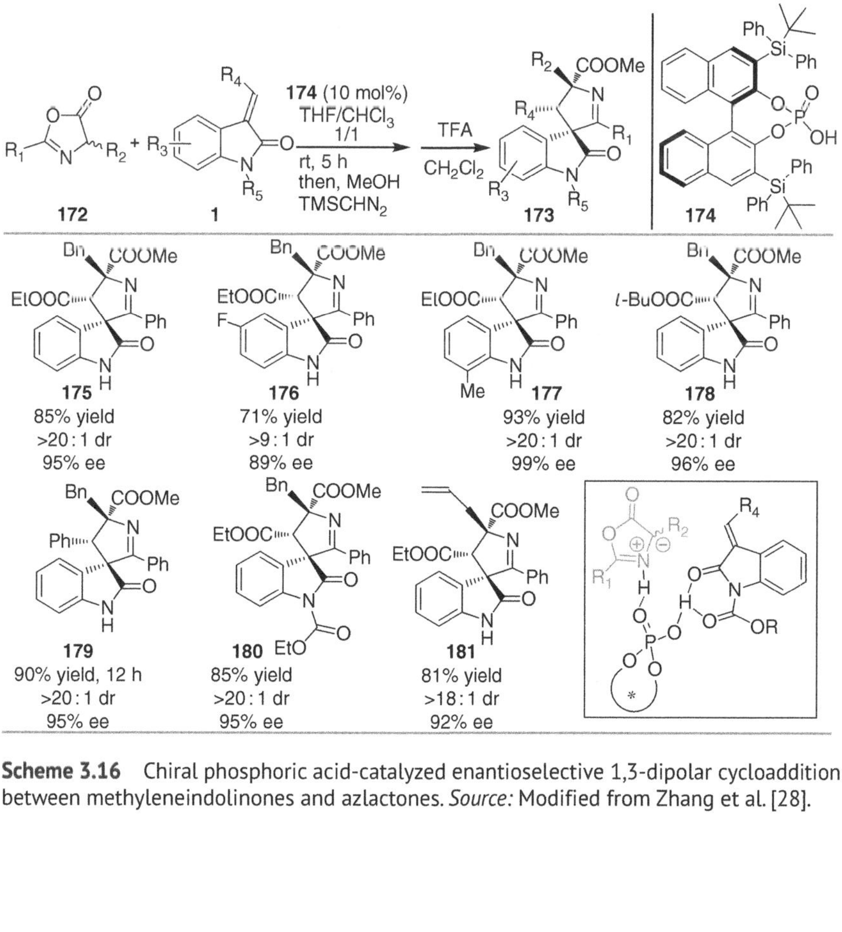

Scheme 3.16 Chiral phosphoric acid-catalyzed enantioselective 1,3-dipolar cycloaddition between methyleneindolinones and azlactones. *Source:* Modified from Zhang et al. [28].

access to highly substituted 3,3-pyrrolidonyl spirooxindole scaffolds **173**, which are present in a number of natural products. The authors take advantage of the high versatility of azlactones, which under the effect of Brønsted acids can form *N*-protonated 1,3-dipoles. The dipole reacts with the methyleneindolinone **1** wherein the hydroxyl proton and the phosphoryl oxygen of the chiral phosphoric acid form double H-bond interactions with both substrates, channeling the reaction in an enatioselective manner. The substituents on the binol backbone of the catalysts have a significant influence on the reaction enantiocontrol, where the bulkier catalyst **174** gave the best results. A wide variety of substitution patterns and functional groups on both substrates are well tolerated, affording the products **173** with good to excellent yields, diastereo- and enantioselectivities.

3.3.2 Organocatalytic [4+2] Cycloaddition Strategies to Construct Spiro Compounds

In 2016, Jørgensen and coworkers reported the asymmetric synthesis of spiroindenes **182** containing up to four contiguous stereocenters via trienamine catalysis (Scheme 3.17) [29]. The reaction between benzofulvenes **183** and 2,4-dienals **184** catalyzed by the diphenylprolinol silyl ether catalyst **138** and ortho-fluorobenzoic acid (*o*-FBA) as additive furnish the indenes spiro-fused to cyclohexene compounds **182** through a formal [4+2] cycloaddition. A range of differently substituted spiroindene compounds was formed efficiently in high yields (66–98%) and excellent stereoselectivities (6 : 1–>20 : 1 dr and 79–99% ee). The authors proposed that the all-*E*-configured trienamine intermediate reacts with the benzofulvene via an *endo* transition state where charge stabilization between the nitrile group and the enamine moiety could be a significant factor for the catalytic performance and the observed stereoselectivity. Finally, the Michael acceptor present in products **182** was demonstrated to facilitate further intramolecular cyclizations to yield complex polycyclic products. Thus, reduction of the aldehyde moiety to the alcohol or reductive amination with benzylamine triggered the intramolecular oxa- or aza-Michael additions to yield the tetracyclic spirocompounds **192** and **193** as a single diastereoisomer in 58 and 54% yield, respectively.

The same group reported the first asymmetric organocatalyzed [4+2]-cycloaddition reaction via a tetraenamine intermediate **194a–b** for the construction of highly functionalized spirocyclic oxindoles **195** (Scheme 3.18) [30]. The reaction between 2-(cyclohepta-1,3,5-trien-1-yl)acetaldehyde **196** and 3-alkyliden oxindoles **197** in the presence of diphenylprolinol silyl ether catalyst **198** forms a wide range of spiranic products in high stereoselectivities (86 : 14–>95 : 5 dr and 68–95% ee) containing a highly functionalized six-membered ring with an annulated seven-membered ring. Mechanistic investigations suggested a stepwise mechanism *via* tetraenamine formation, conjugate

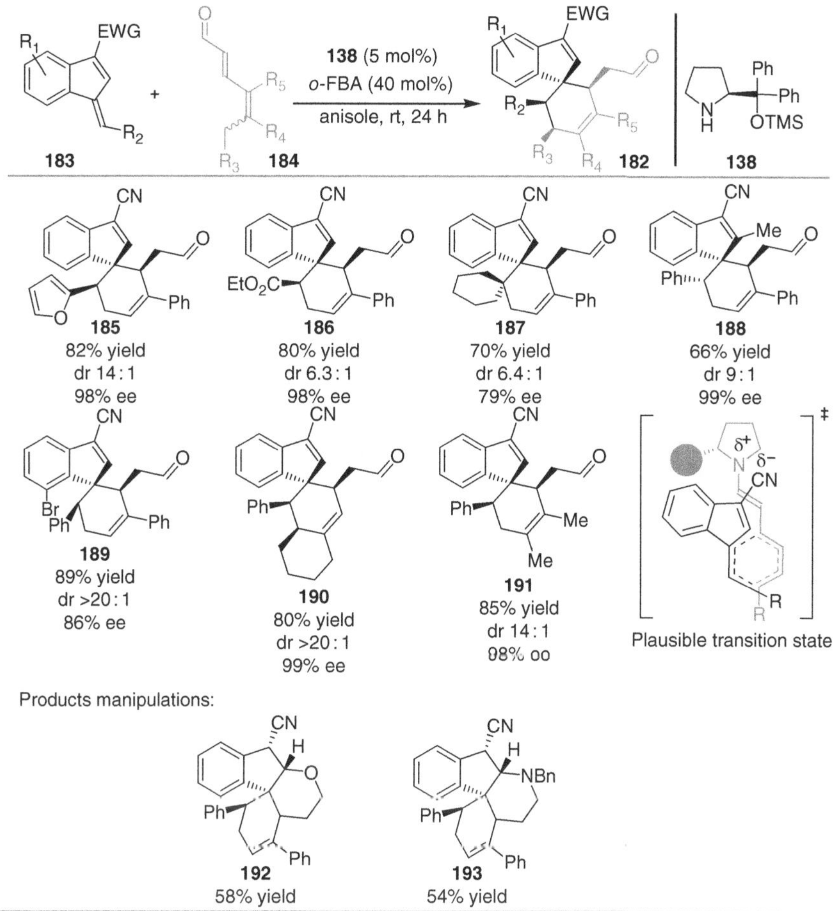

Scheme 3.17 Stereoselective synthesis of spiroindene via trienamine catalysis with benzofulvenes. *Source:* Modified from Donslund et al. [29].

addition, catalyst hydrolysis, and a final cyclization/isomerization sequence. Although the *s-trans* conformer of the nucleophilic enamine **194a** was detected by NMR spectroscopy, the authors proposed that the reactive intermediate is the enamine **194b**, based on computational studies as well as the stereochemical outcome of the reaction.

In 2014, Connon and coworkers developed the first catalytic asymmetric Tamura [4+2] cycloaddition reaction for the construction of spiro compounds **206** (Scheme 3.19) [31]. Alkylidene oxindoles **203** react with enolizable anhydrides **204** by means of a quinine-derived squaramide bifunctional catalyst **205** to generate the densely functionalized spirooxindoles **206** bearing three

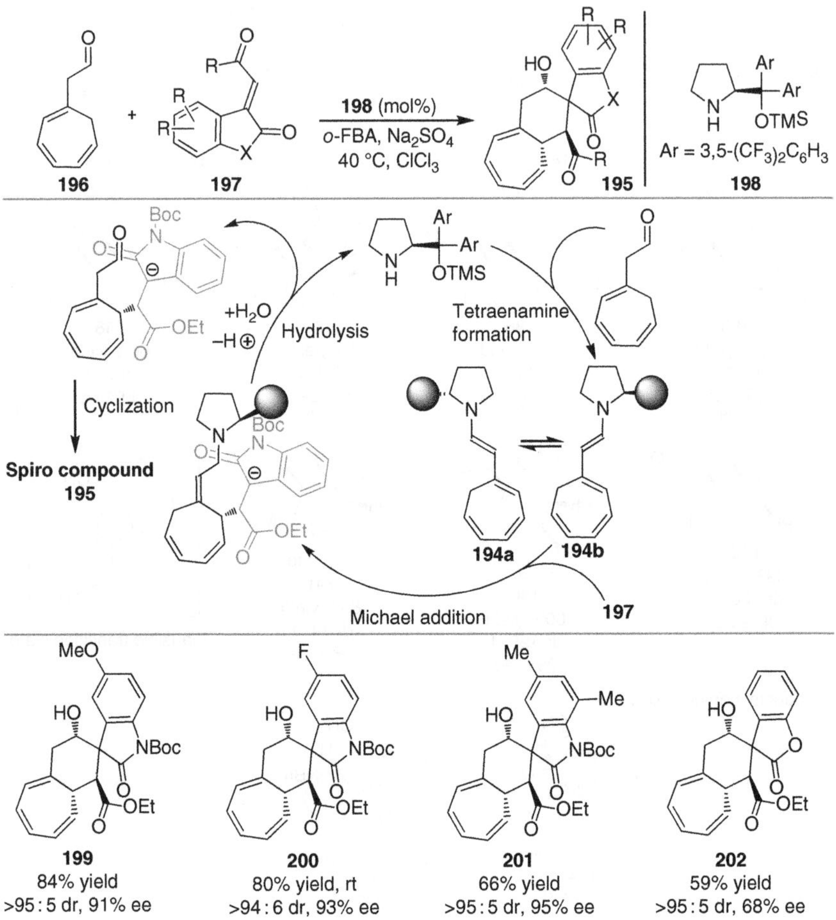

Scheme 3.18 Organocatalytic [4+2] addition reactions *via* tetraenamine intermediate. *Source:* Modified from Stiller et al. [30].

contiguous stereocenters with excellent enantiocontrol (89%–>99% ee). The methodology tolerates a wide range of enolizable anhydrides and alkylidene oxindoles with different stereoelectronic requirements. Interestingly, using 20 mol% of the organocatalyst at −50 °C, the kinetic diastereomer **213** can be isolated in 98% ee. Thus, by simply changing the reaction temperature is possible to switch the stereoselectivity of the process, an important approach in target-oriented synthesis.

In 2017, Yang and coworkers designed a sulphonamide-based H-bond donor organocatalyst derived from *L*-pyroglutamic acid, a cheap chiral feedstock

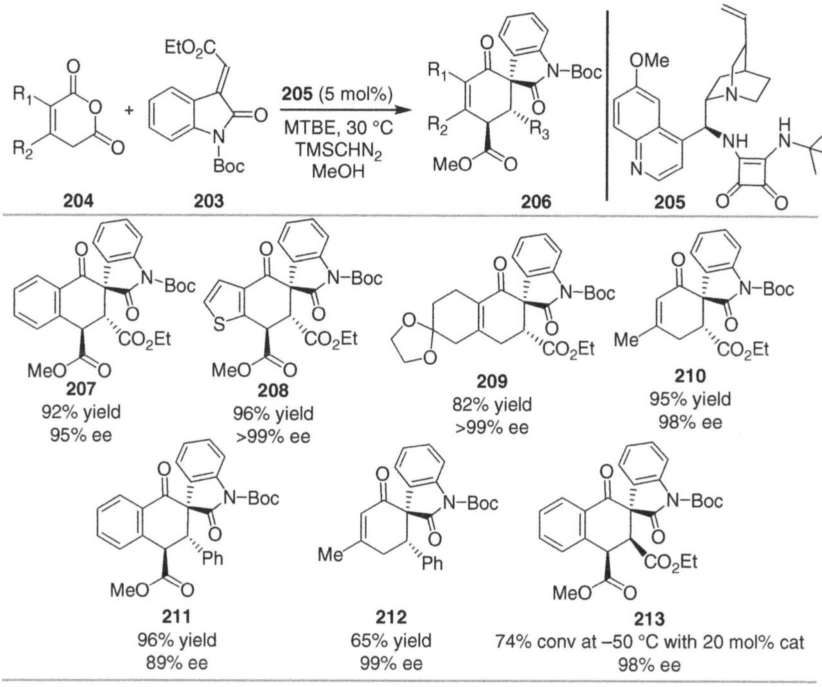

Scheme 3.19 Chiral H-bond donor-catalyzed asymmetric Tamura [4+2]-cycloaddition. *Source:* Modified from Manoni and Connon [31].

available in both enantiomeric forms (Scheme 3.20) [32]. The catalyst **214** was successfully implemented in the [4+2] cyclization between methyleneindolinones **15** and 2-vinyl-1H-indoles **215** to assemble carbazolespirooxindoles **216** in excellent yields (62–99%) and stereoselectivities (81–>99% ee, >20 : 1 dr). The methodology has a broad tolerance to structural changes in both substrates (25 examples). Control experiments suggest that N–H groups in both substrates actively interact with the organocatalyst **214**, having sizable influence on the observed enantioselectivity (Scheme 3.20). Moreover, water plays a critical role in the enantioselectivity via an H_2O-relay H-bond between the catalyst and the 2-vinyl-1H-indole.

3.3.3 Organocatalytic [4+3] and [2+2] Cycloaddition and Switchable Strategies to Construct Spirocompounds

Very recently, the group of Fu and Huang developed an NHC-enabled formal [4+3] annulation of flexible oxotryptamines **223** with enals **224** under mild

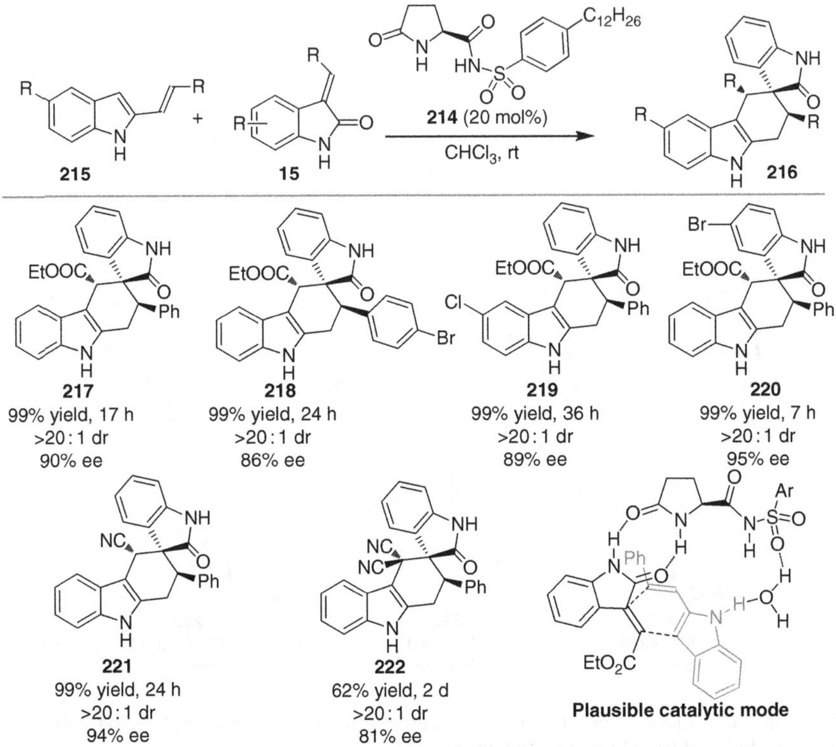

Scheme 3.20 Enantioselective Diels–Alder cyclization for the synthesis of carbazolespirooxindoles. *Source:* Modified from Ren et al. [32].

reaction conditions (Scheme 3.21) [33]. A diverse set of enantioenriched seven-membered heterocycles attached to spiro-ε-lactam oxindoles **225** was obtained with excellent enantioselectivities (90–>99% ee), and moderate to good diastereomeric ratios (3.2 : 1 to 13 : 1 dr). The aminoindanol-derived triazolium pre-catalyst **226** in the presence of DMAP and the oxidant **227** revealed the optimal catalytic system for the transformation. Different oxotryptamine derivatives and enals were successfully used as reaction substrates (22 examples) including a β-alkyl enal, although in the latter case, the corresponding product **235** was obtained in lower yield and diastereoselectivity maintaining excellent asymmetric induction.

In 2015, Jørgensen and coworkers reported an elegant catalytic methodology for the synthesis of biologically relevant complex spirocyclobutaneoxindole derivatives **236** (Scheme 3.22) [34]. The chemistry is based on the organocatalytic enamine-based activation of cyclopropanes **237**, forming a reactive donor–acceptor cyclopropane intermediate that in the presence of 3-alkylidene oxindoles **238**

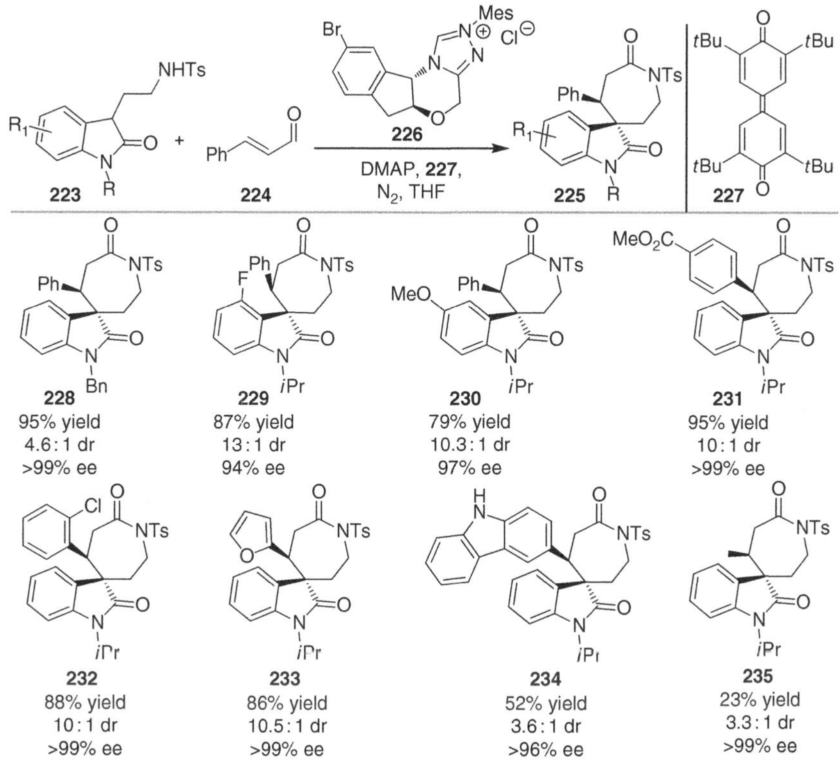

Scheme 3.21 NHC-catalyzed [4+3] annulation of oxotryptamines with enals to access enantioenriched spiro-ε-lactam oxindoles. *Source:* Modified from Liu et al. [33].

promotes the formal [2+2] cycloaddition reaction to construct the spiranic products **236** in good yields, high diastereomeric ratios, and excellent enantiomeric excesses (15 examples, 3.2 : 1->25 : 1 dr,70–97% ee). In fact, the reaction outcome is surprising since the traditional Lewis acid activation of donor–acceptor cyclopropanes typically promotes the [3+2]-cycloaddition pathway. After mechanistic studies, the authors propose two possible reaction mechanisms for the spirocyclobutaneoxindole formation. The first plausible mechanism involves the organocatalytic activation of the cyclopropyl functionality leading to the formation of the iminium-ion intermediate **239** by ring opening. From this intermediate, the dienamine **240** can be formed by deprotonation. Then, this dienamine intermediate reacts in a formal [2+2]-cycloaddition with the 3-olefinic oxindole **238** to form the spirocyclobutaneoxindole product **236**. An alternative mechanistic possibility considers an initial [3+2]-cycloaddition between the 3-alkylidene oxindole and the organocatalytically activated cyclopropane, followed by a ring-contracting

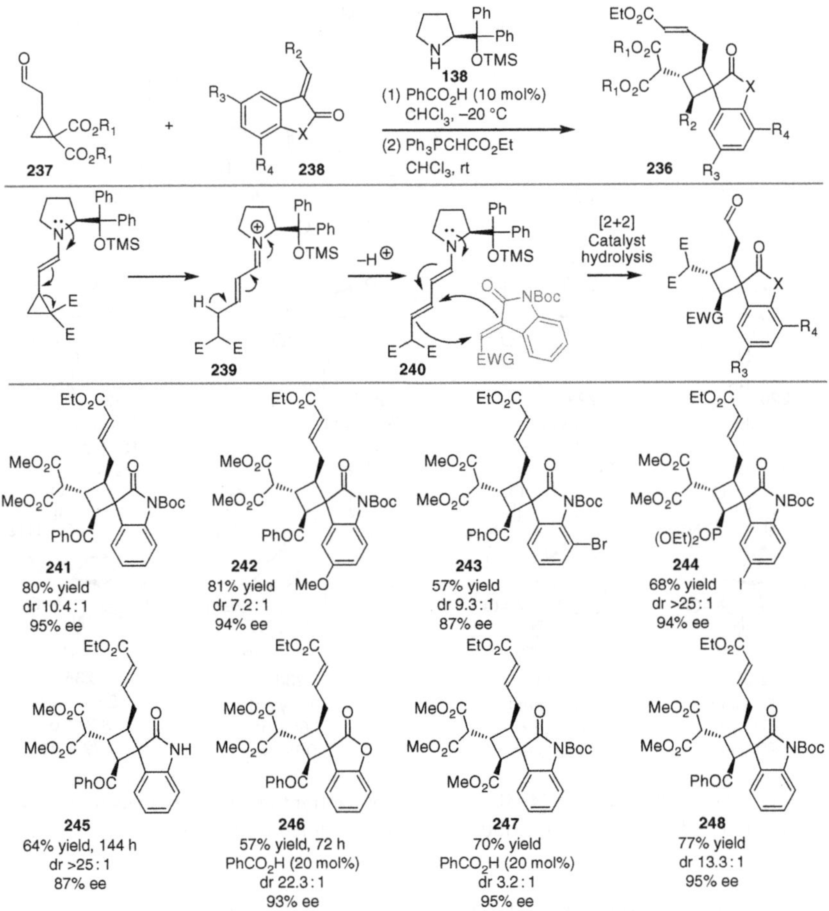

Scheme 3.22 Organocatalytic enamine-activation of cyclopropanes for highly stereoselective formation of spiranic cyclobutanes. *Source:* Modified from Halskov et al. [34].

rearrangement. However, the lack of a rational driving force for this rearrangement makes the latter reaction pathway more unlike.

The development of switchable intermolecular reactions that afford divergent reaction pathways is an attractive and powerful strategy to build up structural and stereogenic diversity in drug discovery. In this context, Ouyang, Chen, and coworkers reported an aminocatalytic divergent cycloaddition toward the formation of two different types of spirocompounds **249** and **250** using the same set of substrates (Scheme 3.23) [35]. By simply changing the cocatalyst, the reaction pathway is switched from trienamine to dienamine activation mode. On the one hand, the *in situ*-generated 4-aminofulvenes act as 6π partners in the γ,β′-regioselective

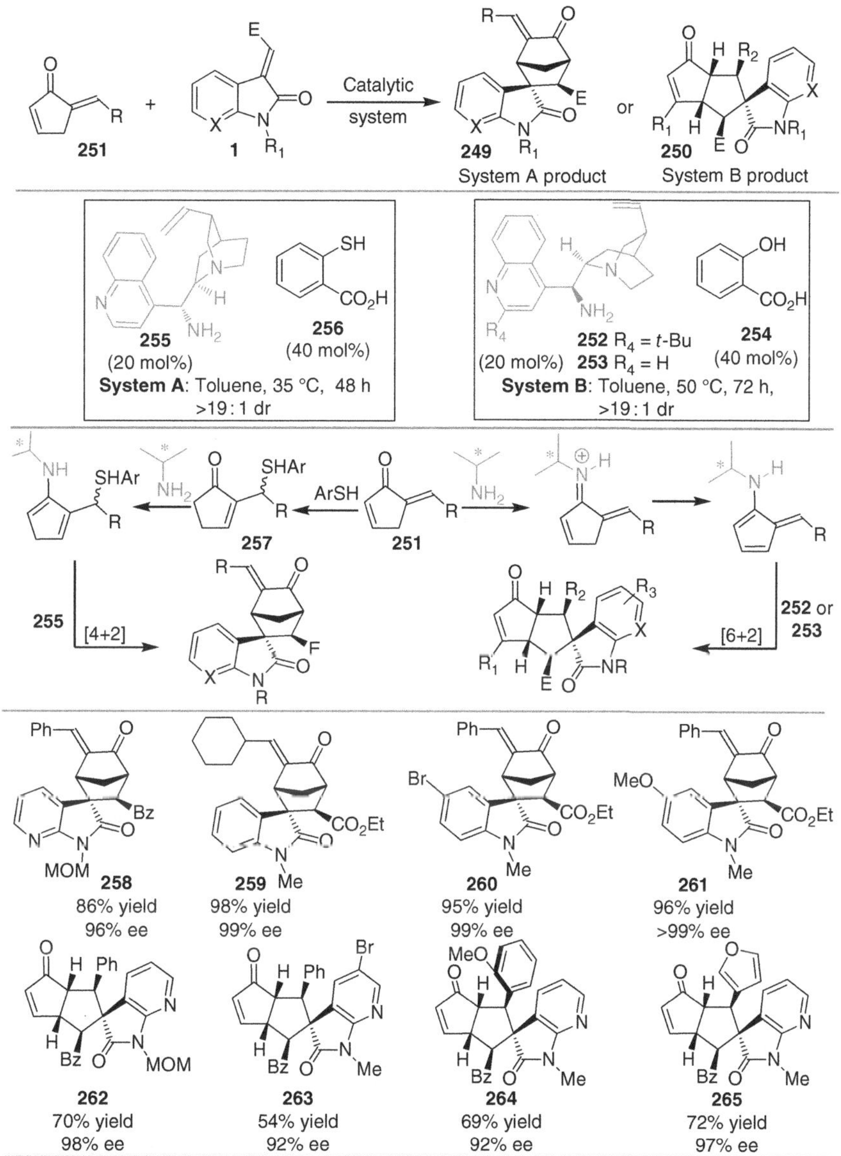

Scheme 3.23 Chiral primary amine-catalyzed regiodivergent asymmetric cycloadditions. *Source:* Modified from Zhou et al. [35].

[6+2] cycloaddition with activated 2π alkenes under the catalysis of **252** or **253** and salicylic acid **254**. On the other hand, the α,γ-regioselective [4+2] cycloaddition reaction is promoted with the catalytic system **255**/2-mercaptobenzoic acid **256**. In the latter case, the β′-regioselective sulfur addition of thiol **256** to **251** gives access to the enone **257** that after condensation with the chiral amine forms the 4π dienamine component in the reaction with activated alkenes. The [6+2] cycloaddition reaction mode provides fused bicycles linked to a pyrrolidinone ring **250** with five contiguous stereogenic centers (28 examples) while the [4+2] pathway affords spiro-bridged cyclic scaffold **249** (22 examples).

3.3.4 Organocatalytic Miscellaneous Strategies to Construct Spiro Compounds

The group of Tu reported the enantioselective synthesis of spiro[4.4]nonane-1,6-diones **266**, which represents the first direct example for the asymmetric synthesis of cyclopentanones using a Nazarov cyclization/semipinacol-rearrangement sequence (Scheme 3.24) [36]. Substrate design revealed key for the successful development of the present strategy, where a cyclobutanol motif is connected to the α-position of the cyclization precursor to induce the 1,2-rearrangement. This tandem process is promoted by means of a chiral *N*-triflylphosphoramide catalyst **268** to construct complex spirocompounds with four consecutive stereocenters in excellent stereocontrol (15 examples, 93 : 7–>99 : 1 dr, 84–97% ee). Computational calculations were performed to understand the reaction mechanism as well as the stereochemical outcome. The reaction begins with the formation of a hydrogen-bond complex between the substrate and the catalyst. Then, the substrate is protonated by the catalyst and can be regarded as a pentadienyl cation, which then undergoes 4π-conrotatory electrocyclization via TS1. The Nazarov reaction step is irreversible, since the transition state TS2 following the ring expansion event is much lower in energy. Then, the intermediate **269** undergoes a [1, 2] migration via TS2 to generate the intermediate **270** that is finally protonated to afford the spiranic product **266**.

In other example of Bronsted acid catalysis, Tan and coworkers developed an enantioselective strategy for the construction of axially chiral compounds **277** (Scheme 3.25) [37]. This important class of spirocompounds is widely found in materials, organocatalysts, and ligands. The enantioselective intramolecular cyclization of ketals **278** to afford the SPINOL derivatives **277** is successfully catalyzed by the chiral phosphoric acid **279**. Various ketals with different substitution patterns, including electron-donating and electron-withdrawing groups at the aryl ring, performed smoothly, affording the corresponding products **281–284** in high enantioselectivities and good to excellent yields (90–96% ee, 62–95% yield,

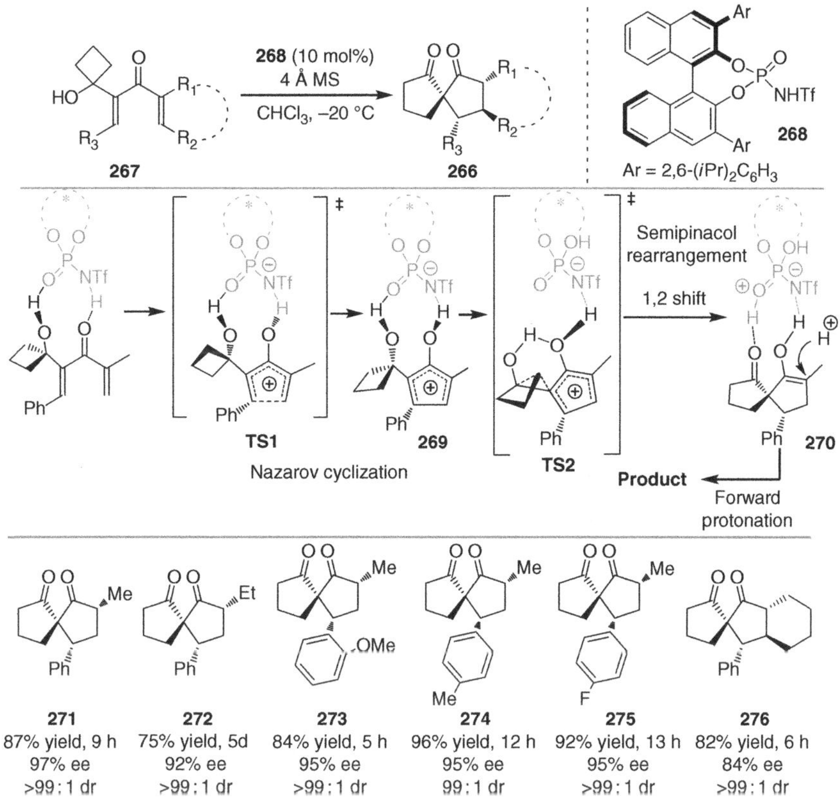

Scheme 3.24 Organocatalytic asymmetric tandem Nazarov cyclization/semipinacol rearrangement. *Source:* Modified from Yang et al. [36].

conditions a). In the case of ketals bearing aromatic groups at the *ortho* position, the catalyst **280** (conditions b) was necessary to obtain the corresponding products. Remarkably, the catalytic system can be used at gram-scale and the catalyst loading can be decreased as little as 0.1 mol%, despite higher temperature and prolonged reaction time are necessary.

On the basis of mechanistic experiments, the authors propose that the reaction may initially form the intermediate **289**, in situ generated from **288**, which further proceed through the active *o*-QM intermediate **290** to deliver the SPINOL derivative. The excellent stereocontrol was attributed to the simultaneous interaction between the bifunctional phosphoric acid and intermediate **291** via hydrogen-bonding interactions.

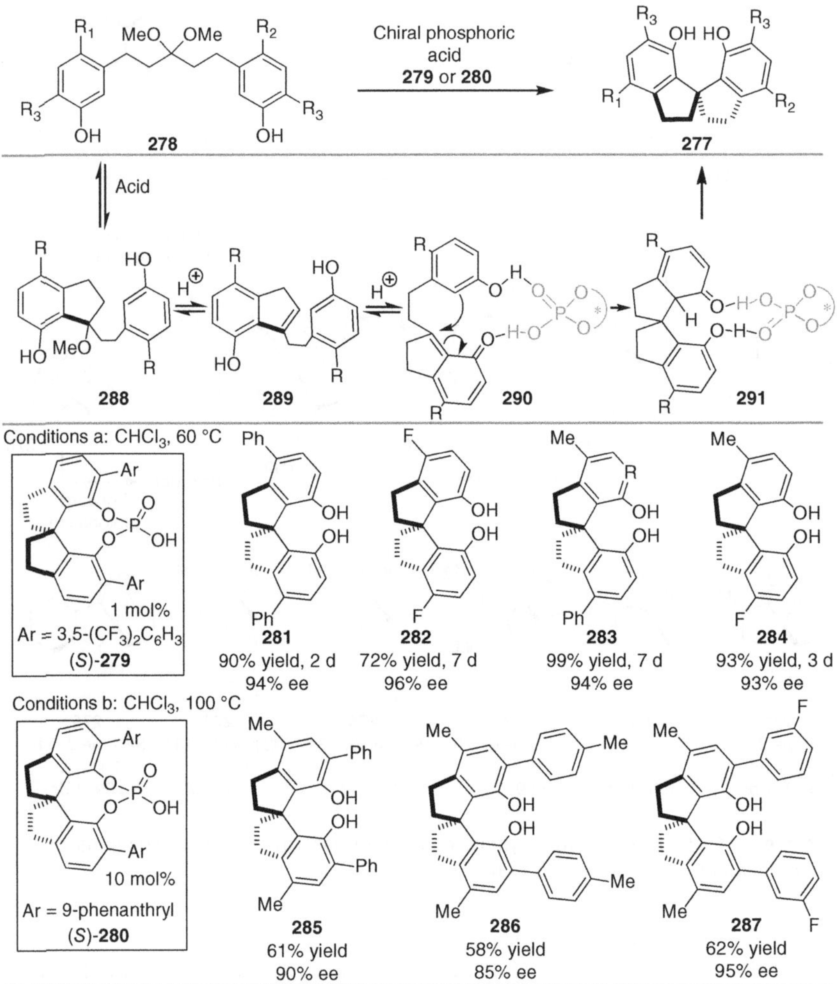

Scheme 3.25 Chiral phosphoric acid-catalyzed asymmetric synthesis of SPINOL derivatives. *Source:* Modified from Li et al. [37].

References

1 Baeyer, A. (**1900**). *Ber. Dtsch. Chem. Ges.* 33: 3771–3775.

2 (a) Lin, H. and Danishefsky, S.J. (**2003**). *Angew. Chem. Int. Ed.* 42: 36–51. (b) Galliford, C.V. and Scheidt, K.A. (**2007**). *Angew. Chem. Int. Ed.* 46: 8748. (c) Undheim, K. (**2014**). *Synthesis* 46: 1957–2006. (d) Smith, L.K. and Baxendale, I.R. (**2015**). *Org. Biomol. Chem.* 13: 9907–9933.

3 Zheng, Y., Tice, C.M., and Singh, S.B. (**2014**). *Bioorg. Med. Chem. Lett.* 24: 3673–3682.

4 For selected reviews, see: (a)Rios, R. (**2012**). *Chem. Soc. Rev.* 41: 1060–1074. (b) Xie, X., Huang, W., Peng, C., and Han, B. (**2018**). *Adv. Synth. Catal.* 360: 194–228. (c) Ding, A., Meazza, M., Guo, H. et al. (**2018**). *Chem. Soc. Rev.* 47: 5946–5996. (d) Xu, P.-W., Yu, J.-S., Chen, C. et al. (**2019**). *ACS Catal.* 9: 1820–1882.

5 For selected reviews, see: (a)Frühauf, H.-W. (**1997**). *Chem. Rev.* 97: 523–526. (b) Reymond, S. and Cossy, J. (**2008**). *Chem. Rev.* 108: 5359–5406. (c) Pellissier, H. (**2011**). *Adv. Synth. Catal.* 353: 189–218. (d) Anand, A., Singh, P., Kumar, V., and Bhargava, G. (**2019**). *RSC Adv.* 9: 25554–25568.

6 For selected reviews, see: (a)Melchiorre, P., Marigo, M., Carlone, A., and Bartoli, G. (**2008**). *Angew. Chem. Int. Ed.* 47: 6138–6171. (b) Bertelsen, S. and Jørgensen, K.A. (**2009**). *Chem. Commun.*: 2178–2189. (c) Valero, G., Companyó, X., Bravo, N. et al. (**2010**). *Synlett* 13: 1883–1908. (d) Cherubini-Celli, A., Mateos, J., Bonchio, M. et al. (**2018**). *ChemSusChem* 11: 3056–3070. (e) Silvi, M. and Melchiorre, P. (**2018**). *Nature* 554: 41–49.(f) Vega-Peñaloza, A., Paria, S., Bonchio, M. et al. (**2019**). *ACS Catal.* 9: 6058–6072.

7 For selected reviews, see: (a)Alba, A.-N.R., Companyó, X., and Rios, R. (**2009**). *Curr. Org. Chem.* 13: 1432–1474. (b) Moyano, A. and Rios, R. (**2011**). *Chem. Rev.* 111: 4703–4832. (c) Hepburn, H.B., Dell'Amico, L., and Melchiorre, P. (**2016**). *Chem. Rec.* 16: 1787–1806. (d) Meazza, M., Companyó, X., and Rios, R. (**2018**). *Asian J. Org. Chem* 7: 1934–1956.

8 Ball-Jones, N.R., Badillo, J.J., Tran, N.T., and Franz, A.K. (**2014**). *Angew. Chem. Int. Ed.* 53: 9462–9465.

9 Arai, T., Ogawa, H., Awata, A. et al. (**2015**). *Angew. Chem. Int. Ed.* 54: 1595–1599.

10 Arai, T., Mishiro, A., Yokoyama, N. et al. (**2010**). *J. Am. Chem. Soc.* 132: 5338–5339.

11 Yang, W.-L., Liu, Y.-Z., Luo, S. et al. (**2015**). *Chem. Commun.* 51: 9212–9215.

12 Yang, W.-L., Tang, F.-F., He, F.-S. et al. (**2015**). *Org. Lett.* 17: 4822–4825.

13 Yin, C., Lin, L., Zhang, D. et al. (**2015**). *J. Org. Chem.* 80: 9691–9699.

14 Zhang, D., Yin, C., Zhou, Y. et al. (**2017**). *Chem. Commun.* 53: 7925–7928.

15 Pham, M.V. and Cramer, N. (**2016**). *Chem. Eur. J.* 22: 2270–2273.

16 Ma, C., Huang, Y., and Zhao, Y. (**2016**). *ACS Catal.* 6: 6408–6412.

17 Shu, T., Zhao, L., Li, S. et al. (**2018**). *Angew. Chem. Int. Ed.* 57: 10985–10988.

18 Jia, Z.-J., Shan, G., Daniliuc, C.G. et al. (**2018**). *Angew. Chem. Int. Ed.* 57: 14493–14497.

19 Zheng, H., Liu, X., Xu, C. et al. (**2015**). *Angew. Chem. Int. Ed.* 54: 10958–10962.

20 Zhou, Y., Lin, L., Liu, X. et al. (**2018**). *Angew. Chem. Int. Ed.* 57: 9113–9116.

21 Zheng, J., Lin, L., Fu, K. et al. (**2015**). *J. Org. Chem.* 80: 8836–8842.

22 Hao, X., Liu, X., Li, W. et al. (**2014**). *Org. Lett.* 16: 134–137.

23 Yoshida, T., Tajima, Y., Kobayashi, M. et al. (**2015**). *Angew. Chem. Int. Ed.* 54: 8241–8244.

24 Yoshizaki, S., Nakamura, Y., Masutomi, K. et al. (**2016**). *Org. Lett.* 18: 388–391.

25 Meazza, M., Kamlar, M., Jašíková, L. et al. (**2018**). *Chem. Sci.* 9: 6368–6373.

26 Li, J.-L., Sahoo, B., Daniliuc, C.-G., and Glorius, F. (**2014**). *Angew. Chem. Int. Ed.* 53: 10515–10519.

27 Sun, Q.-S., Zhu, H., Chen, Y.-J. et al. (**2015**). *Angew. Chem. Int. Ed.* 54: 13253–13257.

28 Zhang, Z., Sun, W., Zhu, G. et al. (**2016**). *Chem. Commun.* 52: 1377.

29 Donslund, B.S., Nielsen, R.P., Mønsted, S.M.N., and Jørgensen, K.A. (**2016**). *Angew. Chem. Int. Ed.* 55: 11124–11128.

30 Stiller, J., Poulsen, P.H., Cruz Cruz, D. et al. (**2014**). *Chem. Sci.* 5: 2052–2056.

31 Manoni, F. and Connon, S.J. (**2014**). *Angew. Chem. Int. Ed.* 53: 2628–2632.

32 Ren, J.-W., Wang, J., Xiao, J.-A. et al. (**2017**). *Org. Chem.* 82: 6441–6449.

33 Liu, D., Hu, Z., Zhang, Y. et al. (**2019**). *Chem. Eur. J.* 25: 11223–11227.

34 Halskov, K.S., Kniep, F., Lauridsen, V.H. et al. (**2015**). *J. Am. Chem. Soc.* 137: 1685–1691.

35 Zhou, Z., Wang, Z.-X., Zhou, Y.-C. et al. (**2017**). *Nat. Chem.* 9: 590–594.

36 Yang, B.-M., Cai, P.-J., Tu, Y.-Q. et al. (**2015**). *J. Am. Chem. Soc.* 137: 8344–8347.

37 Li, S., Zhang, J.-W., Li, X.-L. et al. (**2016**). *J. Am. Chem. Soc.* 138: 16561–16566.

4

Design and Synthesis of Spirocycles *via* Olefin Metathesis

Sambasivarao Kotha, Vikas R. Aswar and Yellaiah Tangella

Department of Chemistry, Indian Institute of Technology Bombay Mumbai, India

4.1 Introduction

Design and synthesis of novel spirocyclic compounds [1a–f] is a challenging task in organic and medicinal chemistry because of the generation of quaternary center, which holds central chirality or an axial chirality. Architecturally, the spirocycles, also termed as spiranes, fused with two rings through a single atom (e.g. carbon, silicon, and phosphorous) that is recognized as spiro-atom. The major advantage of spirocycles is their inherent three-dimensional arrangement of functional groups that induces the rigidity to control the stereochemistry of these molecules and bond rotation, which offers several benefits to drug discovery programs due to significant interactions of a ligand with a three-dimensional binding site. Spirocycles are privileged structures found in a plethora of natural products. Few marketed drugs and medicinally important molecules containing spiro-linkage exhibit a broad spectrum of biological activities. For instance, elatol is a well-known antitubercular drug to treat tuberculosis and diminishes the spread of infection. Similarly, rolapitant is used to avert nausea and vomiting, and tofogliflozin is an antidiabetic agent (Figure 4.1).

Researchers have focused on the assembly of spirocycles by application of a variety of strategies such as cycloaddition reactions [e.g. (2+2), (4+2), and (2+2+2)] along with other key reactions like Claisen rearrangement (CR), Grignard addition, Fischer indolization (FI), Suzuki–Miyaura (SM) cross-coupling, Diels–Alder (DA) reaction, and retro Diels–Alder (rDA) reaction. Besides, ring-closing metathesis (RCM), ring-opening cross-metathesis (ROCM), ring-closing enyne metathesis (RCEM), and ring-rearrangement metathesis (RRM) strategies are used in recent years for the synthesis of several natural products containing spiro center. In this chapter, we have included various strategies that utilize metathesis for the construction of spirocycles in combination with other well-known name reactions, such as CR, Grignard addition, FI, SM

Spiro Compounds: Synthesis and Applications, First Edition. Edited by Ramon Rios Torres.

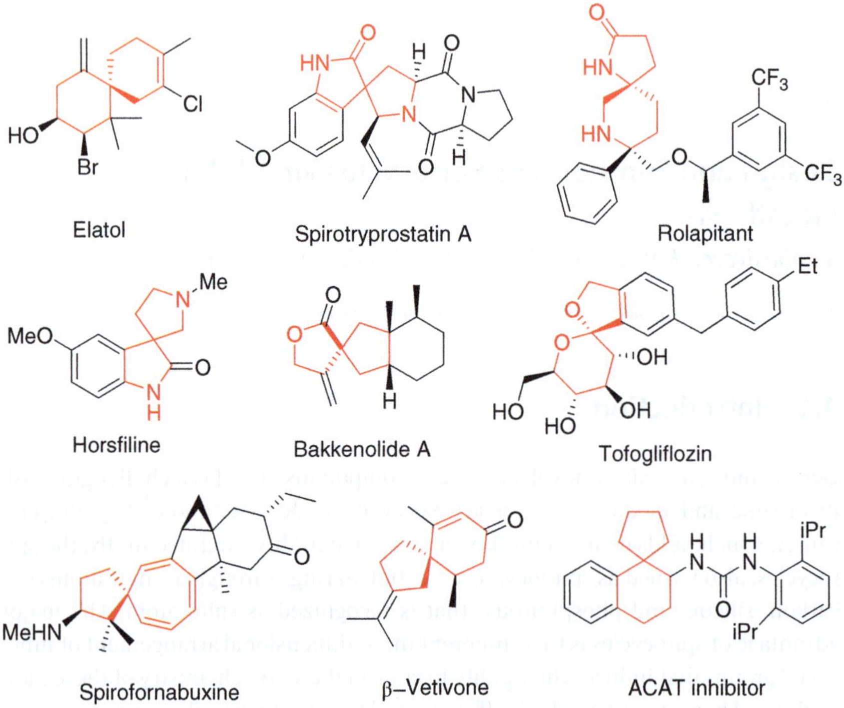

Figure 4.1 Natural and non-natural products containing a spirocenter.

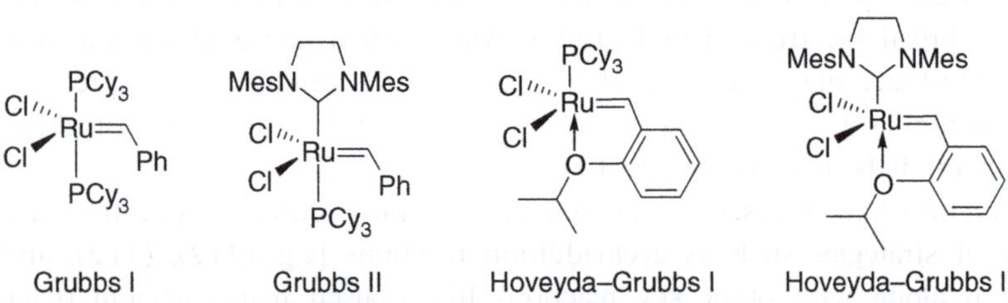

Figure 4.2 List of ruthenium catalysts used in metathesis.

cross-coupling, and rDA reaction. The catalysts used in this chapter to realize the metathesis sequence for the synthesis of spirocycles are Grubbs first-generation catalyst (G-I) **1**, Grubbs second-generation catalyst (G-II) **2**, Hoveyda–Grubbs first-generation catalyst (GH-I) **3**, and Hoveyda–Grubbs second-generation catalyst (GH-II) **4** (Figure 4.2).

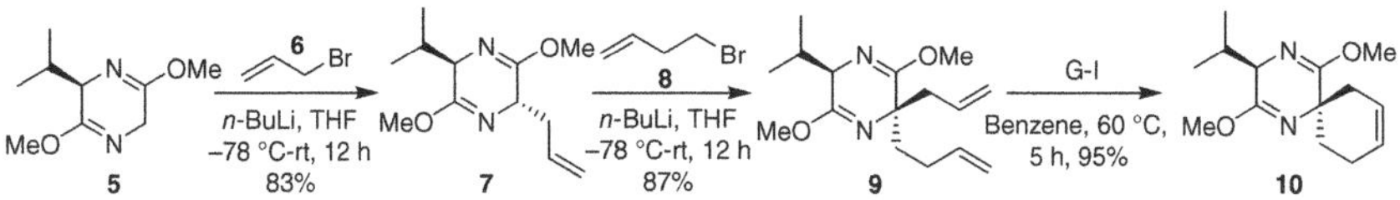

Scheme 4.1 Synthesis of spirocyclic pyrazine. *Sources:* Based on Hammer and Undheim [1g]; Hammer and Undheim [1h]; Andrei et al. [1i]; Jam et al. [1j].

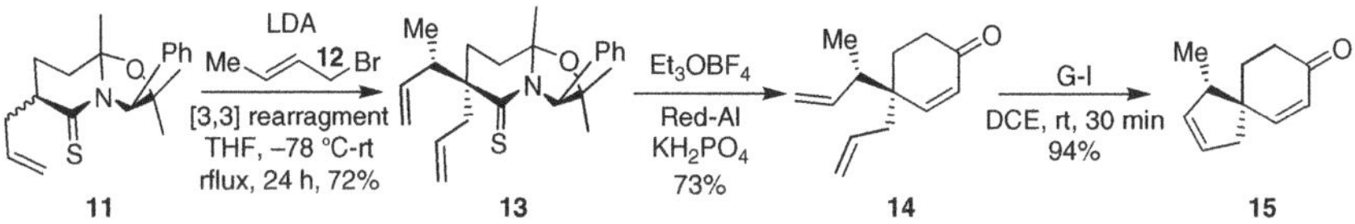

Scheme 4.2 Synthesis of chiral spiro-connected cyclopentene. *Sources:* Based on Lemieux et al. [2a]; Hughes et al. [2b].

Undheim and Hammer reported a stereoselective synthesis of spirocyclic pyrazine derivative **10** *via* RCM by using ruthenium(II)-complex. Initially, they treated the pyrazine derivative **5** with allyl bromide (**6**) using *n*-BuLi to deliver the monoallyl pyrazine derivative **7** in 83% yield. Later, the reaction of **7** with 4-bromo-1-butene (**8**) under similar conditions afforded the diene moiety **9**. Finally, it was subjected to RCM with the aid of G-I (**1**) catalyst to deliver the spirocyclic pyrazine derivative **10**. Later, the spiro derivative **10** undergoes acid hydrolysis to furnish the corresponding amino acid (Scheme 4.1) [1g–j].

Meyers and coworkers reported a versatile approach to the synthesis of chiral cyclohexenone with spiro-linked cyclopentene **15** *via* RCM as a key step. They started with the bicyclic thiolactam **11**, which on treatment with 1-bromo-2-propene (**12**) under basic conditions followed by the [3,3] sigmatropic rearrangement gave the corresponding diallyl bicyclic thiolactam **13**. Later, the diallyl compound **13** reacts with Et_3OBF_4, Red-Al, and KH_2PO_4 sequentially to afford the diallyl cyclohexenone derivative **14**. Finally, the diallyl cyclohexenone **14** undergo RCM with G-I catalyst to yield the corresponding methyl-substituted chiral spiro[4.5]deca-2,6-dien-8-one **15** in excellent yield (Scheme 4.2) [2].

Spitzner and Oesterreich have developed an efficient protocol by using RCM as a key step to synthesize spiro[4.5]–decane derivatives **19** and **20**. Initially, cyclohexanone derivative **16** was treated with methyl acrylate (**17**) to afford vinyl-allyl cyclohexanone **18**, which was subjected to RCM with the aid of G-I catalyst to provide a mixture of regioisomers **19** and **20** (Scheme 4.3) [3].

Colombo and coworkers described the synthesis of spiro-lactam derivative **23** *via* RCM. Reaction of lactam **21** with allyl bromide (**6**) under basic conditions gave the diallyl lactam **22**, which on treatment with G-I catalyst provided the spiro-lactam **23** through RCM sequence (Scheme 4.4) [4].

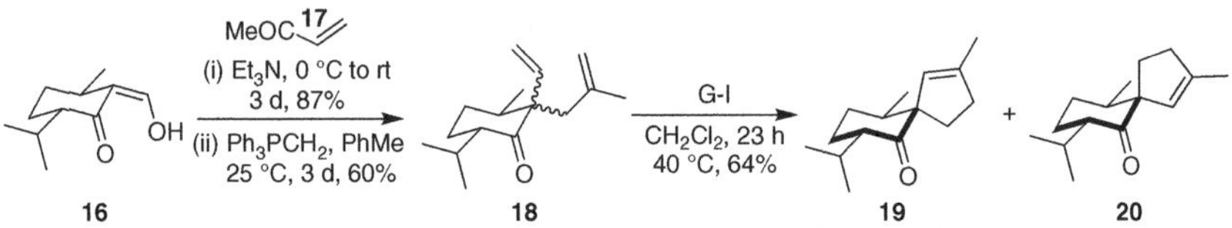

Scheme 4.3 Synthesis of spiro[4.5]-decanes. *Source:* Based on Oesterreich and Spitzner [3].

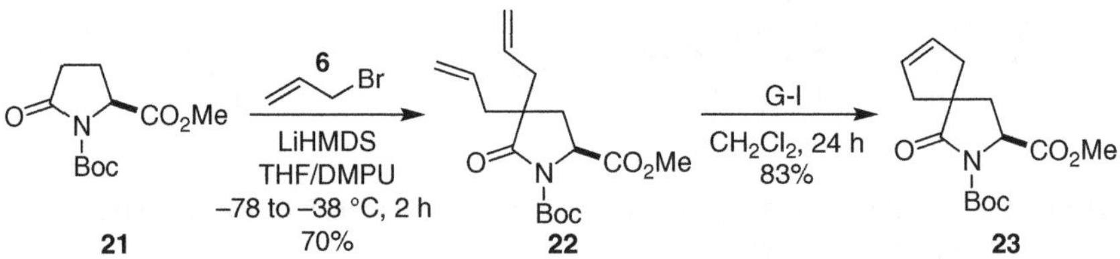

Scheme 4.4 Synthesis of spiro-lactam. *Source:* Based on Colombo et al. [4].

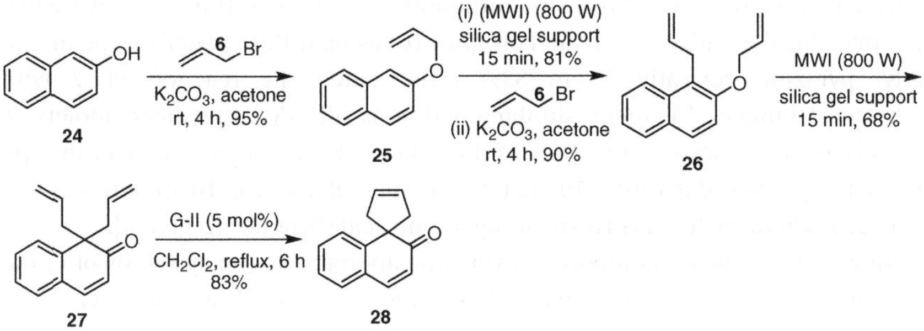

Scheme 4.5 Synthesis of spiro naphthalenone. *Source:* Based on Kotha and Mandal [5].

Kotha and Mandal investigated the application of CR and RCM for the synthesis of spirocycle **28**. In this regard, 2-naphthol (**24**) was treated with allyl bromide (**6**) to deliver the allyl-naphthol ether **25**, which undergo CR under microwave irradiation followed by *O*-allylation furnished the corresponding diallyl naphthol **26**. Further, the naphthol derivative **26** on irradiation under similar conditions provided the desired diallyl enone **27**, which on treatment with G-I catalyst gave the spiro compound **28** in good yield (Scheme 4.5) [5].

Ghosh and coworkers reported the construction of spiroketal **34** by using RCM as a key step. In this regard, the reaction of 3,4-dihydropyran (DHP, **29**) with acrolein (**30**) under basic conditions gave the allyl-THP derivative **31**. Later, allyl derivative **31** was treated with allyl alcohol (**32**) in the presence of PPTS to produce a diallyl compound **33**. Finally, the compound **33** was converted to the corresponding spiroketal **34** through RCM with G-II catalyst (Scheme 4.6) [6].

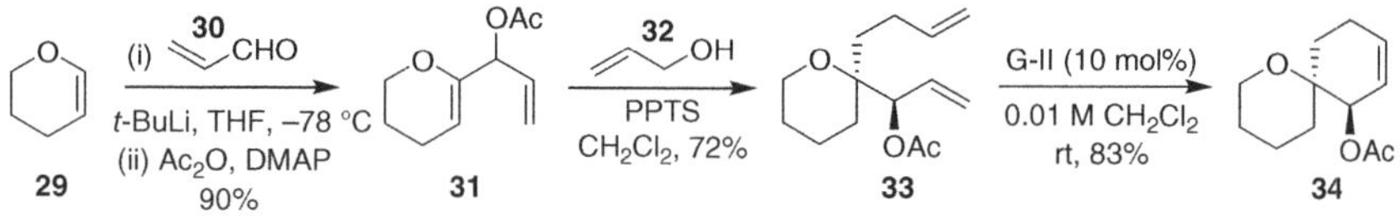

Scheme 4.6 Synthesis of spiroketal **34**. *Source:* Based on Ghosh et al. [6].

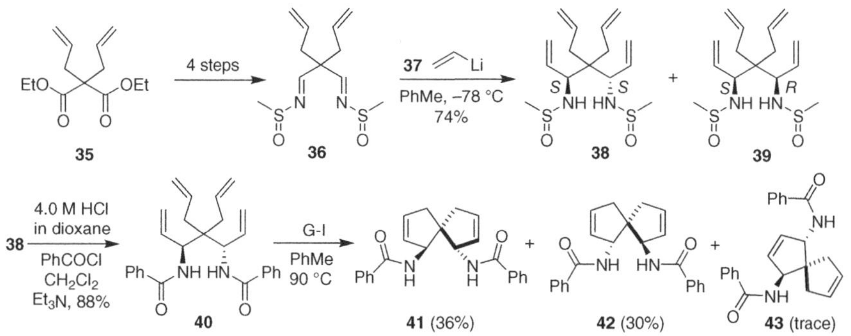

Scheme 4.7 Synthesis of spirocycles. *Source:* Based on Kato et al. [7].

Sasai and coworkers described the synthesis of chiral spirocycles (**41**, **42**, and **43**) by using RCM as a key step. To this end, emine derivative **36** was prepared from the diallyl compound **35** through a four-step sequence. The nucleophilic addition of vinyl lithium (**37**) to the emine **36** afforded a mixture of diastereomers **38** and **39**. Later, the tetra-allyl derivative **38** was treated with 4.0 M HCl followed by benzoyl chloride furnished the RCM precursor **40**, which undergo metathesis with the aid of G-I catalyst to afford various spirocycles **41**, **42**, and **43**. Such types of compounds were further used to synthesize chiral spiro ligand bearing oxazoline moiety (Scheme 4.7) [7].

Schobert and Urbina-Gonzalez reported the synthesis of the spiroannulated butanolide **47** *via* metathesis. For this purpose, the allyl tetronate **44** was treated with allyl acetate (**45**) in the presence of a palladium catalyst to deliver the diallyl lactone **46**. Later, the diallyl derivative **46** undergoes RCM with G-I catalyst to deliver the spiro-butanolide **47** in an excellent yield (Scheme 4.8) [8].

Clive and coworkers reported the synthesis of spiropiperdine derivative *via* the RCM as a key step. In this respect, they carried out the reduction of piperdine derivative **48** with LAH to provide the diol intermediate **49**. To protect a less-hindered OH group, the diol **49** was treated with TBSCl to afford the mono-TBS-protected piperidine derivative **50**. Later, the free OH group of compound **50** was converted to the corresponding aldehyde by Swern oxidation, which on Grignard reaction with vinylmagnesium bromide (**51**) gave the corresponding diallyl

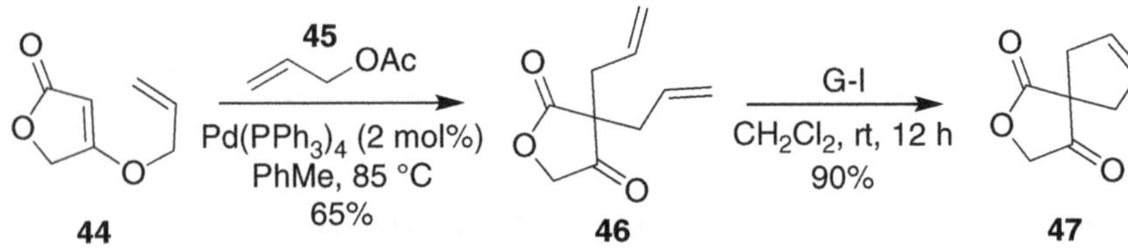

Scheme 4.8 Efficient route to spiroannulated butanolide. *Source:* Based on Schobert and Urbina-González [8].

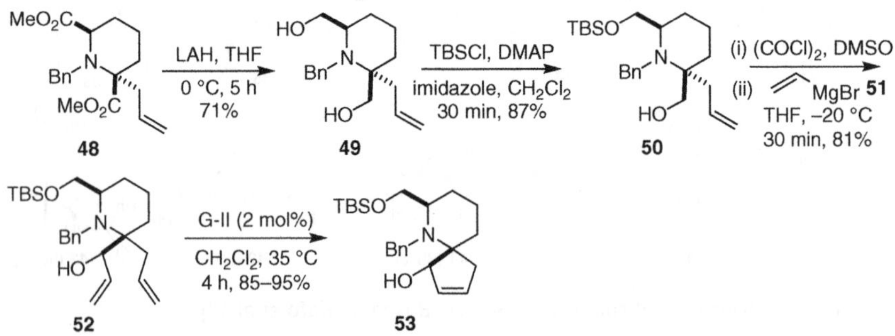

Scheme 4.9 Synthesis of a spirocyclic amine **53**. *Source:* Based on Clive et al. [9].

piperdine derivative **52**. Finally, the diallyl compound **52** was treated with G-II catalyst to produce the spirocyclic piperdine **53** (Scheme 4.9) [9].

Gonzalez-Bello and coworkers reported the synthesis of spiro carba-sugar **58** using the metathesis sequence. Carba-sugars contain a methylene group in place of oxygen, which has biological significance and these compounds are useful to treat many diseases like cancer, viral infection, and diabetes. In this scenario, quinic acid **54** was converted into ketone **55** *via trans*-diol protection followed by oxidation. Later, β-OH group was eliminated by converting to acetate, a better leaving group under basic conditions to generate α,β-unsaturated ketone **56**. Later, the ketone **56** was reacted with allyl bromide (**6**) in the presence of KHMDS to deliver the diallyl derivative **57**, which finally undergo RCM with G-II catalyst (5 mol%) to afford medicinally useful spiro carba-sugar **58** (Scheme 4.10) [10].

Frejd and coworkers reported the synthesis of a novel spiro-cyclohexene **65** using CR and Grignard reaction followed by RCM. They synthesized the bicyclic alcohol **60** from easily accessible bicyclic ketone **59**. Further, the OH group present in compound **60** was protected with TMS under basic conditions to afford the allyl-vinyl ether **61**, which subsequently rearranges to bicyclic aldehyde **62** under thermal conditions through a CR. Later, the aldehyde **62** was treated with allylmagnesium bromide (**63**) to deliver the diallyl bicyclic derivative **64**, which was treated with G-I catalyst to furnish the bicyclic spiro-cyclohexene **65** (Scheme 4.11) [11].

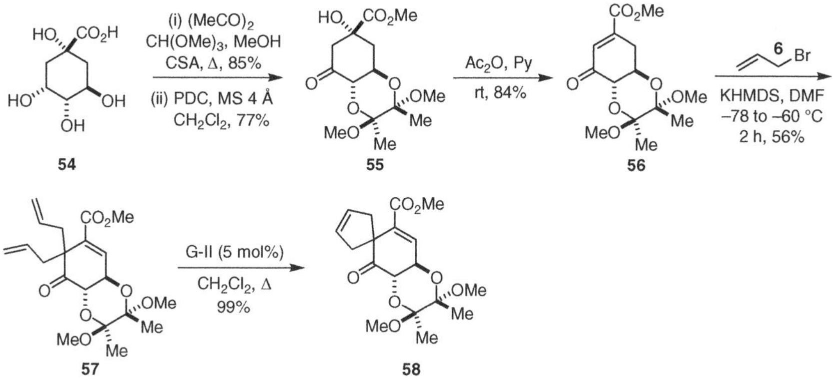

Scheme 4.10 Synthesis of spiro carba-sugar using RCM. *Source:* Based on González-Bello et al. [10].

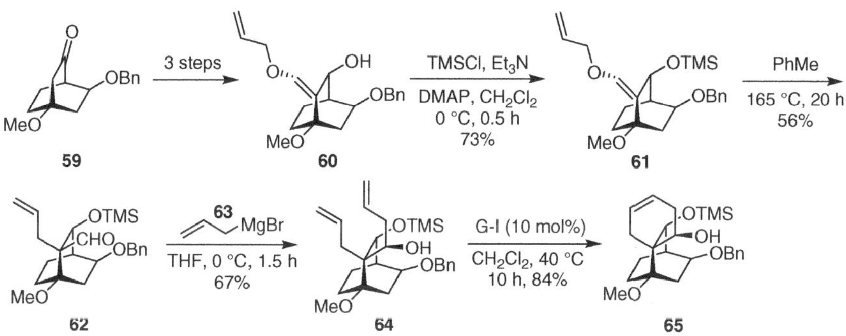

Scheme 4.11 Synthesis of novel spiro-cyclohexene. *Source:* Based on Thornqvist et al. [11].

Majumdar and coworkers reported the spiro-annulation *via* CR and RCM as key steps. In this connection, the substituted 2-pyridone **66** was treated with allyl bromide (**6**) under basic conditions to afford a mixture of *O*-allyl 2-pyridone **67** and C-allyl pyridone derivative **68**. Later, the diallyl compound **68** undergoes RCM with the aid of G-I catalyst to yield the spiro-heterocycle **69** in 96% (Scheme 4.12) [12].

Srikrishna and coworkers developed a new synthetic method to generate spiro-center *via* an Ireland ester CR followed by RCM. In this regard, the cyclohexyl carboxylic acid (**70**) was treated with crotyl bromide (**71**) under DCC and DMAP coupling conditions to obtain the corresponding ester derivative **72**. Then, the ester **72** was treated with trimethylsilyl chloride under basic conditions to provide the enolate **73**, which on *in situ* treatment with dil. HCl followed by treatment with diazomethane gave the corresponding allyl ester **74**. Reduction of the ester **74** using

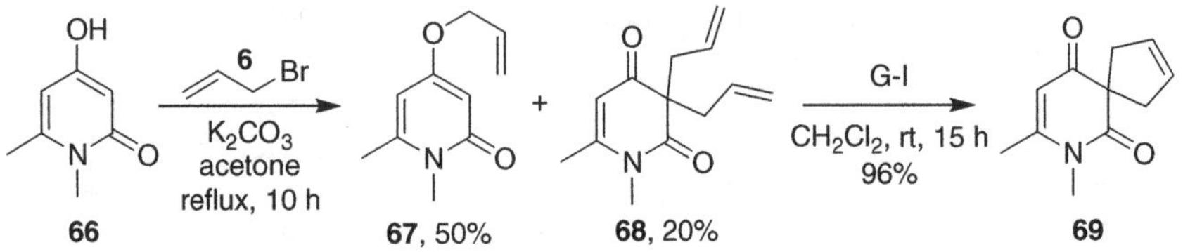

Scheme 4.12 Synthesis of spiro-heterocycle. *Sources:* Based on Majumdar et al. [12a]; Kersten et al. [12b]; Majumdar et al. [12c].

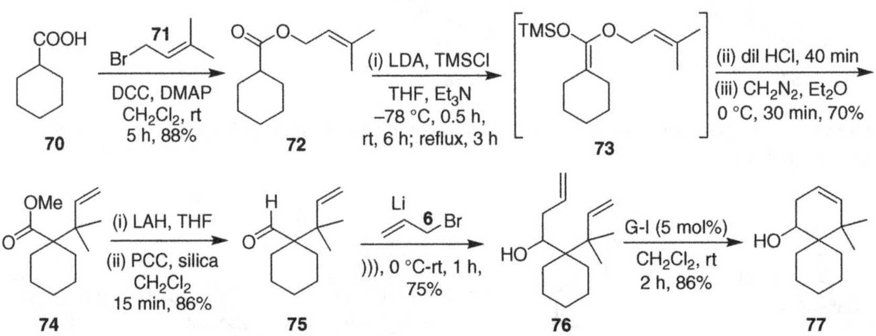

Scheme 4.13 Synthesis of spiro-center *via* an Ireland ester CR. *Sources:* Based on Srikrishna et al. [13a]; Srikrishna et al. [13b].

LAH and oxidation of the alcohol derivative with PCC gave the allyl aldehyde **75**. Further, the aldehyde **75** was treated with an allyl bromide (**6**) in the presence of Li under sonication conditions to produce the di-alkenyl derivative **76**. Next, RCM sequence with G-I catalyst gave the spiro derivative **77** (Scheme 4.13) [13].

We also reported a diversity-oriented approach to generate the spirocyclic derivatives **80** and **83** *via* CR followed by RCM or RCEM as key steps. Initially, allylation reaction of substituted tetralone (**78a-b**) and 1-indanone (**81a-b**) with allyl bromide (**6**) under basic conditions provided the corresponding diallyl derivatives **79a-b** and **82a-b**, respectively. Later, the RCM of these diallyl compounds **79a-b** and **82a-b** using G-I catalyst (10–15 mol%) yielded the spirocycles **80a-b** and **83a-b,** respectively (Scheme 4.14) [14].

Wojciechowski and Modrzejewska reported the synthesis of spiro-benzisothiazoline **86** *via* allylation followed by RCM as key steps. To this end, the benzisothiazoline derivative **84** was treated with allyl bromide (**6**) under basic conditions to deliver the diallyl derivative **85**. The RCM of benzisothiazoline **85** with G-II catalyst provides the corresponding spirocyclopenteno-benzisothiazoline derivative **86** (Scheme 4.15) [15].

Bubnov and coworkers reported the synthesis of the spiro-azetidine derivative **91** *via* RCM. In this regard, the 3-chloropropanenitrile **87** was treated with tri-allyl

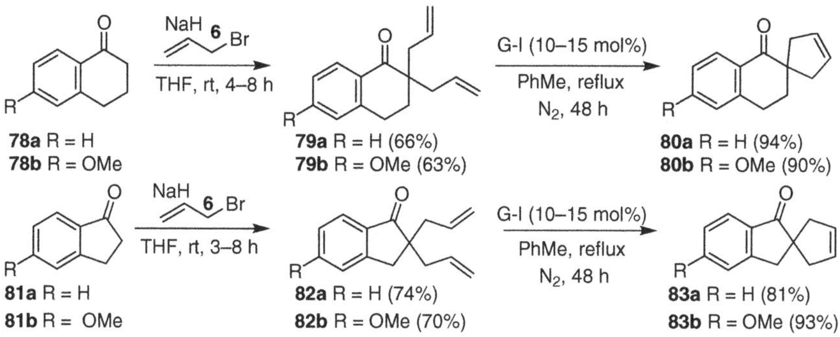

Scheme 4.14 Synthesis of spirocyclic compounds through RCM. *Source:* Based on Kotha et al. [14].

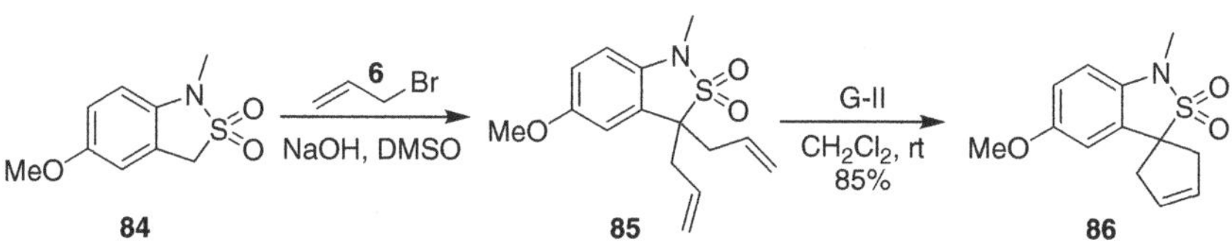

Scheme 4.15 Synthesis of spiro-benzosultams. *Source:* Based on Wojciechowski and Modrzejewska [15].

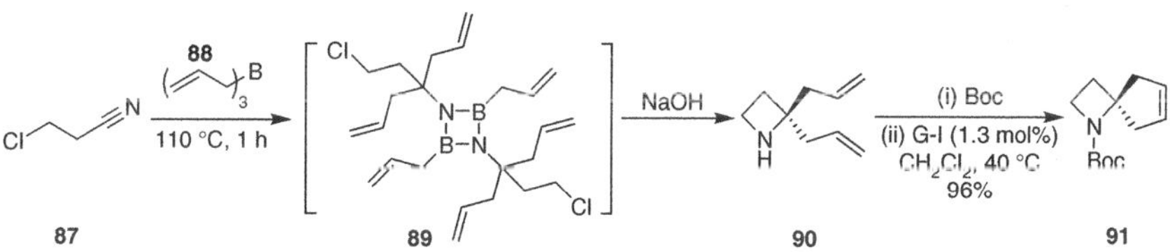

Scheme 4.16 Synthesis of spiro-azetidine **91**. *Source:* Kuznetsov et al. [16].

borane **88** under reflux conditions to provide the diallyl azetidine **90**, which upon RCM with G-I catalyst furnished the spiro-azetidine derivative **91** (Scheme 4.16) [16].

Braun and coworkers reported a simple and efficient protocol for the synthesis of spiroannulated 3-benzofuranone. In this scenario, benzofuran derivative **92** was treated with allyl bromide (**6**) under basic conditions to deliver the diallyl benzofuran **93**. Later, the diallyl derivative **93** was treated with Grubbs first-generation catalyst to afford the spiro-benzofuranone **94** in good yield (66%, Scheme 4.17) [17].

Gais and coworkers reported the asymmetric synthesis of spiroketal, spiroether, and oxabicycle *via* RCM [18]. In this protocol, initially, Z-selective sulfoximines **95**

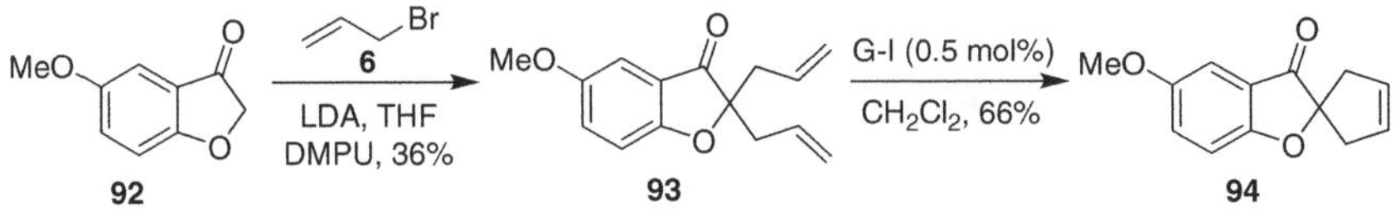

Scheme 4.17 Synthesis of spiroannulated 3-benzofuranone. *Source:* Based on Braun et al. [17].

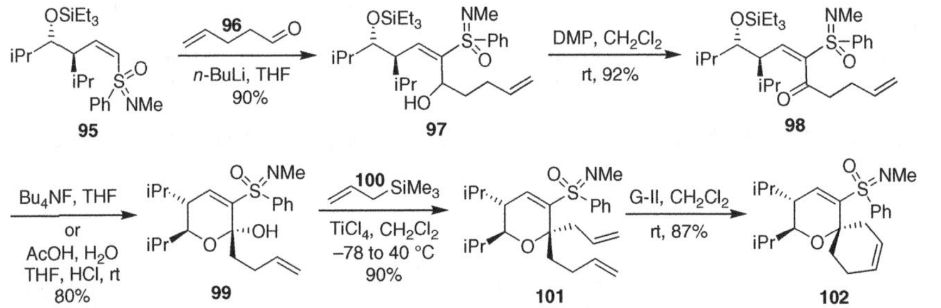

Scheme 4.18 Asymmetric synthesis of spiroether. *Sources:* Based on Kohler et al. [18a]; Lejkowski et al. [18b]; Leikowski et al. [18c].

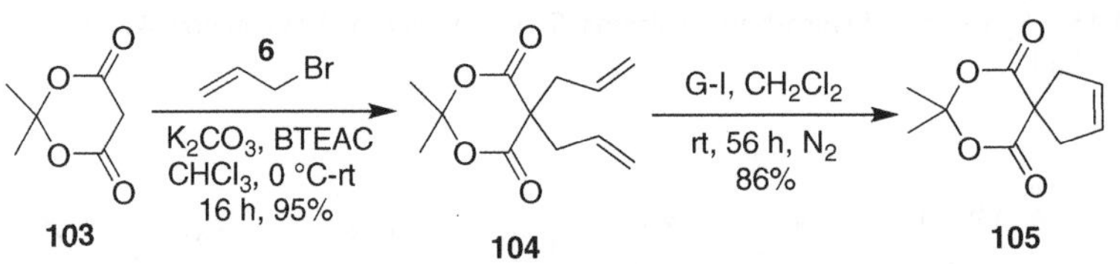

Scheme 4.19 Synthesis of spiro-Meldrum's acid derivative. *Source:* Modified from Kotha and Deb [19].

was treated with pent-4-enal (**96**) in the presence of *n*-BuLi to obtain the hydroxy compound **97**. The Dess–Martin periodinane (DMP) oxidation followed by the deprotection of the silyl group with TBAF gave the corresponding α-OH containing dihydropyran derivative **99**. Further, dihydropyran **99** was treated with allyl trimethylsilane (**100**) in the presence of $TiCl_4$ to furnish the RCM precursor **101**. Finally, the dialkenyl derivative **101** underwent RCM with the aid of Grubbs second-generation catalyst to provide the spiroether **102**. Additionally, this approach is also applicable to the synthesis of spiroketals and oxabicycles (Scheme 4.18) [18].

Likewise, we described a new protocol to the synthesis of spiro-Meldrum's acid derivative **105** by using RCM as one of the key transformations. Reaction of Meldrum's acid (**103**) with allyl bromide (**6**) in the presence of BTEAC under basic conditions provided the diallyl Meldrum's acid **104**. Finally, the RCM of diallyl Meldrum's acid **104** with G-I catalyst yielded the spiro-Meldrum's acid **105** (Scheme 4.19) [19].

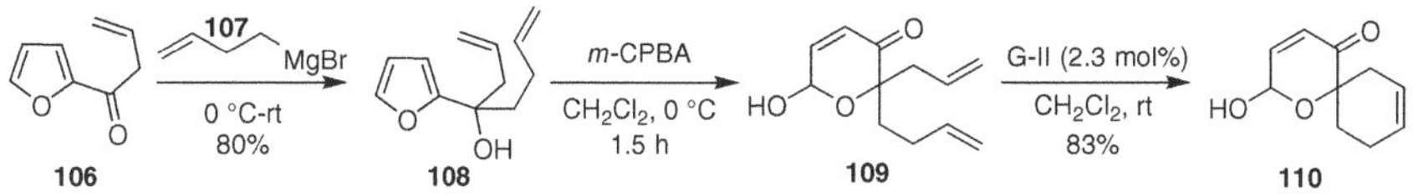

Scheme 4.20 Synthesis of spiro-lactone. *Source:* Based on Singh and Singh [20].

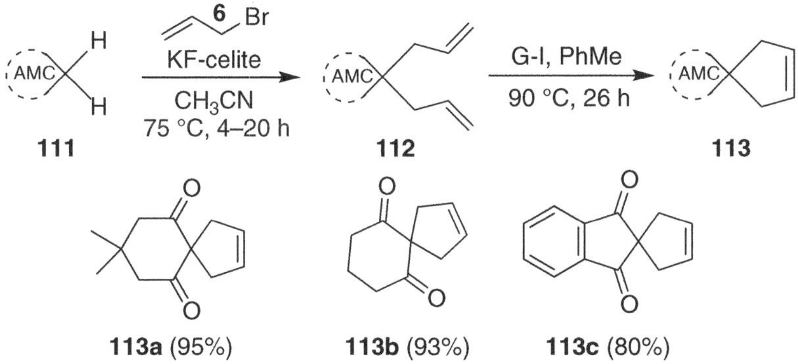

Scheme 4.21 Synthesis of spirocycles starting with active methylene derivative. *Source:* Based on Kotha and Lahiri [22].

Singh and Singh reported the synthesis of spiro-lactone **110** *via* [5+2] cycloaddition followed by RCM. To this end, the furan **106** undergoes a Grignard addition reaction with **107** to afford furyl alcohol **108**. Later, the alcohol **108** undergoes [5+2] cycloaddition reaction in the presence of *m*-CPBA to produce the di-alkenyl pyrone derivative **109**, which on treatment with G-II catalyst afforded the spiro-lactone **110** through RCM sequence (Scheme 4.20) [20].

During the last two decades, we are actively working toward the development of new strategies to assemble spirocycles, heterocycles, unusual amino acids, etc., using metathesis as a key step [21]. In this chapter, we mainly focused on our contribution to the design and syntheses of spirocyclic derivatives. In this regard, we treated the active methylene compound such as **111** with allyl bromide (**6**) to obtain the diallyl compound **112**, which undergo metathesis with G-I catalyst to provide the spirocycle **113**. The same strategy has been extended to synthesize diverse spirocycles **113a-c** (Scheme 4.21) [22].

Along similar lines, we treated the substituted barbituric acid **114** with allyl bromide (**6**) under basic conditions to afford diallyl barbituric acid **115**. Next, RCM of diallyl barbituric acid **115** with G-I catalyst provided the spiro-barbituric acid derivative **116** (Scheme 4.22) [23].

Radhakrishnan and coworkers reported the synthesis of substituted spiroindolin-2-one derivatives **121** and **124** *via* palladium-catalyzed functionalization of isatylidene followed by RCM. For this purpose, the isatylidene **117** was

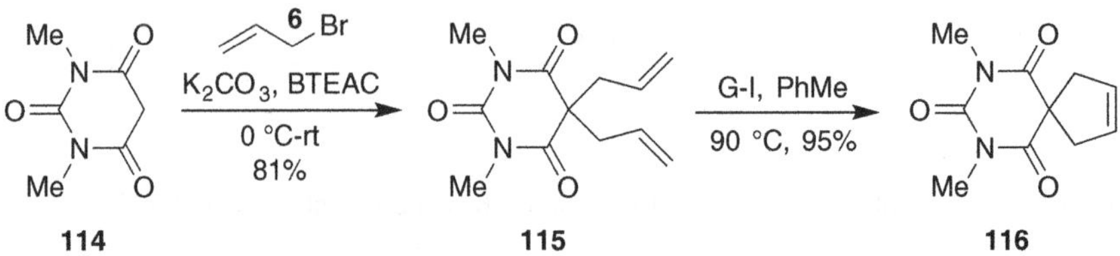

Scheme 4.22 Synthesis of spiro-barbituric acid derivative. *Source:* Based on Kotha et al. [23].

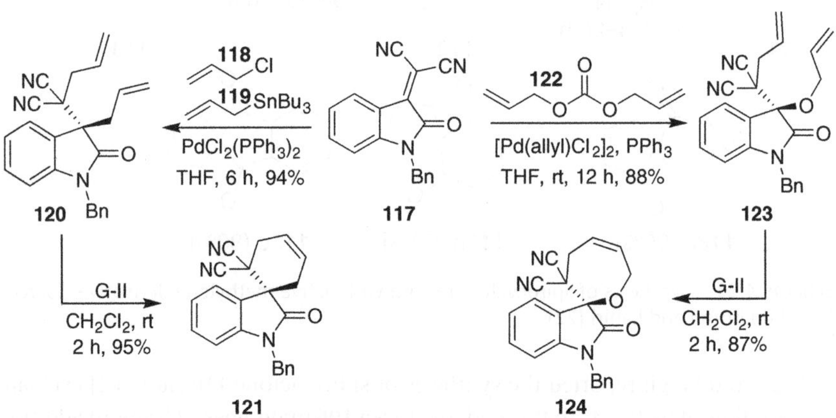

Scheme 4.23 Synthesis of spiroindolin-2-one derivatives. *Sources:* Based on George et al. [24a]; Baiju et al. [24b].

treated with allyl chloride (**118**) followed by allyl tributyltin (**119**) in the presence of a catalytic amount of a palladium catalyst to provide the diallyl indolin-2-one derivative **120** by the generation of a new stereocenter. Later, diallyl compound **120** was treated with G-II catalyst to afford the corresponding RCM product **121**. Subsequently, isatylidene **117** reacts with allyl carbonate (**122**) in the presence of a palladium catalyst to furnish the diallyl indolin-2-one derivative **123**, which then undergo RCM with G-II catalyst to generate spiroindolin-2-one derivative **124** (Scheme 4.23) [24].

Castiglioni and coworkers reported the synthesis of a novel class of spiro-dibenzo[*a*,*d*]cycloheptene **127** *via* allylation followed by RCM. In this regard, the dibenzocycloheptanone derivative **125** was treated with allyl bromide (**6**) under basic conditions to deliver the diallyl compound **126**, which on treatment with G-II catalyst furnished the spiro-dibenzo[*a*,*d*]cycloheptene **127** through RCM sequence (Scheme 4.24) [25].

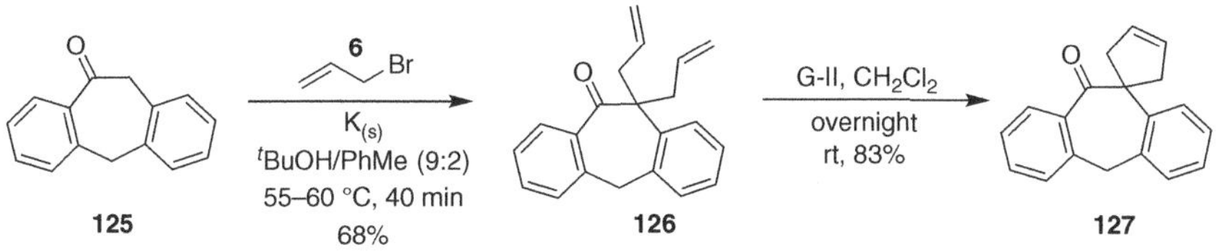

Scheme 4.24 Synthesis of spiro dibenzo[*a*,*d*]cycloheptene. *Source:* Based on Castiglioni et al. [25].

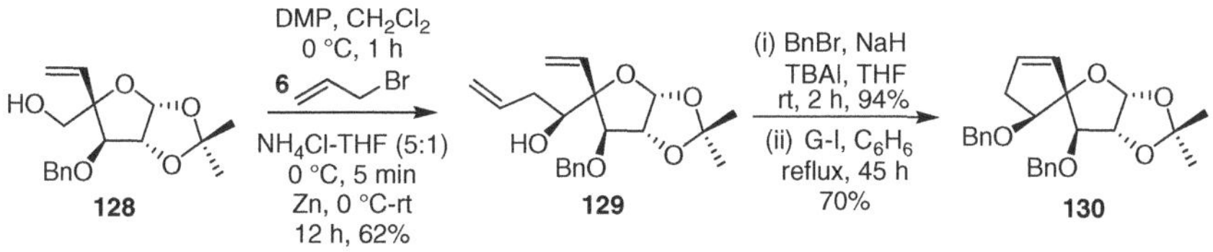

Scheme 4.25 Synthesis of spiro-carbohydrate **130**. *Source:* Based on Maity et al. [26].

Mandal and coworkers described the synthesis of spiro-carbohydrate **130** *via* a RCM as a key step. To this end, oxidation of alcohol **128** with DMP gave the corresponding aldehyde, which on further reaction with allyl bromide (**6**) in the presence of zinc afforded the allyl-derivative **129**. Later, the free hydroxyl group of compound **129** was protected with a benzyl group followed by the RCM with G-I catalyst provided the spiro compound **130** (Scheme 4.25) [26].

Dickschat and Brock have reported the synthesis of sesquiterpenoid acorenone **138** through Wittig reaction, Grignard addition, and RCM sequence. To this end, they prepared the ester **132** from the carboxylic acid derivative **131** *via* a series of known chemical transformations. The reaction of aldehyde **132** with Wittig salt provided the 2-allyl ester derivative **127**. Later, the ester **133** was converted into the corresponding –CN derivative **134** through LAH reduction, protection, followed by reaction with NaCN. Next, the compound **134** on treatment with Grignard reagent **135** provides the diallyl derivative **136**. RCM of compound **136** with G-II catalyst yielded the desired spiro alcohol **137**, which was further oxidized using DMP provided the sesquiterpenoid acorenone **138** (Scheme 4.26) [27].

Ito and coworkers reported the synthesis of (+)-spirocurcasone **144** using RCM as a key transformation. In this connection, the compound **139** undergoes an intramolecular aldol condensation under acidic conditions to afford bicyclic ketone **140**. Later, the bicyclic compound **140** was treated with methyl-substituted acryloyl chloride **141** and subsequently with allyl iodide (**142**) under basic conditions to yield the RCM precursor **143**. Finally, the RCM of **143** with G-II catalyst provided the natural product (+)-spirocurcasone **144** (Scheme 4.27) [28].

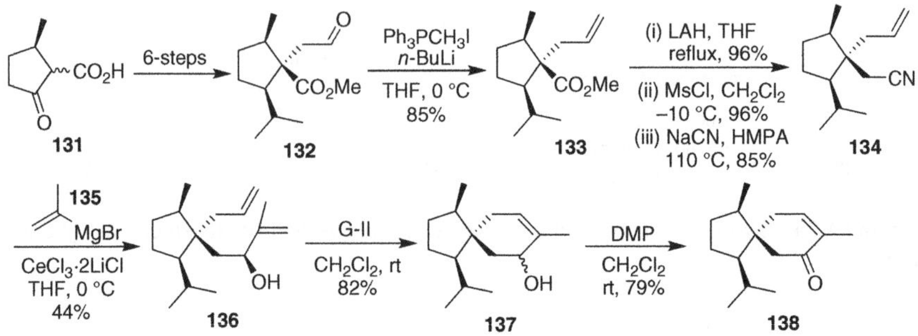

Scheme 4.26 Synthesis of sesquiterpenoid acorenone. *Sources:* Based on Brock and Dickschat [27a]; Kurosawa et al. [27b].

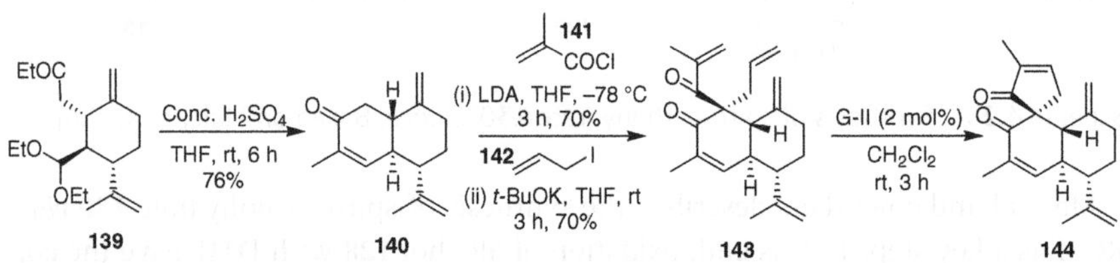

Scheme 4.27 Synthesis of (+)-spirocurcasone. *Source:* Based on Abe et al. [28].

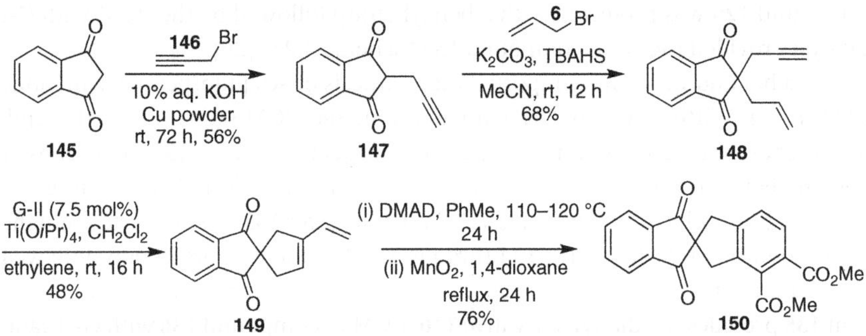

Scheme 4.28 Synthesis of spiroindenes. *Source:* Based on Kotha et al. [29].

Along similar lines, we investigated the application of enyne metathesis (EM) followed by DA reaction for the synthesis of spirocycle **143**. To this end, the 1,3-indandione **145** was treated with propargyl bromide (**146**) to obtain propargyl 1,3-indandione derivative **147**. Further, propargyl 1,3-indandione **147** was reacted with allyl bromide (**6**) to afford enyne building block **148**. Next, the RCM of enyne **148** with G-II catalyst provides the diene derivative **149**, which on treatment with dimethyl acetylenedicarboxylate (DMAD) followed by the MnO_2 oxidation yielded the aromatized spirocyclic compound **150** (Scheme 4.28) [29].

Manner and coworkers reported the synthesis of spiro-bicyclo[2.2.2]octane derivative **156** *via* Grignard addition reaction and RCM as key steps. The key bicyclic compound **152** was prepared from cyclohexenone derivative **151** by a three-step sequence. The bicyclic compound **152** undergoes allylation under basic conditions to deliver the corresponding allylated bicyclic compound **153**. Later, the ester functionality of **153** was interchanged to aldehyde **154** by reduction–oxidation sequence using diisobutylaluminum hydride (DIBAL-H) and TPAP. The Grignard addition reaction of aldehyde **154** with allylmagnesium bromide (**63**) provided the corresponding di-alkenyl derivative **155**. Finally, the di-alkenyl compound **155** on treatment with G-I catalyst afforded the spirobicyclo[2.2.2]octane derivative **156** (Scheme 4.29) [30].

Yang and coworkers reported the construction of oxaspirocycle **161** *via* a rhodium-catalyzed reaction followed by RCM. In this connection, the 1,2,3-triazole derivative **157** was treated with a rhodium catalyst under reflux conditions to obtain the tetrahydrofuran derivative **160**, which on *in situ* treatment with allylmagnesium chloride (**159**) provided the di-alkenyl compound **160**. Later, the di-alkenyl derivative **160** was treated with G-II catalyst to afford the oxaspirocyclic compound **161** (Scheme 4.30) [31].

Sosnicki and coworkers reported the synthesis of spiroquinolizidin-4-ones **165** from substituted 2-pyridones **162** *via* a sequential 1,6-addition, allylation,

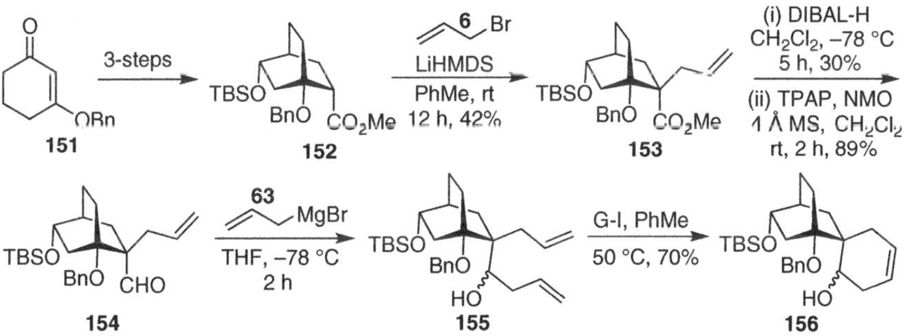

Scheme 4.29 Synthesis of spiro-bicyclo[2.2.2]octane **156**. *Source:* Based on Manner et al. [30].

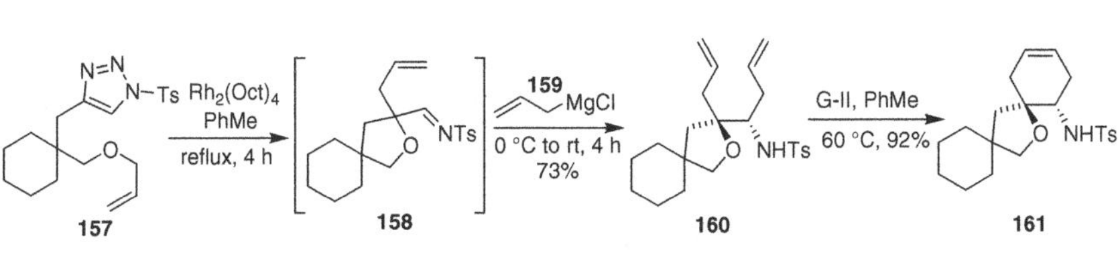

Scheme 4.30 Synthesis of oxaspirocyclic compound. *Source:* Based on Fu et al. [31].

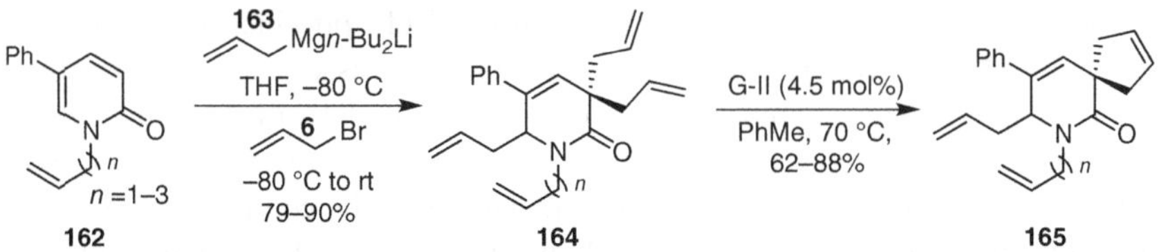

Scheme 4.31 Synthesis of spiro-fused compounds **165**. *Source:* Based on Sośnicki et al. [32].

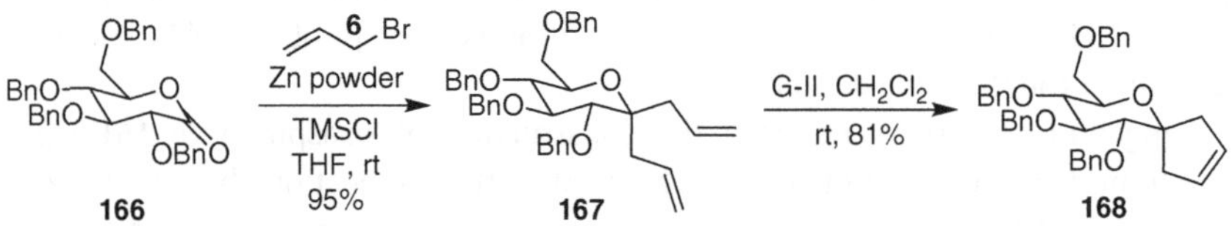

Scheme 4.32 Synthesis of *C*-spiro-glycoconjugate **168**. *Sources:* Based on Lambu et al. [33a]; Chen et al. [33b].

followed by RCM. In this regard, the 2-pyridone **162** was treated with lithium allyl(di-*n*-butyl)magnesate **163** to furnish the corresponding 1,6-addition compounds, which on subsequent treatment with allyl bromide (**6**) provided the tetra-allyl pyridone **164**. Later, the pyridone **164** was treated with G-II catalyst to provide the desired spiro-fused compounds **165** (Scheme 4.31) [32].

Mukherjee and coworkers reported the synthesis of *C*-spiro-glycoconjugate **168** *via* zinc-mediated addition followed by RCM as a key step. To this end, the sugar derivative **166** was treated with allyl bromide (**6**) in the presence of zinc powder to obtain the diallyl sugar **167**. Later, the sugar **167** undergoes a RCM with the aid of G-II catalyst to afford the C-spiro-glycoconjugate **168** (Scheme 4.32) [33].

We have developed a new strategy to synthesize the isomeric spiro-indole derivatives **172** and **176** *via* FI and RCM as key steps. In this connection, the indanone derivatives **169** and **173** had undergone FI with *N*-methyl phenyl hydrazine to obtain the corresponding indole derivatives **170** and **174**. Later, indole derivatives **170** and **174** were treated with allyl bromide (**6**) to produce the corresponding diallyl indoles **171** and **175**. Finally, the RCM reaction of the diallyl derivatives **171** and **175** with G-I catalyst provides the isomeric indole derivatives **172** and **176** in good yields (Scheme 4.33) [34].

We also reported the preparation of bis-spirocycle **180** and bis-spiropropellane **182** *via* allylation followed by RCM as key steps. In this connection, the bicyclicdione **178** was prepared from 1,8-cyclo-octadiene **177** by a sequence of chemical transformations. Later, the dione **178** was treated with allyl bromide (**6**) under basic conditions at room temperature to obtain the tetra-allyl derivative **179**, which upon treatment with G-I catalyst provides the spiro-dione **180**. On the

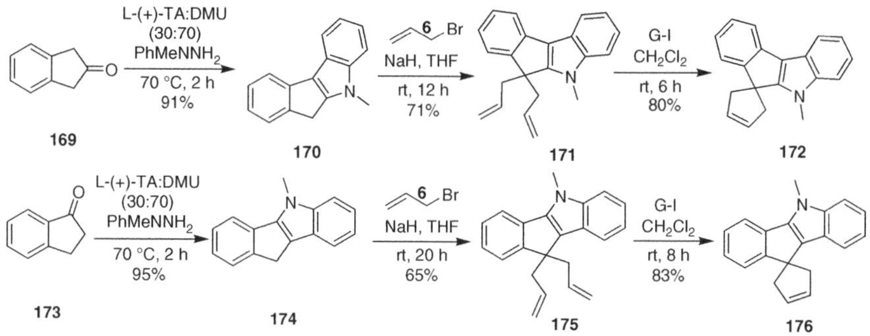

Scheme 4.33 Synthesis of spiro-indole derivatives. *Source:* Modified from Kotha et al. [34].

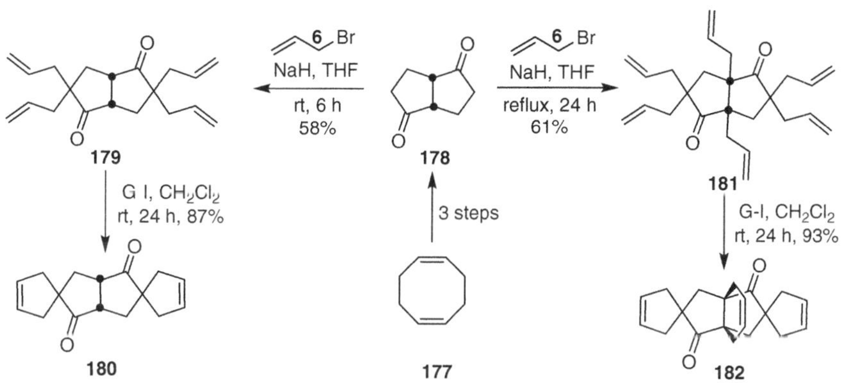

Scheme 4.34 Synthesis of bis-spirocycles. *Source:* Based on Kotha et al. [35].

other hand, the di-keto compound **178** reacts with allyl bromide (**6**) under basic conditions at reflux temperature to deliver the hexa-allyl derivative **181**, which on reaction with the aid of G-II catalyst furnished the corresponding spirocycle containing propellane **182** (Scheme 4.34) [35].

Along similar lines, the tetralone **183** was treated with an excess amount of allyl bromide (**6**) under basic conditions to generate the tetra-allyl tetralone **184**. Later, the metathesis reaction of **184** with G-I catalyst provides the bis-spiro tetralone derivative **185** in an excellent yield (Scheme 4.35) [35].

Singh and Das reported a stereoselective approach to the synthesis of tetracyclic core of daphmanidin A through Grignard addition followed by RCM. For this purpose, the diallyl compound **187** was prepared from bicyclic compound **186** by a sequence of chemical transformations [36b]. The alcohol **187** was oxidized with TPAP/NMO to deliver bicyclic ketone **188**, which on further treatment with G-I

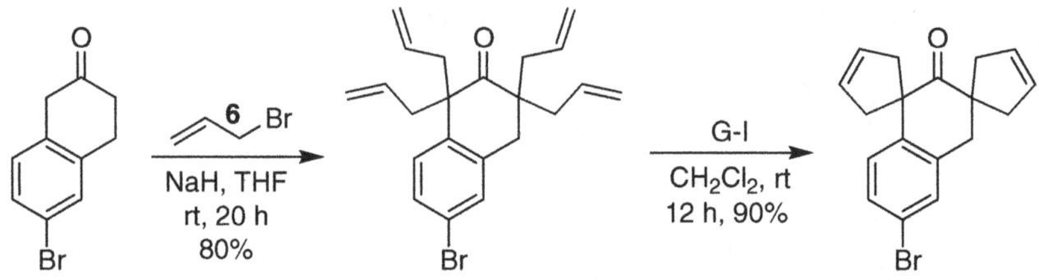

183 **184** **185**

Scheme 4.35 Synthesis of bis-spiro tetralone. *Source:* Based on Kotha et al. [35].

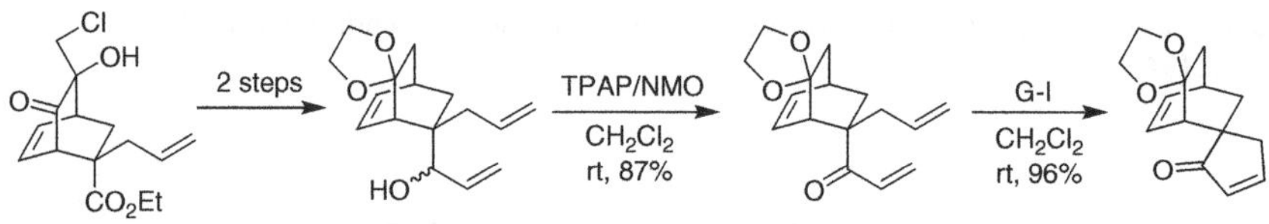

186 **187** **188** **189**

Scheme 4.36 Synthesis of spiro-bicyclic compound **189**. *Sources:* Based on Singh and Das [36a]; Singh et al. [36b].

catalyst yielded the bicyclic spiro derivative **189**. The spiro compound **189** was further utilized successfully for the construction of tetracyclic core of daphmanidin A (Scheme 4.36) [36].

Additionally, we synthesized the spirodione **194** *via* Grignard addition followed by RCM as a key step. To this end, the diethyl malonate **190** was reacted with allyl bromide (**6**) to obtain the diallyl malonate derivative **191**. Later, the diallyl malonate **191** was treated with vinyl magnesium bromide (**51**) to obtain the tetra-ene derivative **192** *via* Grignard followed by 1,4-addition. Finally, the tetra-ene **192** on treatment with G-I catalyst provided the spirocyclic compound **193**. Later, hydrogenation with Pd/C yielded the saturated product **194** (Scheme 4.37) [37].

Along similar lines, the di-ketones **196** and **199** were synthesized from cyclopentadiene (**195**) through a series of chemical transformations. Next, the di-keto derivative **196** was treated with allyl bromide (**6**) under basic conditions to afford the hexa-allyl derivative **197**, which on treatment with G-I catalyst followed by the hydrogenation gave the corresponding saturated spiro-polyquinane derivative **198**. Similarly, a related reaction of di-ketone **199** with allyl bromide (**6**) leads to the tetra-allyl derivative **200**, which undergoes RCM with G-I catalyst followed by hydrogenation with Pd/C provided the saturated polycyclic compound **201** (Scheme 4.38) [38].

Likewise, we have realized the synthesis of di-bromofluorene **202** by alkylation with allyl bromide (**6**) in the presence of sodium hydroxide to produce the diallyl

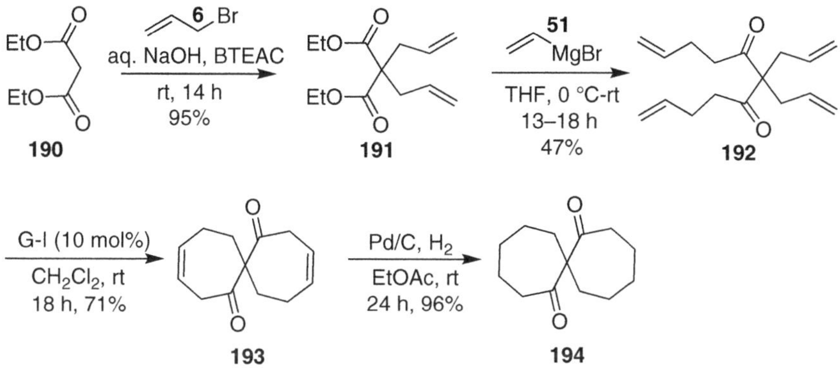

Scheme 4.37 Synthesis of spirodione. *Source:* Kotha et al. [37].

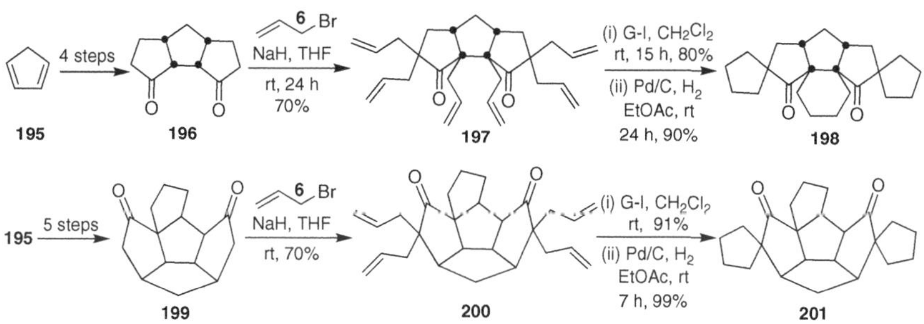

Scheme 4.38 Synthesis of spiro-polycyclic compounds. *Source:* Based on Kotha et al. [38].

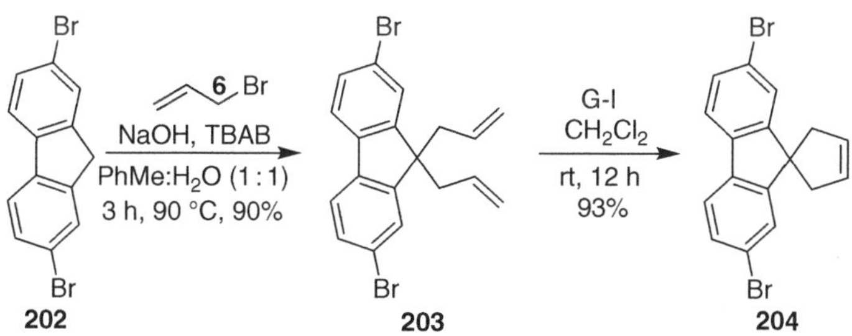

Scheme 4.39 Synthesis of spiro-fluorene **204**. *Source:* Based on Sośnicki et al. [32].

fluorene derivative **203**, which will further generate the spiro-fluorene **204** with the aid of G-I catalyst by RCM sequence (Scheme 4.39) [22].

Acid-mediated condensation of 1-indanone (**173**) under reflux conditions gave the truxene **205**. Nitration of truxene **205** with HNO_3 provides the corresponding

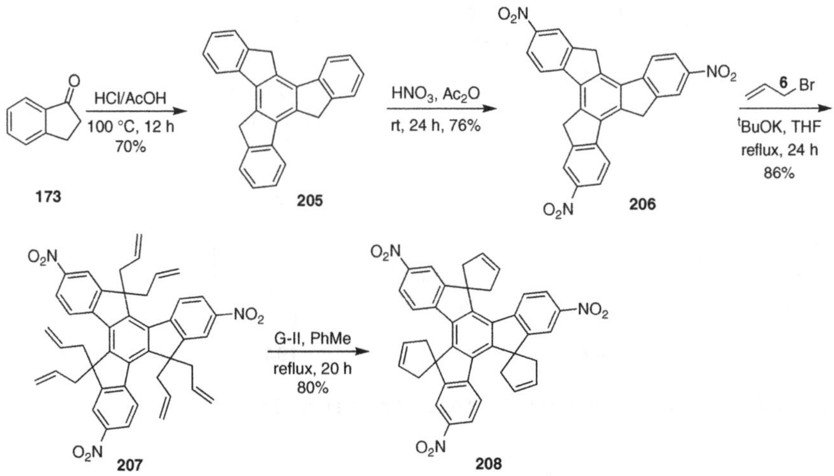

Scheme 4.40 Synthesis of spiro-truxene **208**. *Source:* Based on Kotha et al. [39].

trinitro-truxene **206**. Next, the alkylation reaction of trinitro-truxene with allyl bromide (**6**) under basic reaction conditions provides the hexa-allyl truxene derivative **207**, which upon treatment with G-II catalyst produced the spiro-truxene derivative **208** through RCM (Scheme 4.40) [39].

4.2 Formation of Aza-spirocycles

Tanner and coworkers reported an efficient method to construct (±)-perhydrohistrionicotoxin. In this regard, a useful scaffold, vinyl cyclohexylamine derivative **210**, was synthesized from 2-epoxy cyclohexanone **209** using a five-step sequence. *n*-BuLi-mediated *N*-alkylation of cyclohexylamine **210** with 3-butenyl triflate (**211**) provides the butenyl-vinyl cyclohexane derivative **212**, which undergo RCM with G-I catalyst to afford the spiro derivative **213**. The spiro derivative **213** is an important intermediate for the synthesis of a natural product (±)-perhydrohistrionicotoxin (Scheme 4.41) [40].

Gmeiner and coworkers reported the synthesis of spirocyclic compound **218** from commercially available amino acid proline (**214**). Initially, the allyl derivative of proline **215** was synthesized through LDA-mediated alkylation of proline with allyl bromide (**6**) followed by the Boc-protection. Further, the acid-amine coupling of **215** was accomplished out in the presence of HATU/DIPEA with allyl amine **216** to provide the diallyl proline derivative **217**, which upon RCM with G-I catalyst gave the spirocyclic compound **218** in 67% (Scheme 4.42) [41].

Silvani and coworkers demonstrated the synthesis of spiropiperidine-3,3-oxindole **222** by using RCM as a key step. In this regard, the free NH_2 group of

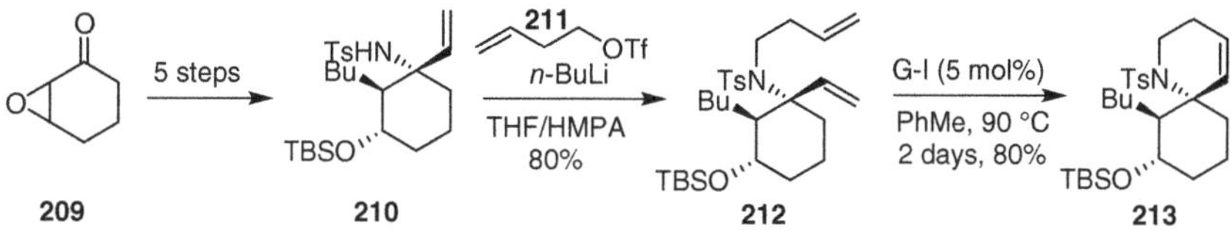

Scheme 4.41 Efficient route to (+)-perhydrohistrionicotoxin. *Source:* Based on Tanner et al. [40].

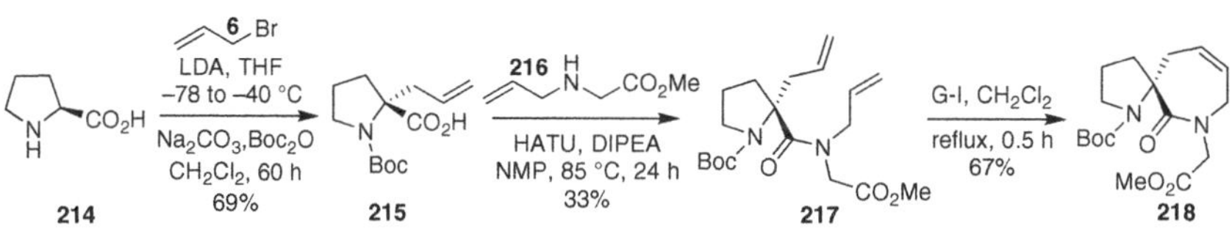

Scheme 4.42 Synthesis of spirocyclic compound **218**. *Source:* Based on Bittermann et al. [41].

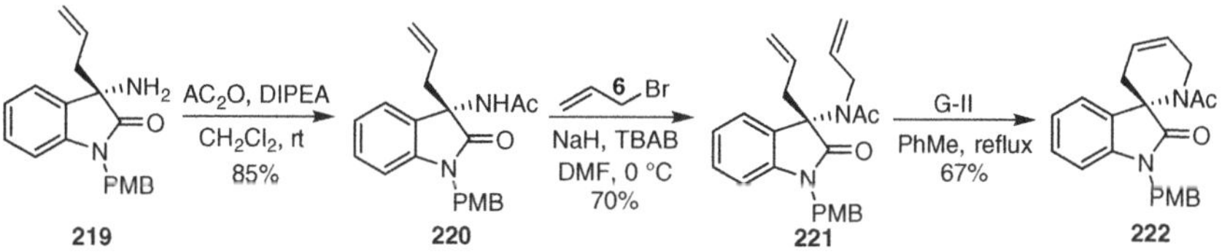

Scheme 4.43 Synthesis of spiropiperidine-3,3-oxindole. *Source:* Based on Lesma et al. [42].

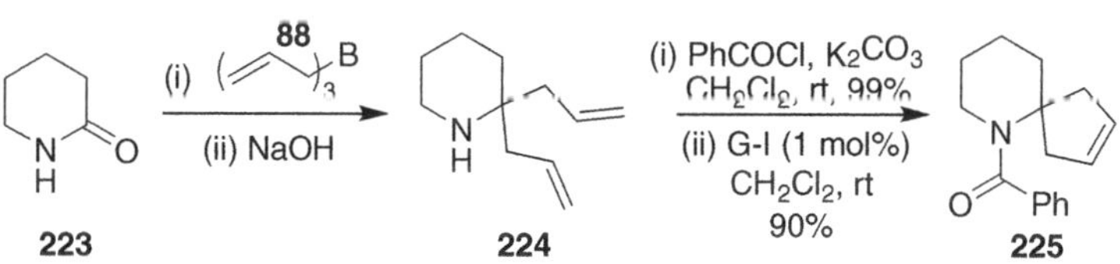

Scheme 4.44 Construction of spiro-mononitrogen heterocycle. *Source:* Based on Kuznetsov et al. [43].

oxindole **219** was protected by acylation using Hunig's base. Next, *N*-allylation delivered diallyl oxindole **221**. Later, the diallyl oxindole derivative **221** was treated with G-II catalyst to obtain the spirooxindole derivative **222** (Scheme 4.43) [42].

Bubnov and coworkers developed a novel method to construct the spiro heterocycle **225** using RCM. To this end, the pipredi-2-one (**223**) was treated with tri-allyl borane (**88**) under reflux conditions to afford the diallyl piperdine **224**. Then, the free NH of **224** was protected with a benzoyl group followed by RCM with G-I catalyst furnished the spiropiperidine derivative **225** in 90% (Scheme 4.44) [43].

Waters and Cheng reported the synthesis of natural product nankakurine (**233**) *via* RCM as a key step. In this connection, the cyclohexenone derivative **226** underwent

DA reaction with diene **227** in the presence of Lewis acid to deliver the bicyclic ketone **228**. Subsequently, the keto derivative **228** was treated with methyl ammonium chloride followed by reduction gave the amine derivative **229**. Next, the amine **229** on treatment with aq. formaldehyde gave luciduline (**230**). Further, the reaction of **230** with allyl amine (**231**) followed by Grignard addition with allylmagnesium bromide (**63**) provides the diallyl derivative **232**. Finally, the diallyl compound **232** reacted with G-II catalyst to afford the (±)-nankakurines (**233**, Scheme 4.45) [44].

Vankar and coworkers reported the synthesis of fused oxa-aza spiro-carbohydrate **239** by Grignard reaction followed by RCM as key transformations. To this end, the carbohydrate derivative **234** was reacted with vinyl magnesium bromide (**51**) followed by reaction with TMSCN provided the cyano-sugar **236**. The –CN group of a carbohydrate **236** selectively reduced using LAH to form amino compound **237**, which on further protection by acylation followed by *N*-alkylation with allyl bromide (**6**) under basic conditions gave the di-alkenyl sugar **238**. Finally, the diallyl compound **238** undergoes RCM reaction with G-II catalyst to yield the spiro compound **239** (Scheme 4.46) [45].

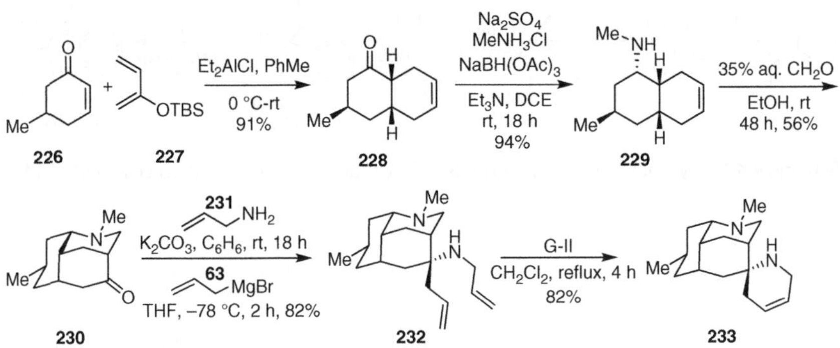

Scheme 4.45 Synthesis of natural product nankakurine. *Source:* Based on Cheng and Waters [44].

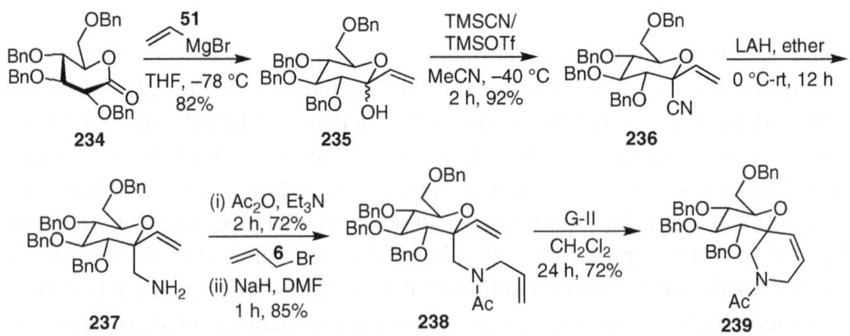

Scheme 4.46 Synthesis of oxa-aza spiro-carbohydrate. *Source:* Based on Pal et al. [45].

4.3 Formation of Oxa-spirocycles

Ghosh and coworkers described the synthesis of annulated carbohydrate **243** *via* Grignard addition and RCM. In this regard, the keto derivative **240** was treated with allylmagnesium bromide (**63**) under reflux conditions to deliver the alcohol **241**, which was then reacted with allyl bromide (**6**) to afford diallyl carbohydrate **242**. Finally, the diallyl compound **242** was reacted with G-I catalyst to give the spiroannulated carbohydrate **243** *via* RCM (Scheme 4.47) [46].

Schmidt and Wildemann reported the synthesis of dihydropyran **248** *via* the Grignard reaction followed by RCM. In this connection, *O*-allyl ester **245** was obtained through a silver oxide-mediated alkylation of α-hydroxy ester **244** with allyl bromide (**6**). Grignard reaction of **245** with vinyl magnesium bromide (**51**) gave the alcohol derivative **246**, which was further *O*-alkylated with allyl bromide (**6**) produced the tetra-ene **247**. Finally, the RCM sequence of tetra-ene derivative **247** with G-I catalyst yielded the corresponding spiro-dihydropyran derivative **248** (Scheme 4.48) [47].

Boom and coworkers reported the synthesis of carbohydrate-containing spiroacetal **252** *via* EM as a key step. In this connection, the *O*-propargylated compound **251** was achieved by the treatment of allyl-carbohydrate derivative **249** with

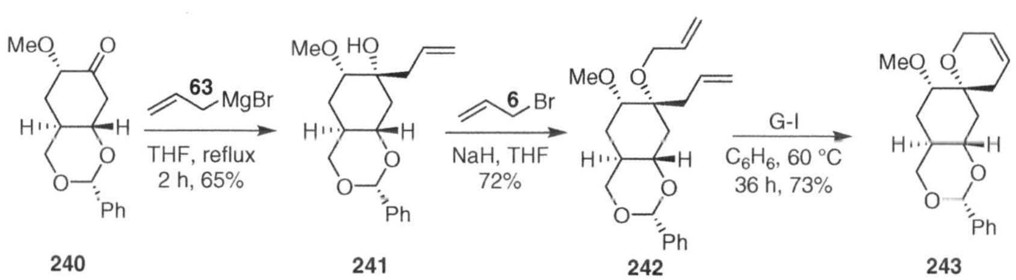

Scheme 4.47 Synthesis of carbohydrate derivative **243**. *Source:* Based on Holt et al. [46].

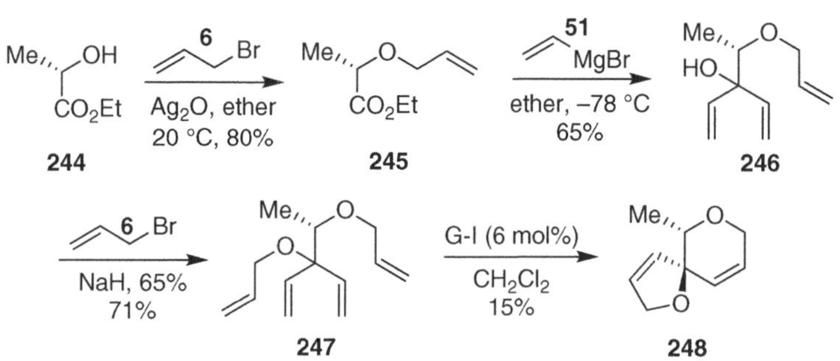

Scheme 4.48 Synthesis of spiro-dihydropyran. *Source:* Based on Schmidt and Wildemann [47].

but-2-yn-1-ol (**250**). Later, the compound **251** has undergone EM with G-I catalyst to provide the spiro-diene **252** (Scheme 4.49) [48].

Walters and coworkers reported the preparation of spirocyclic 3,6-dihydro-2*H*-pyran using RCM as a key step. They started with the Grignard addition reaction of keto-derivative **253** with allylmagnesium bromide (**63**) to obtain the alcohol **254**, which on further treatment with allyl bromide (**6**) produced the corresponding RCM precursor, diallyl derivative **255**. Finally, the diallyl compound **255** was treated with G-I catalyst to yield the spirocyclic 3,6-dihydro-2*H*-pyran **256** (Scheme 4.50) [49].

Along similar lines, Alcaide and coworkers reported the assembly of spirocyclic oxindole **261** *via* RCM as a key step. For this purpose, the oxindole **257** was treated with allyl bromide (**6**) in the presence of an indium (In)/$HfCl_4$ to obtain the allyl indole derivative **258**. Then, allyl indole **258** was treated with acryloyl chloride (**259**) to furnish the RCM precursor **260**, which was then reacted with G-I catalyst to afford the spirooxindole **261** (Scheme 4.51) [50].

Leino and Silver reported the synthesis of chiral spirocyclic ether **267** by using RCM as a key step. In this connection, the cyclohexanone **262** was treated with

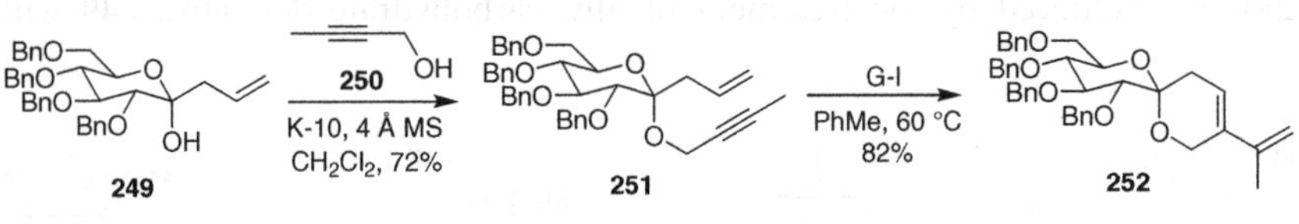

Scheme 4.49 Synthesis of carbohydrate-derived spiroacetal **252**. *Source:* Based on Leeuwenburgh et al. [48].

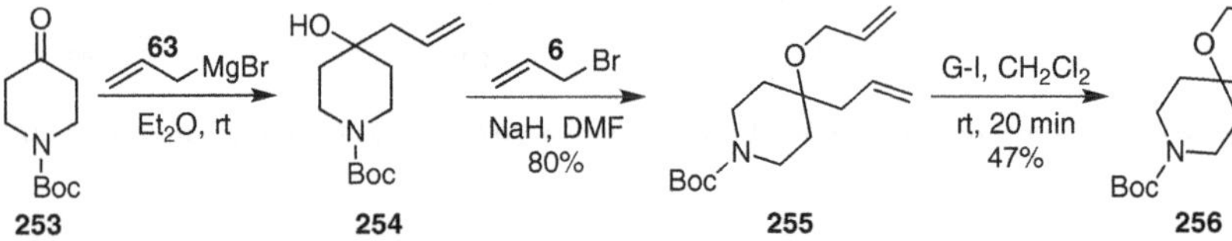

Scheme 4.50 Synthesis of spirocyclic 3,6-dihydro-2*H*-pyran **250**. *Sources:* Based on Walters et al. [49a]; (b) Prusov and Maier [49b].

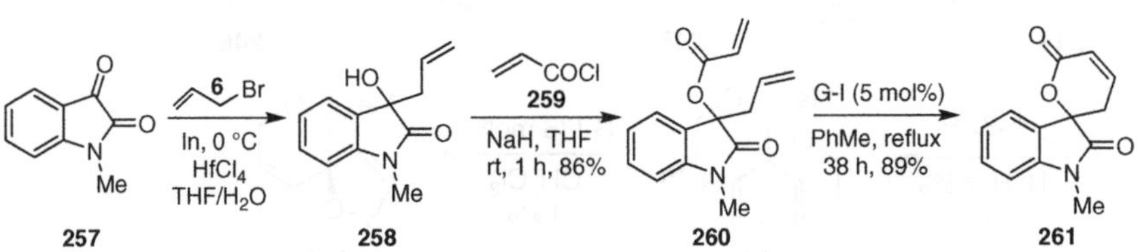

Scheme 4.51 Synthesis of spirocyclic oxindole **261**. *Source:* Based on Alcaide et al. [50].

phenylboronic acid in the presence of a rhodium catalyst to deliver 3-phenyl cyclohexanone **264**. Later, cyclohexanone **264** was treated with allyl bromide (**6**) in the presence of an indium metal to afford the cyclohexanol derivative **265**. Further, the cyclohexanol **265** was treated again with allyl bromide (**6**) under the basic conditions to obtain the diallyl compound **266**, which on treatment with G-I catalyst produced the spirocyclic ether **267** *via* RCM (Scheme 4.52) [51].

Mandal and coworkers developed a useful method for assembling spiroannulated carbanucleoside **271** *via* RCM as a key step. To this end, the keto-derivative **268** was treated with a Grignard reagent **63** to deliver the allylated alcohol **269**. Later, the NaH-mediated *O*-allylation of alcohol **269** with allyl bromide (**6**) under reflux conditions produced the diallyl carbanucleoside derivative **270**. Further, on treatment with G-I catalyst provided the spiroannulated sugar **271** (Scheme 4.53) [52].

Alcaide and coworkers reported the synthesis of enantiopure spirocyclic-2-azetidinone **276** *via* metal-catalyzed reaction followed by RCM. Then, the

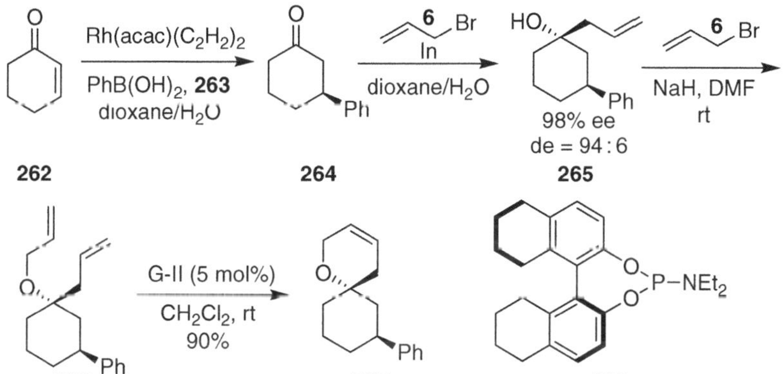

Scheme 4.52 Synthesis of spirocyclic ether **267**. *Source:* Based on Silver and Leino [51].

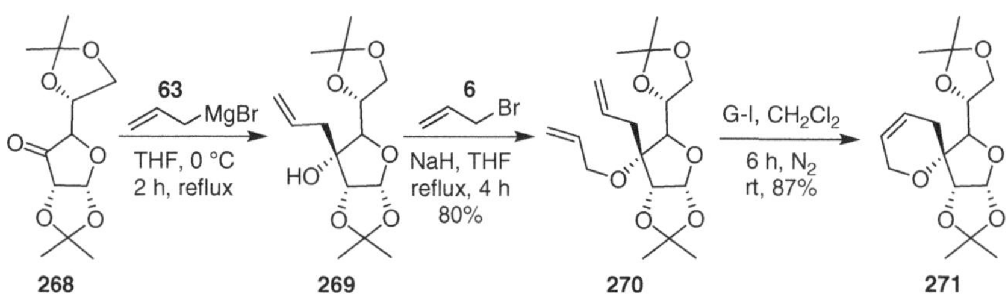

Scheme 4.53 Synthesis of spiroannulated carbanucleoside. *Source:* Modified from Sahabuddin et al. [52].

azetidine-2,3-dione **272** was treated with 1,4-dibromo-2-butyne (**273**) in the presence of a zinc metal to deliver the diene **274**. Later, the diene **274** was reacted with allyl bromide (**6**) under basic conditions to yield the tri-ene **275**. Finally, the triene **275** was treated with G-I catalyst to furnish the spirocyclic 2-azetidinone **276** (Scheme 4.54) [50, 53].

Albrecht and Langer reported the synthesis of spirocyclic butenolide **280** *via* Grignard reaction and RCM as key steps. To this end, 2-adamantanone (**277**) was treated with vinyl magnesium bromide (**51**) followed by *O*-acylation with acryloyl chloride (**259**) to deliver a mixture of alcohol **278** and *O*-acylated compound **279**. Later, the compound **279** was treated with G-I catalyst to afford the spirobutenolide **280** (Scheme 4.55) [54].

Horino and coworkers described the synthesis of spiro-2-oxazolidinone **284** *via* a gold-catalyzed intermolecular addition of allyl alcohol (**32**) followed by RCM. Allene **281** was treated with allyl alcohol (**32**) in the presence of a gold catalyst to deliver a mixture of two different addition products **282** and **283**. The compound **283**, on RCM with G-II catalyst, gave the spiro-2-oxazolidinone **284** (Scheme 4.56) [55].

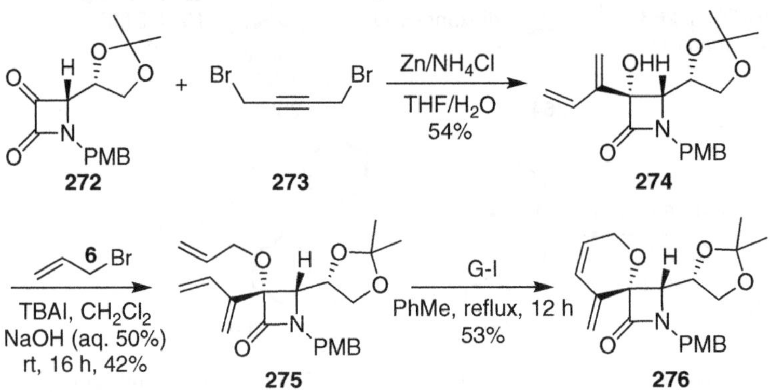

Scheme 4.54 Synthesis of spirocyclic-2-azetidinone. *Sources:* Based on Alcaide et al. [50]; Alcaide et al. [53].

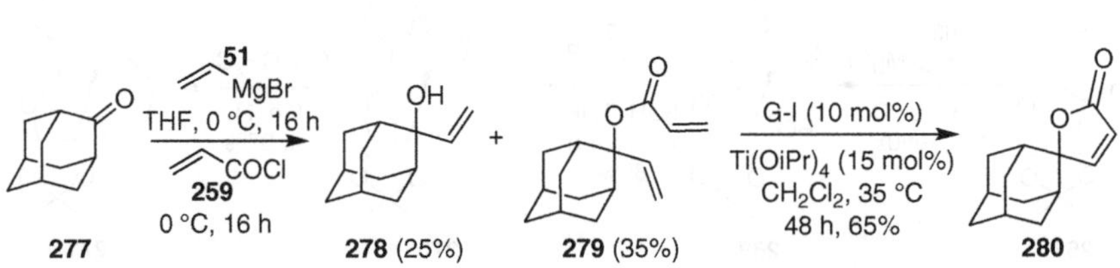

Scheme 4.55 Synthesis of spirocyclic butenolide **280**. *Source:* Based on Albrecht and Langer [54].

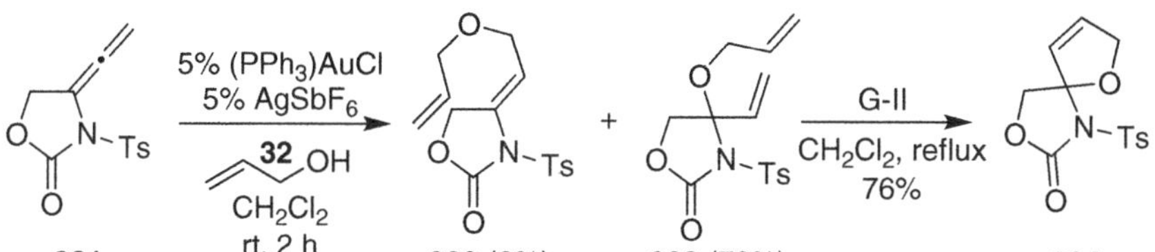

Scheme 4.56 Synthesis of spiro-2-oxazolidinone. *Source:* Based on Horino et al. [55].

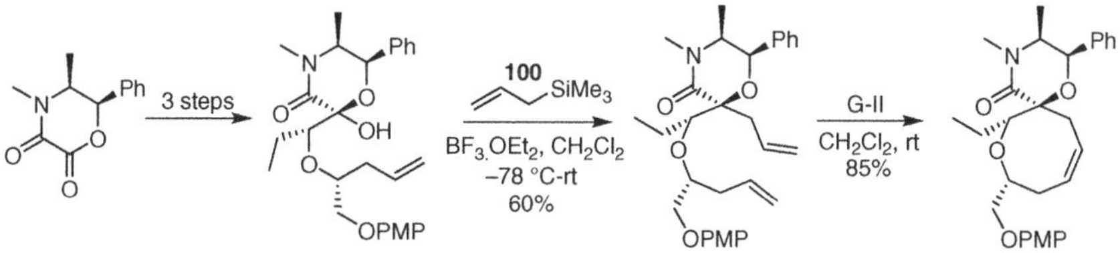

Scheme 4.57 Synthetic approach to (+)-laurencin. *Source:* Based on Adsool and Pansare [56].

Adsool and Pansare reported a RCM as a key step toward the synthesis of (+)-laurencin. In this connection, alcohol **286** was prepared from morpholine dione **285** involving a three-step sequence. Then, the hydroxy derivative **286** was treated with allyl trimethylsilane (**100**) under acidic conditions to afford the key RCM precursor **287**. Later, the dialkene **287** was made to undergo RCM with the aid of G-II catalyst to produce the spirocyclic ether **288**. It is worth mentioning that, the intermediate **288** was utilized for the synthesis of (+)-laurencin involving a series of chemical transformations (Scheme 4.57) [56].

We have developed a new protocol to synthesize spiro oxacycles such as **292** and **296** starting with norbornene derivative **289** containing naphthoquinone moiety by RRM strategy. To this end, the DA adduct **289** was treated with allylmagnesium bromide (**63**) followed by *O*-allylation to afford the *O*-allyl building block **291**. Next, the compound **295** was reacted with G-II catalyst (10 mol%) in the presence of $Ti(OiPr)_4$ to provide the RRM product **292**. Along similar lines, we also studied RRM/EM cascade of *O*-propargylated derivative **293**. The RCM building block **293**, on treatment with HG-II catalyst (10 mol%), produced the expected diene **294**. Finally, the resulting diene **294** was reacted with tetracyanoethylene (TCNE, **295**) to generate the corresponding cycloaddition product **296** (Scheme 4.58) [57].

Metathesis has frequently been used for the construction of key intermediates to synthesize natural products. In 2010, Ghosh and coworkers reported the synthesis of a key intermediate nortriterpenoid **302** by metathesis. In this context, the ketone **298** was prepared from the aldehyde **297** involving a three-step sequence.

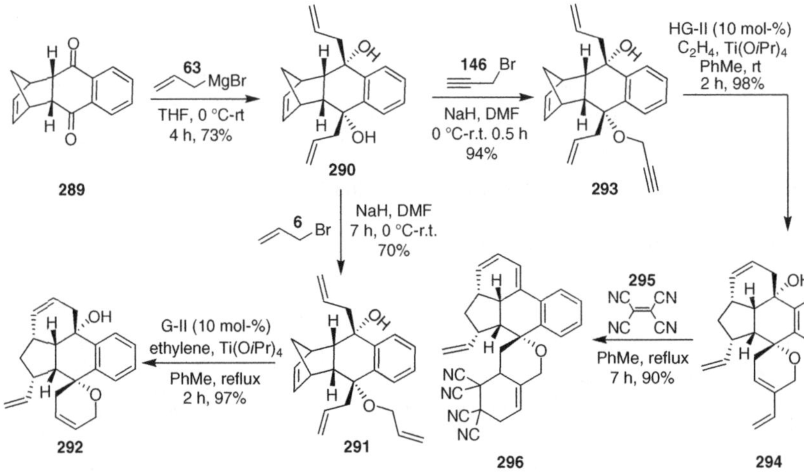

Scheme 4.58 Synthesis of spiro oxacycles. *Source:* Based on Kotha et al. [57].

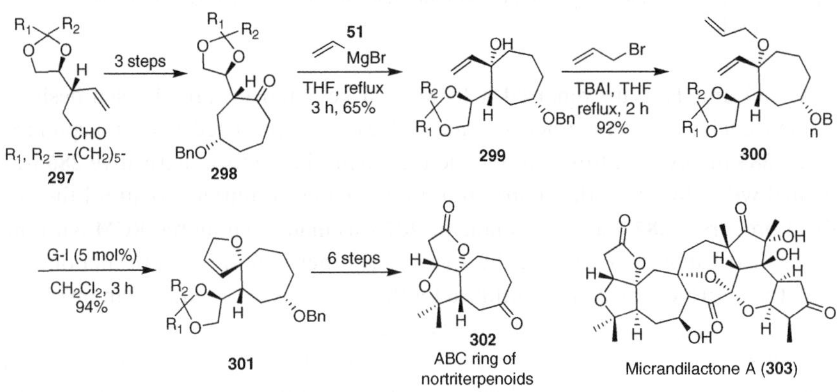

Scheme 4.59 Synthesis of ABC ring of nortriterpenoid. *Source:* Based on Maity et al. [58].

Later, it was treated with vinyl magnesium bromide (**51**) followed by *O*-allylation provided the RCM precursor **300**. The RCM of di-alkenyl precursor **300** with aid of G-I (5 mol%) catalyst afforded the spirocyclic compound **301**. Further, several organic transformations of **301** resulted the enantiopure C-5-*epi* ABC ring system of micrandilactone A (**303**, Scheme 4.59) [58].

Kaliappan and coworkers reported the synthesis of spiro-*C*-aryl glycoside **310** *via* EM followed by DA reaction. In this connection, the lactone **304** was treated with TMS-acetylene under basic conditions to obtain the 1,2-addition product **306**. Then, the resulting –OH group was protected by allylation to afford the enyne building block **307**. The silyl group was deprotected to deliver the enyne building block **308**, which undergoes EM to give diene **309**. Finally, the diene **309** undergoes DA reaction with DMAD to deliver the spiro-*C*-aryl glycoside derivatives **310** (Scheme 4.60) [59].

Elias and coworkers reported the synthesis of spiro phosphazene **313** from terminal alkene by using RCM as a key step. Hexa-chlorophosphazene **311** was reacted with allyl alcohol (**32**) to deliver the hexa(allyloxy)phosphazene **312**, which on RCM reaction with G-I catalyst affords the spiro phosphazene derivative **313** (Scheme 4.61) [60].

Blanchard and coworkers reported the synthesis of spiroketal **318** *via* RCM as a key step. They started with furan derivative **314**, which on reduction with LAH gave the corresponding alcohol. Further, protection with TBSCl under basic conditions gave the furan derivative **315**. Later, the DA reaction of furan derivative **315** and ethyl acrylate (**316**) gave the cycloadduct **317**. Finally, the bicyclic compound **317** undergoes RRM in the presence of G-II catalyst produced the spiroketal **318** (Scheme 4.62) [61].

In 2012, Wulff and coworkers reported the synthesis of anomeric spiroketal involving [2+2] cycloaddition and RCM sequence. Cyclopetenone **319** was reacted with triethylsilyl hydride under basic conditions to afford the enolate **320**, which on [2+2] cycloaddition with methyl propionate (**321**) in the presence of Lewis

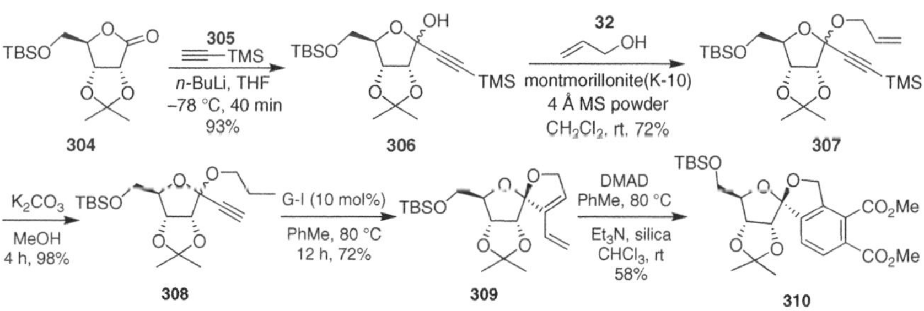

Scheme 4.60 Synthesis of spiro-*C*-aryl glycoside. *Sources:* Based on Subrahmanyam et al. [59a]; (b) S. Kotha et al. [59b].

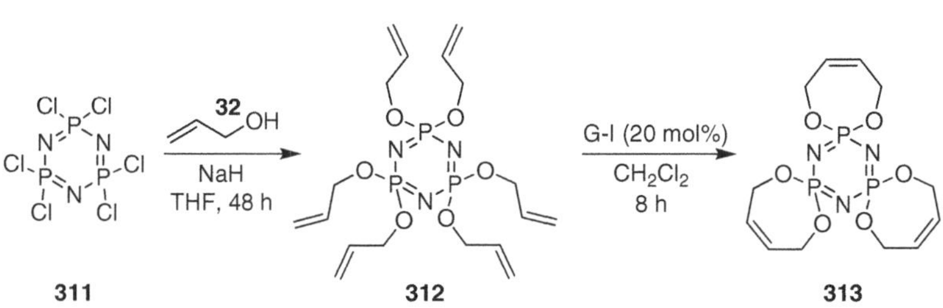

Scheme 4.61 Synthesis of cyclic phosphazene. *Source:* Kumar et al. [60].

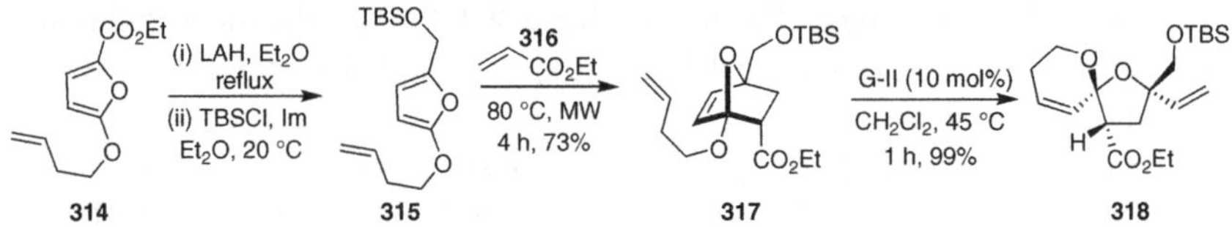

Scheme 4.62 Synthesis of spiroketal **318**. *Source:* Based on Mandel et al. [61].

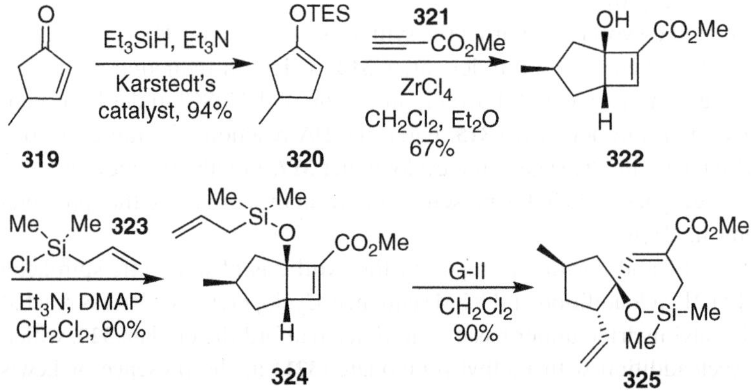

Scheme 4.63 Synthesis of anomeric spiroketal **325**. *Source:* Based on Davy et al. [62].

acid and subsequent deprotection of a silyl group afforded the cyclobutene derivative **322**. Later, the cyclobutene **322** was reacted with allyldimethylsilylchloride (**323**) to deliver the RRM precursor **324**. Further, the RRM of cyclobutene derivative **324** with G-II catalyst provides the anomeric spiroketal **325** (Scheme 4.63) [62].

Hsung and coworkers reported the synthesis of spiroketal **331** *via* RCM as a key step. The aldehyde **326** was transformed to pyran derivative **327**, which on reaction with triflimide formed an oxonium ion **328**. Then, the oxonium ion **328** was treated with alcohol **329** to deliver the RCM precursor **330**. The di-ene **330** was reacted with G-II catalyst to furnish the spiroketal **331** (Scheme 4.64) [63].

Busto and coworkers reported the spiro pyrrolo-azepine core from 7-azanorbornene *via* RRM as a key step. In this connection, 7-azanorbornene derivative **332** was reacted with TFA followed by *N*-acylation with acid **333** to afford the azanorbornene derivative **334**. Later, a selective reduction of the ester group of azanorbornene derivative **334** was carried out using $Ca(BH_4)_2$ to obtain the alcohol **335**, which on further protection with allyl bromide (**6**) gave the RRM precursor **336**. Finally, the azanorbornene derivative **336** was treated with G-II catalyst to yield the rearranged azepine derivatives **337** and **338** (Scheme 4.65) [64].

Cooper and Wright reported the synthesis of oxa-spirocycle **343** *via* Wittig reaction, DA reaction, and RCM as key steps. The substituted benzaldehyde **339**

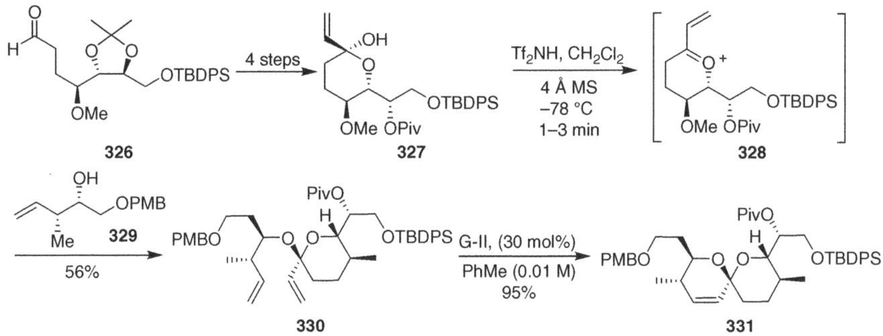

Scheme 4.64 Synthesis of spiroketal **331**. *Source:* Tang et al. [63].

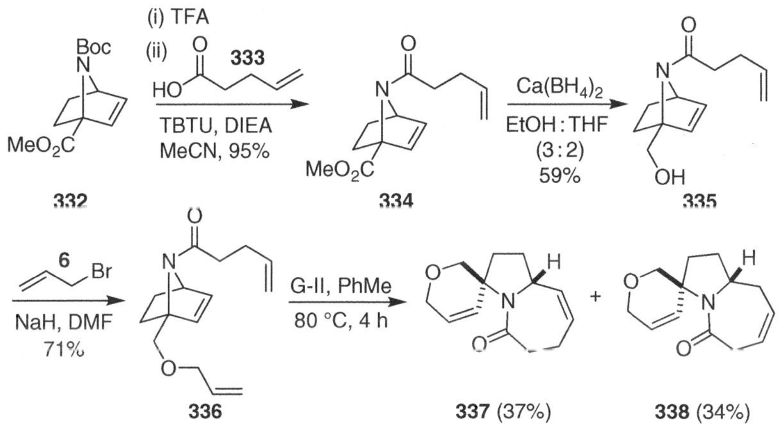

Scheme 4.65 RRM approach to azepine derivatives. *Source:* Based on Rojas et al. [64].

undergoes a Wittig reaction with **340** under basic conditions to deliver the styrene derivative **341**. The styrene derivative **341** on DA reaction with DMAD gave the RRM precursor **342**. Later, the RRM precursor **342** was treated with G-II catalyst to afford the oxa-spirocycle **343** (Scheme 4.66) [65].

Yamamoto and coworkers reported the synthesis of oxa-spirocycle **346** through interceptive decarboxylative addition of allyl carbonate (**122**) followed by RCM. Dibutyl squarate **344** was treated with diallyl carbonate (**122**) under the influence of a palladium catalyst to deliver the bisallyloxy cyclobutenone **345**. Later, the reaction of **345** with G-II catalyst provided the spirocyclic compound **346** *via* the allyloxy rearrangement (bisallyloxy cyclobutane transformed into vicinal diallyloxy compound) followed by RCM (Scheme 4.67) [24b].

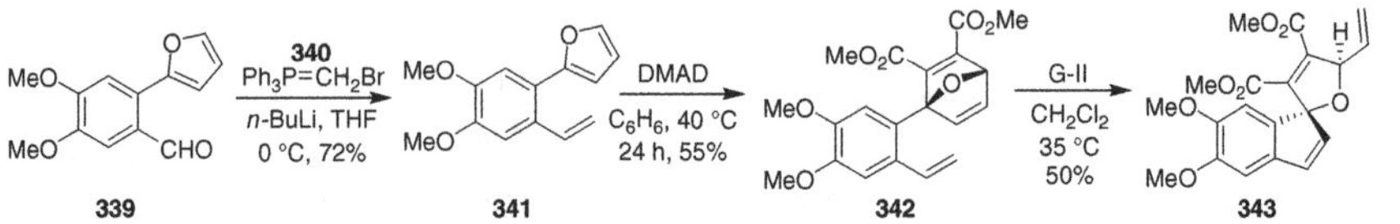

Scheme 4.66 Synthesis of oxa-spirocycle. *Source:* Modified from Cooper and Wright [65].

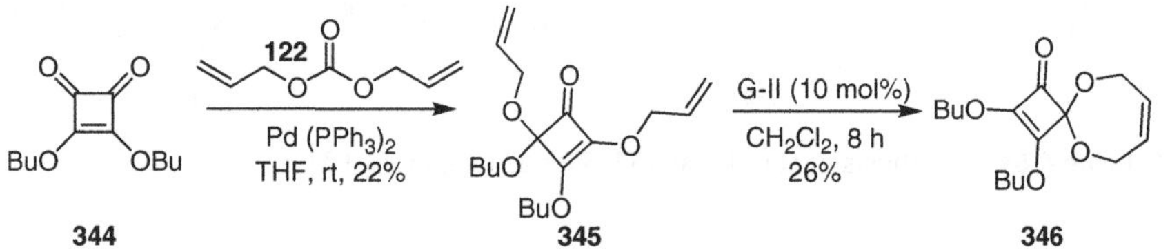

Scheme 4.67 RCM of bisallyloxy cyclobutenone to oxa-spirocycle. *Source:* Based on Baiju et al. [24b].

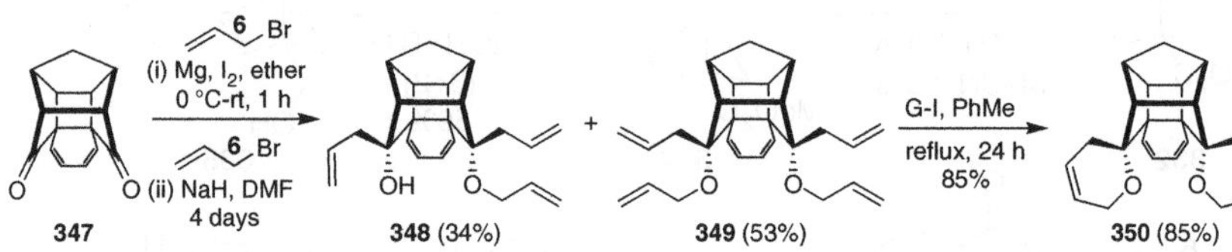

Scheme 4.68 Synthesis of spiro-cage derivative. *Source:* Modified from Kotha et al. [66].

The strategy to assemble the spiro-cage derivative **350** has developed by Kotha and coworkers *via* Grignard addition followed by RCM. The cage-dione **347**, on Grignard addition followed by *O*-allylation, gave a mixture of tri-allyl **348** and tetra-allyl **349** cage derivatives. Later, the tetra-allyl cage derivative **349** was reacted with G-I catalyst to afford the spiro compound **350** (Scheme 4.68) [66].

Along similar lines, Kotha and Ali reported synthesis of bis-oxaspirocyclic compound **354** *via* Grignard addition followed by RCM as key steps. For this purpose, the cyclohex-1,4-dione **351** was treated with allylmagnesium bromide (**63**) to deliver the diol **352**, which upon reaction with allyl bromide (**6**) under basic conditions gave the tetra-allyl compound **353**. Finally, the tetra-allyl compound **353** was treated with G-I catalyst to produce the bis-oxaspirocycles **354** (Scheme 4.69) [67].

Kotha and coworkers developed a simple and atom-economic process involving RRM for the construction of fused and oxaspirocycles **358** and **360** starting with norbornene diol **355**. RRM precursors **357** and **359** are derived from **356** by Swern oxidation, Grignard reaction with **63**, followed by allylation or propargylation sequence. Next, the tri-allyl derivative **357** was subjected to RRM in the presence

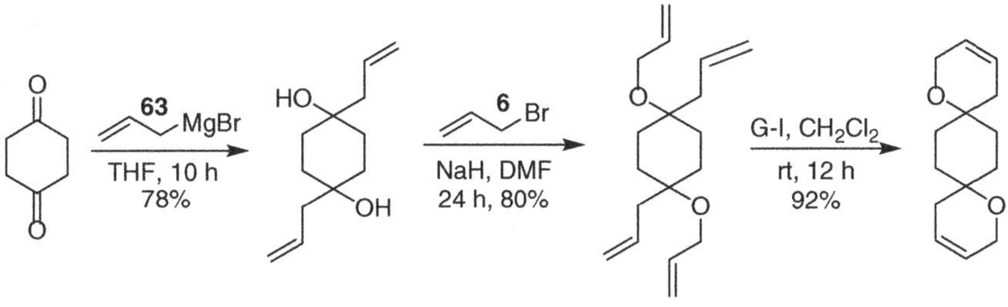

351 352 353 354

Scheme 4.69 Synthesis of bis-oxaspirocycle. *Source:* Based on Kotha and Ali [67].

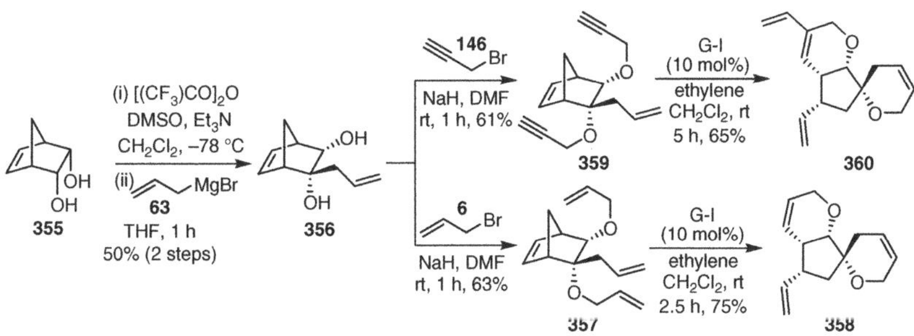

Scheme 4.70 Synthesis of oxaspirocycles. *Source:* Based on Kotha et al. [68].

of G-I catalyst to afford the oxaspirocycle **358**. Similarly, exposure of dipropargyl derivative **359** to G-I catalyst produced the rearranged spiro tri-vinyl system **360** (Scheme 4.70) [68].

4.4 Miscellaneous Examples

Silver and Leino reported the synthesis of spiro-indene **364** starting with indene **361** using RCM as a key step. In this connection, the reaction of indene **361** with allyldimethylsilylchloride (**323**) in the presence of *n*-BuLi resulted in a mixture of 1,3-diallylsiyl substituted indene **362** and 1,1-diallyl silyl indene derivative **363**. Later, 1,1-diallyl silyl indene derivative **363** treated with G-I catalyst to produce the spiro-indene derivative **364** (Scheme 4.71) [51].

Lovchik and Kraft reported the synthesis of spirocycle containing a silicon atom at spioro junction *via* RCM sequence. In this context, the nitrile **366** was prepared from the corresponding allyl alcohol compound **365** by the literature-known multistep synthesis. The nitrile derivative **366** was treated with 1-allyl-1-chlorosilolane (**367**) under basic conditions to deliver the di-ene **368**. Later, the silicon containing

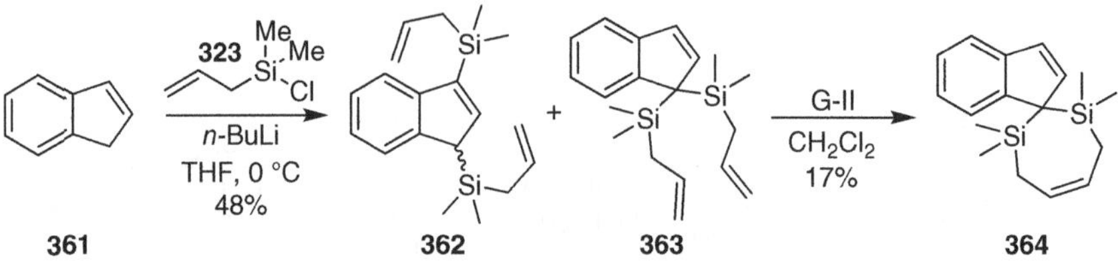

Scheme 4.71 Synthesis of spiro indene. *Source:* Based on Silver and Leino [51].

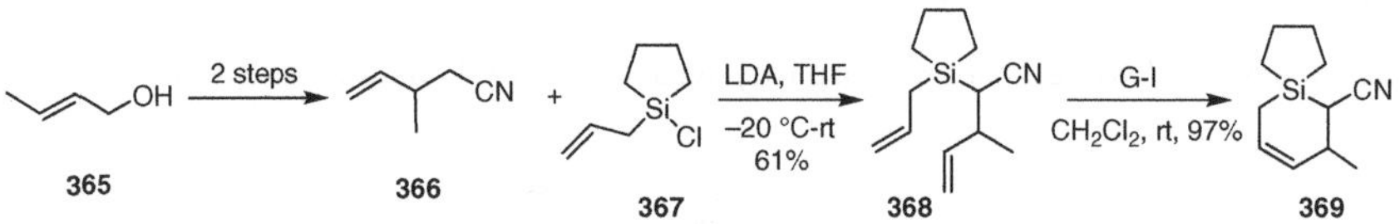

Scheme 4.72 Synthesis of spirocycle containing silicon at spioro junction. *Source:* Based on Lovchik and Kraft [69].

compound **368** was treated with G-I catalyst to afford spirocyclic compound **369** containing the silicon at spioro junction. The compound **369** has been used as an intermediate to synthesize natural product δ-damascone (Scheme 4.72) [69].

References

1 (a) Kotha, S., Deb, A.C., Lahiri, K., and Manivannan, E. (**2009**). *Synthesis* 2: 165. (b) Kotha, S. and Sreevani, G. (**2018**). *Eur. J. Org. Chem.* 2018: 5935. (c) Quintavalla, A. (**2018**). *Curr. Med. Chem.* 25: 917. (d) Kotha, S., Cheekatla, S.R., and Mandal, B. (**2017**). *Eur. J. Org. Chem.* 2017: 4277. (e) Rios, R. (**2012**). *Chem. Soc. Rev.* 41: 1060. (f) Reddy, C.R., Prajapti, S.K., Warudikar, K. et al. (**2017**). *Org. Biomol. Chem.* 15: 3130. (g) Hammer, K. and Undheim, K. (**1997**). *Tetrahedron* 53: 5925. (h) Hammer, K. and Undheim, K. (**1997**). *Tetrahedron* 53: 2309. (i) Andrei, M., Efskind, J., and Undheim, K. (**2007**). *Tetrahedron* 63: 4347. (j) Jam, F., Tullberg, M., Luthman, K., and Grotli, M. (**2007**). *Tetrahedron* 63: 9881.

2 (a) Lemieux, R.M., Devine, P.N., Mechelke, M.F., and Meyers, A.I. (**1999**). *J. Org. Chem.* 64: 3585. (b) Hughes, R.C., Dvorak, C.A., and Meyers, A.I. (**2001**). *J. Org. Chem.* 66: 5545.

3 Oesterreich, K. and Spitzner, D. (**2002**). *Tetrahedron* 58: 4331.

4 Colombo, L., Di Giacomo, M., Vinci, V. et al. (**2003**). *Tetrahedron* 59: 4501.

5 Kotha, S. and Mandal, K. (**2004**). *Tetrahedron Lett.* 45: 1391.

6 Ghosh, S.K., Hsung, R.P., and Wang, J.S. (**2004**). *Tetrahedron Lett.* 45: 5505.

7 Kato, T., Marubayashi, K., Takizawa, S., and Sasai, H. (**2004**). *Tetrahedron Asymmetry* 15: 3693.

8 Schobert, R. and Urbina-Gonzalez, J.M. (**2005**). *Tetrahedron Lett.* 46: 3657.

9 Clive, D.L.J., Wang, H., and Yu, M.L. (**2005**). *Tetrahedron Lett.* 46: 2853.

10 Gonzalez-Bello, C., Castedo, L., and Canada, F.J. (**2006**). *Eur. J. Org. Chem.* 2006: 1002.

11 Thornqvist, V., Manner, S., Wendt, O.F., and Frejd, T. (**2006**). *Tetrahedron* 62: 11793.

12 (a) Majumdar, K.C., Rahaman, H., and Roy, B. (**2006**). *Lett. Org. Chem.* 3: 526. (b) Kersten, L., Taylor, R.H., and Felpin, F.X. (**2009**). *Tetrahedron Lett.* 50: 506. (c) Majumdar, K.C., Islam, R., Rahaman, H., and Roy, B. (**2006**). *Org. Biomol. Chem.* 4: 2393.

13 (a) Srikrishna, A., Lakshmi, B.V., and Mathews, M. (**2006**). *Tetrahedron Lett.* 47: 2103. (b) Srikrishna, A. and Babu, R.R. (**2007**). *Tetrahedron Lett.* 48: 6916.

14 Kotha, S., Mandal, K., Tiwari, A., and Mobin, S.M. (**2006**). *Chem. Eur. J.* 12: 8024.

15 Modrzejewska, H. and Wojciechowski, K. (**2008**). *Synlett* 2008: 2465.

16 Kuznetsov, N.Y., Kolomnikova, G.D., Khrustalev, V.N. et al. (**2008**). *Eur. J. Org. Chem.* 2008: 5647.

17 Braun, M., Hessamian-Alinejad, A., de Lacroix, B.F. et al. (**2008**). *Molecules* 13: 995.

18 (a) Kohler, A., Raabe, G., Runsink, J. et al. (**2014**). *Eur. J. Org. Chem.* 2014: 3355. (b) Lejkowski, M., Banerjee, P., Raabe, G. et al. (**2014**). *Eur. J. Org. Chem.* 2014: 529. (c) Leikowski, M., Banerjee, P., Runsink, J., and Gais, H.J. (**2008**). *Org. Lett.* 10: 2713.

19 Kotha, S. and Deb, A.C. (**2008**). *Indian J. Chem. Sect. B* 47: 1120.

20 Singh, V. and Singh, V. (**2009**). *Tetrahedron Lett.* 50: 3092.

21 Kotha, S., Panguluri, N.R., and Ali, R. (**2017**). *Eur. J. Org. Chem.* 2017: 5316.

22 Kotha, S. and Lahiri, K. (**2007**). *Synlett* 2007: 2767.

23 Kotha, S., Deb, A.C., and Kumar, R.V. (**2005**). *Bioorg. Med. Chem. Lett.* 15: 1039.

24 (a) George, S.C., John, J., Anas, S. et al. (**2010**). *Eur. J. Org. Chem.* 2010: 5489. (b) Baiju, T.V., Joseph, N., Ajit, J. et al. (**2014**). *Synlett* 25: 1246.

25 Castiglioni, E., Di Fabio, R., Togninelli, A. et al. (**2010**). *ChemMedChem* 5: 1843.

26 Maity, J.K., Achari, B., Drew, M.G.B., and Mandal, S.B. (**2010**). *Synthesis* 15: 2533.

27 (a) Brock, N.L. and Dickschat, J.S. (**2011**). *Eur. J. Org. Chem.* 2011: 5167. (b) Kurosawa, S., Bando, M., and Mori, K. (**2001**). *Eur. J. Org. Chem.* 2001: 4395.

28 Abe, H., Sato, A., Kobayashi, T., and Ito, H. (**2013**). *Org. Lett.* 15: 1298.

29 Kotha, S., Ali, R., and Tiwari, A. (**2013**). *Synlett* 24: 1921.

30 Manner, S., Oltner, V.T., Oredsson, S. et al. (**2013**). *Org. Biomol. Chem.* 11: 7134.

31 Fu, J.K., Shen, H.J., Chang, Y.Y. et al. (**2014**). *Chem. Eur. J.* 20: 12881.

32 Sosnicki, J.G., Struk, L., Idzik, T., and Maciejewska, G. (**2014**). *Tetrahedron* 70: 8624.

33 (a) Lambu, M.R., Hussain, A., Sharma, D.K. et al. (**2014**). *RSC Adv.* 4: 11023. (b) Chen, G.R., Fei, Z.B., Huang, X.T. et al. (**2001**). *Eur. J. Org. Chem.* 2001: 2939.

34 Kotha, S., Ali, R., Srinivas, V., and Krishna, N.G. (**2015**). *Tetrahedron* 71: 129.

35 Kotha, S., Ali, R., and Chinnam, A.K. (**2014**). *Tetrahedron Lett.* 55: 4492.

36 (a) Singh, V. and Das, B. (**2015**). *Synth. Commun.* 45: 1912. (b) Singh, V., Das, B., Jarhad, D.B., and Mobin, S.M. (**2015**). *Tetrahedron* 71: 560–576.

37 Kotha, S., Ali, R., and Deepak, M. (**2015**). *J. Indian Chem. Soc.* 92: 277.

38 Kotha, S., Chinnam, A.K., and Ali, R. (**2015**). *Beilstein J. Org. Chem.* 11: 1123.

39 Kotha, S., Ali, R., Panguluri, N.R. et al. (**2018**). *Tetrahedron Lett.* 59: 4080.

40 Tanner, D., Hagberg, L., and Poulsen, A. (**1999**). *Tetrahedron* 55: 1427.

41 Bittermann, H., Boeckler, F., Einsiedel, J., and Gmeiner, P. (**2006**). *Chem. Eur. J.* 12: 6315.

42 Lesma, G., Landoni, N., Sacchetti, A., and Silvani, A. (**2010**). *Tetrahedron* 66: 4474.

43 (a) Nieczypor, P., Mol, J.C., Bespalova, N.B., and Bubnov, Y.N. (**2004**). *Eur. J. Org. Chem.* 2004: 812. (b) Kuznetsov, N.Y., Zhun, I.V., Khrustalev, V.N., and Bubnov, Y.N. (**2005**). *Russ. Chem. Bull.* 54: 2229. (c) Kuznetsov, N.Y., Khrustalev, V.N., and Bubnov, Y.N. (**2010**). *Russ. Chem. Bull.* 59: 1393. (d) Agosti, A., Britto, S., and Renaud, P. (**2008**). *Org. Lett.* 10: 1417.

44 Cheng, X.Y. and Waters, S.P. (**2010**). *Org. Lett.* 12: 205.

45 Pal, A.P.J., Gupta, P., Reddy, Y.S., and Vankar, Y.D. (**2010**). *Eur. J. Org. Chem.* 2010: 6957.

46 Holt, D.J., Barker, W.D., Jenkins, P.R. et al. (**2000**). *J. Org. Chem.* 65: 482.

47 Schmidt, B. and Wildemann, H. (**2000**). *J. Org. Chem.* 65: 5817.

48 Leeuwenburgh, M.A., Appeldoorn, C.C.M., van Hooft, P.A.V. et al. (**2000**). *Eur. J. Org. Chem.* 2000: 873.

49 (a) Walters, M.A., La, F., Deshmukh, P., and Omecinsky, D.O. (**2002**). *J. Comb. Chem.* 4: 125. (b) Prusov, E. and Maier, M.E. (**2007**). *Tetrahedron* 63: 10486.

50 Alcaide, B., Almendros, P., and Rodriguez-Acebes, R. (**2006**). *J. Org. Chem.* 71: 2346.

51 Silver, S. and Leino, R. (**2006**). *Eur. J. Org. Chem.* 2006: 1965.

52 Sahabuddin, S., Roy, A., Drew, M.G.B. et al. (**2006**). *J. Org. Chem.* 71: 5980.

53 Alcaide, B., Almendros, P., del Campo, T.M., and Rodriguez-Acebes, R. (**2007**). *Adv. Synth. Catal.* 349: 749.

54 Albrecht, U. and Langer, P. (**2007**). *Tetrahedron* 63: 4648.

55 Horino, Y., Takata, Y., Hashimoto, K. et al. (**2008**). *Org. Biomol. Chem.* 6: 4105.

56 Adsool, V.A. and Pansare, S.V. (**2008**). *Org. Biomol. Chem.* 6: 2011.

57 (a) Kotha, S. and Ravikumar, O. (**2016**). Eur. J. Org. Chem. 2016: 3900. (b) Kotha, S., Meshram, M., and Tiwari, A. (**2009**). Chem. Soc. Rev. 38: 2065.

58 Maity, S., Matcha, K., and Ghosh, S. (**2010**). *J. Org. Chem.* 75: 4192.

59 (a) Subrahmanyam, A.V., Palanichamy, K., and Kaliappan, K.P. (**2010**). *Chem. Eur. J.* 16: 8545. (b) Kotha, S., Sreenivasachary, N., and Brahmachary, E. (**1998**). *Tetrahedron Lett.* 39: 2805.

60 Kumar, D., Singh, N., Keshav, K., and Elias, A.J. (**2011**). *Inorg. Chem.* 50: 250.

61 Mandel, J., Dubois, N., Neuburger, M., and Blanchard, N. (**2011**). *Chem. Commun.* 47: 10284.

62 Davy, J.A., Moreau, B., and Wulff, J.E. (**2012**). *Synthesis* 44: 1854.

63 Tang, Y., Yang, J.H., Liu, J. et al. (**2012**). *Heterocycles* 86: 565.

64 Rojas, V., Carreras, J., Avenoza, A. et al. (**2013**). *Eur. J. Org. Chem.* 2013: 3817.

65 Cooper, H.D. and Wright, D.L. (**2013**). *Molecules* 18: 2438.

66 Kotha, S., Saifuddin, M., Ali, R., and Sreevani, G. (**2015**). *Beilstein J. Org. Chem.* 11: 1367.

67 Kotha, S. and Ali, R. (**2016**). *Indian J. Chem.* 55B: 1099.

68 Kotha, S., Ravikumar, O., and Sreevani, G. (**2016**). *Tetrahedron* 72: 6611.

69 Lovchik, M.A. and Kraft, P. (**2017**). *Chem. Eur. J.* 23: 4590.

5

Spirooxindoles: Synthesis *via* Organocatalytic Processes

Peng-Wei Xu, Xiao-Yuan Cui, Jin-Sheng Yu, and Jian Zhou

School of Chemistry and Molecular Engineering, East China Normal University, Shanghai, China

5.1 Introduction

Spirooxindoles, a special type of spirocyclic compounds that feature a spiro-ring at the C3 position of the oxindole frameworks [1], are the privileged structural scaffolds frequently found in many natural products and pharmaceutical molecules [2]. As exemplified in Figure 5.1, naturally occurring gelsemine (**A**) [3], spirotryprostatin A (**B**) [4], horsfiline (**C**) [5], and rhynchophylline (**D**) [6] all contain such a motif. A wide variety of non-natural chiral spirooxindole derivatives are also found to have various biological and pharmacological activities, such as spirooxindole **G** is a CRTH2 receptor antagonist [7], MI-888 (**E**) [8] and NITD609 (**F**) [9] are currently in preclinical evaluation for the treatment of human cancer and malaria, respectively. Noticeably, the pharmacological profiles are greatly affected by the absolute configuration of spirooxindole scaffolds in some cases [10]. Therefore, the development of efficient protocol for the synthesis of spirooxindoles, especially in a catalytic enantioselective manner, is highly sought after. In this context, ever-increasing attention from chemical community has been paid to access such scaffolds, and various synthetic strategies have correspondingly been developed during the past decades. Among them, asymmetric catalysis has turned out to be the most direct and powerful strategy, and a great number of catalytic enantioselective synthetic approaches have been established for the diverse synthesis of optically active functionalized spirooxindoles [11].

On the other hand, organocatalysis has emerged as the third pillar of modern asymmetric catalysis in the past two decades, standing next to metal catalysis and biocatalysis [12]. In this context, a variety of asymmetric transformations via organocatalytic processes have been developed for the construction of optically active molecules, owing to organocatalysis has many advantages, such as ready

Spiro Compounds: Synthesis and Applications, First Edition. Edited by Ramon Rios Torres.

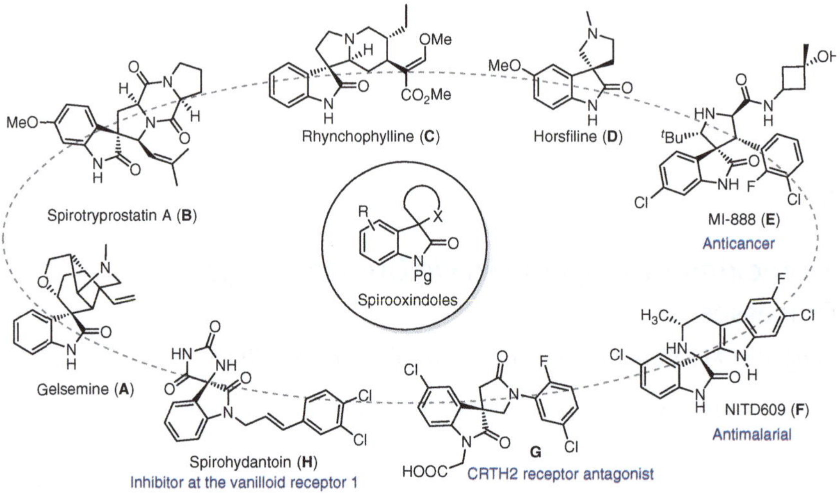

Figure 5.1 Selected natural products and bioactive molecules containing a spirooxindole unit.

availability, nontoxic, no metal residue, and simple reaction conditions. Various activation modes, including enamine catalysis, iminium catalysis, nucleophilic catalysis, NHC catalysis, PTC catalysis, tertiary amine-H bonding donor bifunctional catalysis, phosphoric acid catalysis, and so on, have also been accordingly established during the studies [13]. Especially, the rapid development of various organocatalytic activation modes and chiral organocatalysts enormously aroused the enthusiasm on the exploration of efficient and novel reactions for the diverse synthesis of high value-added chiral spirooxindoles via organocatalytic processes [1].

Consequently, by employing distinct organocatalytic activation modes, a variety of elegant approaches based on the various asymmetric transformations involving different types of oxindole-containing synthons (e.g. isatin, isatin-derived imines, 3-alkenyl oxindoles, and 3-substituted or 3-nonsubstituted oxindoles) have been developed for the construction of optically active spirooxindoles fused with a three-, four-, five-, and six-membered ring in the past decade (Figure 5.2). In this chapter, we will introduce these significant achievements according to the type of activation mode, and each section is further subdivided by the type of oxindole substrates used in the reaction development. We believe this chapter would provide the readers with some inspiration to develop newer and more useful protocols for the diverse synthesis of functionalized spirooxindoles by designing new organocatalysts and/or more suitable substrates.

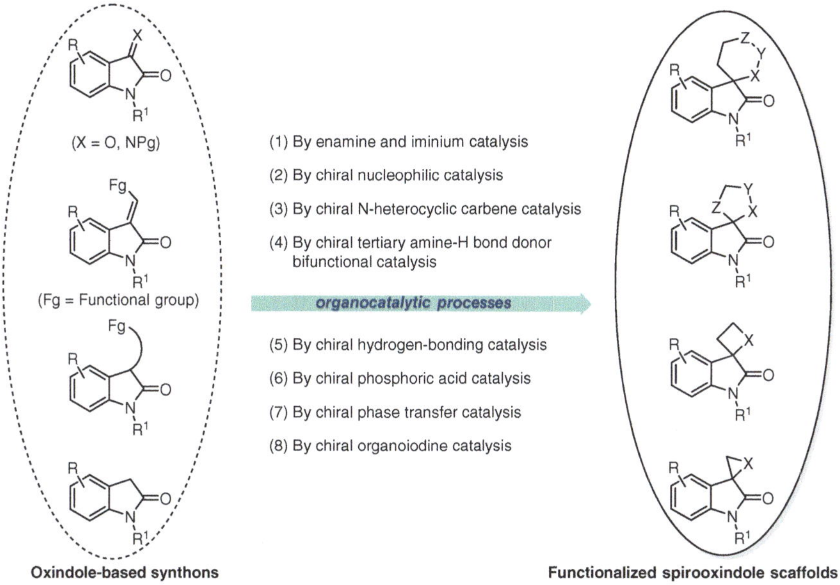

Figure 5.2 Asymmetric synthesis of various spirooxindoles via organocatalytic processes.

5.2 Enamine and Iminium Catalysis

Since the independent landmark work of List, Lerner, and Barbas [14], and MacMillan [15], the enamine catalysis and iminium catalysis have gained ever-increasing attention in the past two decades [16], which are identified as a powerful strategy for the asymmetric transformations of carbonyl compounds via HOMO-raising or LUMO-lowering activation. A variety of reactions have been established well, in particular, the design of new asymmetric domino reactions by enamine and iminium catalysis turned out to be a fruitful tactic for the construction of chiral spirocyclic complex molecules from simple starting materials. In this context, enamine and iminium catalysis also demonstrated their power in the catalytic enantioselective synthesis of spirooxindoles with structural diversity. The typical examples are illustrated in the following.

In 2009, Melchiorre et al. pioneered the use of enamine/iminium catalysis in the synthesis of optically active spirooxindoles. With enamine/iminium sequential catalysis, they developed an asymmetric double Michael cascade reaction of methyleneindolinones **1** and enones **2** (Scheme 5.1) [17]. A diverse set of spirooxindoles **4** were achieved in good yield with high ee values, in the presence of 20 mol% chiral primary amine **3** and 30 mol% *ortho*-fluorobenzoic acid (OFBA).

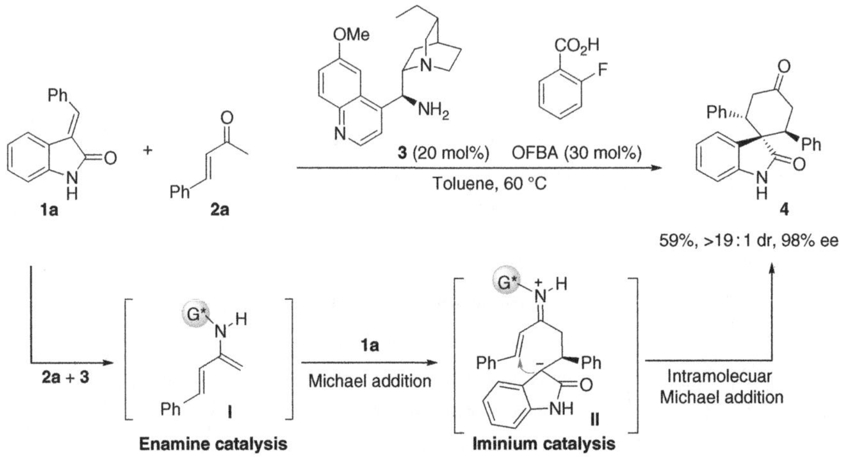

Scheme 5.1 Double Michael cascade reaction enabled by enamine/iminium sequential catalysis. *Source:* Modified from Bencivenni et al. [17].

The reaction pathway was also proposed, in which amine catalyst **3** first reacted with enone **2a** to provide the enamine intermediate **I** that subsequently underwent a Michael addition with methyleneindolinone **1a**. The resulting iminium intermediate **II** proceeded with an intramolecular Michael addition to afford the chiral spirooxindole.

Later on, Wang, Tao, and coworkers combined chiral primary amine **6a** and BINOL-phosphoric acid **7a** to develop an asymmetric double Michael reaction of isatylidene malononitriles **5** with enones **2** through enamine/iminium sequential catalysis, enabling the efficient construction of optically active spiro[cyclohexanone-oxindole] derivatives **8** (eq 1, Scheme 5.2) [18]. Soon after, Ramachary et al. also reported an asymmetric reflexive-Michael reaction of isatylidene malononitriles with α,β-acetylenic ketones by enamine/iminium sequential catalysis (eq 2) [19]. The use of 20 mol% of primary amine **6a** with 30 mol% OFBA provided a number of functionalized drug-like six-membered spirooxindoles with good yields and high to excellent enantioselectivities.

In 2015, Jørgensen et al. described a facile protocol for the synthesis of enantioenriched spirocyclobutaneoxindoles through a stereoselective [2+2] cycloaddition of 3-olefinic oxindole **11a** and activated cyclopropanes **10** by enamine/iminium sequential activation (Scheme 5.3) [20]. A combination of 10 mol% Hayashi–Jørgensen catalyst **12** and 10 mol% $PhCO_2H$ turned out to be an effective catalytic system and provided the biologically relevant spirocyclobutaneoxindoles **13** in good yields with high to excellent stereoselectivities. By mechanistic studies, the authors proposed a plausible reaction pathway shown in Scheme 5.3. The reaction of cyclopropanes **10** with catalyst **12** first occurred to produce an

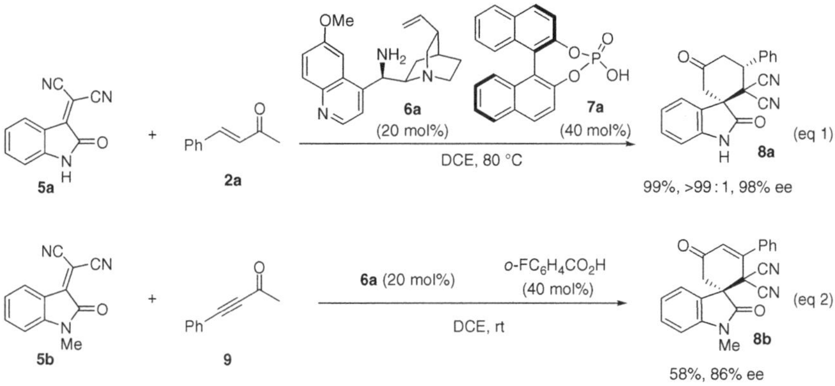

Scheme 5.2 Asymmetric double Michael reactions involving isatylidene malononitriles. *Source:* Lan et al. [18].

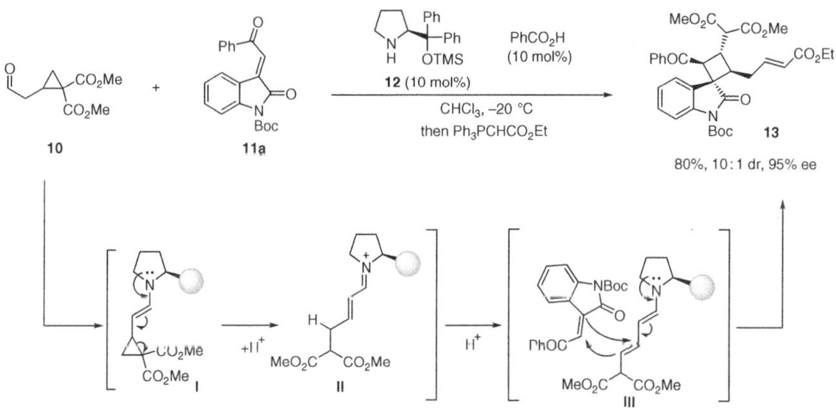

Scheme 5.3 Spirocyclobutaneoxindole synthesis by a stereoselective [2+2] cycloaddition. *Source:* Based on Halskov et al. [20].

iminium intermediate **II**, which went through deprotonation to give the active dienamine intermediate **III**. Subsequently, the dienamine **III** reacted with 3-olefinic oxindoles **11a** via a formal [2+2] cycloaddition to afford the spirooxindole-based cyclobutane **13**.

By employing an enamine/iminium/enamine sequential catalysis, Melchiorre et al. further disclosed a three-component cascade reaction of methyleneindolinones **1**, aliphatic aldehydes **14**, and enals **15**, which allowed to access a variety of spirooxindoles with excellent stereoselectivity under the catalysis of 15 mol% amine catalyst **12** with OFBA (Scheme 5.4) [17]. As illustrated in Scheme 5.4, a similar reaction mechanism was proposed in this case, an activated enamine

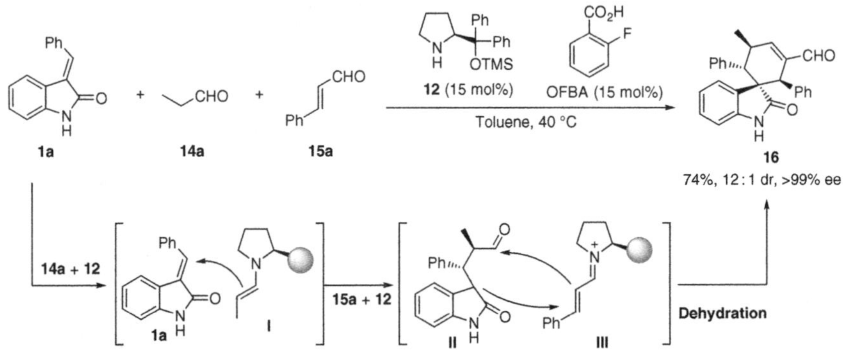

Scheme 5.4 Three-component reaction by enamine/iminium/enamine sequential catalysis. *Source:* Based on Bencivenni et al. [17].

intermediate **I**, generated from aliphatic aldehyde **14a** and catalyst **12**, first reacted with methyleneindolinone **1a**. The resulting 3-substituted oxindolic intermediate **II** subsequently went through a Michael/aldol sequence with α,β-unsaturated iminium **III** that was generated *in situ* from enal **15a** and catalyst **12**, to give the desired spirooxindoles **16** after dehydration. In addition to enals, other electron-deficient olefins also proved to be viable substrates in this type of three-component cascade reaction. For instance, Chen and coworkers demonstrated that nitroolefins, *N*-benzylmaleimide, and aldehyde imide all reacted well, and provided diverse chiral spirooxindole derivatives [21]. Later, Zhong and Zeng discovered that intermediate **II** in the cascade reaction could be trapped by another molecule of aliphatic aldehyde, thus enabling the construction of highly functionalized 5′,6′-dihydro-2′*H*, 4′*H*-spiro[indoline-3,3′-pyran]-2-one derivatives containing multiple stereogenic centers [22].

Furthermore, the Chen group utilized a quadruple iminium/enamine/iminium/enamine catalysis to develop a three-component domino reaction of methyleneindolinones **17** with two molecules of enals (eq 1, Scheme 5.5) [23]. The use of 20 mol% Hayashi–Jørgensen catalyst **12** and benzoic acid enabled an array of optically active spirooxindoles-based polycyclic molecules **18** with excellent diastereo- and enantioselectivities. Later in 2017, by utilizing the same quadruple iminium/enamine/iminium/enamine catalysis, Ren and Wang realized an organocatalytic asymmetric quadruple-cascade reaction of (*E*)-3-(2-hydroxybenzylidene)oxindoles **19** and two molecules of enals (eq 2), which provided a facile entry to access biologically important functionalized polycyclic spirooxindoles **21** [24].

Besides, Chen and Jørgensen designed a novel trienamine catalysis strategy and applied it to the construction of spirooxindoles by a stereoselective Diels–Alder reaction of methyleneindolinones **1** with dienals **22** (Scheme 5.6) [25]. In the presence of 20 mol% secondary amine **12** and OFBA, a wide range of enantioenriched

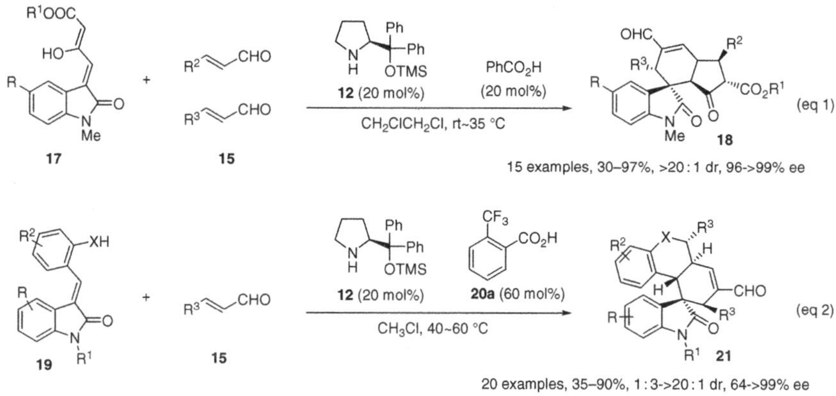

Scheme 5.5 Multicomponent domino reactions by quadruple iminium/enamine/iminium/enamine catalysis. *Source:* Modified from Jiang et al. [23].

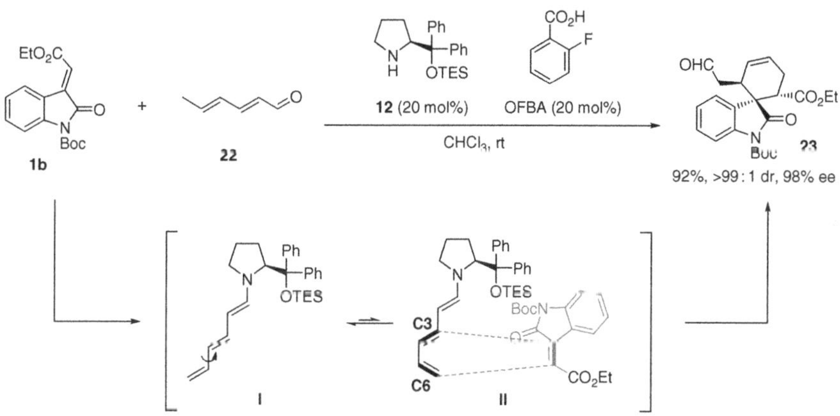

Scheme 5.6 Stereoselective Diels–Alder reaction enabled by trienamine catalysis. *Source:* Based on Jia et al. [25].

spirooxindoles **23** were prepared with high yields and excellent stereoselectivities. On the basis of NMR studies and DFT calculations, the authors proposed the resulting trienamine intermediate **I** underwent a single bond rotation to afford the intermediate **II**, which then reacted with 3-olefinic oxindole **1b** to afford the targeted spirooxindoles **23**. Inspired by this work, the cycloaddition reactions of methyleneindolinones with various polyenal analogues were explored for the diverse synthesis of chiral spirooxindoles by the groups of Melchiorre [26], Chen [27], and Jørgensen [28]. Remarkably, Jørgensen and coworkers further developed a tetraenamine catalysis, which turned out to be a powerful strategy for access of highly functionalized cyclohexane-based spirooxindoles with four stereocenters through the asymmetric [4+2]-cycloaddition of methyleneindolinones with 2-(cyclohepta-1,3,5-trien-1-yl)acetaldehyde [29].

The synthesis of chiral spirooxindoles by enamine and iminium catalysis strategy is not limited to the use of 3-olefinic oxindoles as the starting materials. A variety of other oxindole-based reacting partners (e.g. isatins, 3-substituted oxindoles) have also been exploited over the past few decades. For instance, Tanaka et al. developed an asymmetric hetero-Diels–Alder reaction of α,β-unsaturated ketones **2** with isatins **24** by enamine/iminium sequential catalysis for the synthesis of spirooxindoles featuring an oxa-ring (Scheme 5.7) [30]. A combination of 10 mol% amine catalyst **25a**, 80 mol% amino acid **26**, and 40 mol% thiourea **27** enabled efficiently the aldol/Michael cascade reaction to furnish the desired functionalized spirooxindole tetrahydropyrans **28** in excellent yield with excellent dr and ee values.

In 2010, by using 3-unsubstituted oxindoles as nucleophiles, Rios and coworkers accomplished an asymmetric Michael/Michael/aldol cascade reaction between 3-unsubsituted oxindoles **29** and enals **15** by iminium catalysis, which constitutes a practical method to access spirooxindoles featuring three continuous stereocenters (Scheme 5.8) [31]. A combination of 20 mol% Hayashi–Jørgensen catalyst **12** with $PhCO_2H$ enabled the synthesis of various chiral spirooxindoles **30** with up to 90% yield, >20 : 1 dr, and 99% ee. As illustrated in Scheme 5.8, two sequential Michael addition reactions between oxindole **29a** and two molecules of enal **15a** first occurred through iminium catalysis. The thus obtained intermediate **I** underwent an intramolecular aldol reaction, followed by an irreversible dehydration to produce the target **30**. It should be noted that the addition of $PhCO_2H$ was essential, as no product was observed in the absence of $PhCO_2H$. This is likely because the existence of $PhCO_2H$ facilitated the irreversible dehydration step, thus avoiding a *retro*-Michael reaction.

Later in 2013, also by iminium activation, Melchiorre et al. developed an asymmetric vinylogous cascade reaction of 2,4-dienals **32** with 3-hydroxy-2-oxindoles **31** as reactants (Scheme 5.9) [32]. They found that the sequential 1,6- and 1,4-addition proceeded smoothly to furnish an array of optically active

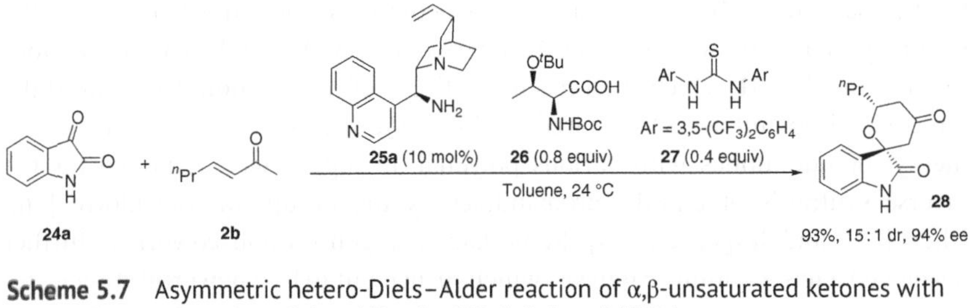

Scheme 5.7 Asymmetric hetero-Diels–Alder reaction of α,β-unsaturated ketones with isatins. *Sources*: Based on Cui and Tanaka [30a]; Cui et al. [30b].

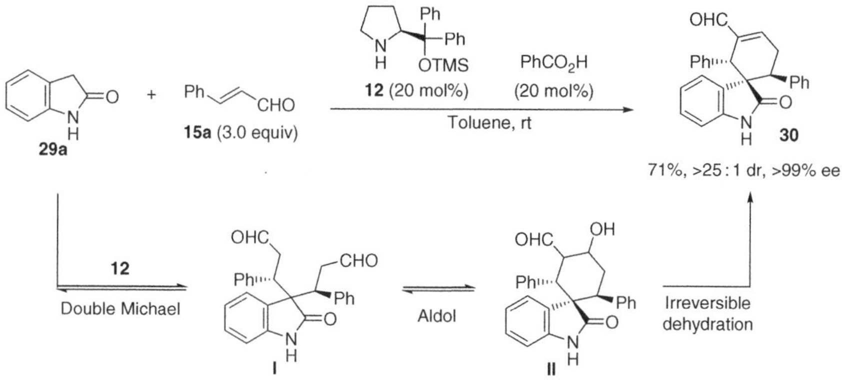

Scheme 5.8 Asymmetric Michael/Michael/aldol reaction of 3-unsubsituted oxindoles with enals. *Source:* Modified from Companyó et al. [31].

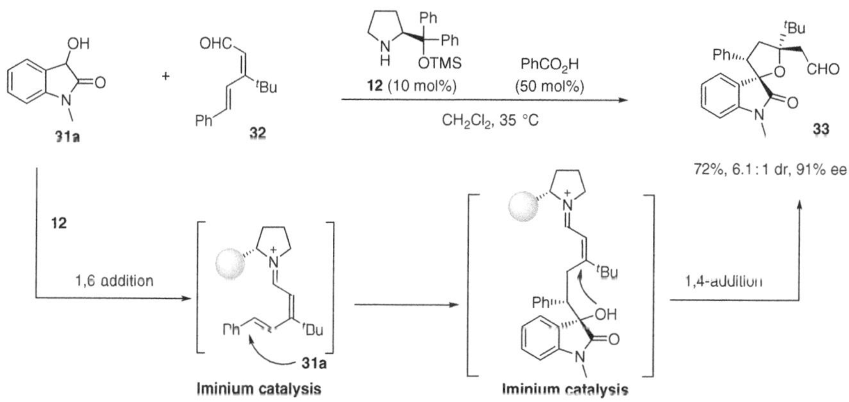

Scheme 5.9 Asymmetric vinylogous cascade reaction of 2,4-dienals with 3-hydroxy-2-oxindoles. *Source:* Based on Silvi et al. [32].

tetrahydrofuran spirooxindoles **33** with good to high yields and good to excellent stereoselectivities, under the influence of 10 mol% secondary amine **12** with 50 mol% $PhCO_2H$. Notably, it represents a rare example of stereoselective and highly δ-site selective 1,6-addition to linear 2,4-dienals. The key to success is the rational design of the substrates employed, as the steering group at the β-position of dienals ensured molecular preorganization of active vinylogous iminium intermediate. Additionally, the same group reported the synthesis of enantioenriched spirooxindole-γ-butyrolactones through a Michael addition/semiacetalization/oxidation cascade reaction involving 3-hydroxy-2-oxindoles by iminium catalysis [33].

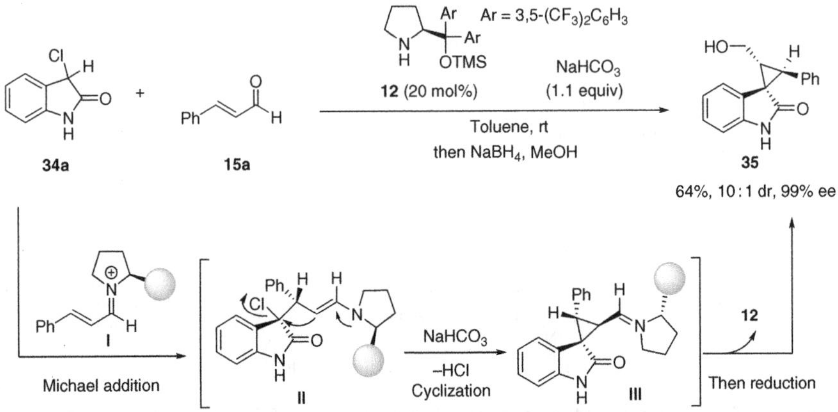

Scheme 5.10 Michael/cyclization sequence of 3-chlorooxindoles to spirocyclopropyl oxindoles. *Source:* Based on Noole et al. [34].

Additionally, 3-chlorooxindoles were identified as the useful synthons for the construction of cyclopropane spirooxindoles. For instance, Kanger, and Malkov provided a Michael/cyclization cascade reaction of 3-chlorooxindoles **34a** and enals **15a** through iminium/enamine sequential catalysis (Scheme 5.10) [34]. With 20 mol% of chiral amine **12** and 1.1 equiv of $NaHCO_3$, a variety of chiral spirocyclopropyl oxindoles **35** could be isolated with good yields and stereoselectivities. As demonstrated in Scheme 5.10, a possible reaction pathway was proposed, in which 3-chlorooxindole **34a** first underwent a Michael addition with the *in situ*-generated iminium intermediate **I** from chiral amine **12** and enals **15a**. Subsequently, the resulting enamine intermediate **II** proceeded with an intramolecular S_N2-alkylation to give the target **35**. Shortly after, the Melchiorre group realized a stereoselective 1,6-addition/cyclization sequence of 3-chlorooxoindoles with dienals by dienamine activation for the access of cyclopropane spirooxindoles [35].

By iminium/enamine activation strategy, asymmetric conjugate addition/aldol condensation cascade reaction involving functionalized 3-alkyl oxindoles **36a** have been developed for the synthesis of optically active spirooxindoles fused five-membered ring. Barbas et al. reported a domino Michael/aldol condensation of 3-alkyl oxindoles **36a** with enals **15** catalyzed by 10 mol% secondary catalyst **12** and benzoic acid, which furnished various spirocyclopentaneoxindoles **37** in high to excellent yields with up to 99% ee and 17 : 1 dr values (eq 1, Scheme 5.11) [36]. Soon after, Melchiorre and coworkers realized a 1,6-conjugate addition/aldol cascade reaction of 3-alkyl oxindoles **36a** with β-substituted cyclic dienones **38** via iminium/dienamine activation. A number of enantiomerically pure functionalized spirooxindoles **39** were isolated in good to excellent yield, dr, and ee values,

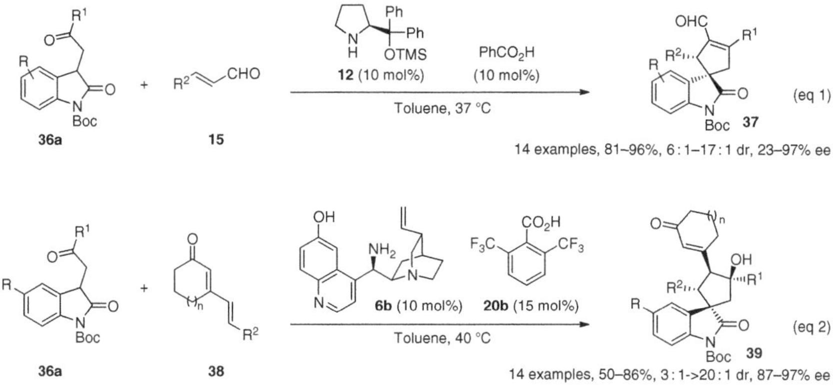

Scheme 5.11. Synthesis of spirocyclopentaneoxindoles with 3-alkyloxindoles **41**. *Source:* Modified from Albertshofer et al. [36].

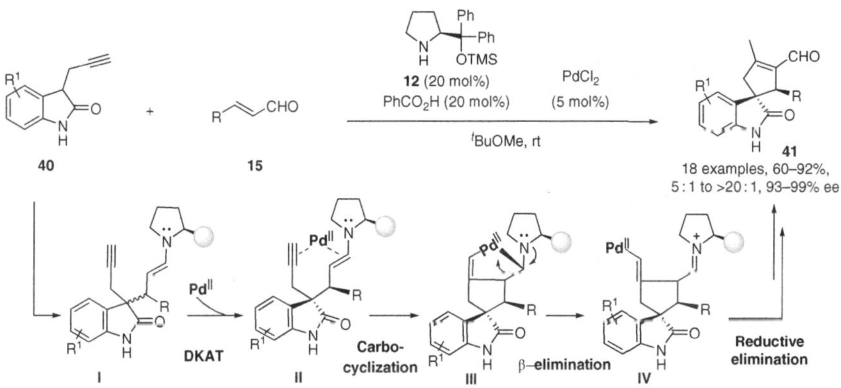

Scheme 5.12 Stereoselective Michael/cyclization sequence of 3-propargyl oxindoles with enals. *Source:* Based on Sun et al. [38].

under the influence of 10 mol% bifunctional primary amine **6b** and 15 mol% bis(trifluoromethyl)benzoic acid **20b** (eq 2) [37].

Wang, Hong, and coworkers developed a highly stereoselective Michael/cyclization sequence between 3-propargyl oxindoles **40** with enals **15** by using iminium/enamine-Pd(II) synergistic catalysis, which provided a facile method to access spirocyclopentene oxindoles **41** (Scheme 5.12) [38]. A range of desired spirooxindoles **41** could be obtained in 60–92% yields with 5 : 1 to >20 : 1 dr, and 93–99% ee, by combining 20 mol% catalyst **12**, $PhCO_2H$ with 5 mol% $PdCl_2$. As depicted in Scheme 5.12, a synergistic reaction pathway was proposed to explain the observed chemoselectivity and stereochemistry, in which a Michael addition

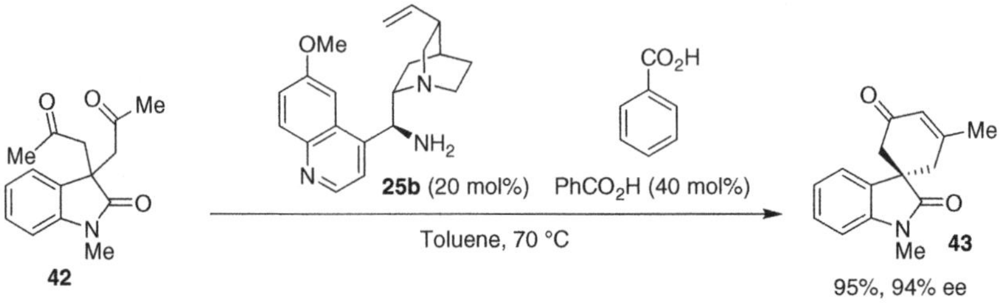

Scheme 5.13 Desymmetric intramolecular aldol condensation of 3,3′-diacetonyloxindoles. *Source:* Based on Sun et al. [39].

of 3-propargyl oxindoles **40** to the generated *in situ* iminium occurred first. The resulting enamine intermediate **I** underwent a dynamic kinetic asymmetric transformation (DKAT), followed by a carbocyclization and β-elimination to produce the intermediate **IV**. Finally, the targets **41** were obtained after undergoing a reductive elimination and cycloisomerization sequence.

By desymmetrization strategy, Wang and coworkers presented a primary amine-catalyzed asymmetric intramolecular aldolization of 3,3′-diacetonyl-modified oxindoles **42** for the synthesis of optically active spirooxindoles **43** (Scheme 5.13) [39]. The use of 20 mol% quinine-derived primary amine **25b** with 40 mol% benzoic acid enabled the preparation of various spiro[cyclohexenone-oxindole] derivatives **43** in high yields and excellent enantioselectivities.

5.3 Chiral Nucleophilic Catalysis with Tertiary Phosphines or Amines

Chiral nucleophilic catalysis with chiral tertiary phosphines or amines, as a powerful strategy via covalent bond activation, has been widely applied in a variety of organic asymmetric reactions in the past few decades [13g, 40]. Accordingly, this strategy has demonstrated their great potency in the construction of chiral spirooxindoles, and considerable progress has been made. Early in 2010, Marinetti and coworkers utilized chiral tertiary phosphine **45** as the catalyst to achieve an asymmetric [3+2] cycloaddition of methyleneindolinones **1c** and allenoates **44a** (Scheme 5.14) [41]. It was found that the use of 10 mol% **45** catalyzed the annulation effectively to give the γ-adduct spirocyclopentane oxindoles **46** in good to excellent yields and stereoselectivities. Moreover, the allenylphosphonate proved to be a competent substrate, and reacted smoothly with **1c** under the catalysis of planar ferrocene-derived chiral phosphine [41].

Later, Lu et al. successfully accomplished an asymmetric [4+2] annulation of isatylidene malononitriles **47a** and α-substituted allenoates **44b** catalyzed by an

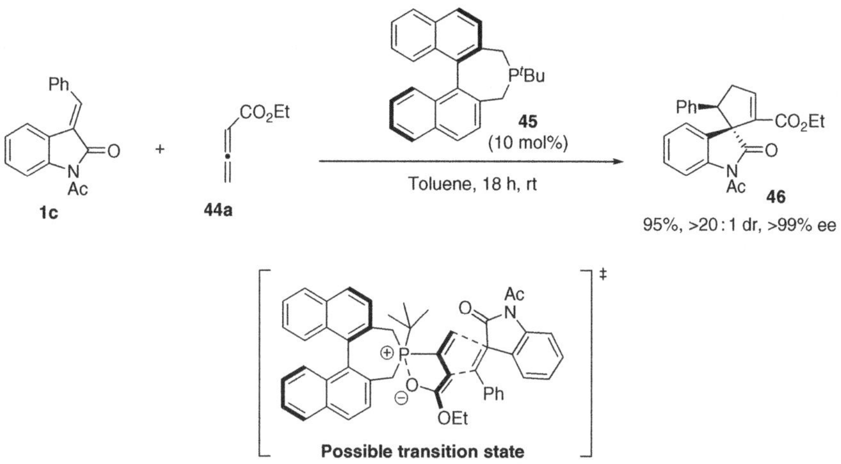

Scheme 5.14 Chiral phosphine-catalyzed [3+2] cycloaddition of allene and methyleneindolinones. *Source:* Based on Voituriez et al. [41].

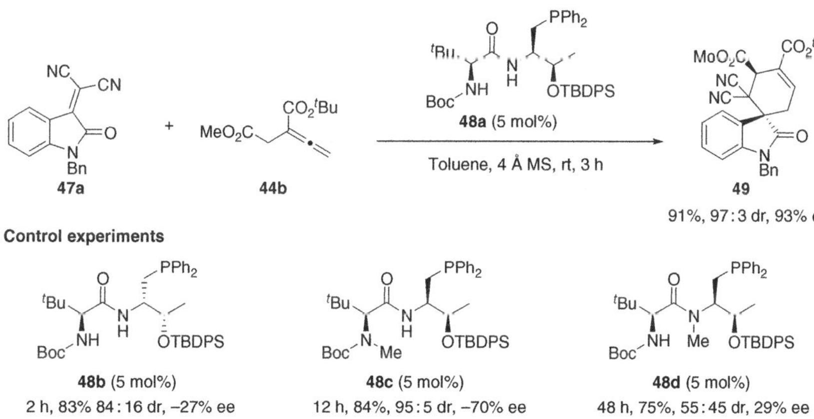

Scheme 5.15 [4+2] Annulation of isatylidene malononitriles and α-substituted allenoates. *Source:* Based on Zhong et al. [42].

amino acid-derived bifunctional phosphine **48a** (Scheme 5.15) [42]. Various 3-spirocyclohexene-2-oxindoles **49** could be produced in 90–96% yields, 95 : 5–99 : 1 dr, and 89–93% ee values when 5–10 mol% phosphine **48a** was used. Notably, the chirality matching was crucial for achieving high diastereo- and enantioselectivity (**48a** *vs* **48b**). The carbamate N–H bond in the catalyst had a great influence on the reaction rate, while the amide N–H bond was important for both stereoselectivity and reactivity, as exemplified by the control experiments shown in Scheme 5.15. Two years later, the group of Lu [43] and Kumar [44] independently reported the enantioselective [3+2] annulation of allenoates with isatin-derived ketimines catalyzed by chiral bifunctional H-bonding donor-phosphines, which afforded a number of 3,2′-dihydropyrrolyl spirooxindoles in high yields with excellent enantioselectivities.

Apart from the use of allenoates, the MBH carbonates have also been used as the versatile synthons in the chiral phosphine-catalyzed annulations for the enantioselective construction of spirooxindoles. In 2011, Barbas and coworkers developed an asymmetric [3+2] cycloaddition of methyleneindolinones **11** with MBH carbonates **50a** by chiral phosphine catalysis (eq 1, Scheme 5.16) [45]. Under the influence of 10 mol% chiral tertiary phosphine (*R*, *R*)-Ph-BPE (**51**), the annulation reaction proceeded well to give a range of spiro cyclopenteneoxindole derivatives **52** with 47–91% yield, 2 : 1 to >20 : 1 dr, and 46–99% ee. By mechanistic studies, the authors proposed a possible transition state **I** shown in the left of Scheme 5.16, in which one phosphine substituent of phosphine **51** reacted with MBH carbonates **50a** to generate the activated 1,3-dipolar intermediate, while the second phosphine moiety might have interacted with the carbonyl group of **11**. Simultaneously,

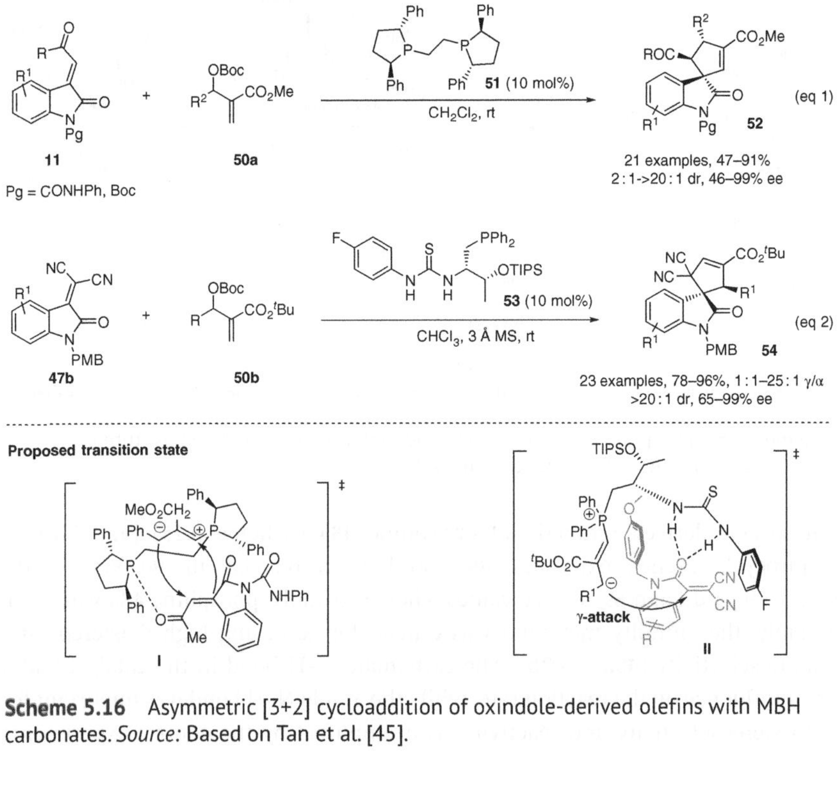

Scheme 5.16 Asymmetric [3+2] cycloaddition of oxindole-derived olefins with MBH carbonates. *Source:* Based on Tan et al. [45].

Lu et al. demonstrated a stereoselective [3+2] cycloaddition of MBH carbonates **50b** with isatin-derived tetrasubstituted alkenes **47b** by employing 10 mol% of their developed threonine-derived phosphine **53** (eq 2) [46]. A plausible dual activation transition state **II** was proposed where the N–H bond of thiourea activated the alkenes **47b** via coordination with the carbonyl group, the tertiary phosphine interacted with MBH carbonates **50b** to generate the active ylide species. Subsequently, a γ-addition of ylide species to activated alkenes occurred to furnish the products **54** after an intramolecular cyclization.

Isatin-derived MBH carbonates **55a** are also viable substrates toward various spirooxindole derivatives by chiral phosphine catalysis. For example, Chen and Ouyang developed a phosphine-catalyzed stereoselective [3+2] annulation of isatin-derived MBH carbonates **55a** with 2-alkylidene-1*H*-indene-1,3(2*H*)-diones **56** (Scheme 5.17) [47]. In the presence of 10 mol% phosphine **48e**, the reaction occurred smoothly to furnish various polycyclic spirooxindoles **57** with excellent diastereo- and enantioselectivities. Intriguingly, the diastereomers **59** were readily obtained in 83–95% yield and 89–99% ee with >19 : 1 dr values when using the tertiary phosphine catalyst **58**. Additionally, by using chiral (2*R*,5*R*)-Me-DuPhos as the catalyst, Chen and coworkers realized an asymmetric [3+2] annulations of isatin-derived MBH carbonates with *N*-phenylmaleimide, which provided a series of spirocyclopentaneoxindoles in good yields with excellent stereoselectivities [48].

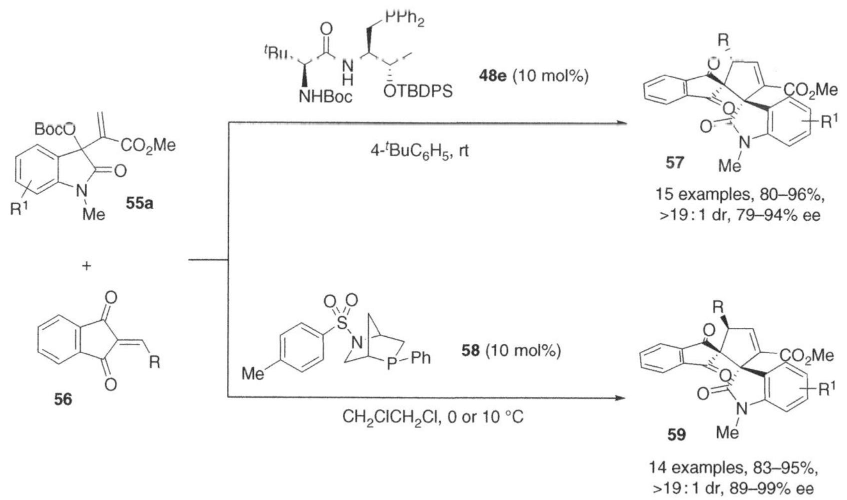

Scheme 5.17 Phosphine-catalyzed [3+2] annulation of isatin-derived MBH carbonates. *Source:* Based on Zhan et al. [47].

In addition to the use of chiral tertiary phosphines as nucleophilic catalysis, chiral tertiary amines have also been efficiently applied for the diverse access of enantioenriched spirooxindoles through the asymmetric annulations between isatin-derived MBH carbonates and various electron-deficient alkenes. In 2011, the Chen group developed a chemo- and enantioselective [3+2] annulation of **55b** with propargyl sulfones **60** by employing a newly synthesized β-ICD *O*-MOM ether **61a** as the catalyst, which delivered spirooxindoles **62** featuring an unusual cyclopentadiene moiety with excellent enantioselectivities (eq 1, Scheme 5.18) [49]. Several years later, the same group established an α-regio-, stereoselective [3+2] annulation of isatin-derived MBH carbonates **55c** with steric hindrance cyclic 1-azadienes **63** (eq 2) [50]. 10 mol% of β-ICD (**61b**) was proven to be efficient and enabled the generation of various spirooxindoles **64** with excellent dr and ee values. By DFT calculation studies, the authors proposed that the observed exclusive α-regioselectivity was likely attributed to the steric hindrance of ammonium salt and bulky electron-withdrawing group of 1-azadienes **63**. Shortly after, they reported an asymmetric [3+2] annulation of isatin MBH carbonates **55d** and α-cyano-α,β-unsaturated ketones **65** catalyzed by tertiary amine **61c** (eq 3) [51].

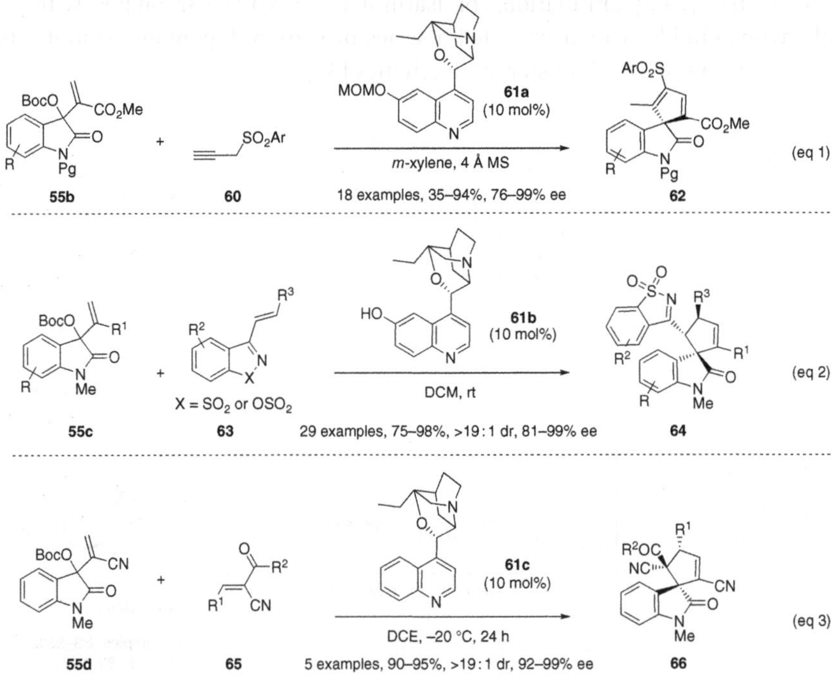

Scheme 5.18 Chiral tertiary amines-catalyzed [3+2] annulation of isatin-derived MBH carbonates. *Source:* Based on Peng et al. [49].

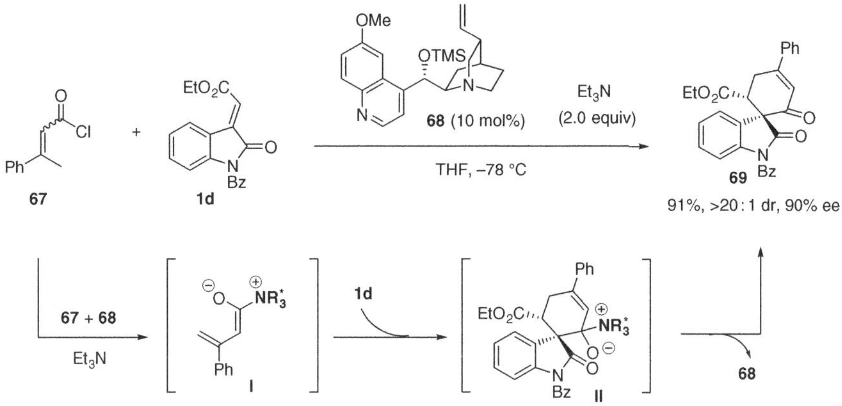

Scheme 5.19 [4+2]-Cycloaddition of α,β-unsaturated acyl chlorides by nucleophilic catalysis. *Source:* Based on Shen et al. [52].

In 2013, Ye et al. developed a chiral tertiary amine-catalyzed asymmetric [4+2] cycloaddition reaction of α,β-unsaturated acyl chlorides **67** and methyleneoxindoles **1d** (Scheme 5.19) [52]. In the presence of 10 mol% *O*-trimethylsilyl (TMS) quinidine **68** and 2.0 equiv of Et_3N, a wide series of functionalized spirooxindoles **69** with six-membered ring were obtained in 72–91% yields, 10 : 1 to >20 : 1 dr, and 65–93% ee. A proposed reaction pathway was proposed in which tertiary amine **68** activated α,β-unsaturated acyl chlorides **67** under Et_3N condition, the resulting active diene species **I** subsequently reacted with methyleneoxindoles **1d** via a Diels–Alder process to give targets **69**.

By chiral tertiary amine nucleophilic catalysis, Zhou and coworkers accomplished an asymmetric [3+2] cycloaddition of various ketones **71a** with brominated MBH adducts **70**, allowing access to spirooxindoles **72a** featuring three continuous stereocenters (eq 1, Scheme 5.20) [53]. A combination of 10 mol% BnO-protected β-isocupreidine **61d** and K_2CO_3 enabled the reaction well to give an array of the desired spirooxindoles **72a** with excellent stereoselectivities. As illustrated in Scheme 5.20, the reaction was initiated by the attacking of tertiary amine **61d** to the brominated MBH adducts **70**, the generated active 1,3-dipole **I** went through a γ-regioselective [3+2] cycloaddition with ketones **71a** to furnish the products **72a**. Based on their work on a β-ICD-catalyzed enantioselective MBH reaction of isatins with acrolein [54], Zhou et al. further developed a one-pot MBH/bromination/asymmetric [3+2] annulation sequence from simple isatins **24b** and acrolein for the synthesis of spirooxindoles **72** (eq 1).

In the same year, Wang and Xu realized the synthesis of functionalized spirooxindole lactones **74** by means of a chiral tertiary amine-catalyzed asymmetric allylic alkylation–cyclization of 3-hydroxyoxindoles **31a** with aldehyde-derived

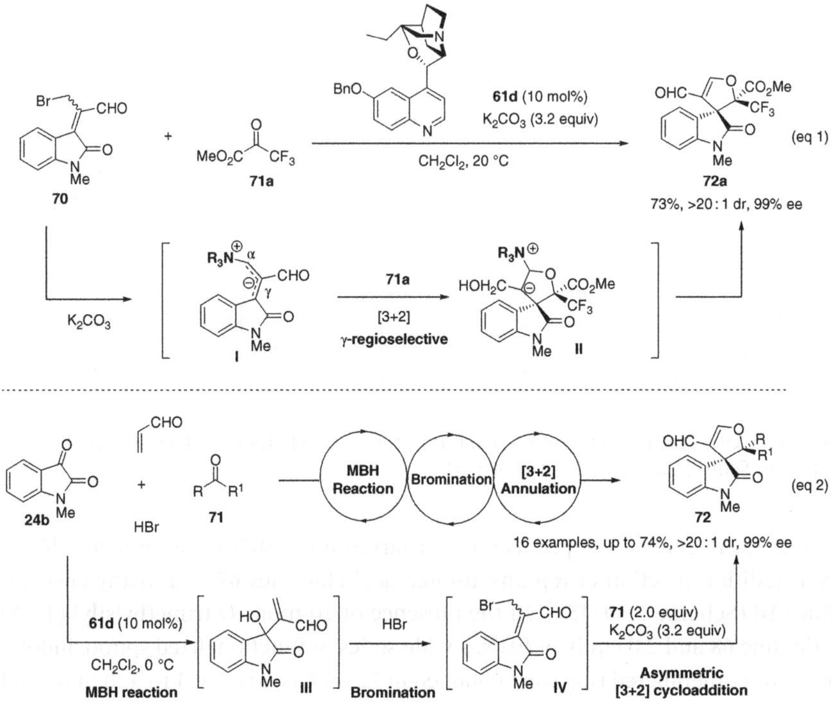

Scheme 5.20 Asymmetric [3+2]-cycloaddition of brominated MBH adducts and ketones. *Source:* Modified from Liu et al. [53].

MBH carbonates **50c** (Scheme 5.21) [55]. The use of 15 mol% quinidine **73a** and 7.5 mol% hydroquinone mediated the synthesis of various spirooxindoles **74** in 25–85% yields with >20 : 1 dr and 60–94% ee. Similarly, quinidine **73a** reacted with MBH carbonates **50c**, and the resulting active intermediate then worked with 3-hydroxyoxindoles **31a** through an allylic alkylation–cyclization process to provide targets **74**.

5.4 Chiral *N*-Heterocyclic Carbene Catalysis

With the rapid development of chiral *N*-heterocyclic carbene (NHC) catalysis in organic synthesis over the past few decades [56], the use of chiral NHC catalysis for the enantioselective assembly of spirooxindoles has gained much attention. As a consequence, a variety of transformations have been developed toward functionalized spirooxindoles. Early in 2010, Ye and coworkers pioneered the asymmetric synthesis of spirooxindoles by chiral NHC catalysis [57]. They found that

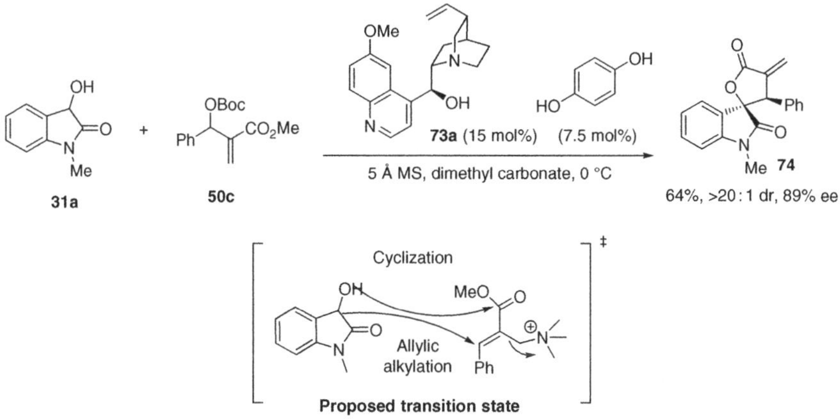

Scheme 5.21 Allylic alkylation–cyclization of 3-hydroxyoxindoles with MBH carbonates. *Source:* Based on Wang et al. [55].

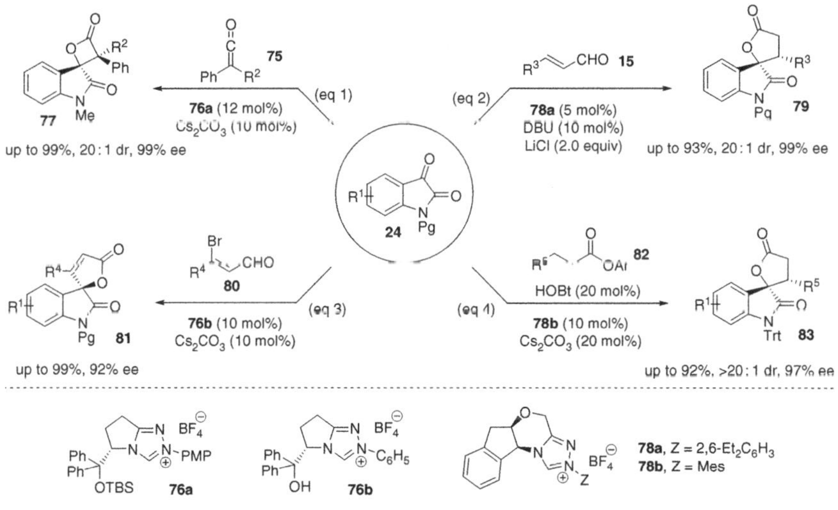

Scheme 5.22 Chiral NHC-catalyzed asymmetric annulation involving isatins.

the use of 10 mol% NHC catalyst **76a** enabled effectively the asymmetric cycloaddition of isatins **24** and ketenes **75**, and afforded a series of optically active spirocyclic oxindole-β-lactones **77** in good to excellent yields and ee with moderate dr values (eq 1, Scheme 5.22). Shortly after, Scheidt et al. accomplished the stereoselective preparation of spirocyclic oxindole-γ-lactones **79** through a homoenolate annulation of isatins with α,β-unsaturated aldehydes catalyzed by a combination of 5 mol% chiral NHC **78a** with 2.0 equiv of LiCl as Lewis acid (eq 2) [58]. Later,

Ma et al. developed a chiral NHC-catalyzed formal [3+2] annulation of isatin **24** with 3-bromoenals **80**, which enabled the assembly of chiral spirooxindole-butenolides **81** (eq 3) [59]. In addition, Xu and coworkers realized a chiral NHC **78b**-catalyzed asymmetric [3+2] annulation of isatin **24** and saturated carboxylic esters **82**, allowing the synthesis of optically active spirooxindole lactones **83** with the assistance of 1-hydroxybenzotriazole (HOBt) (eq 4) [60]. It should be noted that HOBt was proposed to play two important roles: as an ester activating agent and as a hydrogen-bonding donor, which rationalized the enhancement of stereoselectivity.

Chiral NHC-catalyzed asymmetric annulation reactions involving isatins are not only applied for the construction of chiral spirooxindole fused four- or five-membered lactones, but also have been used to access optically active six-membered fused spirooxindole-δ-lactones (Scheme 5.23). Yao and coworkers developed a stereoselective [4+2] annulation of isatin **24** with α-bromo-α,β-unsaturated aldehydes **85** by combining 20 mol% chiral NHC **78c** with 20 mol% of $La(OTf)_3$, which provided a practical entry to access spirocyclic oxindole-dihydropyranones **84** [61]. Later on, the same group reported a chiral NHC-catalyzed formal [4+2] cyclization of isatins **24** with preformed or *in situ*-generated HOBt esters **86** [62]. In 2015, the group of Zhong and Yang [63], and Liu and Zhou [64] independently published the chiral NHC-catalyzed formal [4+2]

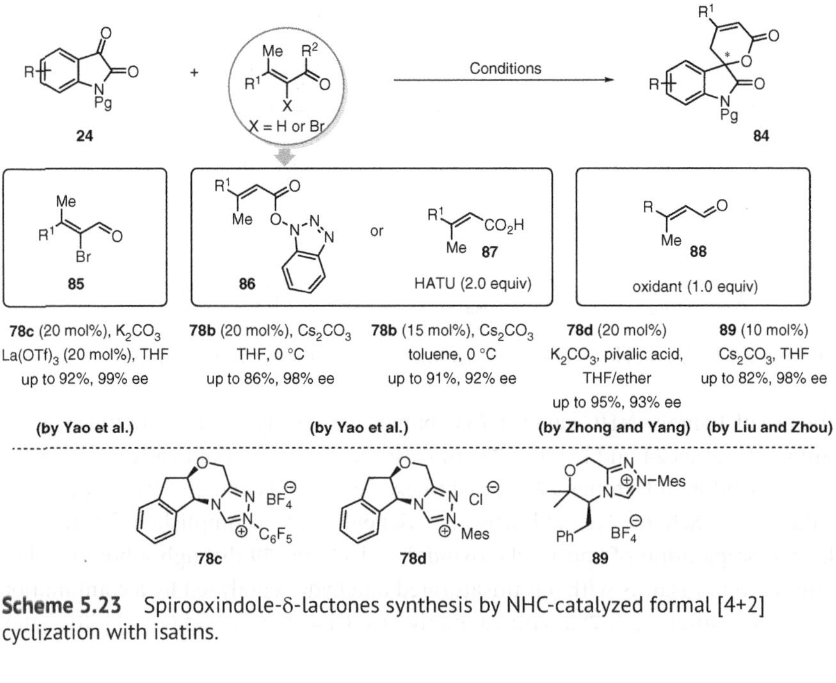

Scheme 5.23 Spirooxindole-δ-lactones synthesis by NHC-catalyzed formal [4+2] cyclization with isatins.

annulations of isatins **24** with β-methyl α,β-unsaturated aldehydes **88** for the construction of spirooxindole-δ-lactones **84**.

In addition to isatins as the electrophiles, the asymmetric cycloadditions involving isatin-derived ketimines have been presented for the access of spirooxindole lactams with structural diversity as well. In 2014, Ye et al. developed a highly efficient chiral NHC-catalyzed Staudinger reaction of isatin-derived ketimines **90** with ketenes **75**, which allowed to access spirooxindole-β-lactams **91** with high to excellent yields, dr, and ee values in the presence of a bifunctional NHC **76c** featuring a free hydroxyl group (eq 1, Scheme 5.24) [65]. Later, Xu, Ren, and coworkers also realized the synthesis of spirooxindole-β-lactams **92** by taking advantage of a NHC **78b**-catalyzed oxidative [2+2] annulation of isatin-derived ketimines **90** with simple aliphatic aldehydes **14** (eq 2) [66]. The Chi group established a highly enantioselective annulation reaction of isatin imines **90** with enals **15**, which afforded the various spirooxindole-γ-lactams **93** with up to >20 : 1 dr and 99% ee values under the influence of 10 mol% NHC **78b** (eq 3) [67]. Four year later, Hui et al. presented a facile method to synthesize optically active five-membered spiro[indoline-3,2′-pyrrole] derivatives through a chiral NHC-catalyzed formal [3+2] annulation between isatin imine **90** and β-bromo-α,β-unsaturated aldehydes **80** (eq 4) [68]. In 2016, by chiral NHC catalysis, the Ye group further reported a formal [4+2] annulation of isatin imine **90** and α,β-unsaturated carboxylic acids **87** for the synthesis of chiral spirooxindole-based dihydropyridinones **95** (eq 5) [69].

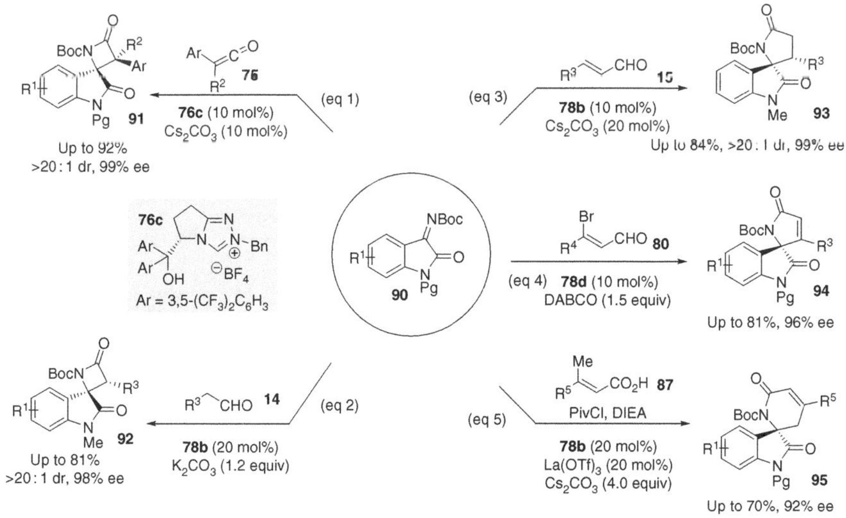

Scheme 5.24 Spirooxindole lactams synthesis by NHC-catalyzed annulation with isatin imines. *Source:* Based on Zhang et al. [65].

3-Olefinic oxindole derivatives, such as isatin-derived enals **96**, are also employed in the chiral NHC-catalyzed enantioselective reactions toward spirooxindoles in sufficient structurally diversity. In 2015, Yang, Zeng, and Zhong developed an asymmetric chiral NHC-catalyzed [3+3] cycloaddition of isatin-derived enals **96** with ketimines **97** [70]. The use of 20 mol% NHC **78d** with 50 mol% DMAP enabled efficiently the reaction to furnish optically active six-membered spirooxindoles **99** in high yields with excellent ee under 1.0 equiv of oxidant **98** conditions (eq 1, Scheme 5.25). One year later, Enders et al. published a novel NHC-catalyzed formal [3+4] cycloaddition of isatin-derived enals **96** with aza-*o*-quinone methides **II** that generated *in situ* from *N*-(ortho-chloromethyl)aryl amides **100**, which provided a facile approach for the assembly of seven-membered spirobenzazepinones **101** (eq 2, Scheme 5.25) [71]. With 10 mol% of chiral NHC **78e**, a range of spirooxindoles **101** could be provided in good to excellent stereoselectivities. A possible reaction pathway was proposed, the NHC homoenolate intermediate **I** first generated from the reaction of enals **96** with the chiral active NHC species, and aza-*o*-quinone methides **II** would form *in situ* from **100** under the influence of Cs_2CO_3. Subsequently, the homoenolate **I** attacked the active aza-*o*-quinone methides **II** in a Michael addition fashion to provide intermediate **III**, which proceeded with a lactamization to give spirobenzazepinones **101**. Moreover, the authors found that both azoalkenes and

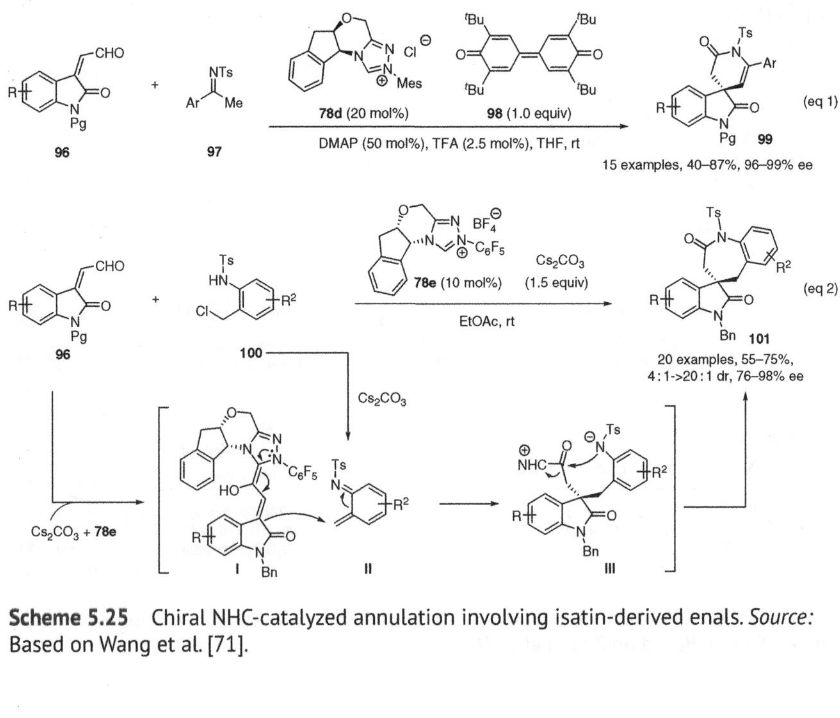

Scheme 5.25 Chiral NHC-catalyzed annulation involving isatin-derived enals. *Source:* Based on Wang et al. [71].

nitrosoalkenes, formed *in situ* from *N*-tosyl hydrazines and α-chloro oximes, could readily undergo such NHC-catalyzed formal [3+4] cycloaddition with isatin-derived enals **96** [71], allowing the synthesis of spirooxindoles featuring pharmaceutically important heterocycles.

Later, Enders et al. developed an NHC-catalyzed stereoselective annulation of isatin-derived enals **96** with α,β-unsaturated *N*-sulfonyl ketimines **102**, enabling the synthesis of chiral spirooxindoles fused five-membered ring (Scheme 5.26) [72]. By slightly varying the structure of chiral NHC catalyst, the same starting materials could produce two different types of spirooxindoles as the products. The use of chiral NHC **103a** gave rise to spirooxindoles **104** bearing an enaminone motif (eq 1), while the spirooxindoles **105** bearing an α,β-lactam motif could be isolated by using a slightly different NHC **78f** (eq 2). To account for the observed stereochemical and divergent annulation, the authors proposed two different pathways as illustrated in Scheme 5.26. Under the catalysis of NHC **103a**, the resulting NHC

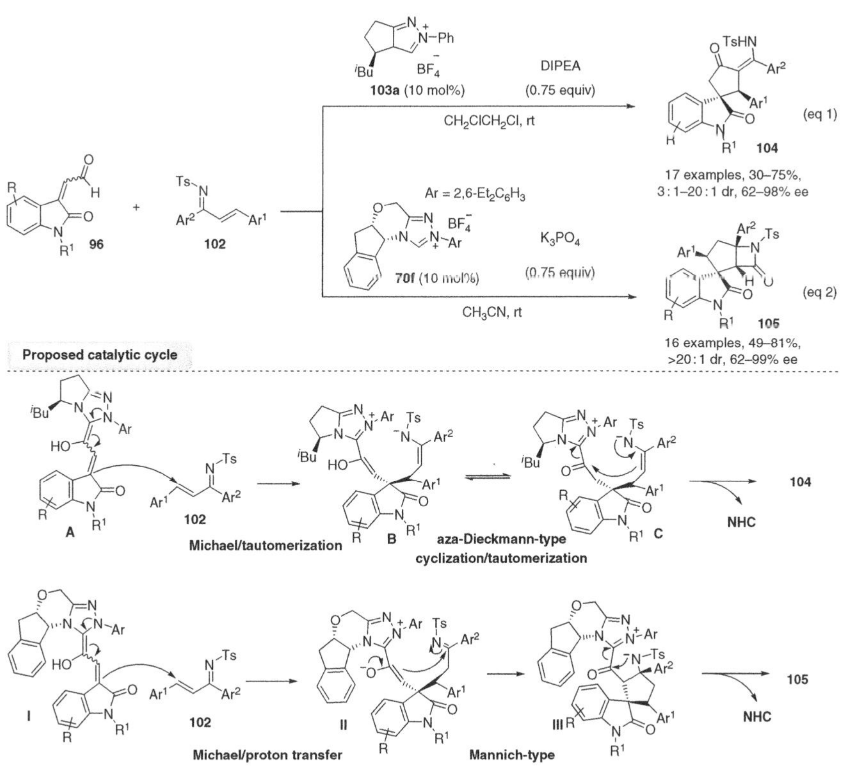

Scheme 5.26 Stereoselective annulation of isatin-derived enals with α,β-unsaturated ketimines. *Source:* Based on Wang et al. [72].

homoenolate **A** proceeded with a Michael addition with α,β-unsaturated ketimines **102** to give the intermediate **B**, which subsequently underwent a tautomerization. The formed intermediate **C** went through an aza-Dieckmann-type cyclization to deliver the desired spirocycle **104**. In the case of NHC **78f**, however, the resulting intermediate **I** also underwent a Michael addition to afford the intermediate **II** followed by a proton transfer process. Subsequently, a Mannich-type reaction of **II** produced the acyl azolium species **III**, which finally generated the target **105** and released the active NHC species through lactamization. Shortly after, the same group discovered that *o*-hydroxyphenylsubstituted *p*-quinone methides were also viable substrates in the NHC-catalyzed annulation with isatin-derived enals, which enabled the access of spirooxindole-ε-lactones [73].

In 2017, Wang et al. developed an asymmetric Michael/intramolecular aldol/lactonization cascade reaction between oxindolyl β,γ-unsaturated α-keto esters **11b** with enals **15a** by chiral NHC catalysis (Scheme 5.27) [74]. Various polycyclic spirooxindoles **106** were prepared in 52–71% yields with >99 : 1 dr and 73–99% ee, in the presence of 10 mol% aminoindanol-derived triazolium-based NHC **78d**. A similar reaction pathway in Scheme 5.27 via a sequential Michael/proton transfer and aldol/lactonization process was proposed to facilitate the readers in understanding the mechanism.

Under chiral NHC catalysis, 3-aminooxindoles and 3-hydroxyoxindoles turned out to be competent substrates for the synthesis of spirooxindole derivatives. In 2015, Du and Lu established a chiral NHC-catalyzed formal [3+2] annulation of α-bromo-α,β-unsaturated aldehydes **108** with 3-aminooxindoles **107a** [75]. The preparation of spirooxindole γ-butyrolactams **109** were achieved with good to high yields with dr and ee values, in the presence of 15 mol% of NHC **103b** and 1.5 equiv of K_2CO_3 (eq 1, Scheme 5.28). Two years later, with the assistance of oxidant nitrobenzene, Ye, Sun, and coworkers reported an oxidative [3+2]

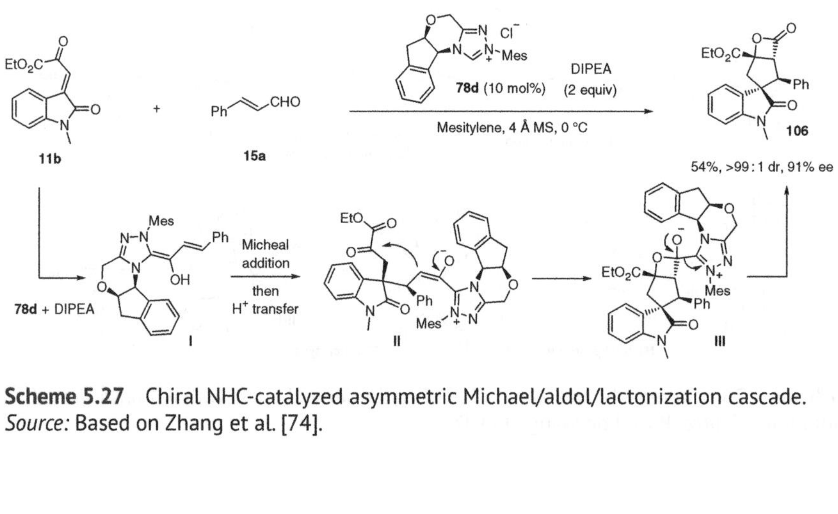

Scheme 5.27 Chiral NHC-catalyzed asymmetric Michael/aldol/lactonization cascade. *Source:* Based on Zhang et al. [74].

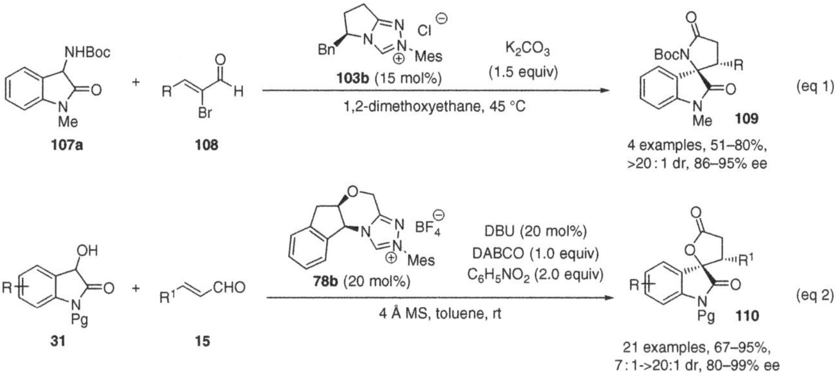

Scheme 5.28. NHC-catalyzed annulation with 3-aminooxindoles or 3-hydroxyoxindoles. *Source:* Based on Zhao et al. [76].

annulation of 3-hydroxyoxindoles **31** with aryl or alkyl enals **15** catalyzed by 20 mol% chiral NHC **78b**, which allowed to access enantioenriched spirooxindole γ-lactons **110** (eq 2) [77]. Later on, the Ye group developed a NHC-catalyzed stereoselective [3+2] annulation of 3-hydroxyoxindoles with β,β-disubstituted enals under the similar conditions [78].

5.5 Chiral Tertiary Amine-H-Bond Donor Bifunctional Catalysis

The tertiary amine H bond donor catalysts have been identified as a privileged catalyst scaffold, as the H-bond donor moiety could activate the electrophilic and/or stabilize the transition state via hydrogen bonding interactions, while the tertiary amine part might serve as a Lewis base or Brønsted base to activate various nucleophiles [79]. In this context, such bifunctional catalysis has recently been widely applied to the enantioselective construction of structurally diverse spirooxindoles from isatins, isatin-derived ketimines, 3-akenyl oxindoles, 3-alkyl or 3-unsubstituted oxindoles, etc. As a consequence, a great number of asymmetric annulation or cascade reactions have been established as a fruitful tactic toward various functionalized chiral spirooxindoles. In the following, we will introduce these achievements according to the type of substrates employed.

In 2013, by chiral tertiary amine-H bonding catalysis, Shi et al. reported an asymmetric [3+2]-cycloaddition of isatins **24** with α-aryl isocyanoacetates **111**. A series of optically active 2-oxo-4′*H*-spiro[indoline-3,5′-oxazole] derivatives **113** were obtained in high yields and good to excellent selectivities, in the presence of 10 mol% of quinine-derived H-bonding catalyst **112a** (eq 1,

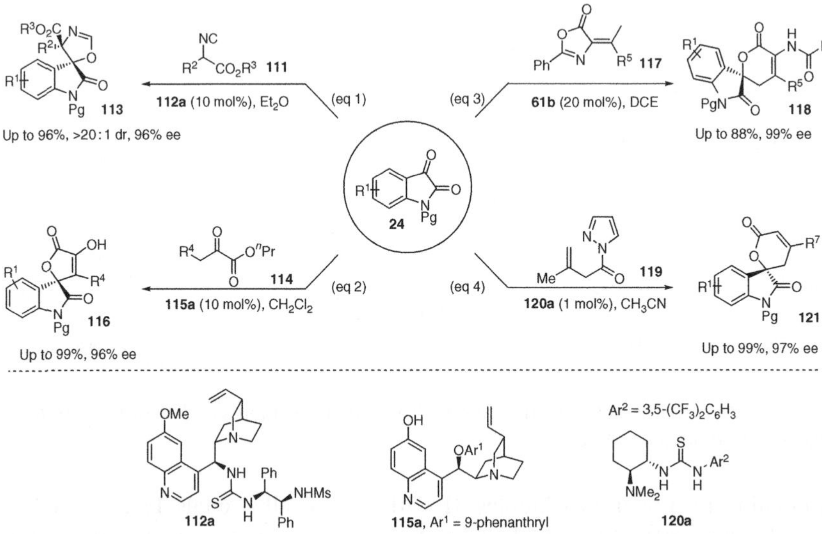

Scheme 5.29. Tertiary amine-H bond donor-catalyzed asymmetric annulation involving isatins. *Sources:* Based on Zhao et al. [76]; Li et al. [80].

Scheme 5.29) [76]. One year later, Li and Guo accomplished the stereoselective synthesis of chiral spirooxindole-butenolides **116** through an aldol–acylation cascade of isatins **24** with α-ketoesters **114** by using 10 mol% bifunctional catalyst **115a** (eq 2) [81]. The Xu group presented a highly enantioselective hetero-Diels–Alder reaction of isatins **24** with olefinic azlactones **117**, which afforded the spirooxindole dihydropyranones **118** with up to 99% ee values under the catalysis of 10 mol% β-ICD **61b** (eq 3) [82]. By using bifunctional tertiary amine **120a**, Wu and Sha developed a highly enantioselective vinylogous aldol–cyclization cascade reaction of isatins **24** with allyl pyrazoleamides **119** for the assembly of chiral spirooxindole dihydropyranones **121** (eq 4, Scheme 5.29) [80]. Later on, Han et al. also developed an asymmetric vinylogous aldol–cyclization cascade reaction of isatin with 3-alkylidene oxindoles catalyzed by bifunctional squaramide catalyst [83].

With chiral tertiary amine-H bonding catalysis, Isatin-derived imines have been also proven to be suitable reacting partners for the synthesis of chiral spirooxindoles. In 2014, Zhou and coworkers realized an asymmetric [1,5]-electrocyclic reaction of functionalized isatin-derived ketimines **122** toward oxindole-based spirocyclic indolines **123a** (Scheme 5.30) [84]. In the presence of 10 mol% of cinchona alkaloid-derived thiourea **112b**, the desired spirooxindoles **123** were provided in 76–98% yield with 94–99% ee. Noticeably, a rare example of a one-pot process that involves asymmetric triple relay catalysis through the sequential use

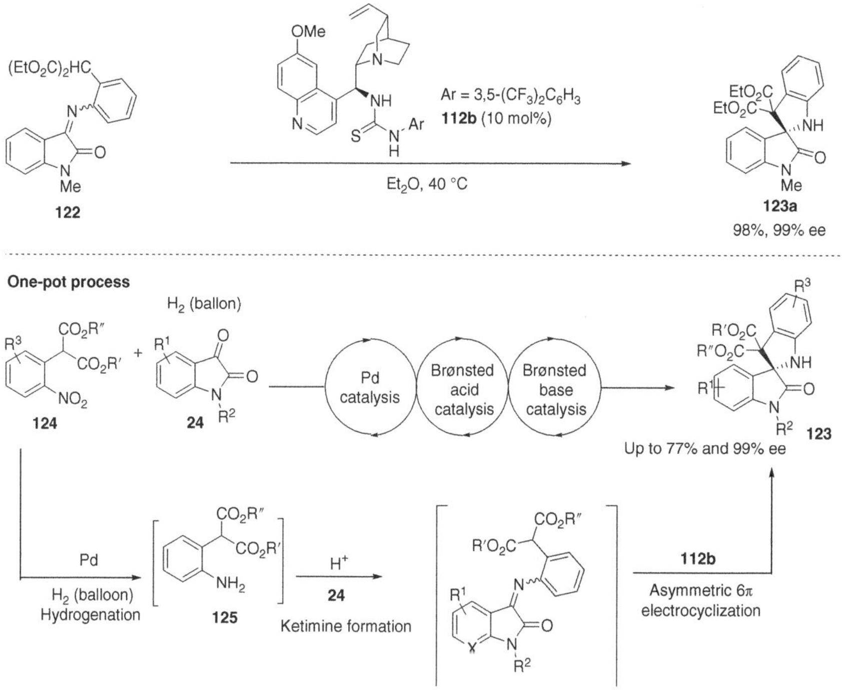

Scheme 5.30. Asymmetric intramolecular [1,5]-electrocyclization of functionalized isatin imines. *Source:* Based on Yin et al. [84].

of palladium, Brønsted acid, and Brønsted base catalysis was further established, which enabled the synthesis of spirooxindoles from isatins directly.

In 2016, Enders et al. developed a Mannich/Boc-deprotection/aza-Michael sequence reaction of isatin imine **90** with functionalized 3-substituted oxindole **126**. With 10 mol% $(DYQD)_2PHYL$, a series of 3,3′-pyrrolidinyldispirooxindole derivatives **127** could be isolated with good to high yields and stereoselectivities (eq 1, Scheme 5.31) [85]. At the same time, Li, Cheng et al. published a bifunctional squaramide **129a**-catalyzed formal [3+2] annulation of isatin imines **90** with 1,4-dithiane-2,5-diol **129a** toward enantioenriched spirooxindole-based 4-thiazolidinone (eq 2) [86].

Under the catalysis of chiral tertiary amine-H bonding catalysts, a variety of enantioselective annulation involving methyleneindolinones have been well-established for the diverse preparation of enantioenriched spirooxindoles. Early in 2010, Gong and coworkers developed a stereoselective formal [4+2] annulation of methyleneindolinones **1** and Nazarov reagents **131** (eq 1, Scheme 5.32) [87]. The use of 10 mol% chiral catalyst **132** mediated the double Michael addition smoothly to give spirooxindoles **133** bearing three continuous stereocenter with

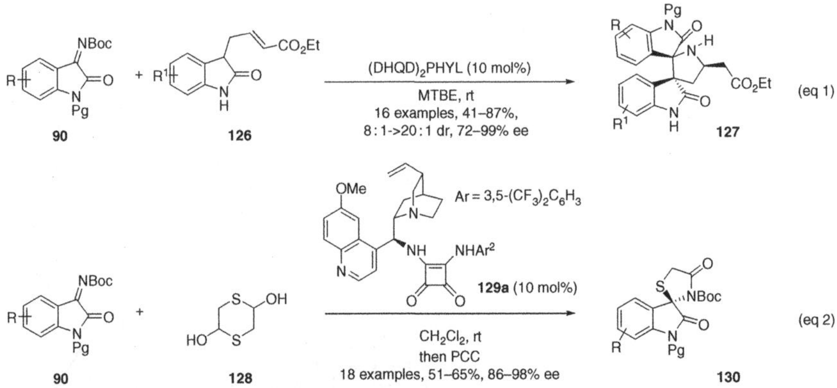

Scheme 5.31 Asymmetric annulation of isatin imine by tertiary amine-H bonding catalysis. *Source:* Based on Zhao et al. [85].

excellent ee values. In 2014, Wang and coworkers realized an asymmetric formal [3+2] cycloaddition of methyleneindolinones **1** with functionalized 3-substituted oxindole **126** catalyzed by bifunctional squaramide **134** (eq 2) [88]. The Quintavalla group achieved a highly enantioselective [4+2] annulation of methyleneindolinones **1** with nitroester **136** catalyzed by 10 mol% catalyst **120b** (eq 3) [89]. In 2016, with bifunctional squaramide **134**, Du et al. developed a formal [3+2] cycloaddition of methyleneindolinones **1** with α-alkylidene succinimide **138**, which produced various polycyclic spirooxindoles **139** bearing five contiguous stereocenters in up to 93% yields with excellent stereoselectivities (eq 4) [90]. Later, a series of chiral five- or six-membered cyclic spirooxindoles were also prepared efficiently through multifunctional chiral tertiary amine-H bonding donor-catalyzed Michael/Michael addition of methyleneindolinones with various reacting partners, such as *ortho*-tosylaminochalcone [91], *ortho*-hydroxychalcone, *trans*-ethyl 4-mercapto-2-butenoate [92], *N*-Ts-2-(*E*)-(2-nitrovinyl)aniline [93], and *N*-tosyl aminomethyl enones [94].

Aside from the double Michael addition process, a plethora of annulation reactions involving methyleneindolinones via Michael addition-initiated cascade process have been established for the access of optically active spirooxindoles with more structural diversity. For example, in 2011, Wang and coworkers reported an asymmetric Michael/Mannich cascade reaction of methyleneindolinones **1** with α-isothiocyanato imide **140** [95] or cyclic imino ester **142** [96] by chiral tertiary amine-thioureas **120c** (eqs 1 and 2, Scheme 5.33). Later, Wang [97], Chen and Xiao [98], and He [99] independently reported the similar annulations of methyleneindolinones with 3-isothiocyanato oxindoles. Sun et al. realized an enantioselective Michael/Mannich cascade reaction of methyleneindolinones **1** with *in situ*-formed ketimine intermediates catalyzed by a quinine-derived squaramide

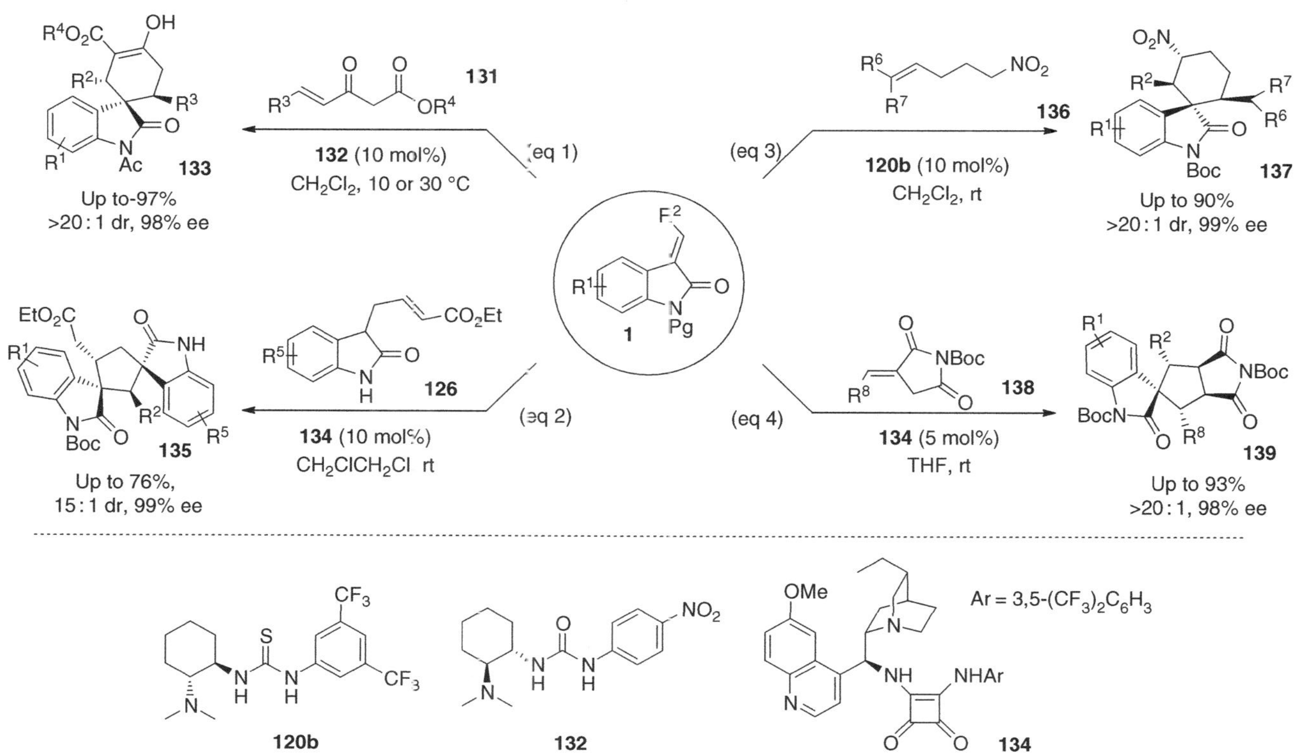

Scheme 5.32 Asymmetric annulation of methyleneindolinones via double Michael additions. *Source:* Based on Wei and Gong [87].

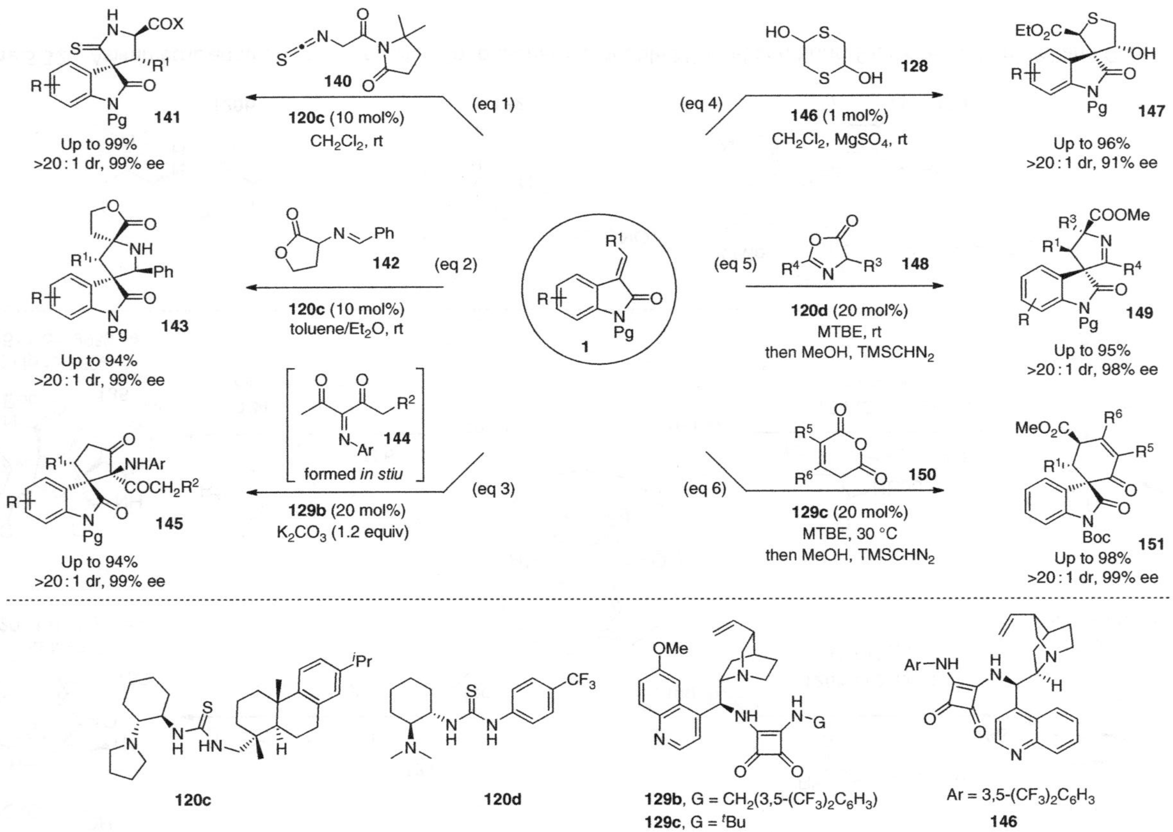

Scheme 5.33 Asymmetric annulation of methyleneindolinones via Michael-initiated cascade process.

144 (eq 3), which allowed to access spirocyclopentanone oxindoles **145** [100]. Xiao et al. developed a highly efficient Michael/aldol cascade reaction of methyleneindolinones **1** with 1,4-dithiane-2,5-diol **128** catalyzed by tertiary amine-squaramide **146** (eq 4) [101]. The asymmetric Michael/acylation cascade reactions of methyleneindolinones **1** with azlactones **148** or enolizable anhydrides **150** were independently accomplished by the group of Wang and Hong [102], and Connon [103] (eqs 5 and 6). The Michael/alkylation cascade process has also demonstrated its potency in the diverse construction of chiral spirooxindoles via chiral tertiary amine-H bonding donor-catalyzed asymmetric annulation of methyleneindolinones with bromonitromethane [104], ethyl 2-chloroacetoacetate [105], 3-bromoethyl substituted oxindoles [106], and γ-halogenated-β-ketoesters [107].

Additionally, isatylidene malononitriles have recently been employed as C2 synthons for the efficient construction of chiral spirooxindoles in the tertiary amine-H bonding multifunctional-catalyzed asymmetric annulation reactions. The Wang group developed an asymmetric domino Michael/cyclization/tautomerization reaction of isatylidene malononitriles **47c** and α,α-dicyanoalkenes **152**, which afforded the polycyclic spirooxindoles **154** with 53–87% yield, 2 : 1–7 : 1 dr, and 70–96% ee in the presence of bifunctional thiourea **153a** (eq 1, Scheme 5.34) [108]. Kesavan and coworkers reported the same reaction by chiral tertiary amine-H bonding bifunctional catalysis [109]. Later on, Du et al. established an asymmetric 1,3-dipolar cycloaddition of isatylidene malononitriles **47d** with the Seyferth–Gilbert reagent **155** catalyzed by 10 mol% of cinchona alkaloid-based catalyst **73b**, which provided a highly efficient entry to access optically active spiro-phosphonylpyrazoline-oxindoles **156** (eq 2) [110].

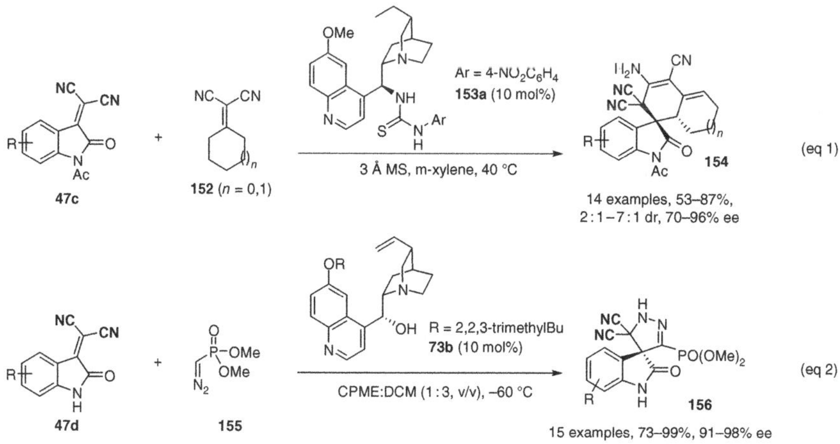

Scheme 5.34 Asymmetric annulation using isatylidene malononitriles as C2 synthons. *Source:* Based on Huang et al. [108].

Isatylidene malononitriles could also be used as C3 synthons in the synthesis of spirooxindole-pyranopyrimidine derivatives, owing to the cyano group (CN) would participate in the annulation process. For instance, by using 10 mol% of bifunctional cupreine **73c** as the catalyst, Yuan and coworkers presented a tandem Michael/cyclization reaction of isatylidene malononitriles **47** and 1,3-diones **157** (eq 1, Scheme 5.35) [111]. After that, the group of Wang and Wu independently reported the synthesis of spirooxindole-pyranopyrimidines via Michael/cyclization sequence of isatylidene malononitriles with α-keto ester **159** (eq 2) [112], and α-cyano ketones **161** (eq 3) [113] or pyrazolones **162** (eq 4) [114].

Similar to isatylidene malononitriles, the isatin-derived β,γ-unsaturated α-ketoesters are found to be a type of efficient C3 synthons as well. In 2016, Kesavan et al. accomplished an enantioselective Michael/hemiketalization reaction of isatin-derived β,γ-unsaturated α-ketoesters **11** with pyrazolones **162** by tertiary amine-H bonding catalysis. A new developed chiral tertiary amine-squaramide **163** was proven to be best choice and enabled the preparation of chiral spirooxindoles **164** with up to 95% yield and 99% ee (Scheme 5.36) [115]. At the same time, Wang and coworkers reported an asymmetric [3+3] annulation reaction of **11** with 3-hydroxy-4*H*-chromen-4-ones or 2-hydroxy-1,4-naphthoquinones via Michael/hemiketalization cascade process, which provided a facile access to optically active spirooxindoles [116].

Apart from the use of 3-alkenyl oxindoles, 3-alkyl oxindoles are competent reacting partners for the enantioselective construction of spirooxindoles. In 2011, by employing a newly designed cinchona alkaloid-based multifunctional catalyst **153b**, the Barbas group developed an asymmetric Michael/aldol cascade reaction of 3-substituted oxindoles **36b** and methyleneindolinones **1e**, allowing to access enantioenriched bis-spirooxindoles (eq 1, Scheme 5.37) [117]. One year later, with nitrostyrenes **166** as the Michael acceptor, they established a stereoselective Michael/Henry cascade reaction involving 3-alkyl oxindoles **36a** by bifunctional tertiary amine **115b**, which provided an entry to highly substituted spirocyclopentane oxindoles **167** (eq 2) [118]. Later in 2016, Wang, Peng, and coworkers reported an enantioselective nitroso aldol intramolecular transesterification cyclization domino reaction between C3 alkyl ester-substituted oxindole with nitrosobenzene by chiral tertiary amine-H bonding catalysis [119].

Meanwhile, Shao et al. presented a highly stereoselective Michael addition/intramolecular silyl nitronate-olefin cycloaddition (ISOC)/fragmentation sequence of 3-allyl oxindoles **168** and nitroolefins **166** (eq 1, Scheme 5.38) [120]. In the presence of 10 mol% of their developed tertiary amine-thiourea catalyst **120f**, the key asymmetric Michael addition of *N*-Boc-3-allyloxindoles **168** to nitroolefins **166** first occurred to afford chiral 3,3′-disubstituted oxindoles **I**, which subsequently underwent a sequential ISOC and fragmentation process to give the desired spirocyclic oxindoles **169**. By using the same catalyst **120f**, the authors

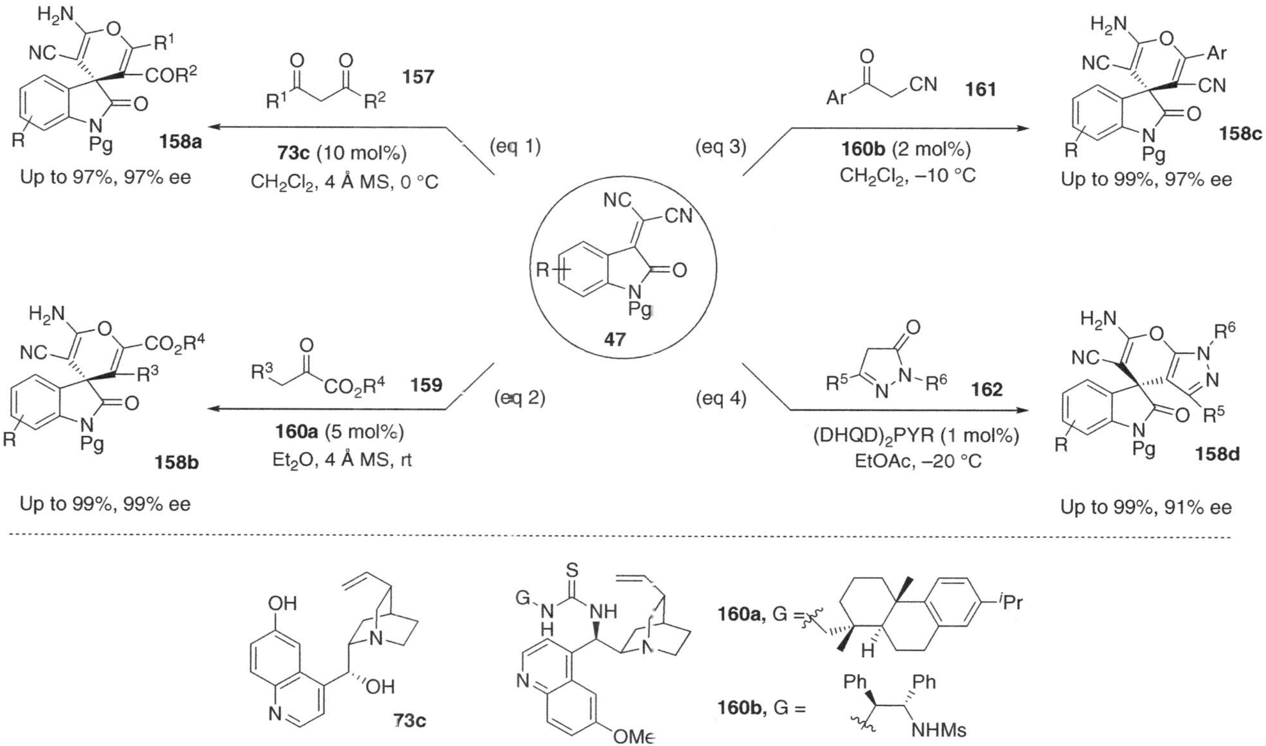

Scheme 5.35 Asymmetric annulation using isatylidene malononitriles as C3 synthons. *Source:* Based on Chen et al. [111].

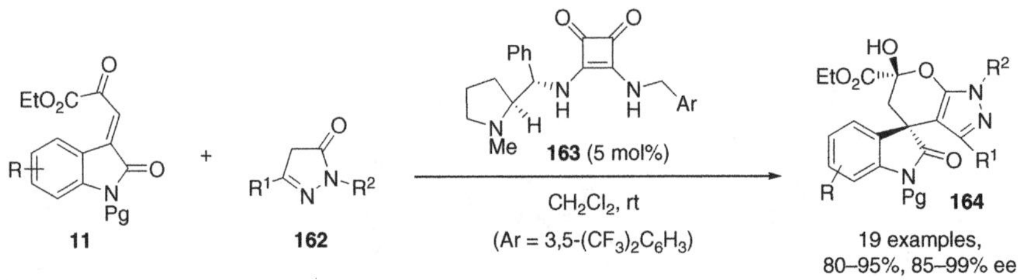

Scheme 5.36 Asymmetric annulation involving isatin-derived β,γ-unsaturated α-ketoesters. *Source:* Based on Kumarswamyreddy and Kesavan [115].

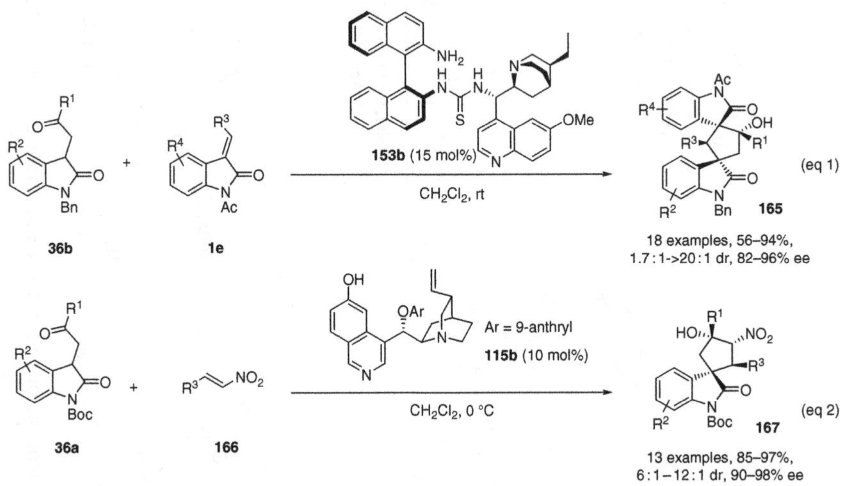

Scheme 5.37 Asymmetric annulation involving C3 alkyl ester-substituted oxindole. *Source:* Based on Tan et al. [117].

further described a highly efficient approach toward chiral spirocyclopentaneoxindoles **172** through an asymmetric sequential cross-metathesis/Michael/Michael cascade reaction of 3-allyl oxindoles **168**, ethyl acrylate, and nitroolefins **166** (eq 2) [121].

With 3-isothiocyanato oxindoles **173** as C3 synthons, a wide range of asymmetric annulation with electrophiles have been developed for the synthesis of optically active spirooxindole fused five-membered *N*-heterocycles [122]. For instance, in 2013, Yuan et al. developed an asymmetric [3+2] annulation of 3-isothiocyanato oxindoles **173** with electron-deficient olefin **174** catalyzed by 1 mol% of quinine, which enabled to access various 3,3′-thiopyrrolidonyl spirooxindoles (eq 1, Scheme 5.39) [123]. The Shi group disclosed that an enantioselective [3+2] annulation of 3,3′-thiopyrrolidonyl spirooxindoles **173** with allenic esters **44** or 2-butynedioic acid diesters **177** by using bifunctional tertiary amine-squaramide

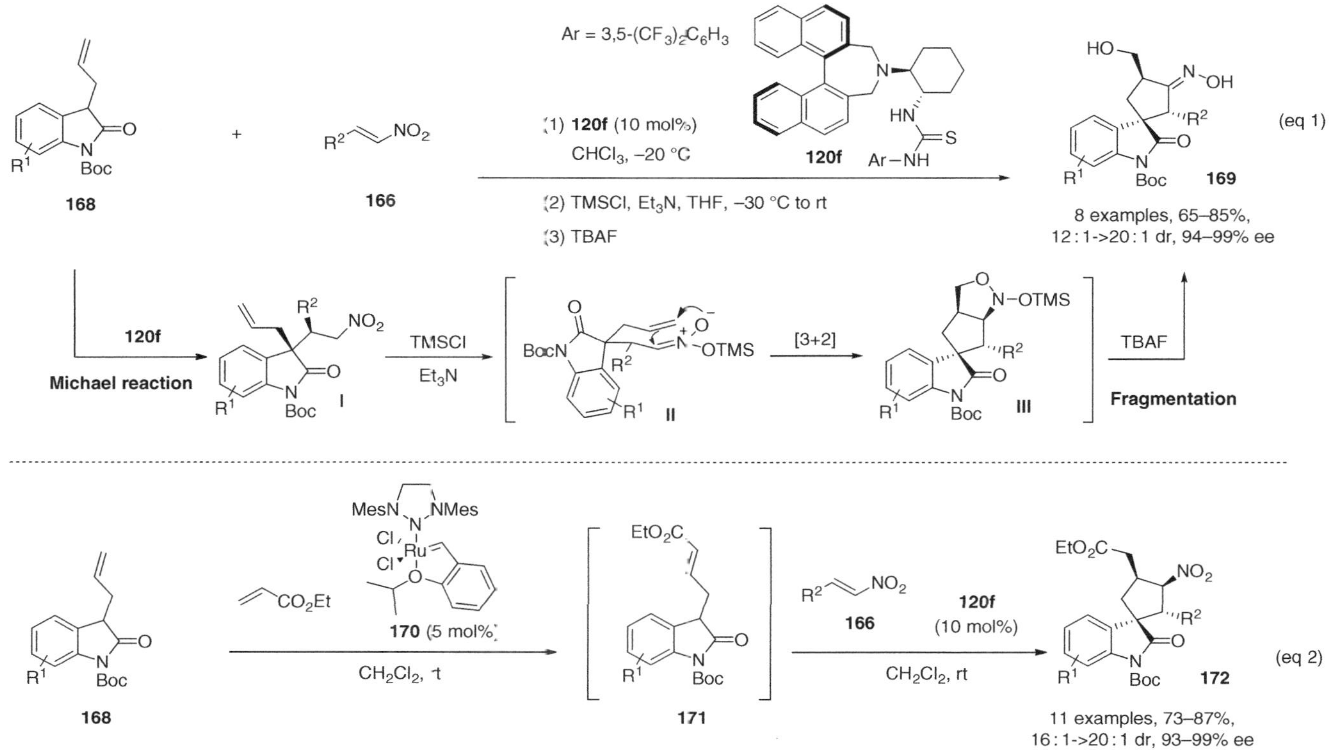

Scheme 5.38 Asymmetric annulation involving 3-allyloxindoles. *Source:* Based on Li et al. [120].

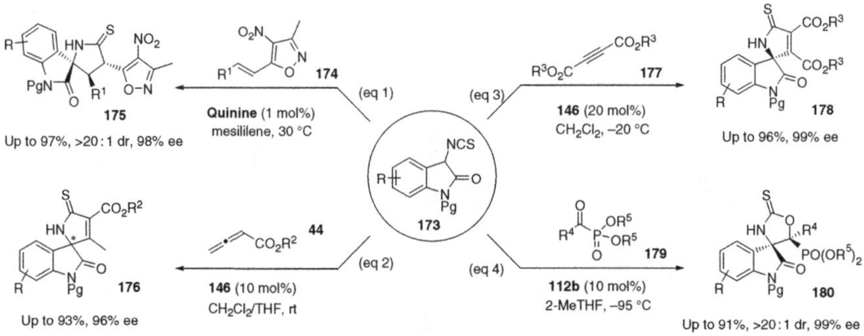

Scheme 5.39 Asymmetric annulation involving 3-isothiocyanato oxindoles. *Source:* Based on Liu et al. [123].

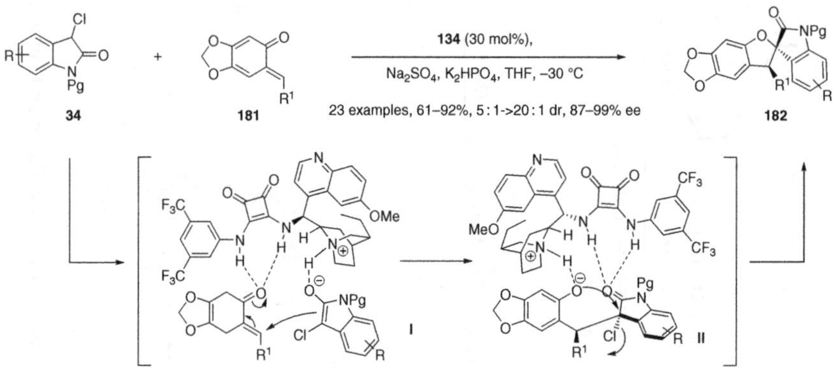

Scheme 5.40 Asymmetric [4+1] cyclization of 3-chlorooxindoles with *o*-QMs. *Source:* Based on Jiang et al. [126].

146 (eqs 2 and 3) [124]. In 2015, Mukherjee et al. reported an asymmetric [3+2] annulation of 3-isothiocyanato oxindoles **173** and α-ketophosphonates **179** (eq 4) [125].

Also by chiral tertiary amine-H bonding bifunctional catalysis, Shi and coworkers developed a stereoselective asymmetric [4+1] cyclization of 3-chlorooxindoles **34** with *ortho*-quinone methides (*o*-QMs) **181** via a Michael addition/S_N2 sequence (Scheme 5.40) [126]. The use of 30 mol% bifunctional squaramide **134** enabled efficiently the sequence to provide various spirooxindole-based 2,3-dihydrobenzofuran derivatives **182** with 61–62% yield, 5 : 1->20 : 1 dr and 87–99% ee values. Notably, this represents the first highly enantioselective [4+1] cyclization involving *o*-QMs. Furthermore, they reported the preparation of spirooxindole-based 2,3-dihydrobenzofurans through a catalytic asymmetric domino oxidation/[4+1] cyclization reaction of 3-chlorooxindoles and 2-alkylphenols.

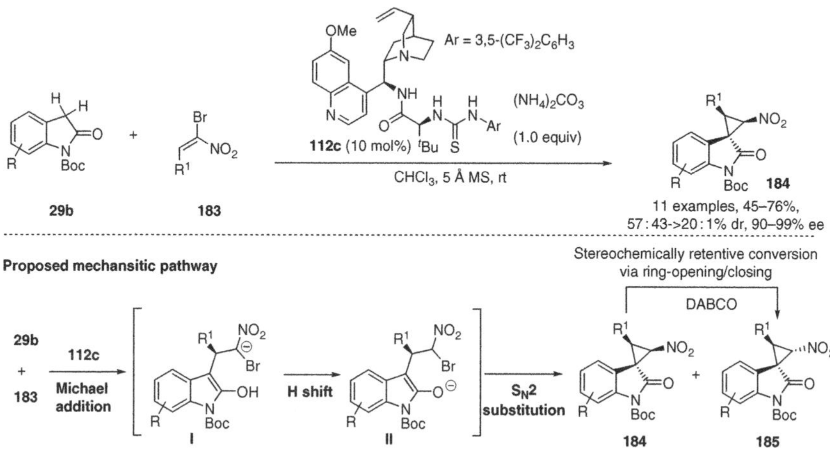

Scheme 5.41 Asymmetric cyclopropanation of 3-unsubstituted oxindoles with α-bromonitroolefins. *Source:* Based on Dou and Lu [127].

Besides, the asymmetric reactions with dinucleophilic 3-unsubtituted oxindoles for the synthesis of optically active spirooxindoles have also been realized by employing chiral tertiary amine-H-bonding multifunctional catalysis. In 2012, Dou and Lu developed a stereoselective cyclopropanation of *N*-Boc 3-unsubstituted oxindoles **29b** with α-bromonitroolefins **183** via a cascade Michael/S_N2 substitution process (Scheme 5.41) [127]. A series of 3-spirocyclopropyl-2-oxindoles **184** were obtained in good to excellent yield and dr with excellent ee values, in the presence of 10 mol% of their developed multifunctional catalyst **112c**. In addition, by a ring opening/closing process, the stereoisomer of products **185** could be easily obtained through a stereochemically retentive conversion of **184** under the influence of DABCO. Moreover, the cyclopropanation/isomerization sequence of 3-unsubstituted oxindoles **29b** with α-bromonitroolefins **183** has been realized for the direct synthesis of stereoisomer **185** as well.

One year later, Li and coworkers reported the enantioselective synthesis of spirooxindoles through a one-pot, asymmetric Michael/Michael/aldol tandem reaction of *N*-Boc 3-unsubstituted oxindoles **29b**, nitroolefins **166**, and enal **15** (Scheme 5.42) [128]. The combination of chiral tertiary amine-thiourea catalyst **112b** with Hayashi–Jørgensen catalyst **12** proved to be effective and provided various highly functionalized spirocyclic oxindoles **186** with up to 94% yield, 7 : 2 : 1 dr, and >99% ee. As shown in Scheme 5.42, the thiourea part of catalyst **112b** activated the nitroolefins through intermolecular H-bonding interaction, while the 3-unsubstituted oxindoles **29b** was deprotonated by the tertiary amine moiety, thus facilitating the first Michael addition of 3-unsubstituted oxindoles and nitroolefins. The resulting Michael adduct **I** underwent the second

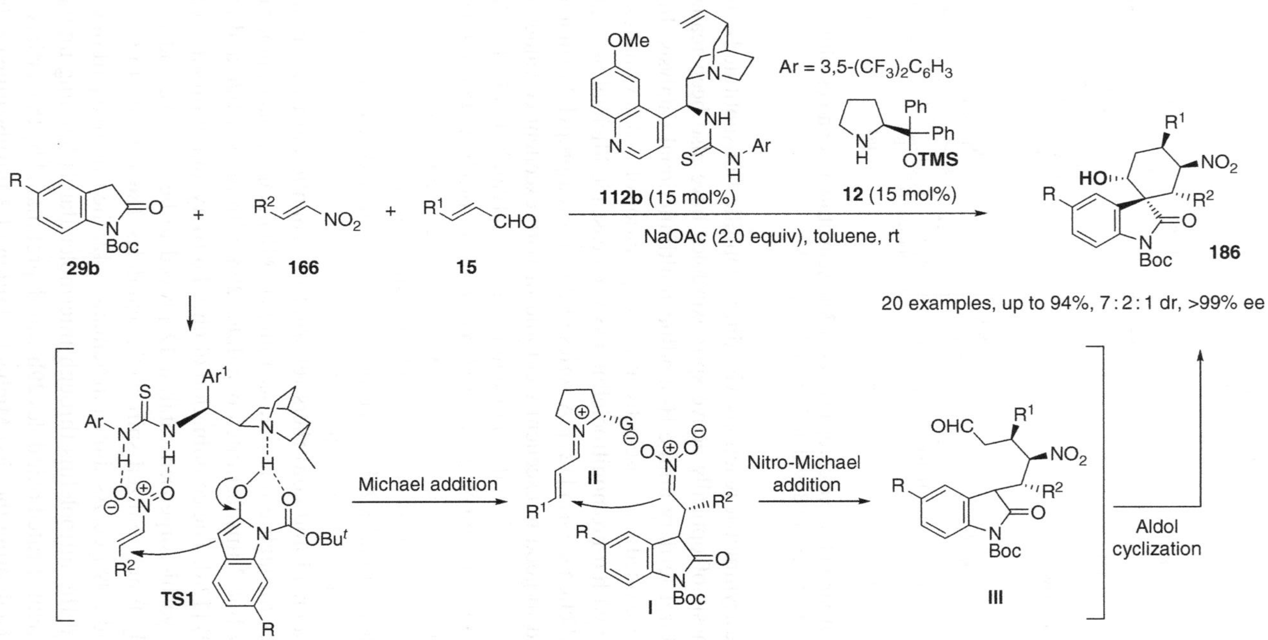

Scheme 5.42 One-pot Michael/Michael/aldol tandem reaction involving 3-unsubstituted oxindoles. *Source:* Based on Zhou et al. [128].

nitro-Michael reaction with *in situ*-generated iminium intermediate **II**, followed by an intramolecular aldol reaction to afford the corresponding products **186**. Later, by chiral tertiary amine-thiourea-catalyzed Michael/Michael cascade process, Xu, Liang, and coworkers developed a stereoselective formal [5+1] cyclization of 3-unsubtituted oxindoles with ester-linked bisenones toward chiral spirooxindoles [129].

By chiral tertiary amine-H bond donor catalysis, Wang and coworkers developed an asymmetric desymmetrization of prochiral spiro cyclohexadienone oxindoles **187** through an organocatalyzed enantioselective sulfa-Michael addition (Scheme 5.43) [130]. The sulfa-Michael addition was efficiently facilitated by 5 mol% multifunctional catalyst **120e**, which gave the desired spirooxindoles **189** with good to excellent yields, dr and ee values. It should be noted that this represents the first application of desymmetrization strategy in the synthesis of chiral spirooxindoles. Additionally, the enone functionality in the spirooxindoles **189** could be selectively reduced by $NaBH_4$ to allylic alcohol, and $Pd(OH)_2/C$ to ketone, or readily underwent secondary sulfa-Michael reaction to provide spirooxindoles with structural diversity.

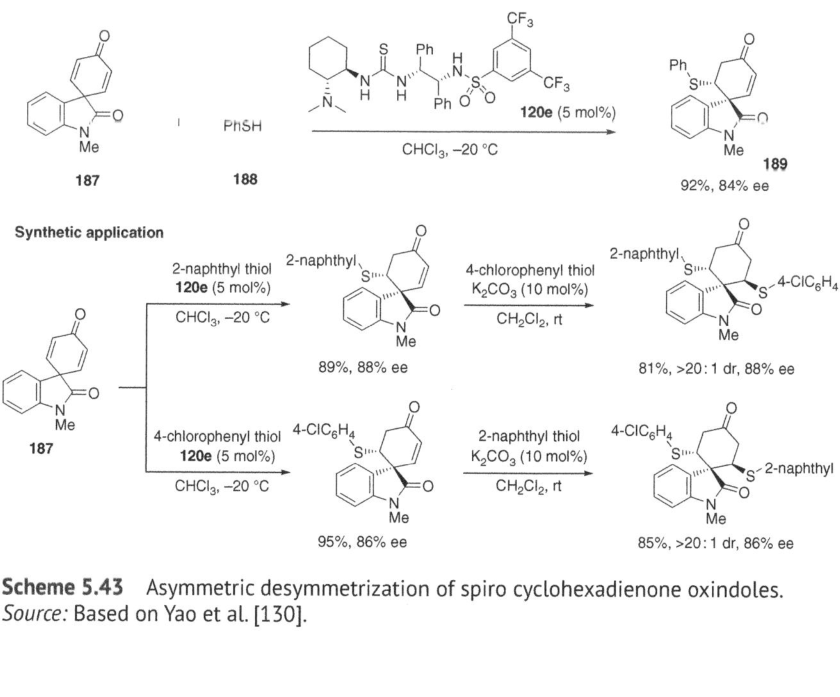

Scheme 5.43 Asymmetric desymmetrization of spiro cyclohexadienone oxindoles. *Source:* Based on Yao et al. [130].

5.6 Chiral Hydrogen-Bonding Catalysis

In parallel with fruitful achievements for the construction of spirooxindoles by chiral tertiary amine-H-bonding donor bifunctional catalysis, chiral hydrogen-bonding donors, as a kind of efficient non-covalent bond organocatalysts, have exhibited their potency in the stereoselective forging of chiral spirooxindoles. In 2011, Barbas et al. developed a highly efficient asymmetric Diels–Alder reaction of methyleneindolinones **1b** with 3-vinylindoles **190** by utilizing a chiral bisthiourea **191a**, which allowed access to enantioenriched carbazolespirooxindoles featuring three or four stereocenters with up to 99% yields, >20 : 1 dr, and 99% ee values (Scheme 5.44) [131]. Intriguingly, a simple centrifugation could provide the pure products, which was attributed to the much poorer solubility of product **192** in hexane. A possible transition state was proposed on the basis of initial NMR studies and the stereochemistry of product, shown in Scheme 5.44. The thiourea activated with the carbonyl group of methyleneindolinones **1b** through H-bonding interactions, and the generated chiral intermediate underwent a Diels–Alder reaction with 3-vinylindoles **190** to afford the targets. Six years later, Yang and coworkers realized the same enantioselective Diels–Alder reaction by devising a novel *L*-pyroglutamic sulphonamides as chiral hydrogen-bonding catalyst [132].

Also with chiral bisthiourea **191a** as hydrogen-bonding catalyst, Zhu, Cheng, and coworkers demonstrated a highly stereoselective [3+2] cycloaddition of methyleneindolinone **1f** with nitrones **193** (Scheme 5.45) [133]. The use of

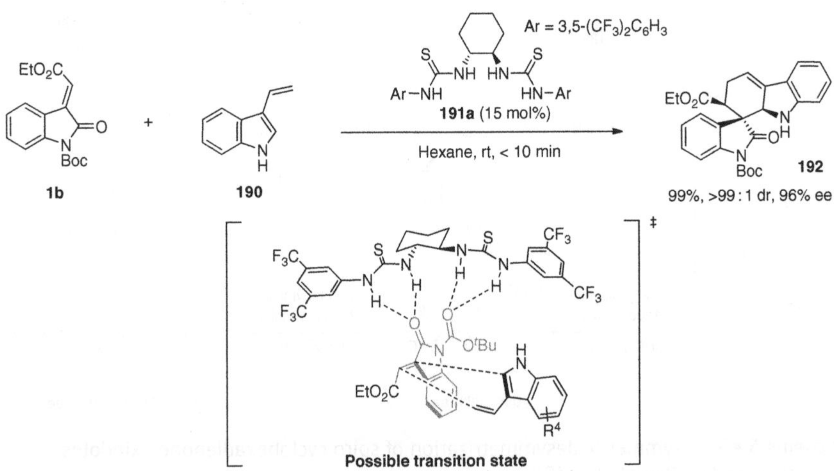

Scheme 5.44 Enantioselective Diels–Alder reaction of methyleneindolinones with 3-vinylindoles. *Source:* Based on Tan et al. [131].

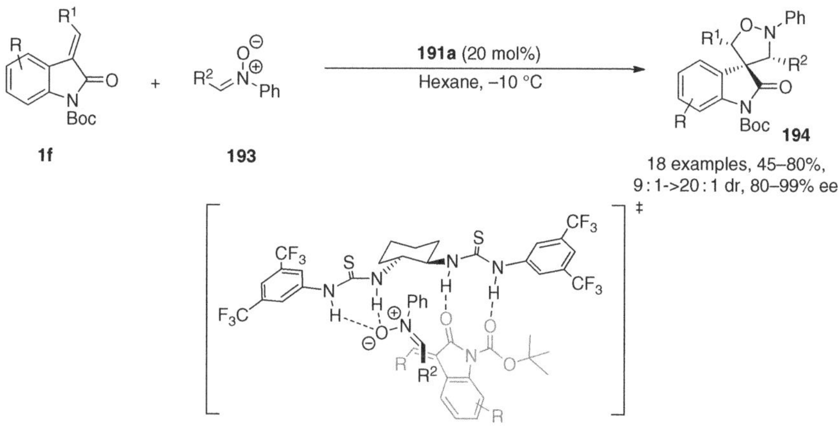

Scheme 5.45 Asymmetric [3+2] cycloaddition of methyleneindolinone and nitrones. *Source:* Based on Shi et al. [133].

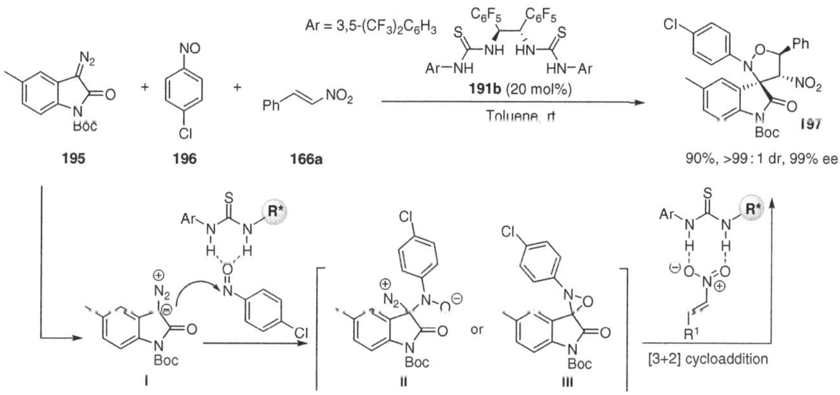

Scheme 5.46 Asymmetric MCR of diazooxindoles, nitrosoarenes, and nitroalkenes. *Source:* Based on Wu et al. [134].

20 mol% bisthiourea **191a** enabled a range of enantioenriched spiro-[isoxazolidine-3,3′-oxindole] derivatives **194** featuring three contiguous stereocenters with good to high yields and high to excellent stereoselectivities. In this case, the multiple hydrogen-bonding catalyst **191a** activated methyleneindolinone and nitrones simultaneously via H-bonding interactions, as described in the transition state of Scheme 5.45.

Later in 2015, Tan and Liu employed chiral hydrogen-bonding catalysis to establish an asymmetric multicomponent reaction (MCR) of diazooxindoles, nitrosoarenes, and nitroalkenes for the assembly of optically active spirooxindoles fused five-membered heterocycle (Scheme 5.46) [134]. 20 mol% of chiral bisthiourea catalyst **191b** turned out to be efficient and furnished the desired spirooxindoles **197** in 57–99% yields, >20 : 1 dr, and 89–99% ee, with excellent functionality tolerance. A plausible reaction pathway was proposed in which the active intermediate **II** or oxaziridine intermediate **III** would generated *in situ* from diazooxindole **195** and nirosoarene **196** under the action of bisthiourea, which subsequently reacted with the activated nitroalkene **166a** by bisthiourea, thus delivering the targets **197** under a well-organized chiral environment. Notably, this represents a successful example of the use of diazo compounds under metal-free conditions.

5.7 Chiral Phosphoric Acid Catalysis

Chiral phosphoric acids have been widely used as a type of powerful organocatalysts in a plenty of asymmetric reactions [135], since the seminar independent work of Akiyama [136] and Terada [137] in 2004. In this context, many efforts have been made to the construction of chiral spirooxindoles through phosphoric acid-catalyzed enantioselective annulation reactions, and considerable achievements have been made. Early in 2008, List et al. reported the first organocatalytic enantioselective cyclization of isatin **24a** and 2-aminobenzamide **198** by using (*S*)-TRIP, which allowed for the synthesis of spirooxindole **199** in 85% yield and 84% ee (Scheme 5.47) [138].

In 2009, by chiral phosphoric acid catalysis, the Gong group developed an asymmetric three-component 1,3-dipolar cycloaddition of aldehydes **200a**, amino esters **201a**, and methyleneindolinones **1c** (Scheme 5.48) [139]. Under the action of 10 mol% chiral phosphoric acid **7b**, the reaction worked well to furnish a variety of 3,3′-pyrrolidinyl spirooxindoles **202** with unusual regioselectivity in good to excellent stereoselectivity. As shown in Scheme 5.48, the possible reaction pathway and transition state were proposed that aldehydes **200a** reacted with amino esters **201a** to generate azomethine ylides **203**, which were activated by chiral phosphoric acid **7b** through H-bonding interactions. Meanwhile, methyleneindolinones **1c** approached the formed chiral intermediate and also interacted with

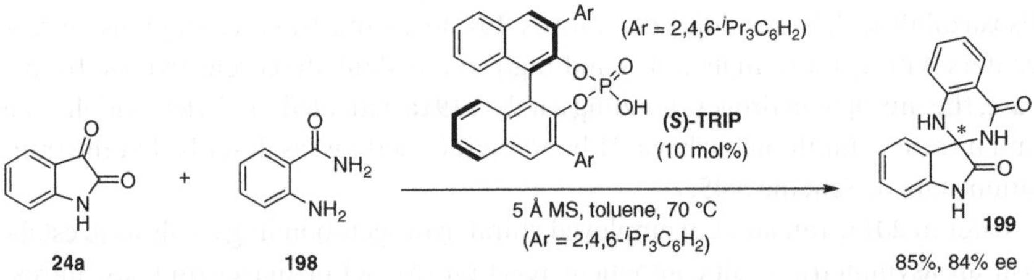

Scheme 5.47 Enantioselective cyclization reaction of isatin and 2-aminobenzamide. *Source:* Based on Cheng et al. [138].

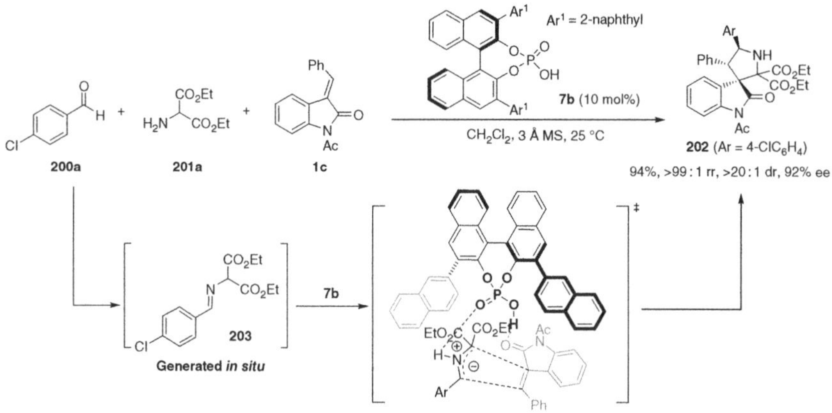

Scheme 5.48 Asymmetric three-component 1,3-dipolar cycloaddition by phosphoric acid catalysis. *Source:* Based on Chen et al. [139].

phosphoric acid via H-bonding. Finally, the desired spirooxindoles **202** were generated followed by a cycloaddition process. By DFT calculations, the authors indicated that the unusual regioselectivity was likely attributed to the favorable π–π stacking interaction between oxindole skeleton and the conjugated esters. Later on, Shi and coworkers reported a similar chiral phosphoric acid-catalyzed asymmetric chemoselective 1,3-dipolar cycloaddition of isatin-derived imines with the *in situ*-generated azomethine ylides from aldehydes and amino esters [140].

Later in 2012, Gong and Luo documented a stereoselective three-component 1,3-dipolar cycloaddition of isatin **24c**, amino esters **201**, and dimethyl maleate **204** catalyzed by chiral bisphosphoric acids **205**, providing an efficient approach for the assembly of biologically important spiro[pyrrolidin-3,2′-oxindole] **206** with structural diversity. Noticeably, this represents the first catalytic 1,3-dipolar cycloaddition reaction with azomethine ylides formed *in situ* from unsymmetrical cyclic ketones (eq 1, Scheme 5.49) [141]. Shortly after, Shi and Tu reported an asymmetric three-component 1,3-dipolar cycloaddition of isatin **24**, amino esters **201a**, and ethynyl ketones **207** which enabled to access spiro[indoline-3,2′-pyrroles] **208** with high yields and excellent enantioselectivities under phosphoric acid **7c** catalysis (eq 2) [142]. In 2015, the Shi group further employed chiral bisphosphoric acid catalysis to achieve the efficient synthesis of spiro[pyrrolidin-3,2′-oxindole] through asymmetric three-component 1,3-dipolar cycloadditions of isatin, amino esters, and methyleneindolinones [143] or allenoates [144].

With 10 mol% of (*R*)-TRIP, Wang et al. developed a catalytic asymmetric three-component 1,3-dipolar cycloaddition of aldehydes **200**, nitroolefins **166**, and 3-aminooxindoles **107b**, which allowed to construct densely substituted

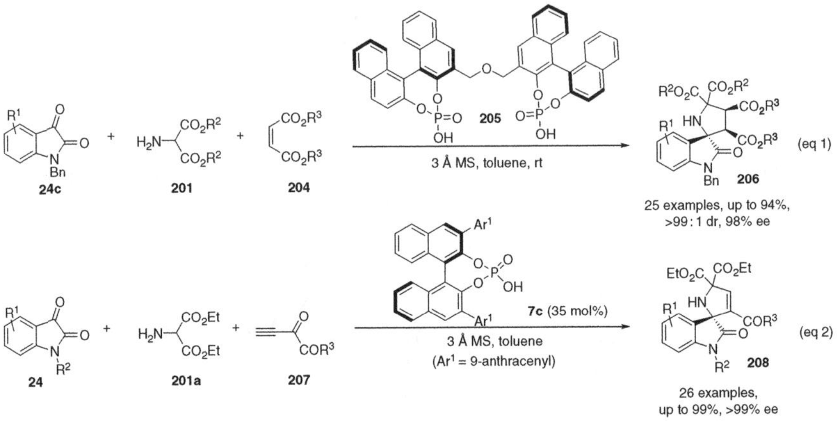

Scheme 5.49 Asymmetric three-component 1,3-dipolar cycloaddition. *Source:* Based on Shi et al. [141].

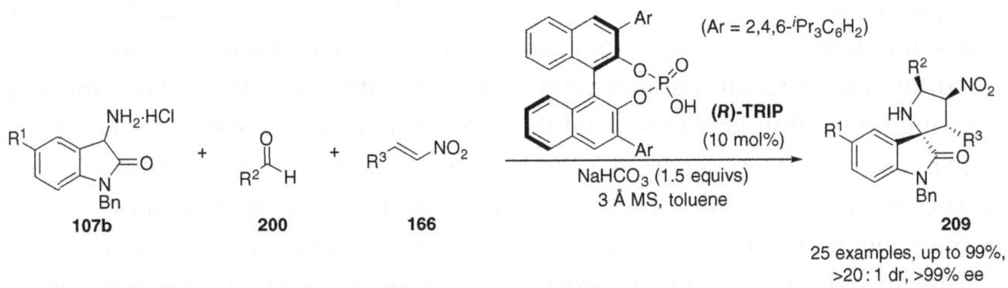

Scheme 5.50 Asymmetric three-component 1,3-dipolar cycloaddition. *Source:* Based on Zhu et al. [145].

spiro[pyrrplidine-2,3′-oxindole] derivatives **209** with high to excellent yields, dr, and ee values (Scheme 5.50) [145].

Hong and Wang designed a novel chiral bisphosphoric acid **211** bearing triple axial chirality and applied it to an asymmetric [3+2] cycloaddition of methyleneindolinones **1f** and *N,N′*-cyclic azomethine imines **210** (eq 1, Scheme 5.51) [146]. The use of 10 mol% of acid **211** enabled the cycloaddition effectively to provide various polycyclic spiro[pyrazolidin-3,3′-oxindoles] **212** with 68–94% yield, 6 : 1–20 : 1 dr, and 91–99% ee values. A dual activation model was proposed in the transition state **I** of Scheme 5.51. Later in 2015, Shi et al. reported a highly stereoselective Diels–Alder reaction of methyleneindolinones **1** and 2-vinylindoles **213a** catalyzed by 10 mol% phosphoric acid **7d**, allowing the synthesis of various spiro[tetrahydrocarbazole-3,3-oxindole] **214** with good to excellent stereoselectivities (eq 2) [147]. As depicted in the transition state **II** of Scheme 5.51, the authors proposed that 2-vinylindole and methyleneindolinone were activated

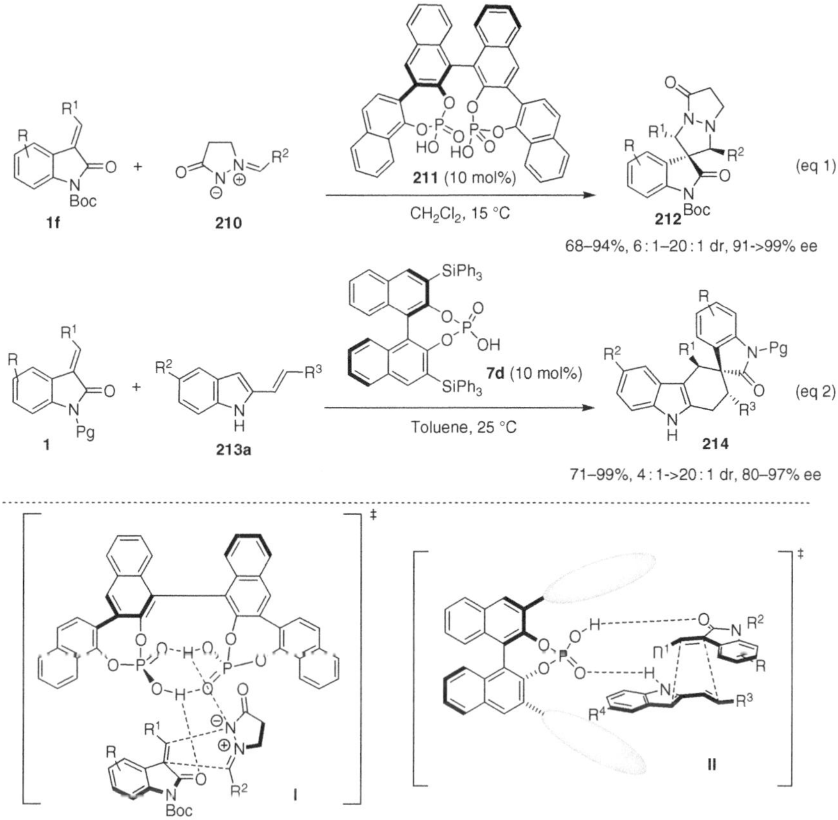

Scheme 5.51 Chiral phosphoric acid-catalyzed cycloaddition with methyleneindolinones. *Source:* Based on Hong et al. [146].

simultaneously by phosphoric acid via H-bonding, thus rendering the Diels–Alder reaction occur under well-organized chiral environment.

In addition, Shi and coworkers realized the construction of optically active polycyclic spirooxindoles through a phosphoric acid-catalyzed asymmetric [3+2] cycloaddition of 3-methyl-2-vinylindoles **213b** with isatin-derived 3-indolylmethanols **215** (Scheme 5.52) [148]. A series of spiro[cyclopenta[β]indole-1,3′-oxindole] **216** were obtained in up to 99% yield, >20 : 1 dr, and 98% ee, under the catalysis 10 mol% of chiral phosphoric acid **7e**. A similar reaction transition state was proposed shown in Scheme 5.52, 3-indolylmethanols **215** underwent an acid-catalyzed dehydration to form a vinyliminium intermediate, while the phosphoric acid interacted with 3-methyl-2-vinylindoles **213b** via hydrogen-bonding. Subsequently, a vinylogous Michael addition to produce the transient intermediate **II** was followed by an intramolecular Friedel–Crafts reaction to deliver the spirooxindoles **216**. By chiral

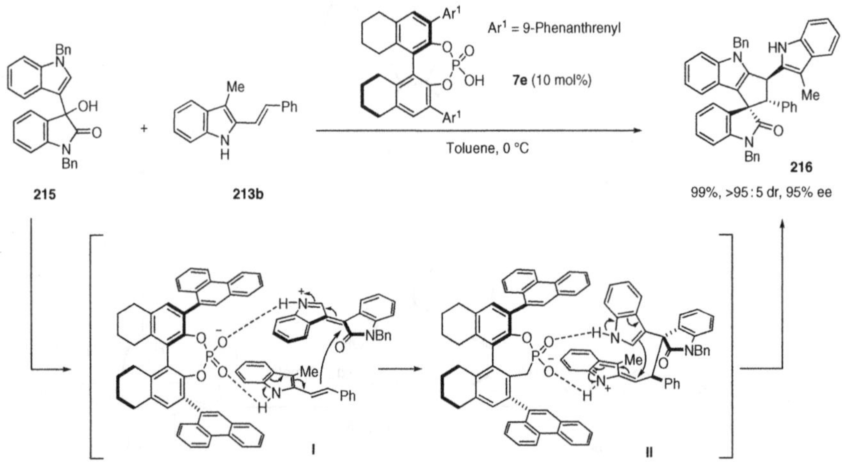

Scheme 5.52 [3+2] Cycloaddition of isatin-derived 3-indolylmethanols and 2-vinylindoles. *Source:* Based on Tan et al. [148].

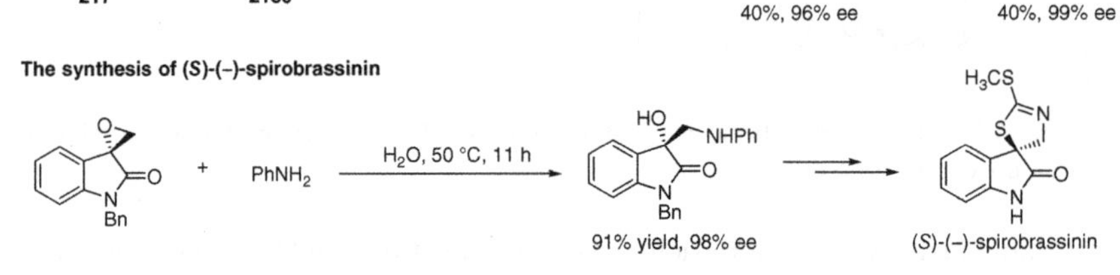

Scheme 5.53 Chiral phosphoric acid-catalyzed kinetic resolution of racemic spiro-epoxyoxindoles. *Source:* Based on Zhu et al. [153].

phosphoric acid catalysis, the same group further developed the asymmetric cascade reactions of isatin-derived 3-indolylmethanols **215** with *in situ*-generated isatin-derived imines [149], azomethine ylides [150], 3-vinylindoles [151], or 7-vinylindoles [152], which enabled the diverse assembly of optically active spirooxindoles.

In 2017, Wang, Hong, Sun, and coworkers developed an efficient kinetic resolution of racemic spiro-epoxyoxindoles **217** through a chiral phosphoric acid-catalyzed regio- and enantioselective Friedel–Crafts alkylation of indoles **213c** and **217** (Scheme 5.53) [153]. Both enantioenriched spiro-epoxyoxindoles (*R*)-**217** and 3-(3-indolyl)-oxindole-3-methanols **219** could be simultaneously produced in good yields with excellent enantioselectivities, in the presence of 20 mol% spirocyclic phosphoric acid **218**. Most importantly, the obtained chiral spiro-epoxyoxindole (*R*)-**217** could be used as a key intermediate for the synthesis of cruciferous phytoalexin (*S*)-(−)-spirobrassinin.

5.8 Chiral Phase-Transfer Catalysis

Phase-transfer catalysis (PTC) with chiral quaternary ammonium or phosphonium salt has become a powerful catalysis model through electrostatic and other non-covalent interactions in many asymmetric reactions [154], since the seminar work of Merck scientists who utilized chiral quaternary ammonium salt to realize a highly enantioselective *C*-methylation of indanones [155]. On the basis of these significant achievements, chiral PTC has been successfully applied in the enantioselective construction of spirooxindoles as well. In 2016, Zhao and Shang employed their developed dipeptide-based quaternary phosphonium salt multifunctional catalyst **221a** to report a highly asymmetric double Michael cascade reaction of 3-alkenyloxindole **1g** with γ-malonate-substituted α,β-unsaturated esters **220**, which furnished a range of optically active spirooxindoles fused five-membered ring **222** in high to excellent yields and stereoselectivities (Scheme 5.54) [156]. Noticeably, both amide N–H bonds and the quaternary ammonium salt moiety of catalyst were indispensable for high reactivity and

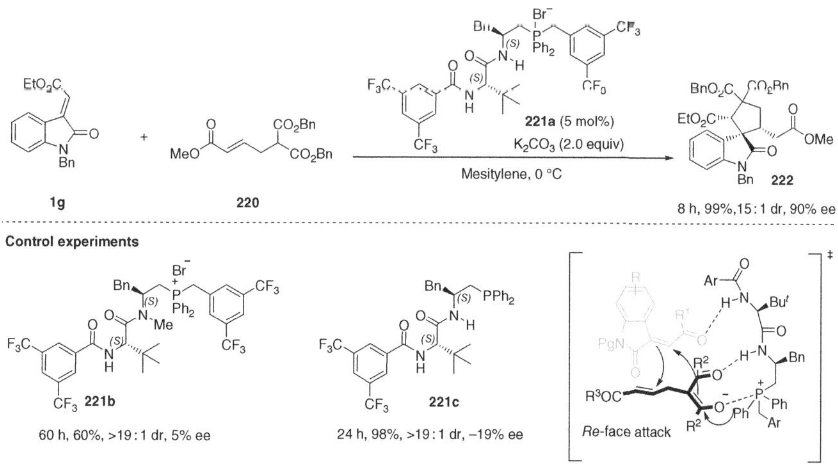

Scheme 5.54 Asymmetric double Michael cascade reaction of 3-alkenyloxindoles by PTC catalysis. *Source:* Based on Zhang et al. [156].

enantioselectivity by control experiments. Accordingly, a plausible reaction transition state was proposed in Scheme 5.54. The *in situ*-formed malonate carbanion of **220** interacted with the phosphonium center and an amide N–H bond in the catalyst **221a** through static electronic and H-bonding interactions, while the electrophile **1g** was activated by another amide N–H bond via coordination with the ester group.

In the same year, by employing thiourea-quaternary ammonium salt **224a** as phase-transfer catalyst, the same group developed a highly enantioselective 1,3-dipolar cycloaddition of methyleneindolinones **1h** with imino esters **223** (Scheme 5.55) [157]. A wide variety of chiral spiro[pyrrolidin-3,3′-oxindoles] **225** were obtained in 80–99% yields, 9 : 1->20 : 1 dr, and 81–99% ee values, in the presence of 5 mol% of **224a** and 50 mol% of K_2CO_3. Similarly, the authors also found that the quaternary phosphonium salt and thiourea moiety all played a crucial role in achieving high reactivity and stereoselectivity.

In addition, Xiang, Yasuda, and coworkers established an intramolecular alkylation of 3-substituted oxindoles **226** bearing a benzyl chloride scaffold by devising a type of novel *N,N'*-disubstituted doubly quaternized cinchona alkaloid **227** as efficient phase-transfer catalysts (Scheme 5.56) [158]. It was found that only 1 mol% **227** enabled the alkylation efficiently to provide the desired optically active spirooxindoles **228** with excellent yields and enantioselectivities, in the presence of 1.1 equiv NaOH. It is worth noting that the leaving group in the substrate **226** had a significant influence on the selectivity, as the ee of product greatly decreased after replacing the Cl atom with a Br or OTs group.

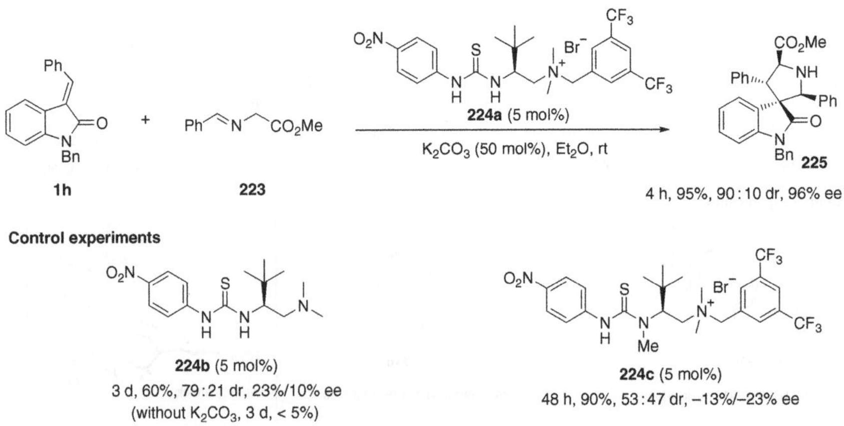

Scheme 5.55 Enantioselective 1,3-dipolar cycloaddition of 3-alkenyloxindoles by PTC catalysis. *Source:* Based on Zhang et al. [157].

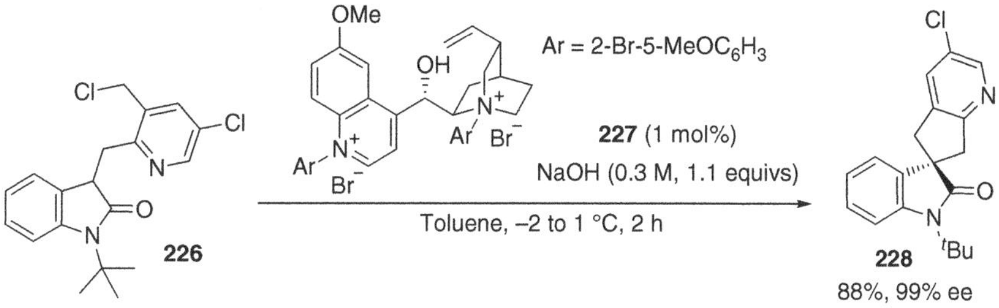

Scheme 5.56 Intramolecular *C*-acylation of 3-substituted oxindoles **320** by PTC catalysis. *Source:* Based on Xiang et al. [158].

5.9 Chiral Organoiodine Catalysis

Aside from the aforementioned several typical organocatalysts, chiral hypervalent iodine reagents have also been employed as efficient organocatalysts for the access of enantioenriched spirooxindoles recently. In 2014, the Gong group pioneered an asymmetric direct intramolecular C(sp^2)-H/C(sp^3)-H oxidative cross-coupling of N^1,N^3-diphenylmalonamides **229** catalyzed by chiral organoiodine **230a**, which provided a facile protocol to access chiral quaternary spirooxindoles **231** (eq 1, Scheme 5.57) [159]. A combination of 15 mol% of **230a** with 2.0 equiv of CF_3CO_2H and 4.0 equiv of $MeCO_3H$ enabled the reaction smoothly to produce the desired spirocyclic bisoxindoles **231** with moderate yields and good enantioselectivities. As illustrated in the transition state of Scheme 5.57, the authors proposed that the formed iodonium intermediate with trifluoroacetate counterion first proceeded with a *syn* addition elimination process to give 3 amide substituted indoles. Subsequently, an intramolecular Friedel–Crafts addition of anilide to the iodoenol in the associative intermediates from the less sterically hindered *Si*-face of the 2-indololate moiety occurred to give the targeted product, after the release of chiral iodine catalyst. Later on, by using the same chiral organoiodine **230a** as the catalyst, Du et al. developed an enantioselective oxidative spirocyclization of alkyl 3-oxopentanedioate monoamide derivatives **232** for the assembly of biologically important optically active spirofurooxindole **233** (eq 2) [160].

Additionally, Gong et al. developed an asymmetric oxidative dearomatization reaction of naphthol derivatives **234** enabled by a chiral hypervalent iodine **230b** (Scheme 5.58) [161]. Under the action of 15 mol% iodobenzene catalyst **230b** and 1.3 equiv *m*-CPBA, a series of 1-hydroxy-*N*-aryl-2-naphthamide derivatives **234** were smoothly converted into the corresponding chiral quaternary spirooxindoles **235** with good to high yields and high to excellent ee values. It should be mentioned that the addition of both trifluoroethanol and H_2O were crucial for achieving high yield and enantioselectivity. A possible catalytic cycle was proposed in

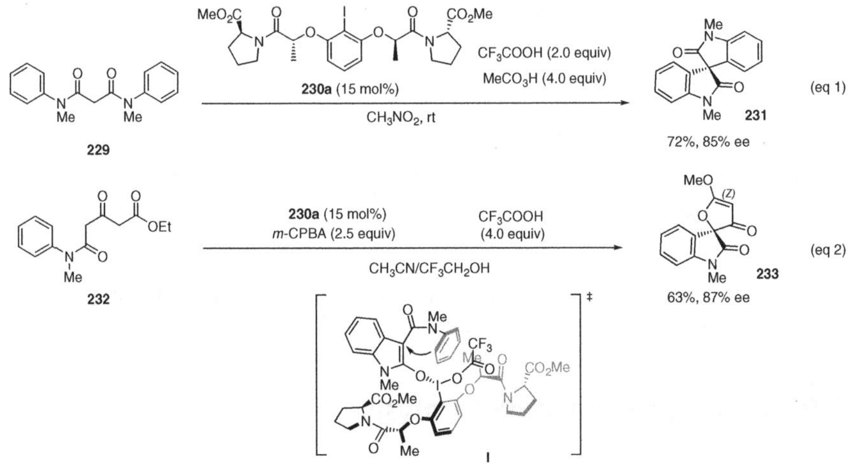

Scheme 5.57 Enantioselective direct intramolecular C(sp^2)–H/C(sp^3)–H oxidative cross-coupling. *Source:* Based on Wu et al. [159].

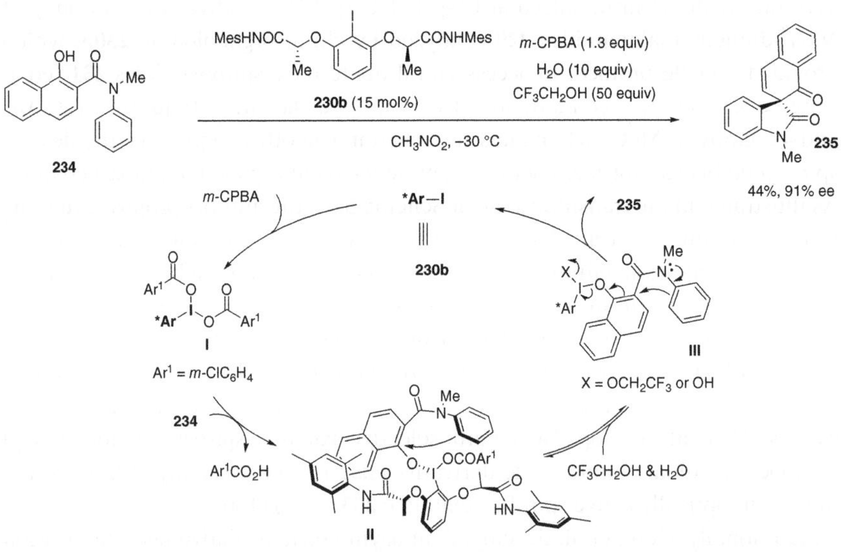

Scheme 5.58 Chiral organoiodine-catalyzed oxidative dearomatization of phenol derivatives. *Source:* Based on Zhang et al. [161].

Scheme 5.58, catalyst **230b** was oxidized by *m*-CPBA to give a hypervalent phenyl-λ^3-iodane **I**, which underwent a substitution with the hydroxy group of **234**. The resulted chiral iodo-enol-type intermediate **II** then proceeded with a ligand exchange with either trifluoroethanol or water to produce an associative intermediate **III**. Subsequently, an intramolecular S_N2'-like Friedel–Crafts substitution occurred via *Si*-face attack to afford the target **235** and release chiral catalyst **230b**.

5.10 Conclusion

The asymmetric organocatalysis proves to be one of the most powerful and versatile tactics for the efficient construction of optically active spirooxindole scaffolds. The past few decades have witnessed the significant progress in constructing such sophisticated spirocycles via the organocatalytic process. As summarized in this chapter, eight major activation modes with various chiral organocatalysts have been successfully applicable for the enantioselective synthesis of functionalized chiral spirooxindoles fused three-, four-, five-, and six-membered rings with high to excellent yields and stereoselectivities. Despite the remarkable achievements, there is still much room for the further improvement and development with regard to the substrate scope and the type of substrates employed, as well as catalyst loadings. Therefore, the more efficient approaches for the construction of spirooxindoles with low catalyst loading, simplicity, and mild conditions, as well as readily available synthons is highly desirable and sought after. Given the development of more and more chiral organocatalysts and synthetic strategies, we believe that more practical and efficient organocatalytic transformations toward such privileged scaffolds with structurally diversity will continue to be explored, thus facilitating their applications in the field of pharmaceuticals, agrochemicals, materials science, and other related areas.

References

1 (a) Singh, G.S. and Desta, Z.Y. (**2012**). *Chem. Rev.* 112: 6104–6155. (b) Hong, L. and Wang, R. (**2013**). *Adv. Synth. Catal.* 355: 1023–1052. (c) Cheng, D., Ishihara, Y., Tan, B., and Barbas, C.F. III (**2014**). *ACS Catal.* 4: 743–762. (d) Galliford, C.V. and Scheidt, K.A. (**2007**). *Angew. Chem. Int. Ed.* 46: 8748–8758. (e) Ball-Jones, N.R., Badillo, J.J., and Franz, A.K. (**2012**). *Org. Biomol. Chem.* 10: 5165–5181.

2 (a) Badillo, J.J., Hanhan, N.V., and Franz, A.K. (**2010**). *Curr. Opin. Drug Discovery Dev.* 13: 758.(b) Zheng, Y., Tice, C.M., and Singh, S.B. (**2014**). *Bioorg. Med. Chem. Lett.* 24: 3673–3682.(c) Gupta, A.K., Bharadwaj, M., Kumar, A., and Mehrotra, R. (**2017**). *Top. Curr. Chem. (Z)* 375: 3.

3 (a) Lin, H. and Danishefsky, S.J. (**2003**). *Angew. Chem. Int. Ed.* 42: 36–51. (b) Zhang, J.-Y. and Wang, Y.-X. (**2015**). *Fitoterapia* 100: 35–43.

4 Ding, K., Lu, Y., Nikolovska-Coleska, Z. et al. (**2005**). *J. Am. Chem. Soc.* 127: 10130–10131.

5 Jossang, A., Jossang, P., Hadi, H.A. et al. (**1991**). *J. Org. Chem.* 56: 6527–6530.

6 (a) Warner, M.L., Kaufman, N.C., and Grundmann, O. (**2016**). *Int. J. Legal Med.* 130: 127–138. (b) Marti, C. and Carreira, E.M. (**2003**). *Eur. J. Org. Chem.* 2003: 2209–2219.

7 Crosignani, S., Page, P., Missotten, M. et al. (**2008**). *J. Med. Chem.* 51: 2227–2243.

8 Zhao, Y., Yu, S., Sun, W. et al. (**2013**). *J. Med. Chem.* 56: 5553–5561.

9 Rottmann, M., McNamara, C., Yeung, B.K.S. et al. (**2010**). *Science* 329: 1175–1180.

10 Yu, B., Yu, D.-Q., and Liu, H.-M. (**2015**). *Eur. J. Med. Chem.* 97: 673–698.

11 (a) Zhou, F., Liu, Y.-L., and Zhou, J. (**2010**). *Adv. Synth. Catal.* 352: 1381–1407. (b) Shen, K., Liu, X., Lin, L., and Feng, X. (**2012**). *Chem. Sci.* 3: 327–334. (c) Dalpozzo, R. (**2017**). *Adv. Synth. Catal.* 359: 1772–1810. (d) Dalpozzo, R. (**2017**). *Org. Chem. Front.* 4: 2063–2078. (e) Cao, Z.-Y., Zhou, F., and Zhou, J. (**2018**). *Acc. Chem. Res.* 51: 1443–1454. (f) Mei, G.-J. and Shi, F. (**2018**). *Chem. Commun.* 54: 6607–6621.

12 Bertelsen, S. and Jørgensen, K.A. (**2009**). *Chem. Soc. Rev.* 38: 2178–2189.

13 (a) Dalko, P.I. and Moisan, L. (**2004**). *Angew. Chem. Int. Ed.* 43: 5138–5175. (b) Seayad, J. and List, B. (**2005**). *Org. Biomol. Chem.* 3: 719–724. (c) MacMillan, D.W.C. (**2008**). *Nature* 455: 304–308. (d) Dondoni, A. and Massi, A. (**2008**). *Angew. Chem. Int. Ed.* 47: 4638–4660. (e) Melchiorre, P., Marigo, M., Carlone, A., and Bartoli, G. (**2008**). *Angew. Chem. Int. Ed.* 47: 6138–6171.(f) Alba, A.N., Companyó, X., Viciano, M., and Rios, R. (**2009**). *Curr. Org. Chem.* 13: 1432–1474.(g) Moyano, A. and Rios, R. (**2011**). *Chem. Rev.* 111: 4703–4832.

14 List, B., Lerner, R.A., and Barbas, C.F. III (**2000**). *J. Am. Chem. Soc.* 122: 2395–2396.

15 Ahrendt, K.A., Borths, C.J., and MacMillan, D.W.C. (**2000**). *J. Am. Chem. Soc.* 122: 4243–4244.

16 List, B. (**2006**). *Chem. Commun.* 42: 819–824.

17 Bencivenni, G., Wu, L.-Y., Mazzanti, A. et al. (**2009**). *Angew. Chem. Int. Ed.* 48: 7200–7203.

18 Lan, Y.-B., Zhao, H., Liu, Z.-M. et al. (**2011**). *Org. Lett.* 13: 4866–4869.

19 Ramachary, D.B., Venkaiah, C., and Madhavachary, R. (**2013**). *Org. Lett.* 15: 3042–3045.

20 Halskov, K.S., Kniep, F., Lauridsen, V.H. et al. (**2015**). *J. Am. Chem. Soc.* 137: 1685–1691.

21 Jiang, K., Jia, Z.-J., Chen, S. et al. (**2010**). *Chem. Eur. J.* 16: 2852–2856.

22 Zhu, L., Chen, Q., Shen, D. et al. (**2016**). *Org. Lett.* 18: 2387–2390.

23 Jiang, K., Jia, Z.-J., Yin, X. et al. (**2010**). *Org. Lett.* 12: 2766–2769.

24 Ren, W., Wang, X.-Y., Li, J.-J. et al. (**2017**). *RSC Adv.* 7: 1863–1868.

25 Jia, Z.-J., Jiang, H., Li, J.-L. et al. (**2011**). *J. Am. Chem. Soc.* 133: 5053–5061.

26 Liu, Y., Nappi, M., Arceo, E. et al. (**2011**). *J. Am. Chem. Soc.* 133: 15212–15218.

27 (a) Zhou, Z., Wang, Z.-X., Zhou, Y.-C. et al. (**2017**). *Nat. Chem.* 9: 590–594. (b) Chen, P.-Q., Xiao, Y.-C., Yue, C.-Z., and Chen, Y.-C. (**2014**). *Org. Chem. Front.* 1: 490–493. (c) Zhou, Q.-Q., Yuan, X., Xiao, Y.-C. et al. (**2013**). *Tetrahedron* 69: 10369–10374.

28 (a) Halskov, K.S., Johansen, T.K., Davis, R.L. et al. (**2012**). *J. Am. Chem. Soc.* 134: 12943–12946. (b) Stiller, J., Kowalczyk, D., Jiang, H. et al. (**2014**). *Chem. Eur. J.* 20: 13108–13112.

29 Stiller, J., Poulsen, P.H., Cruz, D.C. et al. (**2014**). *Chem. Sci.* 5: 2052–2056.

30 (a) Cui, H.-L. and Tanaka, F. (**2013**). *Chem. Eur. J.* 19: 6213–6216.(b) Cui, H.-L., Chouthaiwale, P.V., Yin, F., and Tanaka, F. (**2016**). *Org. Biomol. Chem.* 14: 1777–1783.

31 Companyó, X., Zea, A., Alba, A.-N.R. et al. (**2010**). *Chem. Commun.* 46: 6953–6955.

32 Silvi, M., Chatterjee, I., Liu, Y., and Melchiorre, P. (**2013**). *Angew. Chem. Int. Ed.* 52: 10780–10783.

33 Bergonzini, G. and Melchiorre, P. (**2012**). *Angew. Chem. Int. Ed.* 51: 971–974.

34 Noole, A., Ošeka, M., Pehk, T. et al. (**2013**). *Adv. Synth. Catal.* 355: 829–835.

35 César da Silva, R., Chatterjee, I., Escudero-Adán, E. et al. (**2014**). *Asian J. Org. Chem.* 3: 466–469.

36 Albertshofer, K., Anderson, K.E., and Barbas, C.F. III (**2012**). *Org. Lett.* 14: 5968–5971.

37 Tian, X. and Melchiorre, P. (**2013**). *Angew. Chem. Int. Ed.* 52: 5360–5363.

38 Sun, W., Zhu, G., Wu, C. et al. (**2012**). *Chem. Eur. J.* 18: 13959–13963.

39 Sun, B.-B., Zhang, Y.-F., Zhang, J.-Q. et al. (**2017**). *Eur. J. Org. Chem.* 2017: 2871–2877.

40 (a) Fu, G.C. (**2004**). *Acc. Chem. Res.* 37: 542–547. (b) Wurz, R.P. (**2007**). *Chem. Rev.* 107: 5570–5595. (c) Marinetti, A. and Voituriez, A. (**2010**). *Synlett* 21: 174–194.

41 Voituriez, A., Pinto, N., Neel, M. et al. (**2010**). *Chem. Eur. J.* 16: 12541–12544.

42 Zhong, F., Han, X., Wang, Y., and Lu, Y. (**2012**). *Chem. Sci.* 3: 1231–1234.

43 Han, X., Chan, W.-L., Yao, W. et al. (**2016**). *Angew. Chem. Int. Ed.* 55: 6492–6496.

44 Sankar, M.G., Garcia-Castro, M., Golz, C. et al. (**2016**). *RSC Adv.* 6: 56537–56543.

45 Tan, B., Candeias, N.R., and Barbas, C.F. III (**2011**). *J. Am. Chem. Soc.* 133: 4672–4675.

46 Zhong, F., Han, X., Wang, Y., and Lu, Y. (**2011**). *Angew. Chem. Int. Ed.* 50: 7837–7841.

47 Zhan, G., Shi, M.-L., He, Q. et al. (**2016**). *Angew. Chem. Int. Ed.* 55: 2147–2151.

48 Wang, Y., Liu, L., Zhang, T. et al. (**2012**). *J. Org. Chem.* 77: 4143–4147.

49 Peng, J., Huang, X., Jiang, L. et al. (**2011**). *Org. Lett.* 13: 4584–4587.

50 Wang, K.-K., Jin, T., Huang, X. et al. (**2016**). *Org. Lett.* 18: 872–875.

51 Wang, K.-K., Wang, P., Ouyang, Q. et al. (**2016**). *Chem. Commun.* 52: 11104–11107.

52 Shen, L.-T., Jia, W.-Q., and Ye, S. (**2013**). *Angew. Chem. Int. Ed.* 52: 585–588.

53 Liu, Y.-L., Wang, X., Zhao, Y.-L. et al. (**2013**). *Angew. Chem. Int. Ed.* 52: 13735–13739.

54 Liu, Y.-L., Wang, B.-L., Cao, J.-J. et al. (**2010**). *J. Am. Chem. Soc.* 132: 15176–15178.

55 Wang, Q.-L., Peng, L., Wang, F.-Y. et al. (**2013**). *Chem. Commun.* 49: 9422–9424.

56 (a) Enders, D. and Balensfeier, T. (**2004**). *Acc. Chem. Res.* 37: 534–541. (b) Johnson, J.S. (**2004**). *Angew. Chem. Int. Ed.* 43: 1326–1328. (c) Christmann, M. (**2005**). *Angew. Chem. Int. Ed.* 44: 2632–2634. (d) Zeitler, K. (**2005**). *Angew. Chem. Int. Ed.* 44: 7506–7510. (e) Enders, D., Niemeier, O., and Hensler, A. (**2007**). *Chem. Rev.* 107: 5606–5655.

57 Wang, X.-N., Zhang, Y.-Y., and Ye, S. (**2010**). *Adv. Synth. Catal.* 352: 1892–1895.

58 Dugal-Tessier, J., O'Bryan, E.A., Schroeder, T.B.H. et al. (**2012**). *Angew. Chem. Int. Ed.* 51: 4963–4967.

59 Zheng, C., Yao, W., Zhang, Y., and Ma, C. (**2014**). *Org. Lett.* 16: 5028–5031.

60 Xu, J., Yuan, S., Miao, M., and Chen, Z. (**2016**). *J. Org. Chem.* 81: 11454–11460.

61 Xiao, Z., Yu, C., Li, T. et al. (**2014**). *Org. Lett.* 16: 3632–3635.

62 (a) Que, Y., Li, T., Yu, C. et al. (**2015**). *J. Org. Chem.* 80: 3289–3294. (b) Zhu, L., Yu, C., Li, T. et al. (**2016**). *Org. Biomol. Chem.* 14: 1485–1491.

63 Lin, Y., Yang, L., Deng, Y., and Zhong, G. (**2015**). *Chem. Commun.* 51: 8330–8333.

64 Rong, X., Yao, H., Xia, W. et al. (**2016**). *ACS Comb. Sci.* 18: 220–224.

65 Zhang, H.-M., Gao, Z.-H., and Ye, S. (**2014**). *Org. Lett.* 16: 3079–3081.

66 Xu, J., Yuan, S., Peng, J. et al. (**2017**). *Chem. Commun.* 53: 3430–3433.

67 Lv, H., Tiwari, B., Mo, J. et al. (**2012**). *Org. Lett.* 14: 5412–5415.

68 Wang, C., Zhu, S., Wang, G. et al. (**2016**). *Eur. J. Org. Chem.* 2016: 5653–5658.

69 Jia, W.-Q., Zhang, H.-M., Zhang, C.-L. et al. (**2016**). *Org. Chem. Front.* 3: 77–81.

70 Xie, D., Yang, L., Lin, Y. et al. (**2015**). *Org. Lett.* 17: 2318–2321.

71 Wang, L., Li, S., Blümel, M. et al. (**2016**). *Angew. Chem. Int. Ed.* 55: 11110–11114.

72 Wang, L., Li, S., Blgmel, M. et al. (**2017**). *Chem., Int. Ed.* 56: 8516–8521.

73 Liu, Q., Li, S., Chen, X.-Y. et al. (**2018**). *Org. Lett.* 20: 3622–3626.

74 Zhang, J.-Q., Li, N.-K., Yin, S.-J. et al. (**2017**). *Adv. Synth. Catal.* 359: 1541–1551.

75 Jiang, D., Dong, S., Tang, W. et al. (**2015**). *J. Org. Chem.* 80: 11593–11597.

76 Zhao, M.-X., Zhou, H., Tang, W.-H. et al. (**2013**). *Adv. Synth. Catal.* 355: 1277–1283.

77 Chen, X.-Y., Chen, K.-Q., Sun, D.-Q., and Ye, S. (**2017**). *Chem. Sci.* 8: 1936–1941.

78 Song, Z.-Y., Chen, K.-Q., Chen, X.-Y., and Ye, S. (**2018**). *J. Org. Chem.* 83: 2966–2970.

79 For reviews on bifunctional organocatalysis, see: (a)Bhadury, P.S., Song, B.-A., Yang, S. et al. (**2009**). *Curr. Org. Synth.* 6: 380–399. (b) Lattanzi, A. (**2009**). *Chem. Commun.* 45: 1452–1463. (c) Liu, X., Lin, L., and Feng, X. (**2009**). *Chem. Commun.* 45: 6145–6158. (d) Connon, S.J. (**2008**). *Chem. Commun.* 44:

2499–2510.For reviews on organo/metal catalysis, see: (e) Stegbauer, L., Sladojevich, F., and Dixon, D.J. (**2012**). *Chem. Sci.* 3: 942–958. (f) Paull, D.H., Abraham, C.J., Scerba, M.T. et al. (**2008**). *Acc. Chem. Res.* 41: 655–663.For reviews on multimetallic catalysis: (g) Ma, J.-A. and Cahard, D. (**2004**). *Angew. Chem. Int. Ed.* 43: 4566–4583. (h) Shibasaki, M., Sasai, H., and Arai, T. (**1997**). *Angew. Chem. Int. Ed. Eng.* 36: 1236–1256. (i) Shibasaki, M. and Yoshikawa, N. (**2002**). *Chem. Rev.* 102: 2187–2209. (j) Shibasaki, M., Kanai, M., Matsunaga, S., and Kumagai, N. (**2009**). *Acc. Chem. Res.* 42: 1117–1127. (k) Park, J. and Hong, S. (**2012**). *Chem. Soc. Rev.* 41: 6931–6943. (l) Haak, R.M. and Wezenberg, S.J. (**2010**). *Chem. Commun.* 46: 2713–2723.For book chapter: (m) Yu, J.-S. and Zhou, J. (**2014**, Chap. 3). Asymmetric multifunctional catalysis. In: Multicatalyst System in Asymmetric Catalysis (ed. J. Zhou), 159–289. New York: Wiley-VCH.

80 Li, T.-Z., Jiang, Y., Guan, Y.-Q. et al. (**2014**). *Chem. Commun.* 50: 10790–10792.

81 Guo, W., Wang, X., Zhang, B. et al. (**2014**). *Chem. Eur. J.* 20: 8545–8550.

82 Gao, T.-P., Lin, J.-B., Hu, X.-Q., and Xu, P.-F. (**2014**). *Chem. Commun.* 50: 8934–8936.

83 Han, J.-L. and Chang, C.-H. (**2016**). *Chem. Commun.* 52: 2322–2325.

84 Yin, X.-P., Zeng, X.-P., Liu, Y.-L. et al. (**2014**). *Angew. Chem. Int. Ed.* 53: 13740–13745.

85 Zhao, K., Zhi, Y., Li, X. et al. (**2016**). *Chem. Commun.* 52: 2249–2252.

86 Cheng, P., Guo, W., Chen, P. et al. (**2016**). *Chem. Commun.* 52: 3418–3421.

87 Wei, Q. and Gong, L.-Z. (**2010**). *Org. Lett.* 12: 1008–1011.

88 Sun, W., Hong, L., Zhu, G. et al. (**2014**). *Org. Lett.* 16: 544–547.

89 Monari, M., Montroni, E., Nitti, A. et al. (**2015**). *Chem. Eur. J.* 21: 11038–11049.

90 Zhao, B.-L. and Du, D.-M. (**2016**). *Chem. Commun.* 52: 6162–6165.

91 (a) Yang, W. and Du, D.-M. (**2013**). *Chem. Commun.* 49: 8842–8844.(b) Huang, Y.-M., Zheng, C.-W., and Zhao, G. (**2013**). *RSC Adv.* 3: 16999–17002.

92 Huang, Y., Zheng, C., Chai, Z., and Zhao, G. (**2014**). *Adv. Synth. Catal.* 356: 579–583.

93 Mao, H., Lin, A., Tang, Y. et al. (**2013**). *Org. Lett.* 15: 4062–4065.

94 Zhao, B.-L. and Du, D.-M. (**2015**). *Asian J. Org. Chem.* 4: 1120–1126.

95 Cao, Y., Jiang, X., Liu, L. et al. (**2011**). *Angew. Chem. Int. Ed.* 50: 9124–9127.

96 Wang, L., Shi, X.-M., Dong, W.-P. et al. (**2013**). *Chem. Commun.* 49: 3458–3460.

97 Wu, H., Zhang, L.-L., Tian, Z.-Q. et al. (**2013**). *Chem. Eur. J.* 19: 1747–1753.

98 Tan, F., Cheng, H.-G., Feng, B. et al. (**2013**). *Eur. J. Org. Chem.* 2013: 2071–2075.

99 Wu, C., Jing, L., Qin, D. et al. (**2016**). *Tetrahedron Lett.* 57: 2857–2860.

100 Sun, Q.-S., Zhu, H., Chen, Y.-J. et al. (**2015**). *Angew. Chem. Int. Ed.* 54: 13253–13257.

101 Duan, S.-W., Li, Y., Liu, Y.-Y. et al. (**2012**). *Chem. Commun.* 48: 5160–5162.

102 Sun, W., Zhu, G., Wu, C. et al. (**2013**). *Angew. Chem. Int. Ed.* 52: 8633–8637.

103 Manoni, F. and Connon, S.J. (**2014**). *Angew. Chem. Int. Ed.* 53: 2628–2632.

104 Pesciaioli, F., Righi, P., Mazzanti, A. et al. (**2011**). *Chem. Eur. J.* 17: 2842–2845.

105 Noole, A., Sucman, N.S., Kabeshov, M.A. et al. (**2012**). *Chem. Eur. J.* 18: 14929–14933.

106 Sun, W., Zhu, G., Wu, C. et al. (**2012**). *Chem. Eur. J.* 18: 6737–6741.

107 Zhou, J., Wang, Q.-L., Peng, L. et al. (**2014**). *Chem. Commun.* 50: 14601–14604.

108 Huang, X.-F., Zhang, Y.-F., Qi, Z.-H. et al. (**2014**). *Org. Biomol. Chem.* 12: 4372–4385.

109 Gajulapalli, V.P.R., Vinayagam, P., and Kesavan, V. (**2015**). *RSC Adv.* 5: 7370–7379.

110 Du, T., Du, F., Ning, Y., and Peng, Y. (**2015**). *Org. Lett.* 17: 1308–1311.

111 Chen, W.-B., Wu, Z.-J., Pei, Q.-L. et al. (**2010**). *Org. Lett.* 12: 3132–3135.

112 Jiang, X., Sun, Y., Yao, J. et al. (**2012**). *Adv. Synth. Catal.* 354: 917–925.

113 Xie, J., Xing, W.-L., Sha, F., and Wu, X.-Y. (**2016**). *Eur. J. Org. Chem.* 2016: 3983–3992.

114 Xie, J., Xing, X.-Y., Sha, F. et al. (**2016**). *Org. Biomol. Chem.* 14: 8346–8355.

115 Kumarswamyreddy, N. and Kesavan, V. (**2016**). *Org. Lett.* 18: 1354–1357.

116 Yin, S.-J., Zhang, S.-Y., Zhang, J.-Q. et al. (**2016**). *RSC Adv.* 6: 84248–84254.

117 Tan, B., Candeias, N.R., and Barbas, C.F. III (**2011**). *Nat. Chem.* 3: 473–477.

118 Albertshofer, K., Tan, B., and Barbas, C.F. III (**2012**). *Org. Lett.* 14: 1834–1837.

119 Ji, S.-P., Liu, L.-W., Chen, F. et al. (**2016**). *Eur. J. Org. Chem.* 2016: 5437–5444.

120 Li, X., Li, Y.-M., Peng, F.-Z. et al. (**2011**). *Org. Lett.* 13: 6160–6163.

121 Li, Y.-M., Li, X., Peng, F.-Z. et al. (**2011**). *Org. Lett.* 13: 6200–6203.

122 (a) Han, W.-Y., Li, S.-W., Wu, Z.-J. et al. (**2013**). *Chem. Eur. J.* 19: 5551–5556. (b) Chen, Q., Liang, J., Wang, S. et al. (**2013**). *Chem. Commun.* 49: 1657–1659. (c) Kayal, S. and Mukherjee, S. (**2016**). *Org. Biomol. Chem.* 14: 10175–10179. (d) Zhao, J.-Q., Zhou, M.-Q., Wu, Z.-J. et al. (**2015**). *Org. Lett.* 17: 2238–2241. (e) Zhao, H.-W., Tian, T., Pang, H.-L. et al. (**2016**). *Adv. Synth. Catal.* 358: 2619–2630. (f) Cao, Y.-M., Shen, F.-F., Zhang, F.-T., and Wang, R. (**2013**). *Chem. Eur. J.* 19: 1184–1188. (g) Du, D., Jiang, Y., Xu, Q. et al. (**2016**). *ChemistryOpen* 5: 311–314. (h) Kayal, S. and Mukherjee, S. (**2014**). *Eur. J. Org. Chem.* 2014: 6696–6700. (i) Liu, L., Zhao, B.-L., and Du, D.-M. (**2016**). *Eur. J. Org. Chem.* 2016: 4711–4718. (j) Chowdhury, R., Kumar, M., and Ghosh, S.K. (**2016**). *Org. Biomol. Chem.* 14: 11250–11260. (k) Lin, Y., Liu, L., and Du, D.-M. (**2017**). *Org. Chem. Front.* 4: 1229–1238. (l) Fu, Z.-K., Pan, J.-Y., Xu, D.-C., and Xie, J.-W. (**2014**). *RSC Adv.* 4: 51548–51557.

123 Liu, X.-L., Han, W.-Y., Zhang, X.-M., and Yuan, W.-C. (**2013**). *Org. Lett.* 15: 1246–1249.

124 Du, D., Jiang, Y., Xu, Q., and Shi, M. (**2013**). *Adv. Synth. Catal.* 355: 2249–2256.

125 Kayal, S. and Mukherjee, S. (**2015**). *Org. Lett.* 17: 5508–5511.

126 Jiang, X.-L., Liu, S.-J., Gu, Y.-Q. et al. (**2017**). *Adv. Synth. Catal.* 359: 3341–3346.

127 Dou, X. and Lu, Y. (**2012**). *Chem. Eur. J.* 18: 8315–8319.

128 Zhou, B., Yang, Y., Shi, J. et al. (**2013**). *J. Org. Chem.* 78: 2897–2907.

129 Zhao, S., Lin, J.-B., Zhao, Y.-Y. et al. (**2014**). *Org. Lett.* 16: 1802–1805.

130 Yao, L., Liu, K., Tao, H.-Y. et al. (**2013**). *Chem. Commun.* 49: 6078–6080.

131 Tan, B., Hernández-Torres, G., and Barbas, C.F. III (**2011**). *J. Am. Chem. Soc.* 133: 12354–12357.

132 Ren, J.-W., Wang, J., Xiao, J.-A. et al. (**2017**). *J. Org. Chem.* 82: 6441–6449.

133 Shi, Y., Lin, A., Mao, H. et al. (**2013**). *Chem. Eur. J.* 19: 1914–1918.

134 Wu, M.-Y., He, W.-W., Liu, X.-Y., and Tan, B. (**2015**). *Angew. Chem. Int. Ed.* 54: 9409–9413.

135 (a) Akiyama, T. (**2007**). *Chem. Rev.* 107: 5744–5758. (b) Akiyama, T., Itoh, J., and Fuchibe, K. (**2006**). *Adv. Synth. Catal.* 348: 999–1010. (c) Parmar, D., Sugiono, E., Raja, S., and Rueping, M. (**2014**). *Chem. Rev.* 114: 9047–9153. (d) Rueping, M., Kuenkel, A., and Atodiresei, I. (**2011**). *Chem. Soc. Rev.* 40: 4539–4549. (e) You, S.-L., Cai, Q., and Zeng, M. (**2009**). *Chem. Soc. Rev.* 38: 2190–2201.

136 Akiyama, T., Itoh, J., Yokota, K., and Fuchibe, K. (**2004**). *Angew. Chem. Int. Ed.* 43: 1566–1568.

137 Uraguchi, D. and Terada, M. (**2004**). *J. Am. Chem. Soc.* 126: 5356–5357.

138 Cheng, X., Vellalath, S., Goddard, R., and List, B. (**2008**). *J. Am. Chem. Soc.* 130 (47): 15786–15787.

139 Chen, X.-H., Wei, Q., Luo, S.-W. et al. (**2009**). *J. Am. Chem. Soc.* 131: 13819–13825.

140 Wang, Y.-M., Zhang, H.-H., Li, C. et al. (**2016**). *Chem. Commun.* 52: 1804–1807.

141 Shi, F., Tao, Z.-L., Luo, S.-W. et al. (**2012**). *Chem. Eur. J.* 18: 6885–6894.

142 Shi, F., Zhu, R.-Y., Liang, X., and Tu, S.-J. (**2013**). *Adv. Synth. Catal.* 355: 2447–2458.

143 Dai, W., Jiang, X.-L., Wu, Q. et al. (**2015**). *J. Org. Chem.* 80: 5737–5744.

144 Wang, C.-S., Zhu, R.-Y., Zheng, J. et al. (**2015**). *J. Org. Chem.* 80: 512–520.

145 Zhu, G., Wang, B., Bao, X. et al. (**2015**). *Chem. Commun.* 51: 15510–15513.

146 Hong, L., Kai, M., Wu, C. et al. (**2013**). *Chem. Commun.* 49: 6713–6715.

147 Wang, Y., Tu, M.-S., Yin, L. et al. (**2015**). *J. Org. Chem.* 80: 3223–3232.

148 Tan, W., Li, X., Gong, Y.-X. et al. (**2014**). *Chem. Commun.* 50: 15901–15904.

149 Dai, W., Lu, H., Li, X. et al. (**2014**). *Chem. Eur. J.* 20: 11382–11389.

150 Shi, F., Zhu, R.-Y., Dai, W. et al. (**2014**). *Chem. Eur. J.* 20: 2597–2604.

151 Fan, T., Zhang, H.-H., Li, C. et al. (**2016**). *Adv. Synth. Catal.* 358: 2017–2031.

152 Shi, F., Zhang, H.-H., Sun, X.-X. et al. (**2015**). *Chem. Eur. J.* 21: 3465–3471.

153 Zhu, G., Bao, G., Li, Y. et al. (**2017**). *Angew. Chem. Int. Ed.* 56: 5332–5335.

154 (a) Maruoka, K. and Ooi, T. (**2003**). *Chem. Rev.* 103: 3013–3028. (b) Lygo, B. and Andrews, B.I. (**2004**). *Acc. Chem. Res.* 37: 518–525. (c) Hashimoto, T. and Maruoka, K. (**2007**). 5656–5682. *Chem. Rev.* 107: 12.

155 Dolling, U.-H., Davis, P., and Grabowski, E.J.J. (**1984**). *J. Am. Chem. Soc.* 106: 446–447.

156 Zhang, J., Cao, D., Wang, H. et al. (**2016**). *J. Org. Chem.* 81: 10558–10568.

157 Zhang, J.-X., Wang, H.-Y., Jin, Q.-W. et al. (**2016**). *Org. Lett.* 18: 4774–4777.

158 Xiang, B., Belyk, K.M., Reamer, R.A., and Yasuda, N. (**2014**). *Angew. Chem. Int. Ed.* 53: 8375–8378.

159 Wu, H., He, Y.-P., Xu, L. et al. (**2014**). *Angew. Chem. Int. Ed.* 53: 3466–3469.

160 Cao, Y., Zhang, X., Lin, G. et al. (**2016**). *Org. Lett.* 18: 5580–5583.

161 Zhang, D.-Y., Xu, L., Wu, H., and Gong, L.-Z. (**2015**). *Chem. Eur. J.* 21: 10314–10317.

6

Spirooxindole Synthesis by Organometallic Processes

Albert Moyano

Secció de Química Orgànica, Departament de Química Inorgànica i Orgànica, Facultat de Química, Universitat de Barcelona, Barcelona, Catalonia, Spain

6.1 Introduction

The spirooxindole structural subunit is widely distributed among alkaloid natural products [1] and synthetic compounds [2], usually endowed with interesting biological activities (Figure 6.1). The synthesis of spirocyclic oxindoles has accordingly been the object of considerable attention, centered mainly on the development of both metal- and organocatalyzed methods [3]. In this chapter, we will focus our attention in the former type of methodologies.

We will describe some of the most relevant metal-promoted methodologies for the direct construction of the spirooxindole moiety, classified according to the type and number of the precursors: (i) direct construction of the spirooxindole unit by cyclization reactions, (ii) two-component annulation or cycloaddition reactions of oxindole or isatin derivatives, and (iii) miscellaneous methods, including three-component reactions.

6.2 Direct Construction of the Spirooxindole Unit by Cyclization Methodologies

6.2.1 Intramolecular Heck Reaction and Related Cyclizations

The use of the intramolecular Heck reaction for the synthesis of spirocyclic oxindoles was pioneered by Overman in the 1990s. In the course of their highly innovative synthetic approach to (−)-spirotryprostratin B [4], Overman and Rosen constructed the spirocyclic core of this compound by two Pd-catalyzed tandem reactions. The triene cyclization substrate **1**, upon exposure to a catalytic amount of $Pd_2(dba)_3$ in the presence of $(o\text{-tol})_3P$, underwent an intramolecular Heck

Spiro Compounds: Synthesis and Applications, First Edition. Edited by Ramon Rios Torres.

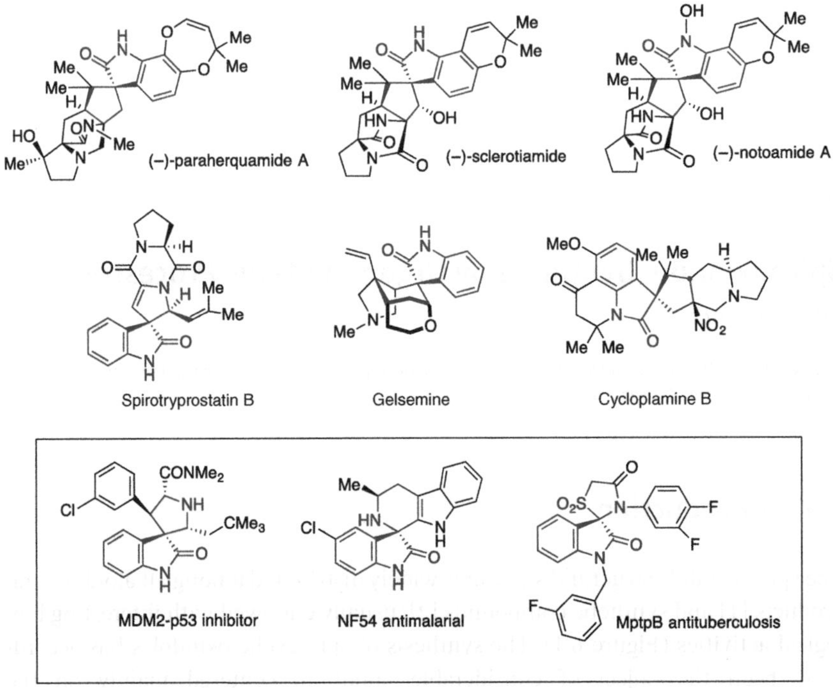

Figure 6.1 Some representative examples of biologically active natural and synthetic spirooxindoles. *Sources:* Based on Marti and Carreira [1a]; Galliford and Scheidt [1b]; Yang et al. [1c]; Trost and Brennan [1d]; Zhou et al. [1e]; Rios [1f]; Fensome et al. [2a]; Tokunaga et al. [2b]; Badillo et al. [2c]; Cao et al. [2d].

insertion, and the diastereomeric pair of intermediate π-allyl Pd-complexes was trapped intramolecularly by the nitrogen atom of the diketopiperazine moiety to afford a 1 : 1 mixture of the pentacyclic compounds **2** and **3**, in 72% overall yield. Deprotection of the oxindole nitrogen transformed **2** into (−)-spirotryprostratin B in 93% yield (Scheme 6.1).

During their exploratory investigations [5] of the intramolecular Heck reactions of (*E*)-α,β-unsaturated 2-iodoanilides, Overman et al. had already demonstrated that treatment of iodoanilides such as **4** with $Pd_2(dba)_3$, (*R*)-BINAP, and Ag_3PO_4 as an HI scavenger provided spirooxindoles (e.g. **5**) in excellent chemical yield and with moderate enantioselectivity (Scheme 6.2).

In 2009, Takemoto et al. [6] developed the synthesis of spiro[indoline-3,3′-pyrrolidine] derivatives using the tandem Pd(0)-catalyzed cycloamidation and the intramolecular nucleophilic addition, starting from 2-(butadienyl)phenylcarbamoyl chlorides (Scheme 6.3).

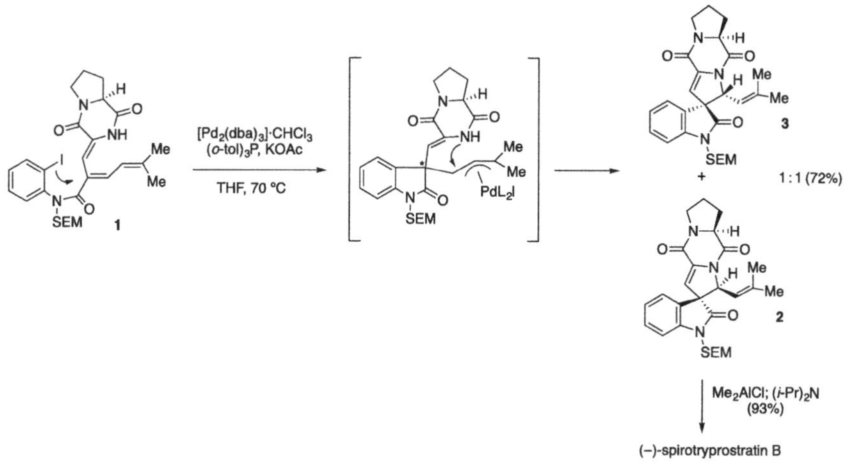

Scheme 6.1 Overman's approach to (−)-spirotryprostratin B.

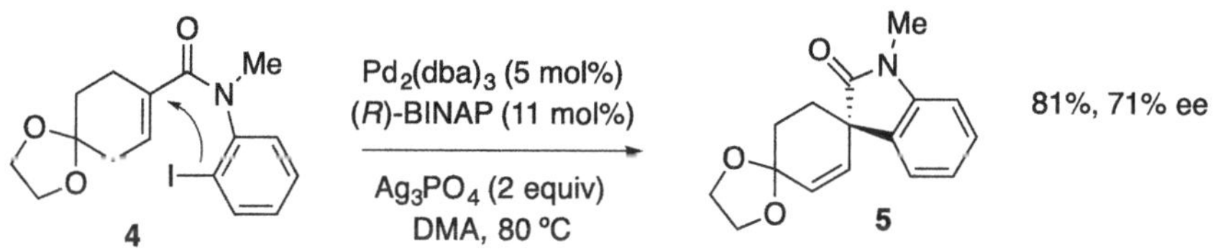

Scheme 6.2 Catalytic asymmetric intramolecular Heck synthesis of a spirocyclic oxindole.

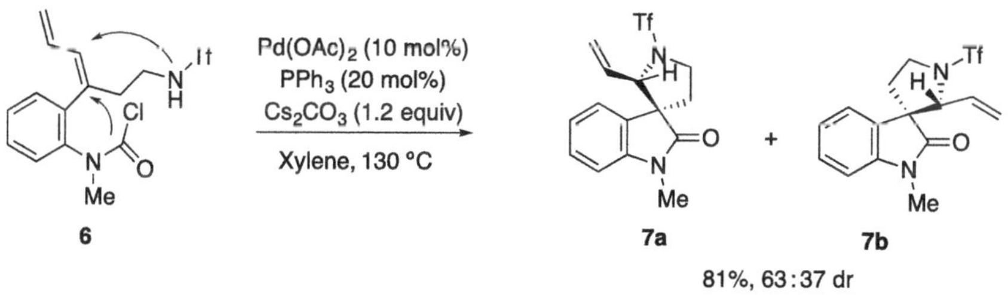

Scheme 6.3 Takemoto's tandem cycloamidation of 2-(butadienyl)phenylcarbamoyl chlorides. *Source:* Modified from Kamisaki et al. [6].

In subsequent studies, Takemoto found that improved conditions for the bicyclization of 2-(butadienyl)phenylcarbamoyl chlorides involved a domino Pd(0)-catalyzed intramolecular Heck reaction and Bi(III)-catalyzed hydroamination, and they applied these tactics to the formal synthesis of the hemiterpene spirooxindole alkaloids elacomine (**8**) and isoelacomine (**9**) in racemic form (Scheme 6.4) [7].

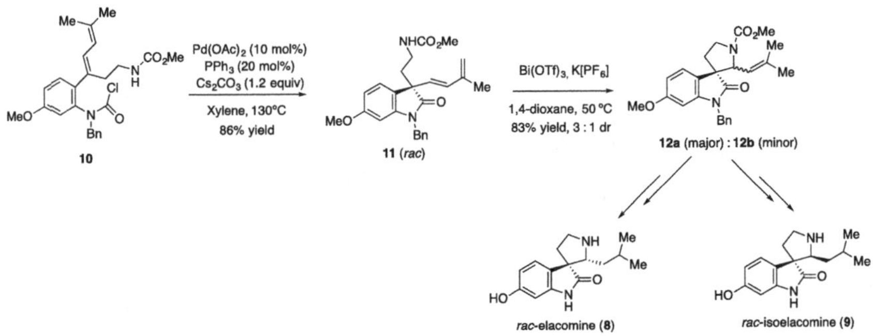

Scheme 6.4 Takemoto's racemic synthesis of spirooxindole alkaloids elacomine and isoelacomine. *Source:* Based on Kamisaki et al. [7].

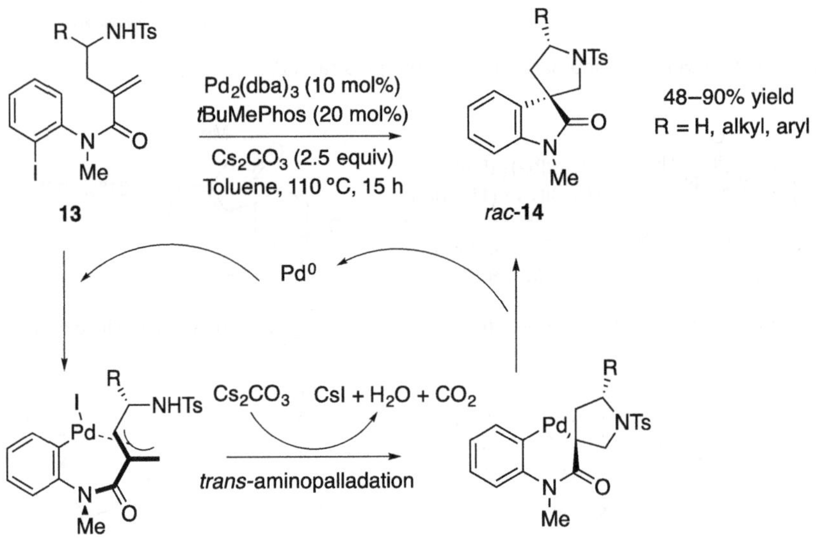

Scheme 6.5 Spirocyclization of α-substituted tosylamines and proposed mechanism.

In 2010, Zhu, Neuville et al. developed a novel palladium-catalyzed, domino spirocyclization process for the formation of biologically relevant spiropyrrolidine-3,3′-oxindoles [8]. Experimental results indicate that both Heck and aminopalladation intramolecular processes were viable pathways from amides and that the preferred route is dependent upon the ligand chosen. However, only the latter process is conducive to the spirocyclization and the use of *t*BuMePhos as the ligand is required for the success of this palladium-catalyzed process. Excellent diastereoselectivity was obtained for cases in which α-substituted terminal sulfonamides **13** were used, a fact that can be accounted for by a *trans*-aminopalladation process (Scheme 6.5).

In the course of a total synthesis of the alkaloid *rac*-gelsemoxonine, Diethelm and Carreira [9] described the successful, totally diastereoselective reductive Heck cyclization of the *N*-(2-bromophenyl)hydroxyamide **15** (Scheme 6.6). It is worth noting that the stereochemical outcome of the addition of the arylpalladium intermediate to the double bond could not be predicted on purely steric grounds, and that the reductive quenching (by formic acid) of the resulting alkylpalladium species could undergo several undesired side reactions, such as reopening of the oxindole ring, β-hydride elimination, and cleavage of the N–O bond. The desired spirocyclic oxindole **16** was subsequently transformed to *rac*-gelsemoxonine after some additional steps involving functional group manipulation.

At the end of 2018, Du, Chen, and coworkers reported an unprecedented auto-tandem cooperative catalysis for Morita–Baylis–Hillman carbonates derived from isatins (**17**) and allylic carbonates using a simple $Pd(PPh_3)_4$ precursor (Scheme 6.7). Dissociated phosphine generates a phosphorus ylide and the Pd leads to a

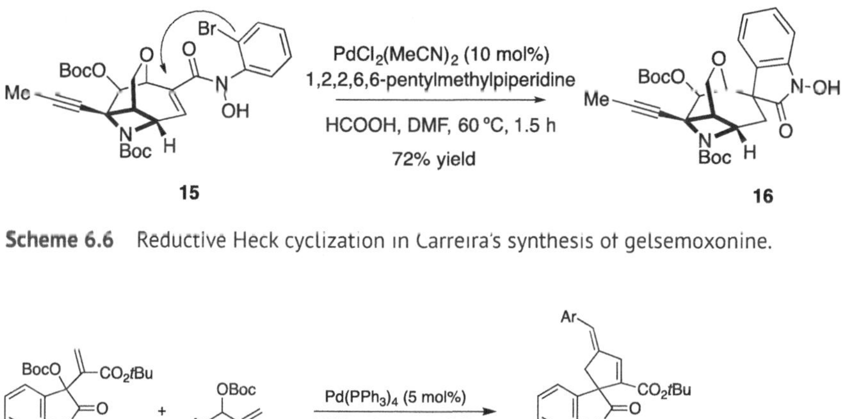

Scheme 6.6 Reductive Heck cyclization in Carreira's synthesis of gelsemoxonine.

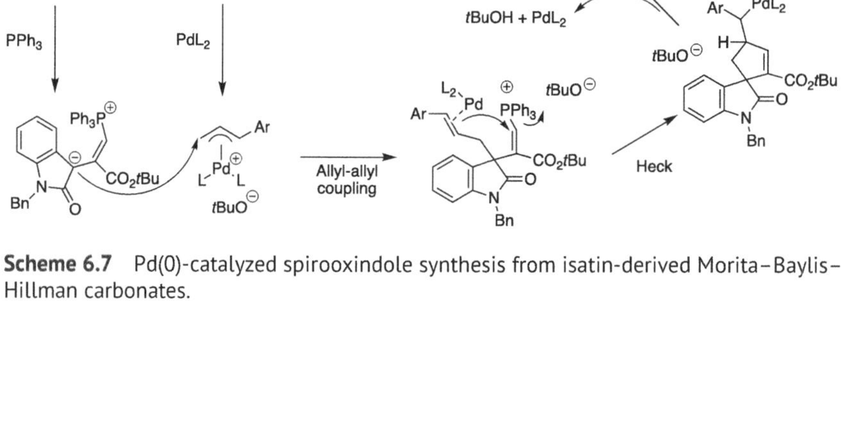

Scheme 6.7 Pd(0)-catalyzed spirooxindole synthesis from isatin-derived Morita–Baylis–Hillman carbonates.

π-allylpalladium complex, and these two intermediates undergo a regioselective allylic alkylation reaction, followed by an intramolecular Heck-type coupling that furnishes spirooxindoles **18** endowed with a 4-methylene-2-cyclopentene subunit [10].

In 2019, Luo, Zhu et al. developed an enantioselective synthesis of 2-oxindole spirofused γ- and δ-lactones (**19**) based on the carbonylative Heck cyclization of *N*-(2-iodophenyl)-*N*-methylacrylamide alcohol derivatives **20** (Scheme 6.8). Interestingly enough, the lactam analogues **21** could also be produced from similar substrates featuring an intramolecular nitrogen nucleophile (**22**, Scheme 6.9). In both instances, yields ranged from moderate to good, and excellent enantioselectivities were achieved with a chiral bidentate phosphine ligand [11].

Charette and coworkers described in 2013 the palladium-catalyzed, silver(I)-mediated intramolecular direct arylation of cyclopropane C–H bonds [12]. Starting from easily accessible *N*-(2-bromoaryl) cyclopropyanecarboxamides **23**, various spiro-3,3′-cyclopropyl oxindoles **24** were obtained in good to excellent yields and with good diastereoselectivities. Mechanistic studies suggest that the formation of a putative Pd-enolate does not take place and that the reaction proceeds via a C–H arylation pathway (Scheme 6.10).

Very recently, Nishino has investigated the Mn(III)-mediated oxidative cyclization of *N*-aryl-2-oxocycloalkane-1-carboxamides **25** [13]. The reaction, that can be described as a formal 5-*exo*-trig cyclization, afforded the expected spiro[cycloalkane-1,3′-indoline]-2,2′-diones **26** in high to quantitative yields (Scheme 6.11).

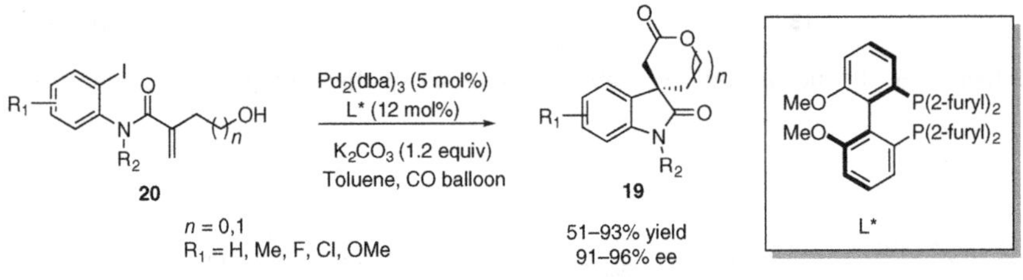

Scheme 6.8 Enantioselective synthesis of 2-oxindole spirofused lactones by a Heck/carbonylative cyclization sequence.

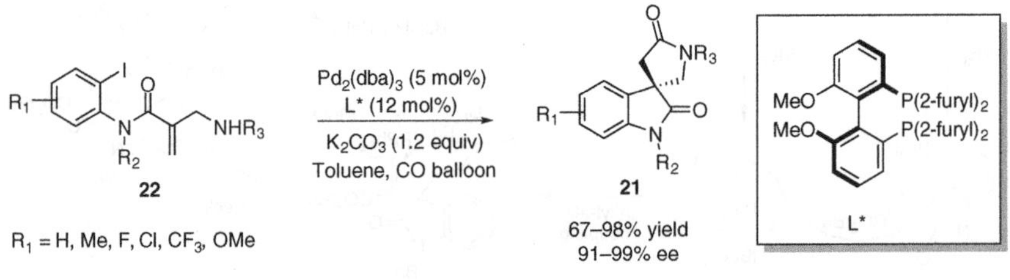

Scheme 6.9 Enantioselective synthesis of 2-oxindole spirofused lactams by a Heck/carbonylative cyclization sequence.

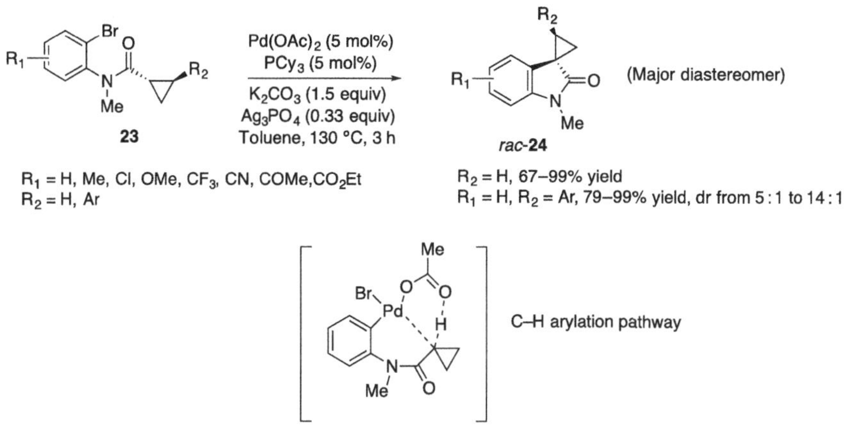

Scheme 6.10 Charette's synthesis of spiro 3,3′-cyclopropyloxindoles by silver-promoted, palladium-catalyzed intramolecular arylation of cyclopropanes.

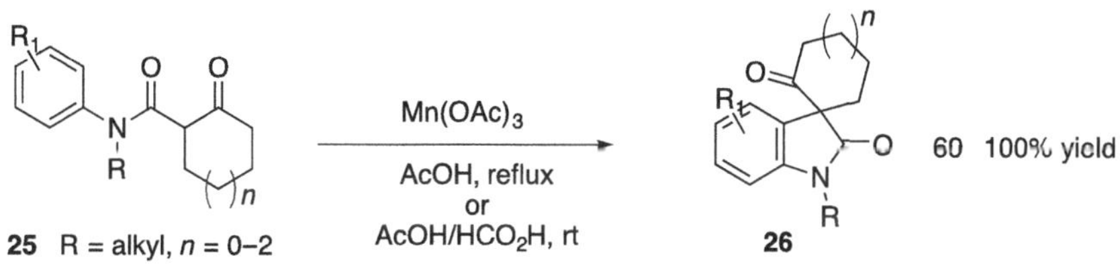

Scheme 6.11 Oxidation of *N*-aryl-2-oxocycloalkane-1-carboxamides with $Mn(OAc)_3$.

6.2.2 Alkylation–Cyclization of 3-Substituted Oxindoles

In 2006, Almendros and coworkers [14] explored the use of 2-indoline-tethered homoallylic alcohols, and (buta-1,3-dien-2-yl)methanols (obtained by addition of unsaturated organoindium compounds to isatins) in the synthesis of oxaspirocyclic oxindoles by Ru-catalyzed ring-closing metathesis. Three 6-membered and one 5-membered oxaspirocyclic oxindoles were obtained in moderate to good yields (57–89% yield).

In the last step of their synthesis of (−)-spirotryprostratin B [15], Trost and Stiles attempted the cyclization of the 3,3′-disubstituted oxindole intermediate **27** (itself obtained by diastereoselective palladium-catalyzed prenylation of a 3-methyleneoxindole) [16] under a variety of conditions using catalytic palladium of copper, to no avail. Finally, they found that the use of an aluminum amide, generated *in situ* by the treatment of **27** with trimethylaluminum, was able to achieve the desired cyclization, affording the target natural product in 45% isolated yield, along with minor quantities of two diastereomers (Scheme 6.12).

At the end of 2014, Córdova and coworkers reported a highly enantioselective synthesis of substituted spirocyclic oxindoles with four new contiguous stereocenters [17]. This remarkable reaction takes place by the highly stereoselective control of a dynamic conjugate addition/intramolecular allylic alkylation relay sequence that relies on a synergistic cooperation between a metallic and an organic catalyst. Thus, when a series of 3-substituted oxindoles **28** were reacted with α,β-unsaturated aldehydes **29** in the presence of catalytic amounts of the L-prolinol derivative **30**, $Pd_2(dba)_3$, and tris(*p*-anisyl)phosphine, the highly functionalized spirocyclic oxindoles **31** were obtained in moderate to good yields, good diastereoselectivities, and excellent enantioselectivities (Scheme 6.13).

The mechanism proposed by the authors (Scheme 6.14) begins with the reversible stereoselective conjugate addition of **28** (*via* its enol tautomer) to the chiral iminium intermediate derived from **29** and **30**, by the less-hindered *Si* face,

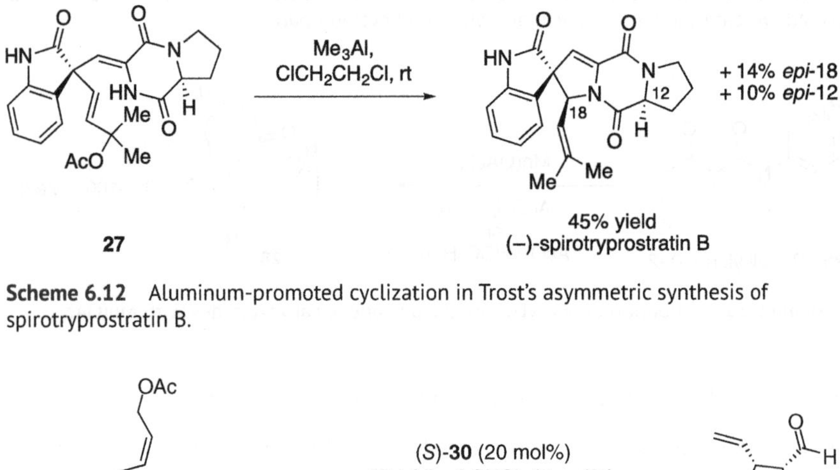

Scheme 6.12 Aluminum-promoted cyclization in Trost's asymmetric synthesis of spirotryprostratin B.

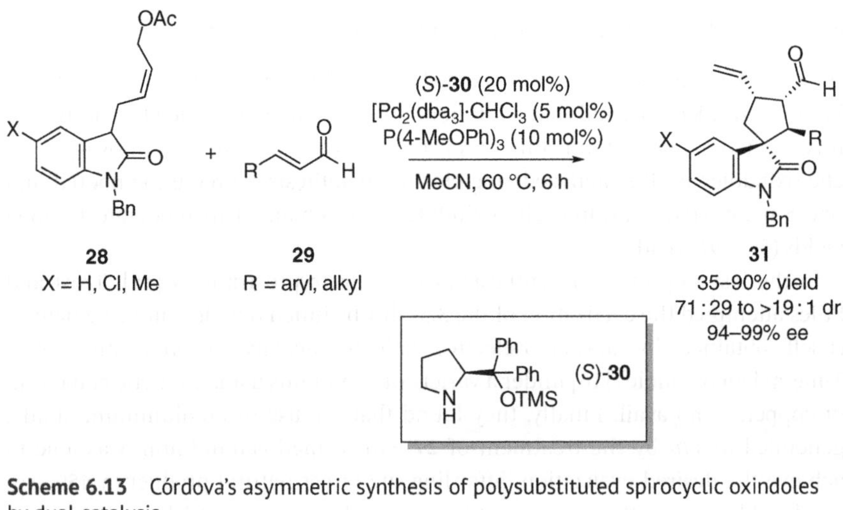

Scheme 6.13 Córdova's asymmetric synthesis of polysubstituted spirocyclic oxindoles by dual catalysis.

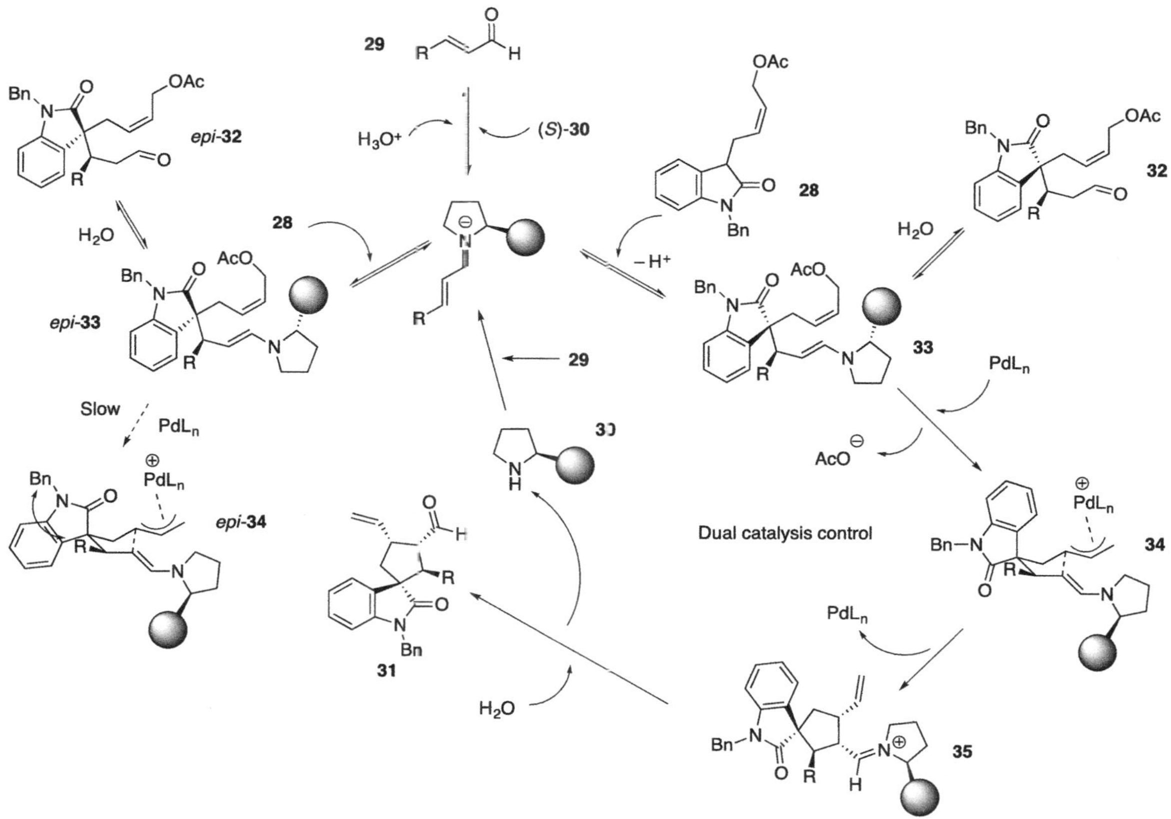

Scheme 6.14 Proposed catalytic cycle for Córdova's asymmetric synthesis of polysubstituted spirocyclic oxindoles.

leading to the predominant formation of the major stereoisomers **32** and *epi*-**32**, that differ in the configuration of the spiranic carbon atom. These two compounds are in equilibrium through the corresponding enamine intermediates **33** and *epi*--**33**, respectively. Next, oxidative addition of the Pd catalyst takes preferentially place on the enamine **33**, delivering the electrophilic π-allyl palladium complex **34**. This complex undergoes an irreversible intramolecular nucleophilic *Si*-facial attack by the chiral enamine, which after Pd decomplexation leads to the spirocyclic iminium intermediate **35**, whose hydrolysis delivers the major isomer of **31** and regenerates the chiral amine catalyst **30**. The attack on **33** is much faster than in *epi*-**33** due to minor steric repulsion experienced by the R substituent in the former intermediate. Thus, the formation of only one of the 16 possible stereoisomers of **31** is kinetically favored, formally through the Pd-catalyzed cyclization of the 3,3′-disubstituted oxindole **32**.

It is worth noting here that Du and Chen's Pd(0)-catalyzed spirooxindole synthesis from isatin-derived Morita–Baylis–Hillman carbonates, discussed above [10], could also have been ascribed to this section, since it takes place through a palladium-catalyzed cyclization of a 3,3′-disubstituted oxindole (Scheme 6.7).

6.2.3 Cyclization of 3-Substituted Indoles

En route to the asymmetric synthesis of 3-aminooxindoles, Xu et al. reported in 2013 an efficient, iron(II)-catalyzed enantioselective intramolecular aminohydroxylation of indoles [18]. Under optimized conditions, *N*-alkoxycarbamates derived from 3-hidroxymethylindoles (**36**) were converted into spirocyclic oxazolidinone-derived 2-alkoxyindolanes **37** with good yields and stereoselectivities (Scheme 6.15). A chiral bis(oxazoline)ligand afforded the highest

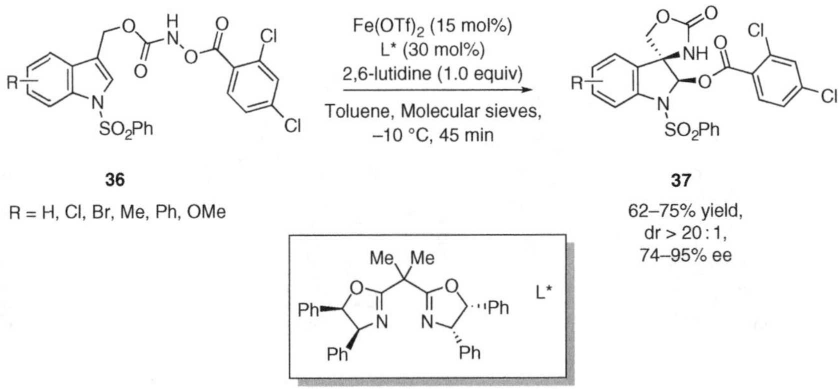

Scheme 6.15 Iron(II)-catalyzed asymmetric intramolecular aminohydroxylation of indoles.

enantiomeric excesses. The racemic version of the reaction, using 1,10-phenantroline as an Fe(II) ligand, also took place in good yields and with essentially complete diastereoselectivity (dr > 20 : 1).

6.3 Two-component Annulation/Cycloaddition Methodologies

6.3.1 Cyclopropyl Spirooxindoles by [2+1] Cycloaddition Reactions

Whereas unsubstituted spiro(cyclopropane-1,3′-oxindoles) are readily obtained by the alkylation of enolates derived from *N*-protected oxindoles with 1,2-dibromopropane [19], the preparation of the corresponding 1,2-disubstituted spirocyclopropanes involves the rhodium-catalyzed cyclopropanation of 3-diazo-2-oxindoles with alkenes. This approach was pioneered by Meyers and Carreira in 2003, in the context of their synthesis of (−)-spirotryprostratin B (Scheme 6.16) [20].

Subsequently, Singh et al. expanded the scope of this reaction [21]. These authors studied the rhodium-catalyzed cyclopropanation of **38** and of other substituted derivatives (**41**) with allyl alcohol and with allyl chloride, and they found that the reaction took place with good yields (Scheme 6.17). The *trans* isomer **42** was formed as the major product in all cases along with a small amount of the *cis*

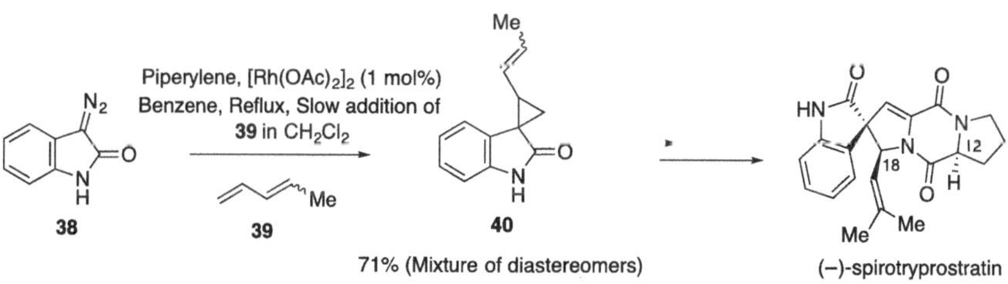

Scheme 6.16 Rhodium(II)-catalyzed cyclopropanation of 3-diazo-2-oxindole **38**. *Source:* Based on Meyers and Carreira [20].

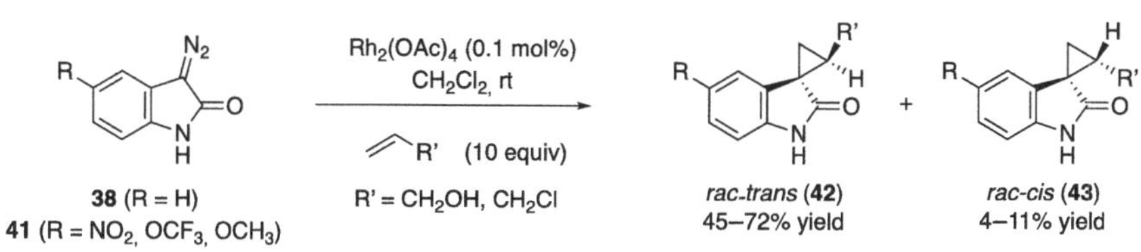

Scheme 6.17 Rhodium(II)-catalyzed cyclopropanation of 3-diazo-2-oxindole **38**.

isomer **43** and alkene insertion product (for allyl alcohol adducts). The pure diastereomers could be easily isolated by silica gel column chromatography.

The catalytic asymmetric synthesis of spiro cyclopropyloxindoles was reported independently and almost simultaneously at the end of 2012 by the research groups of Arai [22] and Zhou [23]. Using a chiral, *tert*-leucine-derived Rh(II) catalyst previously developed by Hashimoto and Ikegami, (Rh_2(S-PPTTL)$_4$) [24], Arai et al. effected the highly diastereoselective *trans*-spirocyclopropanation of **38** (Scheme 6.18). Excellent yields and moderate enantioselectivities (48–66% ee for the *trans* isomer) were achieved with styrenes **44**. In a single example, cyclopropanation of **38** with 1-pentene took place with moderate yield (65%), diminished diastereoselectivity (88 : 12 *trans/cis* dr), and somewhat improved enantioselectivity (74% ee for the *trans* isomer).

Zhou's approach, involving the use of Hg(II) as a catalyst [23], provided both better yields and stereoselectivities. $Hg(OTf)_2$ proved to be the catalyst of choice, which in the presence of equimolar amounts of the chiral bis(phosphine) catalyst **46** mediated the spirocyclopropanation of a variety of 3-diazo-2-oxindoles with styrenes (Scheme 6.19) with moderate to good yields, very high *trans*-diastereoselectivity, and excellent enantioselectivities (up to 99% ee).

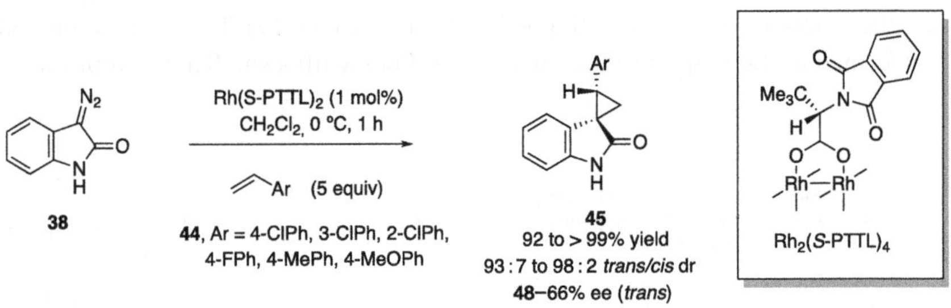

Scheme 6.18 Arai's rhodium(II)-catalyzed asymmetric cyclopropanation of 3-diazo-2-oxindole **38**. *Sources:* Based on Hashimoto et al. [24a]; Tsutsui et al. [24b].

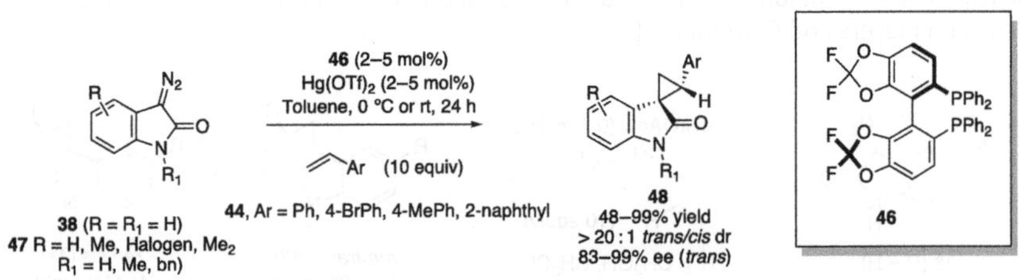

Scheme 6.19 Zhou's mercury(II)-catalyzed asymmetric cyclopropanation of 3-diazo-2-oxindoles.

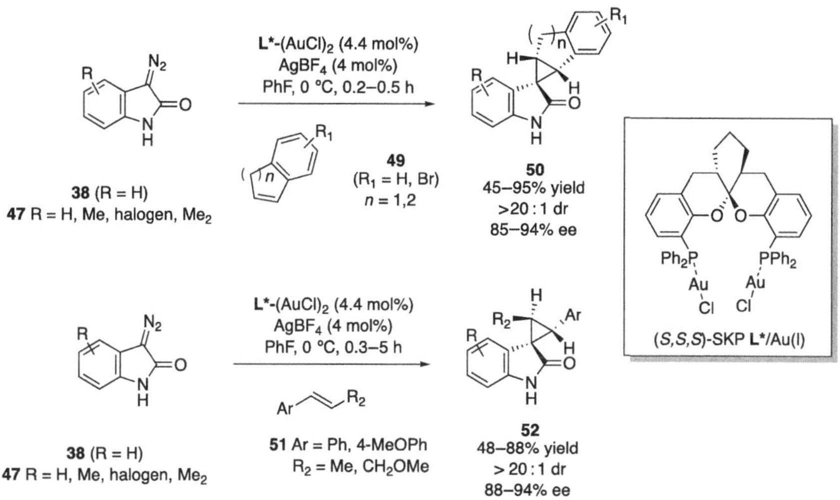

Scheme 6.20 Zhou's gold(I)-catalyzed asymmetric cyclopropanation of 3-diazo-2-oxindoles.

Subsequently, Zhou's group [25] described the use of a spiroketal bisphosphine-derived chiral digold(I) complex [(*S,S,S*)-SKP–$(AuCl)_2$] as a powerful catalyst for the highly diastereo- and enantioselective synthesis of spiro cyclopropyloxindoles from diazooxindoles (**38**, **47**) and a broad range of alkenes, including both *cis*-1,2- (**49**) and *trans*-1,2-disubstituted (**51**) alkenes (Scheme 6.20).

In 2018, Feng and coworkers have shown that the enantioselective cyclopropanation of alkylidene oxindoles **53** with sulfoxonium ylides **54** can be efficiently realized by using a chiral *N,N′*-dioxide/Mg(II) complex as a catalyst [26]. Various chiral spiro cyclopropyl-oxindoles **55**, containing two or three contiguous chiral centers, were obtained in high yields (up to 99%), good diastereoselectivities (up to 97 : 3 dr), and moderate to good enantioselectivities (66–94% ee; Scheme 6.21). The transition state model proposed by the authors explains reasonably well the observed stereochemical outcome of the reaction.

6.3.2 [2+2] Cycloaddition Reactions

Feng and coworkers have also demonstrated that chiral *N,N′*-dioxide/scandium(III) complexes are excellent catalysts for the highly diastereo- and enantioselective [2+2] cycloaddition of disubstituted aryl, alkyl ketenes **56** to *N*-protected isatins **57** (Scheme 6.22) [27]. In the presence of molecular sieves, the reaction takes place in dichloromethane at 30 °C, with very good yields (77–96%) with small amounts of catalyst (0.2–2 mol%). In all cases, only one diastereomer was observed (>99 : 1 dr).

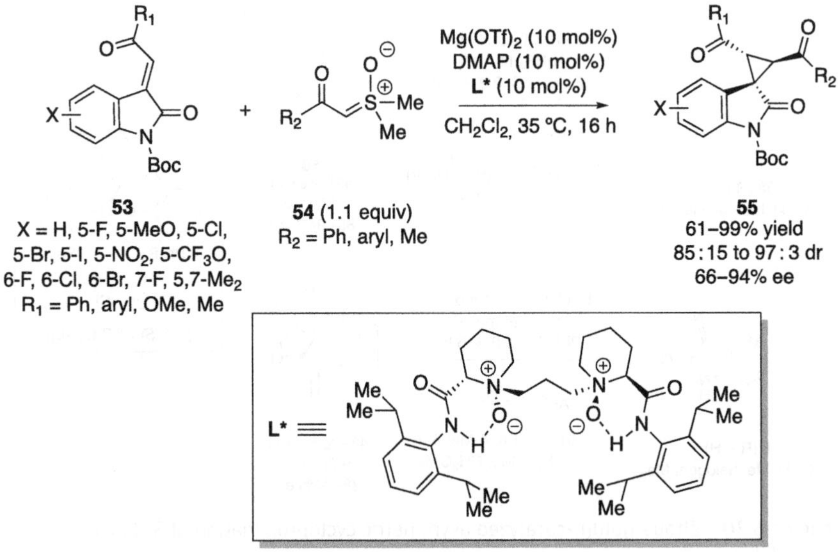

Scheme 6.21 Feng's magnesium(II)-catalyzed asymmetric cyclopropanation of alkylidene oxindoles with sulfoxonium ylides.

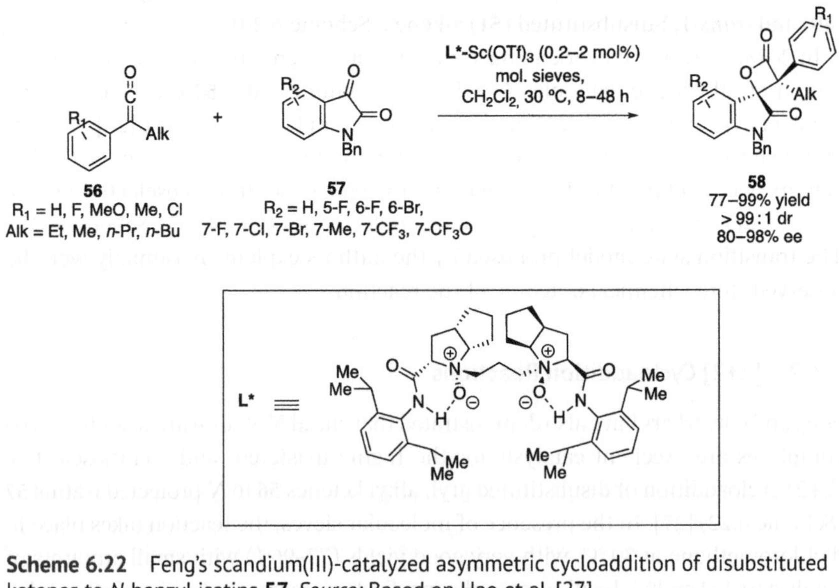

Scheme 6.22 Feng's scandium(III)-catalyzed asymmetric cycloaddition of disubstituted ketenes to *N*-benzyl isatins **57**. *Source:* Based on Hao et al. [27].

For a given ketene, isatin derivatives substituted at the C5 and C6 positions gave somewhat lower levels of enantioselectivity than isatins with substituents at the C7 positions, which with only one exception (7-CF_3 group) afforded the spirocyclic β-lactones **58** with 97–98% ee. The absolute configuration of the adducts, determined by a combination of X-ray diffraction analysis and circular dichroism, was (*S*,*S*) in all instances.

6.3.3 [3+2] Annulation/Cycloaddition Reactions

In 1999, Carreira and coworkers [28] disclosed a facile methodology for the synthesis of spiro[pyrrolidin-3,3′-oxindoles] **59**, based on the magnesium-catalyzed reaction between *N*-benzyl spiro(cyclopropane-1,3′-oxindole) **60** [19] with imines **61** (Scheme 6.23). Whereas *N*-arylsulfonyl imines reacted smoothly at 60 °C, the more electron-rich *N*-allyl imines required reaction temperatures of 80 °C for product formation. The reaction takes place with moderate to excellent yields (57–99% yield), with complete regioselectivity and with variable levels of diastereoselectivity (from 52 : 48 to 98 : 2 dr). In the major diastereomers, the C2 pyrrolidine substituent is *trans* to the oxindole carbonyl, and in general the stereoselectivity can be improved by appropriate selection of the *N*-alkyl protecting group on the oxindole or the *N*-arylsulfonyl group of the aldimine. Carreira's approach appears to be more general than those previously developed by Danishefsky [29] and by Corey [30] to the same type of compounds.

In the most probable mechanistic pathway (Scheme 6.24), iodide opens the cyclopropane ring, furnishing a magnesium enolate that adds to the imine; subsequent cyclization affords the spirocyclic pyrrolidin-oxindole adduct. This ring expansion can be thus formally considered as a [3+2] cycloaddition, in which the cyclopropane ring in **60** acts as an *umpoled* d^2-a^4 synthon, a reactivity pattern that is in turn enabled by selection of MgI_2 as a catalyst possessing dual nucleophilic and electrophilic activity [28].

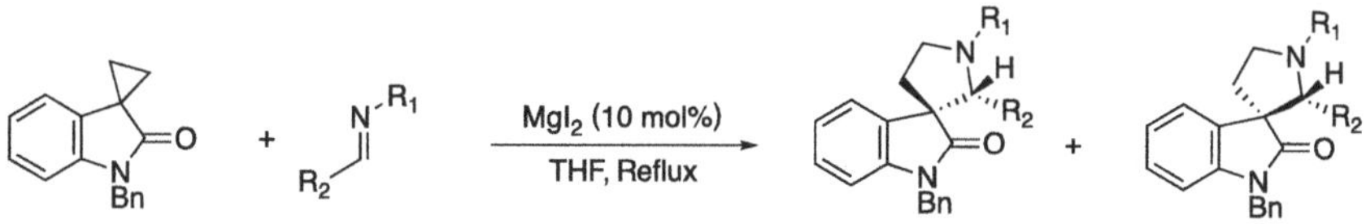

60 **61** **59a** (Major) **59b** (Minor)

R_1,R_2 = –$CH_2CH_2CH_2$–
R_1 = Allyl, *n*Bu, *p*-$TolSO_2$
R_2 = Alkyl, Aryl, Furyl, Alkenyl, Alkynyl

55–99% yield,
52 : 48 to 98 : 2 dr

Scheme 6.23 Carreira's MgI_2-catalyzed ring expansion of spiro(cyclopropane-1,3′-oxindole) **60** to spiro[pyrrolidin-3,3′-oxindoles] **59**.

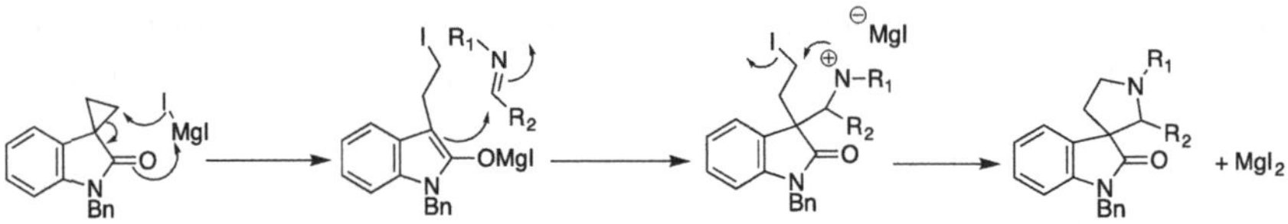

Scheme 6.24 Proposed mechanism for the MgI_2-catalyzed reaction of spiro (cyclopropane-1,3′-oxindoles) with aldimines.

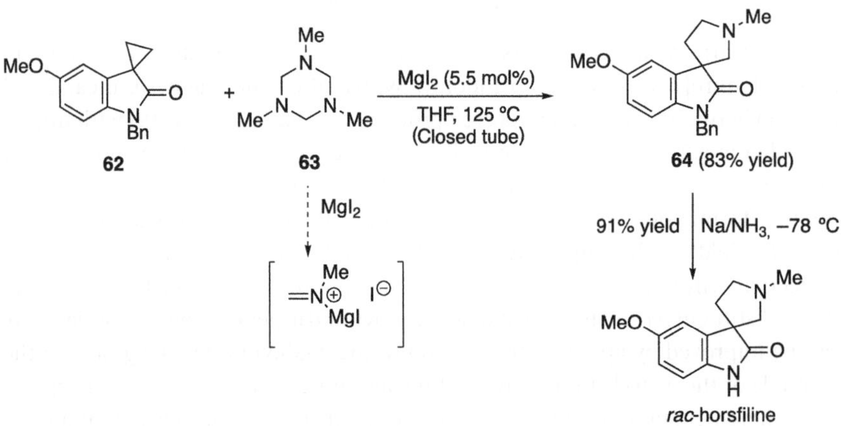

Scheme 6.25 Ring expansion of a spiro(cyclopropane-1,3′-oxindole) in the synthesis of horsfiline. Note the use of triazine **63** as an *in situ* precursor of *N*-methyl formimine. *Source:* Based on Fischer et al. [31].

This methodology was subsequently put to use by Carreira's group in the total syntheses of spiro[pyrrolidin-3,3′-oxindole] alkaloids such as horsfiline (Scheme 6.25) [31] and strychnofoline (Scheme 6.26) [32].

In the course of the synthesis of (−)-spirotryprostratin B, Meyers and Carreira [20, 33] found that although the reaction of aldimines with 1,2-disubstituted spirocyclopropanes such as **40** still takes place with total regioselectivity, leading to 2,5-disubstituted pyrrolidine moieties in the resulting spiro[pyrrolidin-3,3′-oxindoles], although with low diastereoselectivity (Scheme 6.27). The observed regioselectivity implies that the iodide-induced ring-opening of the cyclopropane (see Scheme 6.24) takes place through the breaking of the more substituted bond.

First disclosed by Trost and Chan in 1979 [34], the palladium-catalyzed [3+2] cycloaddition of trimethylenemethane (TMM) is a powerful method for the synthesis of five-membered carbocycles. The carboxylative TMM cycloaddition [35] was used in 2007 by Trost et al. for an efficient approach to the synthesis of *rac*-marcfortine B, a spirocyclic oxindole with a heavily substituted cyclopentane core

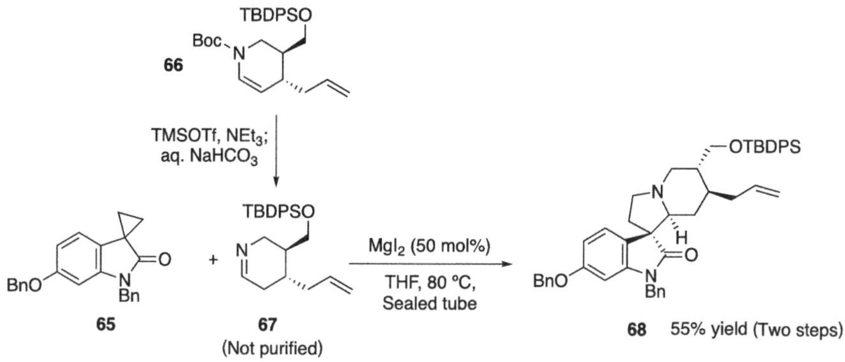

Scheme 6.26 Ring expansion of a spiro(cyclopropane-1,3′-oxindole) in Carreira's synthesis of strychnofoline. *Sources:* Based on Lerchner and Carreira [32a]; Lerchner and Carreira [32b].

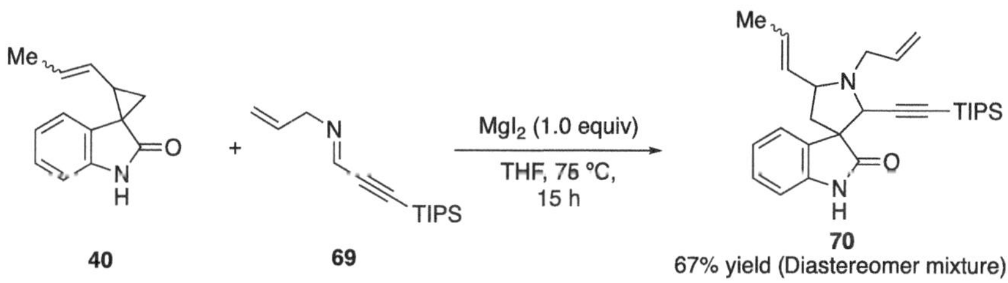

Scheme 6.27 Ring expansion of a spiro(cyclopropane-1,3′-oxindole) in Carreira's synthesis of strychnofoline.

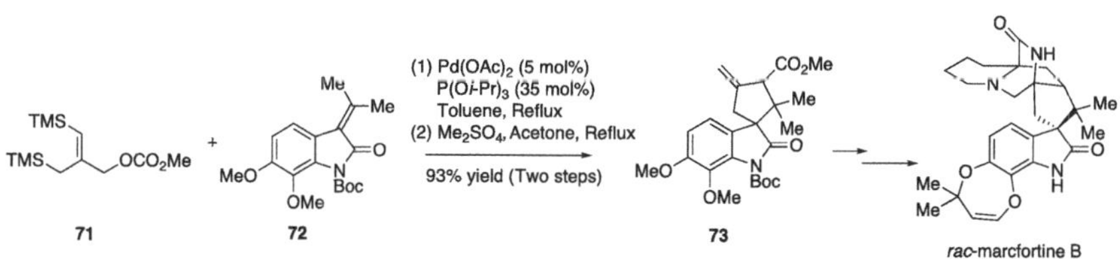

Scheme 6.28 Carboxylative Pd-catalyzed TMM cycloaddition to an isopropylidene oxindole in Trost's synthesis of marcfortine B. *Sources:* Based on Trost et al. [36]; Trost et al. [37].

(Scheme 6.28) [36, 37]. The intermediate carboxylic acid resulting from the [3+2] cycloaddition of the TMM donor **71** to the isopropylidene oxindole **72** was not purified, and was directly converted to the corresponding methyl ester **73** (93% overall yield).

Subsequently, Trost and coworkers developed an enantioselective version of the carboxylative Pd-catalyzed TMM [3+2] cycloaddition that was put to use in the asymmetric synthesis of the related alkaloid (−)-marcfortine C (Scheme 6.29) [37,38].

The first Lewis-acid-catalyzed highly enantioselective synthesis of 3,3′-pyrrolidinyl-spirooxindoles by means of 1,3-dipolar cycloadditions of azomethine ylides to ylidene oxindoles **53** was disclosed in 2010 by Waldmann and coworkers [39]. After extensive exploration, these authors found that optimal conditions involved the use of a combination of Cu(I) with a chiral, ferrocene-derived bidentate (*P,N*)-ligand for the catalysis of the addition of a variety of glycine-derived imines **77** to *N*-unprotected, (*E*)-configured 3-arylidene- or 3-alkylidene oxindoles **53** (Scheme 6.30). The reaction is totally regio- and diastereoselective, and the

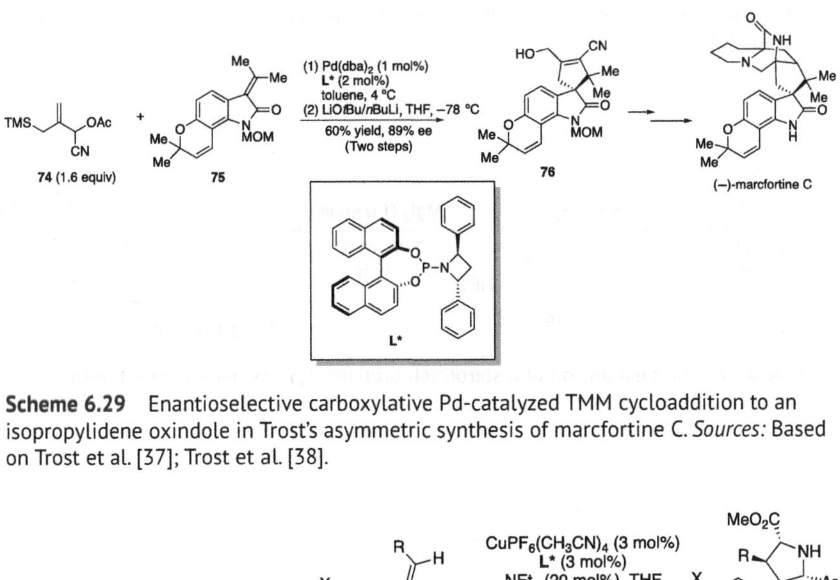

Scheme 6.29 Enantioselective carboxylative Pd-catalyzed TMM cycloaddition to an isopropylidene oxindole in Trost's asymmetric synthesis of marcfortine C. *Sources:* Based on Trost et al. [37]; Trost et al. [38].

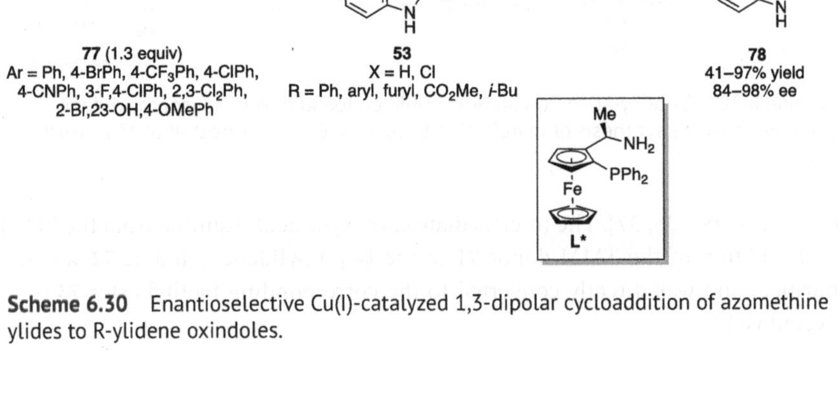

Scheme 6.30 Enantioselective Cu(I)-catalyzed 1,3-dipolar cycloaddition of azomethine ylides to R-ylidene oxindoles.

resulting 3,3′-(2-aryl-5-methoxycarbonyl-pyrrolidinyl)spirooxindoles **80** were obtained with moderate to excellent yields (41–97% yield) and with very good enantioselectivities (84–98% ee).

The regioselectivity of this Cu(I)-catalyzed reaction is opposite to that observed for enantioselective phosphoric-acid-catalyzed 1,3-dipolar cycloadditions of azomethine ylides to *N*-protected methyleneoxindoles [40]. The authors have proposed a working transition-state model that rationalizes both the observed regioselectivity and the steric course of the reaction (Figure 6.2).

Asymmetric 1,3-dipolar cycloadditions of *N*-unprotected (*E*)-alkylidene oxindoles **53** with azomethine ylides **77** were also achieved, almost simultaneously, by Wang and coworkers [41]. These authors used an Ag(I)/TF-BiphamPhos complex as a chiral catalyst. The 3,3′-(5-methoxycarbonyl-pyrrolidinyl)spirooxindole products **78** were obtained with complete diastereoselectivity (>98 : 2 dr), with good yields (up to 95%), but with modest enantioselectivity (Scheme 6.31).

In 2011, Annaliese Franz and coworkers [42] developed a regio- and diastereoselective cyclization between isatins **79** and 5-methoxyoxazoles **80** using titanium (IV) chloride as the catalyst to afford spiro[3,3′-oxindoleoxazolines] **81**/**82** in excellent yield (up to 99%) and diastereoselectivity (up to 98 : 2 dr) (Scheme 6.32). The presence of a secondary alkyl substituent (*i*-Pr) at the C4-position of the oxazole controls nucleophilic attack, providing 3-oxazoline spirocycles (**82**) instead of the 2-oxazoline regioisomers (**71**) produced either with 4-H- or with 4-Me-5-methoxyoxazoles. This reaction can be considered as a formal [3+2] cycloaddition of an azomethine ylide to the C3-carbonyl of the isatin, although the mechanism is stepwise and probably proceeds through nucleophilic attack of the oxazole followed by an intramolecular acyl transfer [41].

Subsequently, the same research group reported the first example of an asymmetric catalytic [3+2] annulation reaction of allyl(triisopropyl)silane **83** with

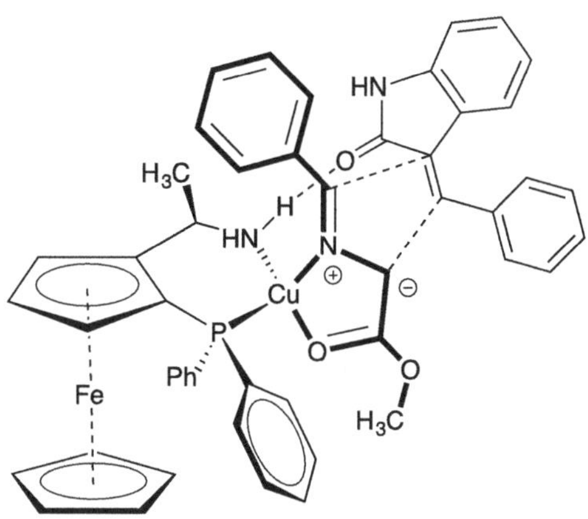

Figure 6.2 Proposed structure of the tetrahedral complex formed by coordination of Cu(I) with the chiral (*P,N*)-ligand and the azomethine ylide, and approach of the dipolarophile (benzylidene oxindole) by the least hindered face.

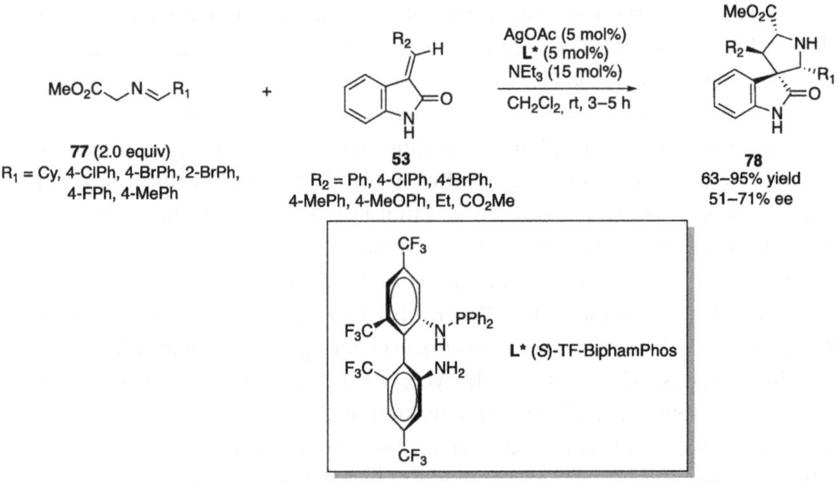

Scheme 6.31 Enantioselective Ag(I)-catalyzed 1,3-dipolar cycloaddition of azomethine ylides to R-ylidene oxindoles.

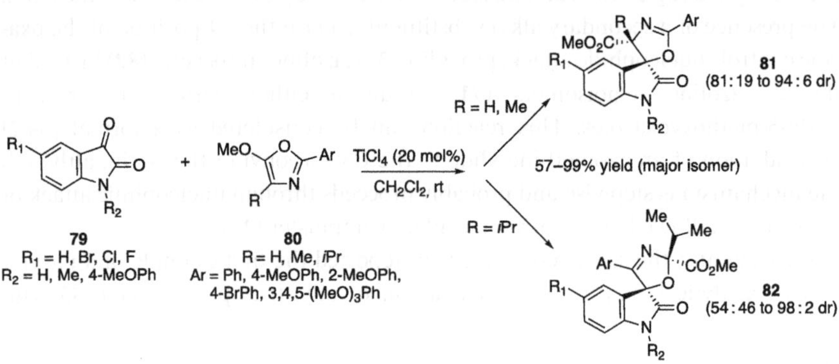

Scheme 6.32 Franz's titanium(IV)-catalyzed stereoselective synthesis of spirooxindole oxazolines.

N-protected isatins **79** [43]. In the presence of a chiral indapybox ligand, the reaction is efficiently catalyzed by scandium(III) salts, furnishing [3′-silyl-2-oxacyclopentane]spirooxindoles **84** with moderate to good yields (up to 82% yield) and with outstanding enantioselectivity (97–99% ee; Scheme 6.33). Interestingly enough, proceeding with the easily oxidable benzhydryl silyl group (allyl benzhydryldimethylsilane **85**), the annulation products can be oxidized *in situ* with TBAF and hydrogen peroxide, furnishing the corresponding 3′-hydroxyspirooxindoles **86** with retention of configuration, in good yields and enantioselectivities (Scheme 6.34).

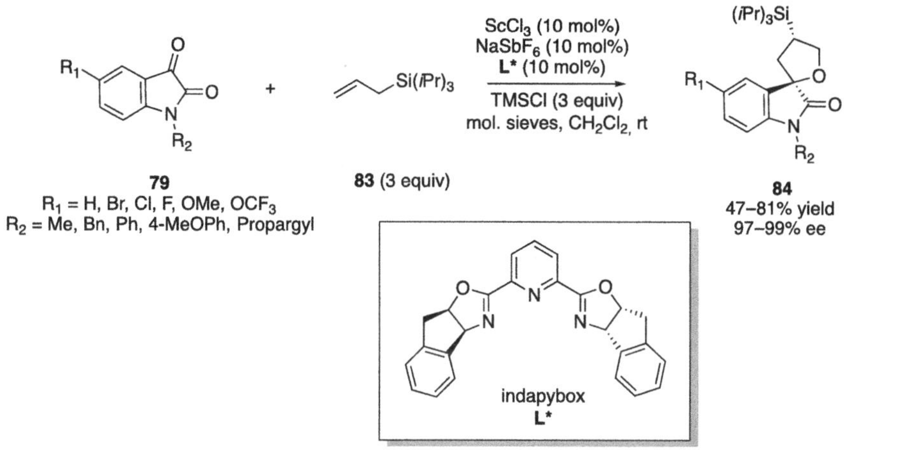

Scheme 6.33 Franz's scandium(III)-catalyzed asymmetric [3+2] annulation of allylsilanes with isatins.

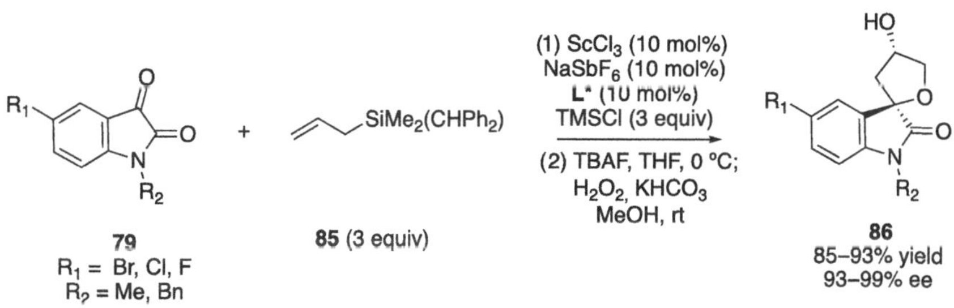

Scheme 6.34 Franz's enantioselective allylsilane annulation and C–Si oxidation (**L*** = indapybox, see Scheme 6.33).

Franz and coworkers have demonstrated that both types of reactions can also be performed with *N*-protected (*E*)-alkylidene oxindoles. In 2014, her research group [44] disclosed that titanium(IV) chloride was an excellent catalyst for the formal [3+2] cyclization between 5-methoxyoxazoles **80** and 3-alkene oxindoles **53**, affording spiro[3,3′-oxindole-1-pyrrolines] **87** in excellent yield (up to 99%) and diastereoselectivity (up to 99 : 1 dr, Scheme 6.35). In one example, the authors showed that the use of a Sc(III)/(*R*,*S*)-indapybox complex (Scheme 6.36) afforded an enantioenriched (up to 72% ee for the major diastereoisomer) spirooxindole-1-pyrroline **88** where a ligand-induced reversal of diastereoselectivity was observed [44].

In the same year, Franz's group reported [45] that the combination Sc(III)/(*R*,*S*)-indapybox ligand also effected the catalytic enantioselective annulation of alkylidene oxindoles **53** with allyl(triisopropyl)silane **83**, providing a convenient

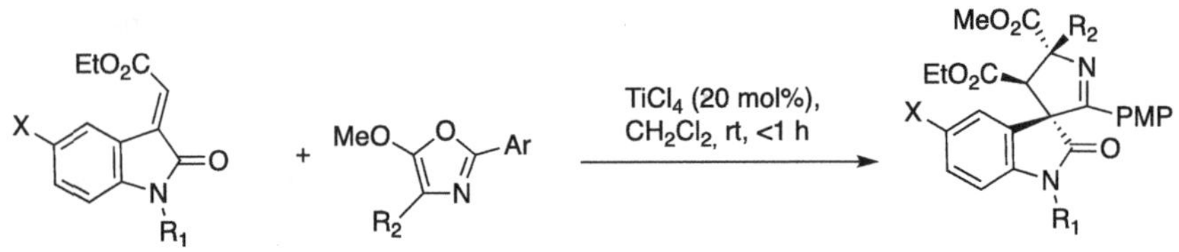

53 (1.5 equiv)
X = H, F
R_1 = Ac, Cbz, Me

80
R_2 = H, Me
Ar = Ph, 4-BrPh, 4-MeOPh

87
69–99% yield
dr generally > 86 : 14

Scheme 6.35 Franz's titanium(IV)-catalyzed diastereoselective synthesis of spirooxindole pyrrolines.

53

80
PMP = 4-MeOPh

88
99% yield
91 : 9 dr
72% ee (Major diastereomer)

Scheme 6.36 Franz's scandium(III)-catalyzed enantioselective synthesis of a spirooxindole pyrroline (**L*** = indapybox, see Scheme 6.33).

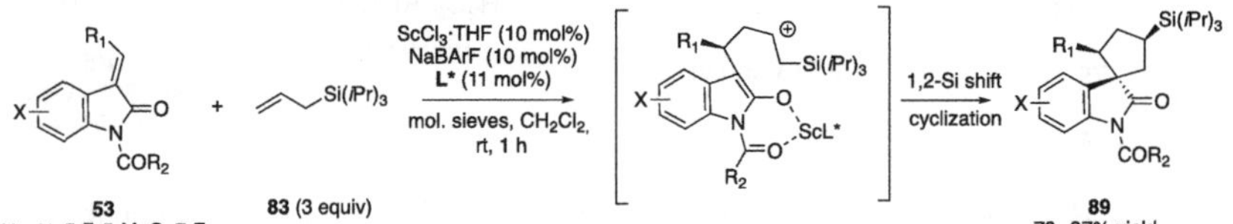

53
X = H, 5-F, 5-MeO, 7-F
R_1 = CO_2Et, CO_2Me, CO_2tBu, Ph, CN
R_2 = Ac, CONHPh, Cbz

83 (3 equiv)

89
72–97% yield
85 : 15 to 99 : 1 dr
76–99% ee (major diastereomer)

Scheme 6.37 Franz's scandium(III)-catalyzed enantioselective [3+2] carboannulation of allylsilanes with alkylidene oxindoles (**L*** = indapybox, see Scheme 6.33).

entry to spiro[3,3′-oxindole-1-cyclopentanes] **89**. The [3+2] carboannulation, that involves a Michael–Sakurai addition followed by a 1,2-allyl shift and cyclization (Scheme 6.37), takes place with remarkable yields (75–97%), diastereoselectivities (>85 : 15 dr), and enantioselectivities (up to 98% ee).

Very recently [46], the same research group has found that the use of allenylsilanes **90** (instead of allylsilanes) in similar reaction conditions gives rise to highly enantioenriched (up to 99% ee) novel cyclopentene-spirooxindoles **91** containing a synthetically useful vinylsilane moiety (Scheme 6.38). In this case, the highest stereoselectivities were obtained with a (*S*)-*tert*-butyl-pybox ligand.

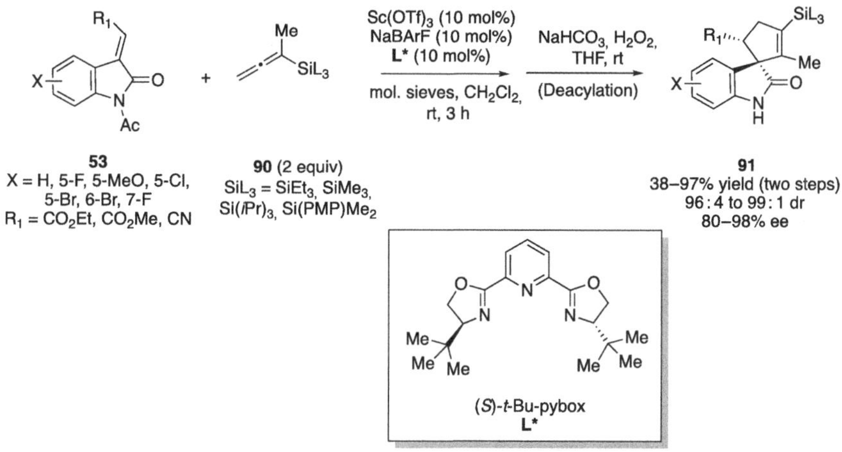

Scheme 6.38 Franz's scandium(III)-catalyzed enantioselective [3+2] carboannulation of allenylsilanes with alkylidene oxindoles.

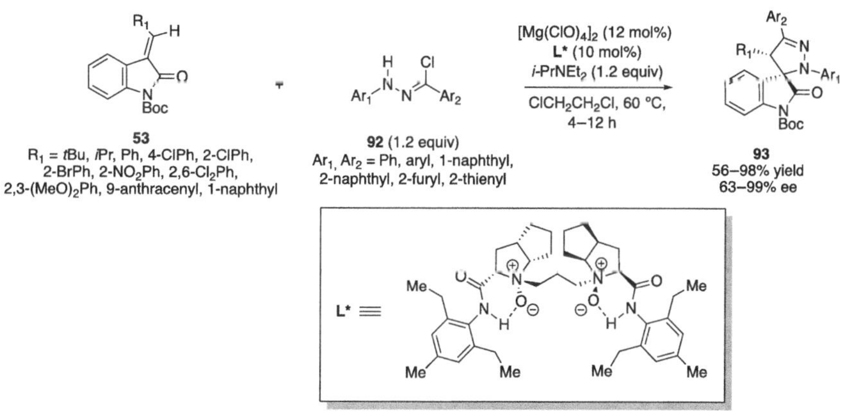

Scheme 6.39 Feng's magnesium(II)-catalyzed enantioselective [3+2] cycloaddition of nitrile imines with alkylidene oxindoles.

(*E*)-3-alkylidene oxindoles **53** (*N*-protected) can undergo an enantioselective 1,3-dipolar cycloaddition with nitrile imines (generated *in situ* from dehydrohalogenation of hydrazonyl halides **92**), affording the pyrazoline-spirooxindoles **93** [47]. After optimization of the reaction conditions, Feng and coworkers found that with a suitable *N,N′*-dioxide ligand, Mg(II) perchlorate was able to catalyze efficiently the [3+2] cycloaddition, which took place with good to excellent (up to 98%) yields, in most instances with complete diastereoselectivity, and with good to excellent enantioselectivity (75–98% ee, Scheme 6.39).

In 2016, Li and coworkers reported an interesting Pd(0)-catalyzed [3+2] cycloaddition between *N*-protected vinyl aziridines **94** and alkylidene oxindoles **53** [48]. The resulting 3,3′-(3-vinyl-pyrrolidinyl)spirooxindole products **95** were obtained with high diastereoselectivity (up to 19 : 1 dr), with good yields (up to 93%), and excellent enantioselectivity (up to 99% ee; Scheme 6.40). The reaction involves the complexation of Pd to the vinyl aziridine, which results in the *in situ* formation of a N^1-1,3-dipole upon aziridine-ring opening and formation of a cationic Pd(II) π-allyl complex. The Michael addition of this dipole to the oxindole **53** is efficiently controlled by a chiral phosphoramidite ligand. Intramolecular asymmetric allylic alkylation of the chiral intermediate furnishes the 3,3′-pyrrolidinyl spirooxindole 95 and regenerates the Pd(0) catalyst (Scheme 6.41).

A mechanistically related transformation, in which vinyl cyclopropanes **96** bearing electron-withdrawing groups react with *N*-protected isatins **79**, has been previously disclosed by Mei et al. [49] In the presence of $Pd_2(dba)_3$ and a chiral imidazoline-phosphine ligand, the resulting 1,3-dipolar species undergoes a [3+2] cycloaddition with the carbonyl group. As a result, highly functionalized oxindole-fused spirotetrahydrofuran derivatives **97** were obtained in good yields (up to 98%), along with good diastereoselectivities (up to 94 : 6 dr) and high enantioselectivities (up to 96% ee), under mild reaction conditions (Scheme 6.42).

In the context of a general rhodium-catalyzed asymmetric [3+2] cycloaddition of vinyl oxirane **98** with imines, Du and coworkers found [50] that *N*-(*p*-methoxyphenyl)isatin imines **99** were excellent substrates for this reaction. A chiral sulfur/alkene ligand was crucial for the success of this reaction and afforded optically active spirooxindole oxazolidinones **100** with moderate to good yields

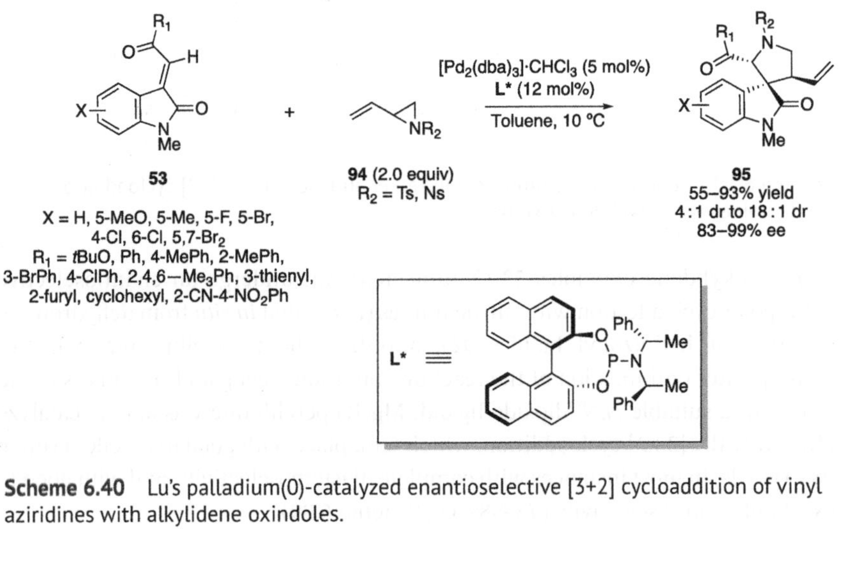

Scheme 6.40 Lu's palladium(0)-catalyzed enantioselective [3+2] cycloaddition of vinyl aziridines with alkylidene oxindoles.

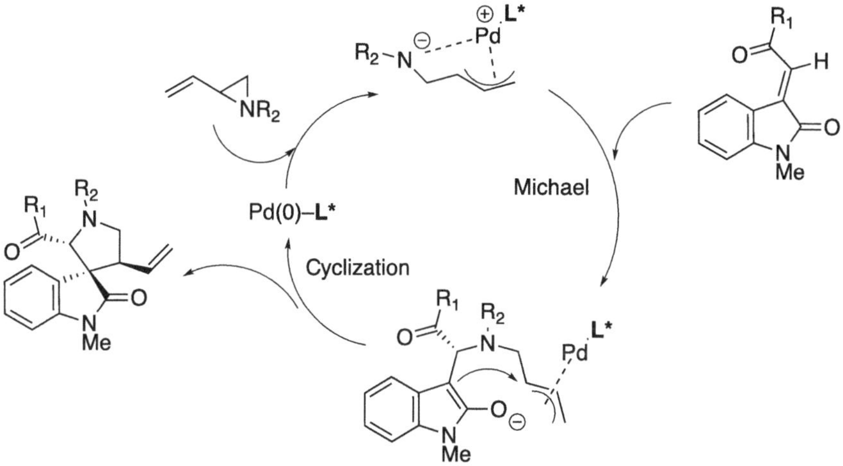

Scheme 6.41 Proposed mechanism for the palladium(0)-catalyzed enantioselective [3+2] cycloaddition of vinyl aziridines with alkylidene oxindoles.

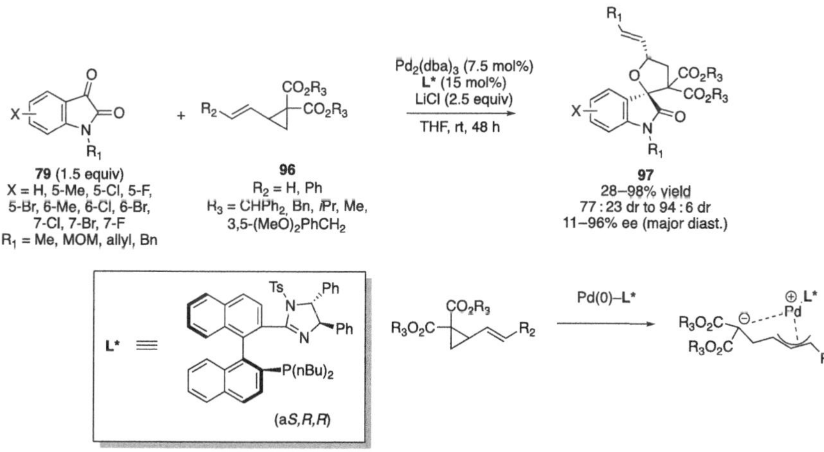

Scheme 6.42 Mei's palladium(0)-catalyzed enantioselective [3+2] cycloaddition of vinyl cyclopropanes with isatins.

(81–91%), good diastereoselectivities (mostly 20 : 1 dr), and good enantioselectivities (78–96% ee; Scheme 6.43). The mechanism of the reaction is unclear, but probably involves a rhodium-complexed O^1-1,3-dipole.

An asymmetric organo- and gold-catalyzed sequential Mannich/hydroamination reaction of *N*-Boc (or *N*-Cbz) imino isatins **101** with propargyl malonodinitrile **102** was disclosed in 2013 by Jiang, Liu, and coworkers [51]. The asymmetric

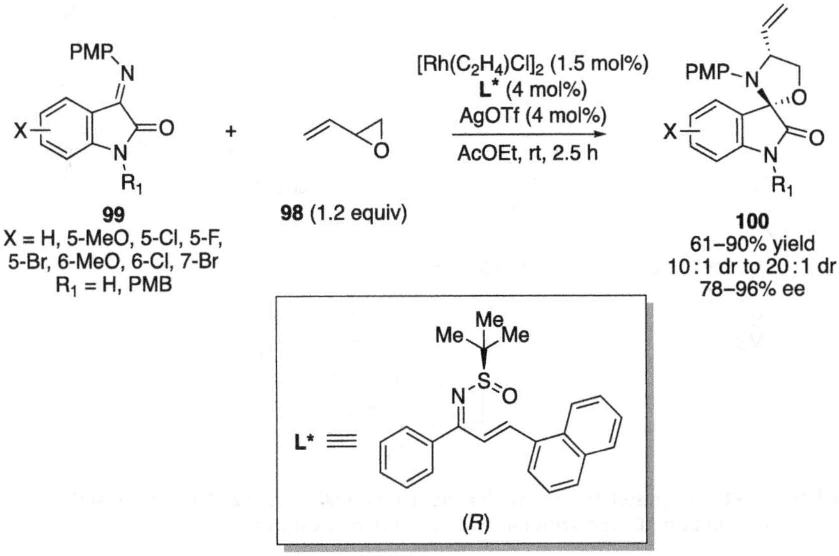

Scheme 6.43 Du's rhodium(0)-catalyzed enantioselective [3+2] cycloaddition of vinyl cyclopropanes with isatins.

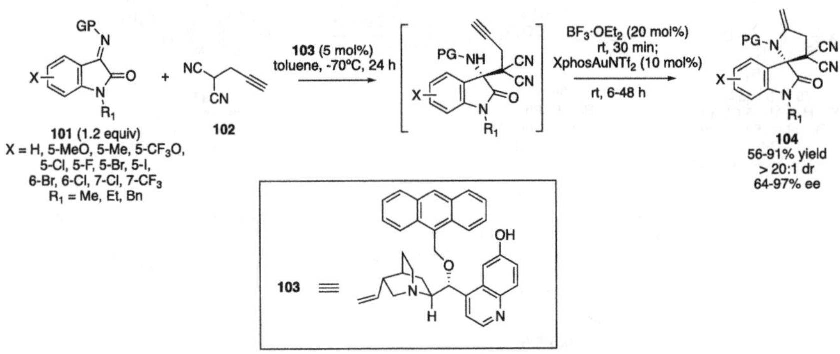

Scheme 6.44 Asymmetric one-pot sequential organo- and gold(I)-catalyzed Mannich/hydroamination reaction of imino isatins, leading to spiro[pyrrolidin-3,2′-oxindole] derivatives. PG = Boc or Cbz.

Mannich addition step was catalyzed by a quinidine derivative **103**, while the intramolecular hydroamination was efficiently promoted by a XPhosAuNTf_2 complex (XPhos = 2-[dicyclohexylphosphino]-2′,4′,6′-triisopropylbiphenyl). The resulting (2-methylen-5-pyrrolidinyl)spirooxindole products **104** were produced in up to 91% yield, with almost complete diastereoselectivity (>20 : 1 dr) and with high enantioselectivity (Scheme 6.44).

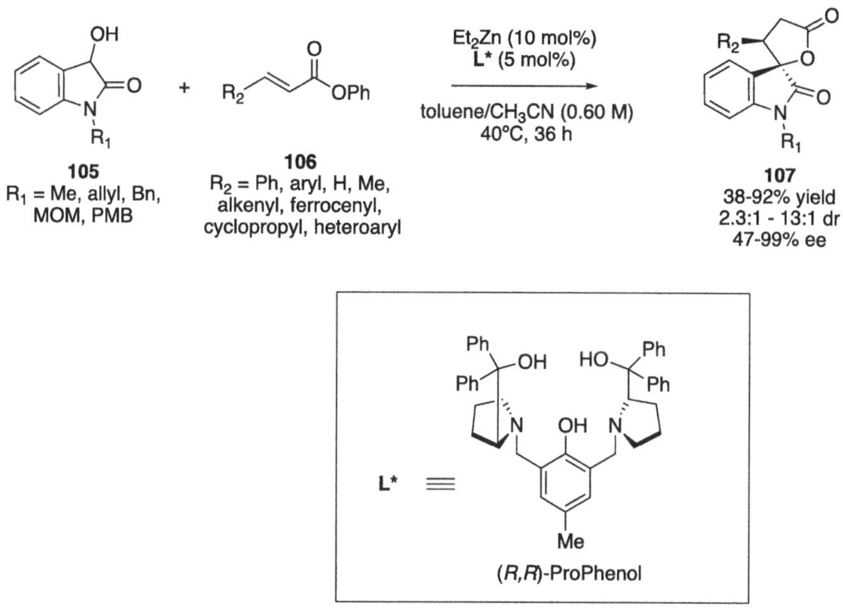

Scheme 6.45 Trost's dinuclear zinc-catalyzed asymmetric spiroannulation reaction of 3-hydroxy oxindoles with α,β-unsaturated esters.

In 2012, Trost and Hirano reported [52] the enantioselective nucleophilic addition of 3-hydroxyindoles **105** (via their Zn-enolates) to a variety of α,β-unsaturated phenyl esters **106**. The reaction involves the dinuclear zinc-ProPhenol-catalyzed Michael addition [53] of **105** to the esters **106**, followed by intramolecular transesterification to afford the spirocyclic δ-lactones **107** in good yields (generally higher than 67%), moderate to good diastereoselectivities (2.3 : 1 dr to 13 : 1 dr), and poor to high enantioselectivities (up to 99% ee; Scheme 6.45).

Kanai, Matsunaga, and coworkers have found [54] that spiro[imidazolidine-4,3′-oxindole] derivatives **108** can be obtained in a highly stereocontrolled way by means of the Michael addition of 3-isothiocyanato oxindoles **109** to *N*-(diphenylphosphinoyl)imines **110**, catalyzed by a chiral strontium(II)/Schiff base complex (Scheme 6.46). The initially formed Mannich adduct cyclizes in the reaction conditions, so that the reaction can be regarded as a formal [3+2] cycloaddition.

More recently, Chen, Xiao, and coworkers [55] have also used 3-isothiocyanato oxindoles **109** in a cascade Michael addition/cyclization with 3-nitro-2*H*-chromenes **111**. The reaction takes place by zinc(II) catalysis, in the presence of a chiral bis(oxazoline) ligand, and provides access to highly functionalized, polycyclic spirooxindoles **112**, in a highly stereoselective manner under mild conditions (72–99% yield, up to >95 : 5 dr and >99% ee). It is worth noting that this process

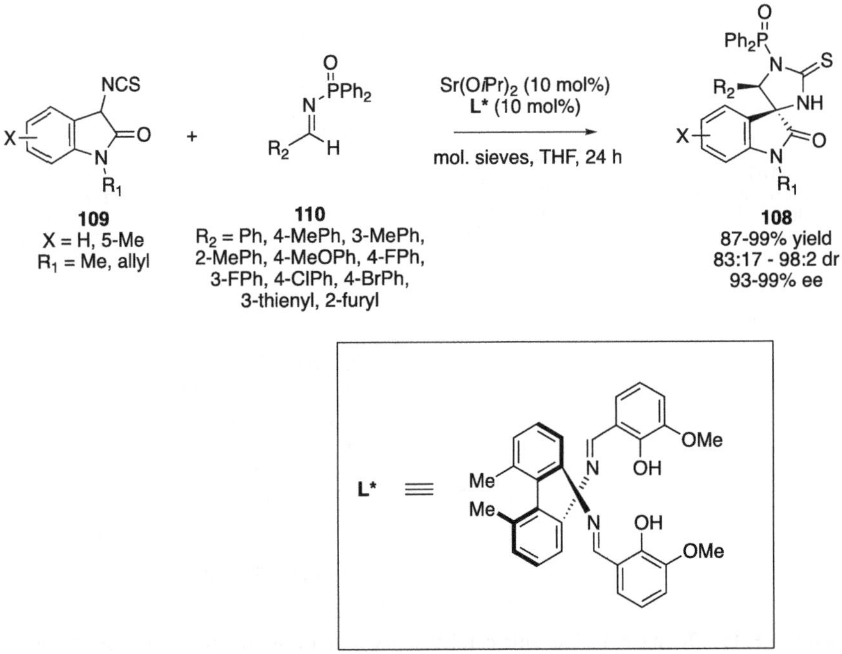

Scheme 6.46 Kanai–Matsunaga's strontium-catalyzed asymmetric synthesis of 3-spirooxindoles by a Mannich-type reaction of isothiocyanato oxindoles.

leads to the formation of three consecutive chiral centers, including a pair of 1,3-nonadjacent tetrasubstituted carbons, in a single synthetic operation (Scheme 6.47). A bifunctional Zn-promoted Lewis acid activation and *N*-promoted Lewis base activation was proposed by the authors.

6.3.4 [3+3] Cycloaddition Reactions

In 2017, a diastereo- and enantioselective [3+3] cycloaddition of racemic spirocyclopropyl oxindoles **113** with nitrones **114** was jointly developed by the groups of Yan (Nanjing) and Zhou (Shanghai) [56]. The key to the reaction development was the activation of the spirocyclopropyl oxindoles by a suitable electron-withdrawing *N*-protecting group ($PO[OEt]_2$). In this way, the unprecedented [3+3] cycloaddition took place at moderate temperatures under catalysis by nickel(II), affording the spiro tetrahydro-1,2-oxazinyl-oxindoles **115** with high yields and excellent diastereoselectivity (>20 : 1 dr). Moreover, when the reaction was performed in the presence of a chiral bis(oxazoline) ligand, a kinetic resolution took place (selectivity factors from 7 up to 116), allowing the recovery of enantiomerically enriched spirocyclopropyl oxindoles **113** and affording the spirocyclic tetrahydro-1,2-oxazines **115** with excellent enantioselectivity (90–98% ee; Scheme 6.48).

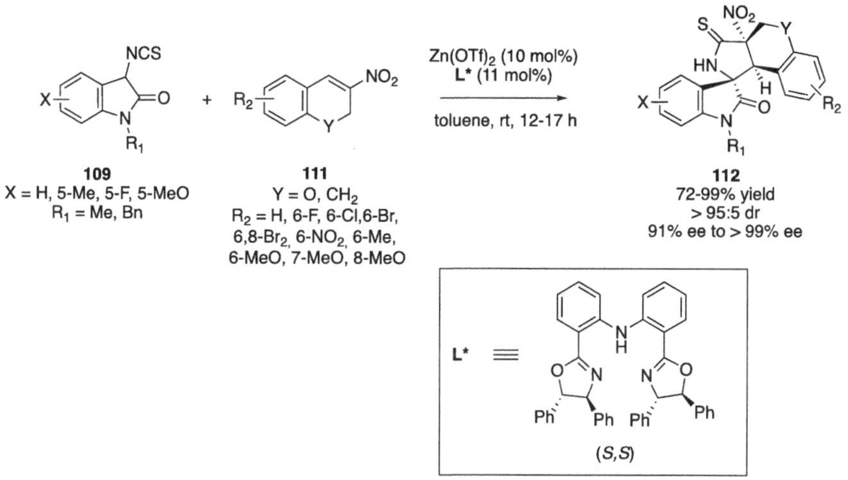

Scheme 6.47 Chen-Xiao's zinc(II)-catalyzed enantioselective cascade Michael addition/cyclization of 3-nitro-2*H*-chromenes with 3-isothiocyanato oxindoles.

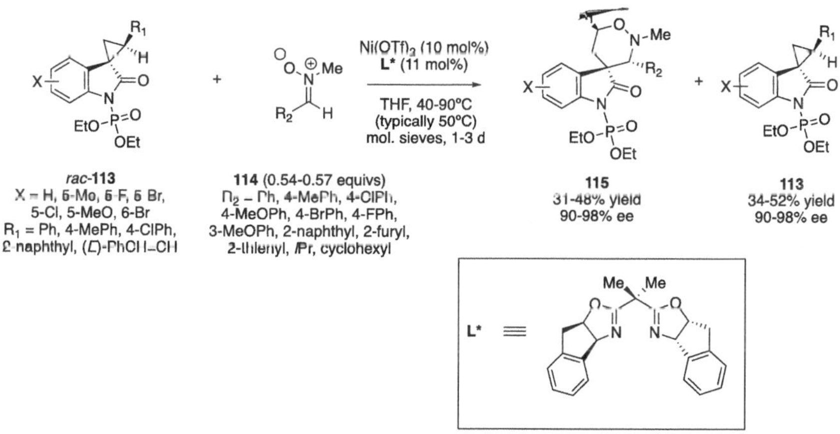

Scheme 6.48 Yan-Zhou's nickel(II)-catalyzed diastereo- and enantioselective [3+3] cycloaddition of spirocyclopropyl oxindoles.

6.3.5 [4+1] Cycloaddition Reactions

In 2017, Ashfeld and Meloche [57] developed a formal [4+1] cycloaddition of vinyl isocoyanates **116** with diazooxindoles **38**, catalyzed by dirhodium tetraacetate, that afforded the spirooxindole pyrrolones **117** with high yields (Scheme 6.49). The reaction mechanism involves the initial formation of a spirocyclopropane intermediate that undergoes ring expansion.

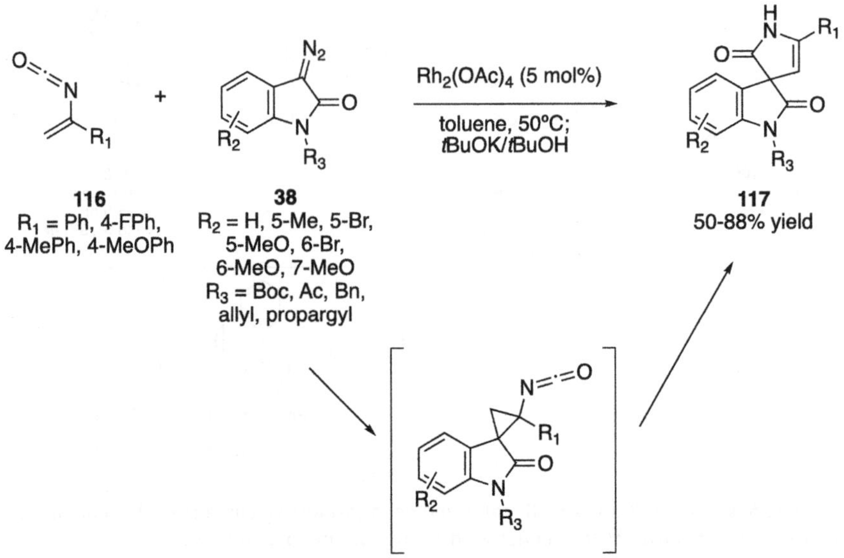

Scheme 6.49 Ashfeld's rhodium-catalyzed formal [4+1] cycloaddition route to spirooxindole pyrrolones. *Source:* Based on Meloche and Ashfeld [57].

Later on, the same research group used α-silyl vinyl ketenes **118** (generated *in situ* from 2-silyl-2-cyclobutenones **119**) to obtain spirooxindole cyclopentenones **120** [58]. A catalytic asymmetric version [59] of the same reaction employed a chiral tetrachlorophthalimide-derived rhodium carboxylate as the catalyst (Scheme 6.50).

6.3.6 [4+2] Cycloaddition/Annulation Reactions

The asymmetric Diels–Alder reaction of *N*-protected (*E*)-3-alkylidene oxindoles **53** with the Brassard-type diene **121** can be efficiently catalyzed by a complex formed *in situ* from zinc(II) triflate and a chiral *N,N*′-dioxide ligand [60]. The spiro[cyclohex[3]ene-1,3′-indoline]-1′-carboxylate-2,2′-dione derivatives **122**, containing three adjacent stereocenters, were obtained in moderate yields (58–68%), but with an outstanding degree of stereocontrol (single diastereomer, 98 to >99% ee). Amide and carbamate *N*-protecting groups in the oxindole are required, and no reaction was observed with a Bn group (Scheme 6.51).

In 2017, Kumar and coworkers [61] reported the asymmetric hetero-Diels–Alder reaction of (*E*)-3-cyanomethylidene oxindoles **53** with 3-silyloxy-2-azadienes **123**. After careful optimization of the reaction conditions, Kumar's group found that dysprosium(III) triflate, again in combination with a Feng *N,N*′-dioxide chiral ligand, offered the best results both in terms of yield and of steric control

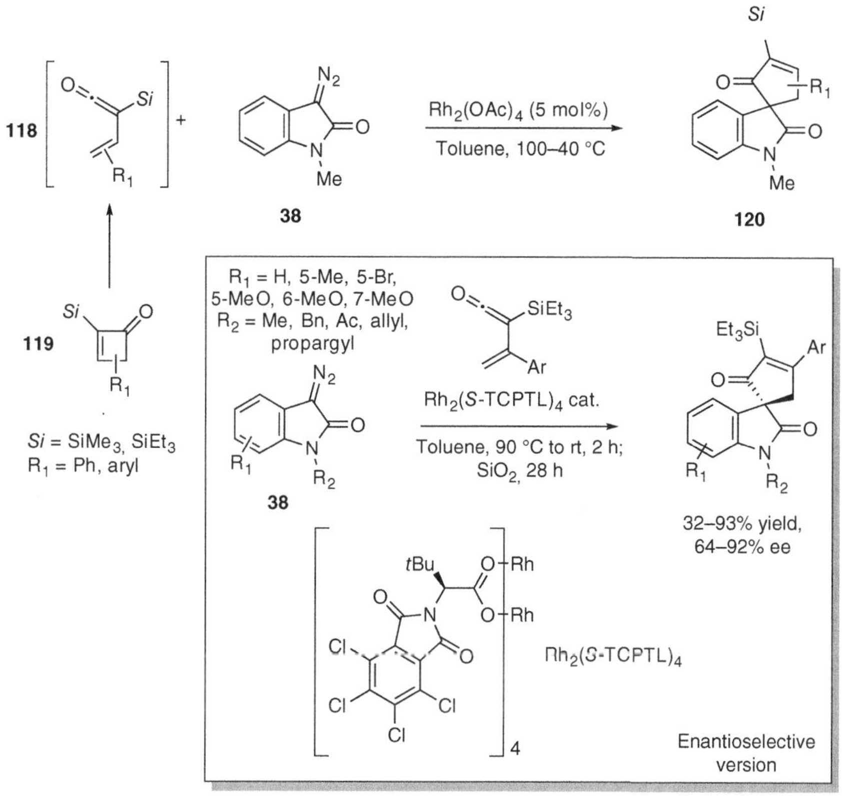

Scheme 6.50 Ashfeld's rhodium-catalyzed formal [4+1] cycloaddition syntheses of spirooxindole cyclopentenones.

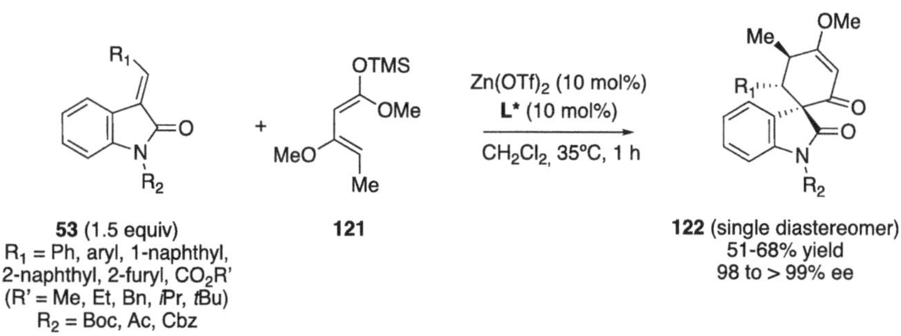

Scheme 6.51 Lin-Feng's zinc-catalyzed asymmetric Diels–Alder reaction of a Brassard diene with alkylidene oxindoles. See Scheme 6.39 for the structure of the chiral *N,N*-dioxide ligand **L***.

(Scheme 6.52). The substituted 3,3′-piperidinoyl-spirooxindoles **124** were obtained in most instances as a single diastereomer (>20 : 1 dr), with excellent yields (>85%) and with variable but usually very good (up to 99% ee) enantioselectivity.

This hetero-Diels–Alder reaction could also be extended to the differently substituted dicyano-methylidene oxindoles **125**, but a re-optimization of the reaction conditions proved to be necessary, in order to control the regioselectivity of the process, that gave rise to regioisomeric mixtures of 3,4′- (**126**) and 3,3′-piperidinoyl spirooxindoles (**127**). Magnesium(II) bis(trifluoromethanesulfonyl) imide emerged as the best metal catalyst. Electron-withdrawing as well as electron-donating substituents in the aryl oxindole ring were well tolerated and afforded majoritarily the 3,4′-piperidinoyl-spirooxindoles **126** in excellent yields, and in most cases with high enantioselectivities (Scheme 6.53).

In 2008, Tanaka's research group developed a [4+2] annulation reaction between 2-alkynylbenzaldehyde derivatives **128** and carbonyl compounds [62]. A cationic rhodium(I) complex with the chiral bisphosphine (*R,R*)-walphos

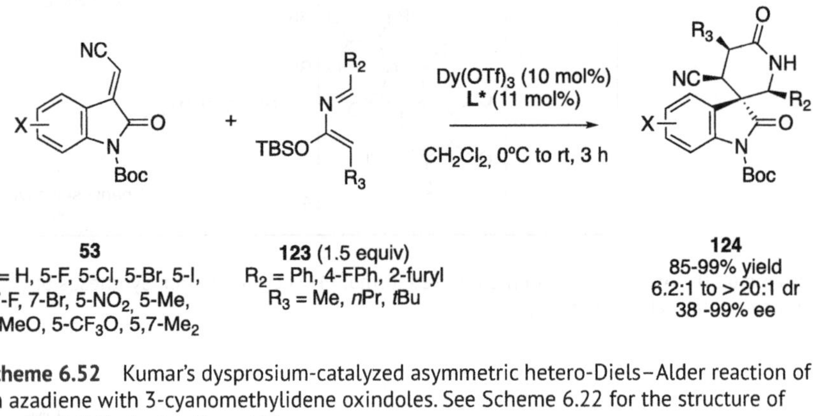

53
X = H, 5-F, 5-Cl, 5-Br, 5-I, 7-F, 7-Br, 5-NO_2, 5-Me, 5-MeO, 5-CF_3O, 5,7-Me_2

123 (1.5 equiv)
R_2 = Ph, 4-FPh, 2-furyl
R_3 = Me, *n*Pr, *t*Bu

124
85-99% yield
6.2:1 to > 20:1 dr
38 -99% ee

Scheme 6.52 Kumar's dysprosium-catalyzed asymmetric hetero-Diels–Alder reaction of an azadiene with 3-cyanomethylidene oxindoles. See Scheme 6.22 for the structure of the chiral *N,N′*-dioxide ligand **L***.

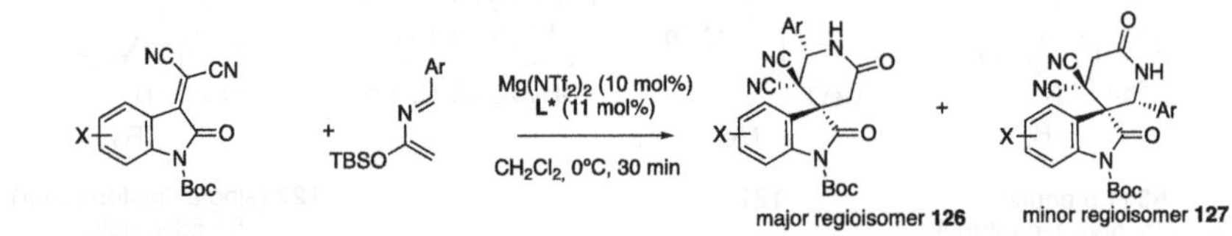

4.2:1 to 8:1 regioisomer ratio

125
X = H, 5-F, 5-Cl, 5-Br, 5-I, 5-Me, 5-MeO, 5-CF_3O, 5,7-Me_2

123 (1.5 equiv)
Ar = Ph, 4-FPh, 2-furyl

126: 90-99% yield
single diastereomers
24 -96% ee

Scheme 6.53 Kumar's magnesium-catalyzed asymmetric hetero-Diels–Alder reaction of an azadiene with 3-dicyanomethylidene oxindoles. See Scheme 6.22 for the structure of the chiral *N,N′*-dioxide ligand **L***.

furnished the desired cross-annulation products in mild conditions and in optically enriched form. Isatins **79** were very good substrates for this annulation and gave rise to the spirocyclic benzopyranones **129** in excellent yield (up to 97%) and enantioselectivity (up to 99% ee). It should be noted that the absolute (*S*) stereochemistry of cycloadducts **129** in Scheme 6.54 is temptative, since it has been assigned by analogy to that determined by X-ray crystallographic analysis for an acenaphthene quinone adduct.

In 2009, Shintani, Hayashi, and coworkers [63] developed a palladium-catalyzed decarboxylative cyclization of γ-methylene-δ-valerolactones **130** with isatins **79**, which led to the racemic spirocyclic oxindole pyrans **131**. The reaction was optimized for a variety of *P,N* ligands, and a phosphoramidite having a bis(diphenylmethyl)amino group on phosphorus produced the best results (up to 94% yield, >95 : 5 dr; Scheme 6.55). The mechanism involves the oxidative addition of the ester moiety of **130** to palladium(0), followed by decarboxylation, to give a 1,4-zwitterionic species that after attack to the isatin carbonyl undergoes a ring closure through nucleophilic attack of the oxygen to the π-allylpalladium moiety (Scheme 6.56) The authors also explored the use of a chiral phosphoramidite ligand for an asymmetric version of this process, with promising results (4 examples, 73–87% ee).

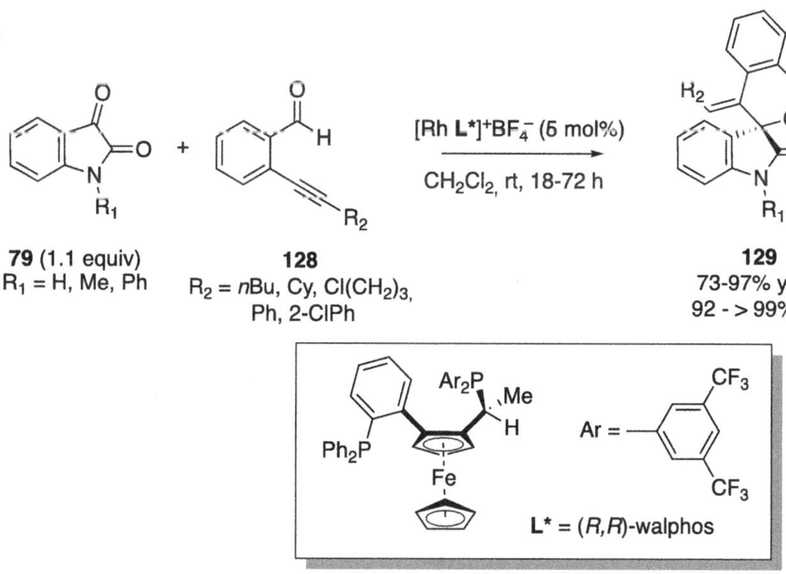

Scheme 6.54 Tanaka's enantioselective synthesis of spirocyclic benzopyranones by rhodium-catalyzed [4+2] annulation.

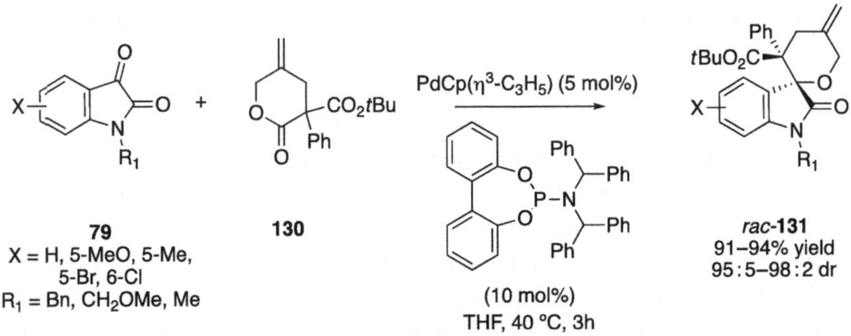

Scheme 6.55 Shintani–Hayashi's stereoselective synthesis of spirocyclic oxindoles by palladium-catalyzed decarboxylative cyclization of γ-methylene-δ-valerolactones with isatins.

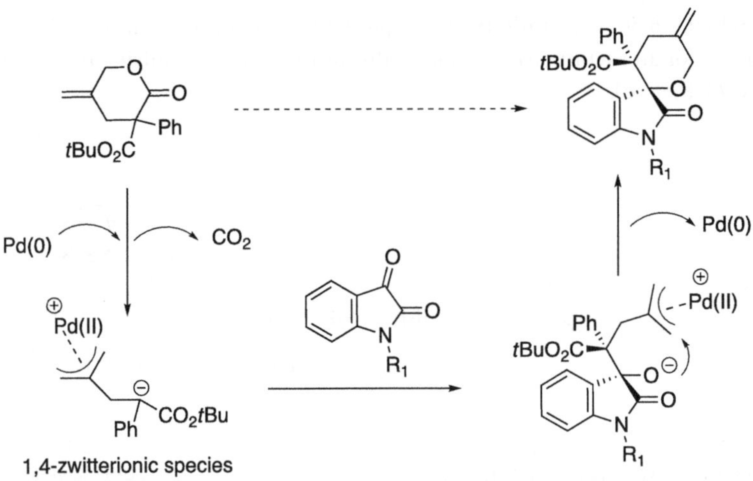

Scheme 6.56 Proposed catalytic cycle for the palladium-catalyzed decarboxylative cyclization of γ-methylene-δ-valerolactones with isatins.

An asymmetric aza-Diels–Alder reaction of 3-vinylindoles **132** with *N*-Boc imino isatins **101** was developed in 2015 by Feng and coworkers [64]. The reaction was efficiently catalyzed by nickel(II) triflate, and the choice of the chiral *N,N'*-dioxide ligand was the key to achieve high enantioselectivities, probably by harnessing the adequate π–π interactions between two indoline rings of the substrates. The resulting *exo*-spiroindolones **133** were obtained in good yields (75–95%) and with excellent enantioselectivities (84–94% ee), with considerable substrate scope (Scheme 6.57).

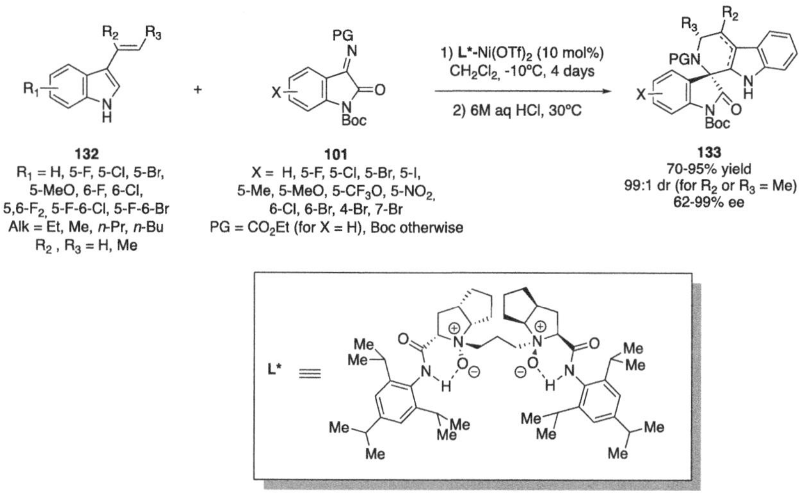

Scheme 6.57 Liu–Feng's nickel-catalyzed asymmetric aza-Diels–Alder reaction of 3-vinylindoles with imino isatins.

6.3.7 [5+2] Cycloaddition Reactions

The [5+2] cycloaddition of the chiral, nonracemic (98% ee) η^3-pyridinylmolybdenum complex **134** with 3-methyleneoxindole **135** was reported in 2000 by Malinakova and Liebeskind [65]. This reaction affords a direct, convergent, and enantiocontrolled synthesis of the core structural unit of the spirooxindole alkaloids but is stoichiometric in molybdenum. The cycloaddition reaction is reversible, and the authors discovered that under catalysis by scandium(III) triflate, a combined 41% yield of the cycloadducts **136**-*exo* and **136**-*endo* (4 : 1 dr) was obtained. Moreover, a 45% yield of starting complex **134** was recovered. After chromatographic separation, the major isomer **136**-*exo* could be oxidatively demetallated with copper(II) chloride to provide the tetracyclic enone **137**. Catalytic hydrogenation followed by treatment by iodotrimethylsilane afforded the known amine **138** (95% ee), an intermediate in a proposed route to *Alstonia* alkaloids (Scheme 6.58).

6.4 Miscellaneous Methods

6.4.1 Darzens Reaction of Isatins

Spiro-epoxindoles **139** have been obtained by asymmetric catalytic Darzens reaction between α-bromomethyl ketones **140** and isatins **79** in Feng's research group [66]. The optically active, almost diastereomerically pure products (99 : 1 dr) were obtained in moderate to good yields (35–99%) and enantioselectivities (47–95% ee) catalyzed by chiral *N,N′*-dioxide-Co(acac)$_2$ complexes (Scheme 6.59).

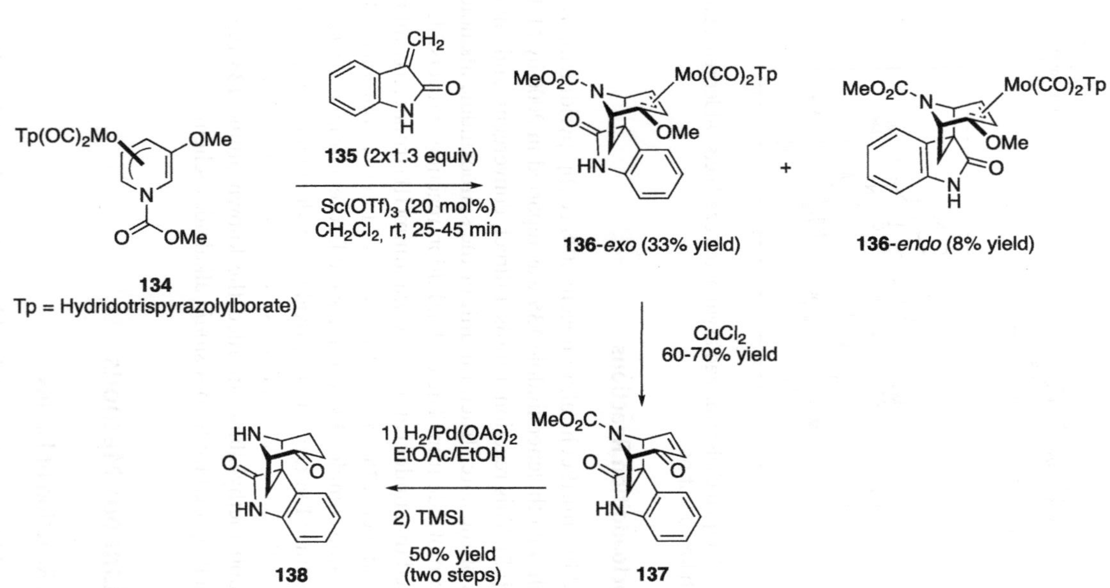

Scheme 6.58 Liebeskind's enantioselective synthesis of spirooxindoles based on the scandium-catalyzed [5+2] cycloaddition of a η^3-pyridinylmolybdenum complex to methyleneoxindole.

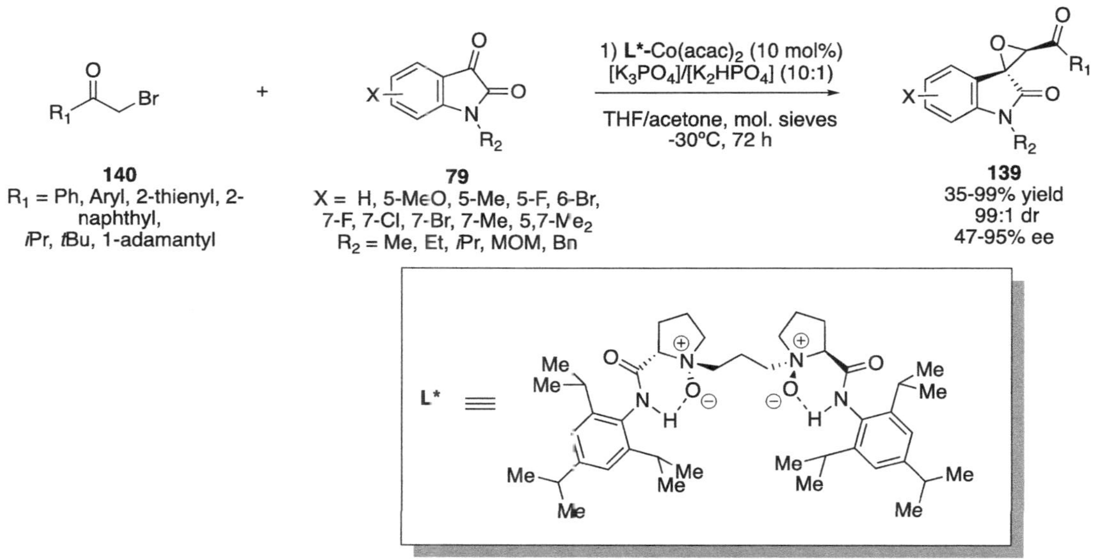

Scheme 6.59 Feng's cobalt-catalyzed asymmetric Darzens reaction of phenacyl bromides with isatins.

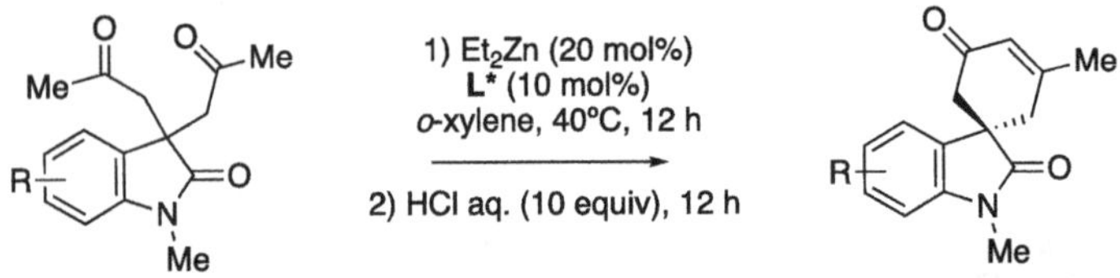

142
R = 5-Me, 5-MeO, 5,7-Me_2,
5-F, 5-Cl, 5-Br, 4-Cl, 4-Br,
6-Br, 6-Cl, 7-Cl, 7-F, 7-Me

141
45-79% yield
63-92% ee

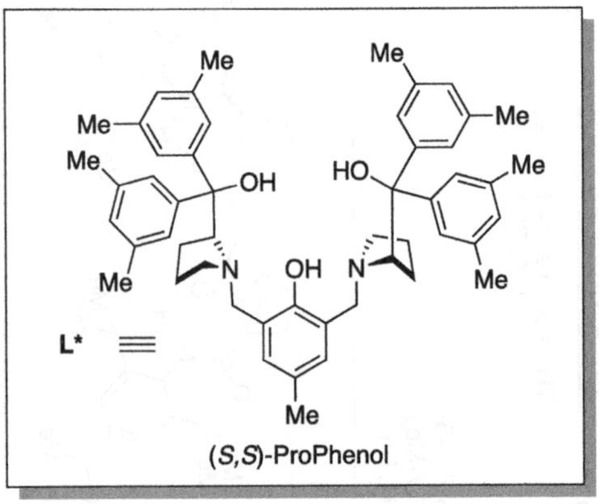

Scheme 6.60 Wang's zinc-catalyzed enantioselective desymmetrization/cyclization of 3,3-bis(2-oxopropyl)oxindoles.

6.4.2 Cyclization of 3,3-Disubstituted Oxindoles

Xin-Wang Wang and coworkers [67] were able to synthesize spiro[cyclohexenone-oxindole] derivatives **141** by the asymmetric intramolecular aldol reaction of achiral diketone-oxindoles **142** with Trost's Pro-Phenol dinuclear zinc catalysts [53]. The reaction led to an inseparable mixture of aldol and dehydrated compounds, so that the authors found it convenient to treat *in situ* the cyclization mixture with hydrochloric acid to attain exclusively the spirocyclohexenone compounds **141** in moderate to good yields (45–79%) and variable enantioselectivities (63–92% ee; Scheme 6.60).

6.4.3 Three-component Reactions

Whereas the vast majority of synthetic approaches to spirocyclic oxindoles through multicomponent reactions rely on the use of organocatalysis [3], Shanmugam and coworkers [68] have shown that 3-spiroazetidinimine-2-oxindo les can be obtained by means of a copper-catalyzed formal [2+2] cycloaddition between a ketenimine **143** and an imino isatin **101**. Since the ketenimines **143** are

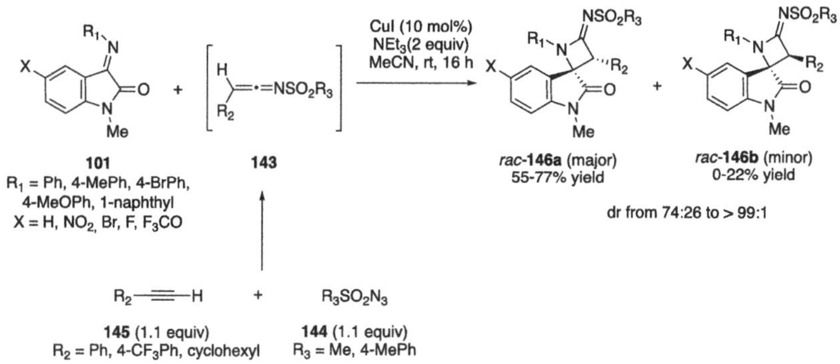

Scheme 6.61 Shanmugam's copper(I)-catalyzed one-pot, three-component diastereoselective synthesis of 3-spiroazetidinimine-2-oxindoles.

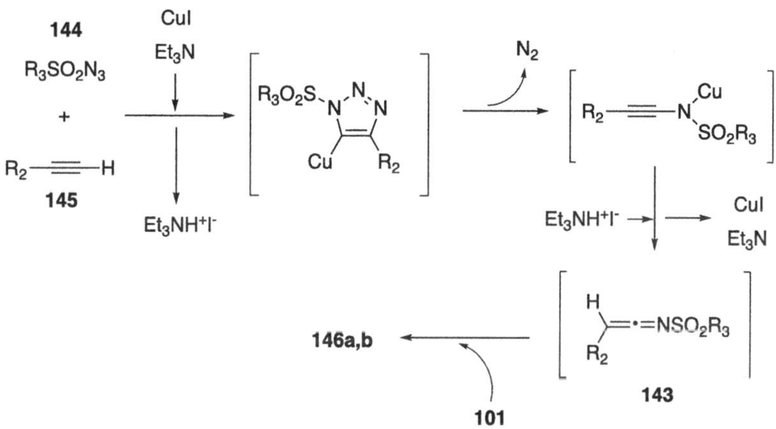

Scheme 6.62 A plausible mechanism of the copper(I)-catalyzed three-component diastereoselective synthesis of 3-spiroazetidinimine-2-oxindoles. *Sources:* Based on Whiting and Fokin [69]; Bae et al. [70].

generated *in situ* from sulfonyl azides **144** and terminal acetylenes **145**, the overall procedure is a one-pot, three-component cascade reaction (Scheme 6.61). The 3-s piroazetidinimine-2-oxindoles **146a,b** were obtained in good to excellent combined yields (55–9%), and with good diastereoselectivity (from 26 : 74 to >99 : 1 dr).

The probable mechanism of the reaction involves the initial formation of an alkynyl-copper species, which undergoes a [3+2] cycloaddition with the azide to form an unstable copper-triazolyl intermediate. This intermediate is transformed into the ketenimine **143** upon protonation, followed by loss of nitrogen and regeneration of the Cu(I) [69]. Subsequent [2+2] ketenimine-imine cycloaddition affords the final products **146** with the expected [70] stereochemical outcome (*exo* approach with regard to R_2, Scheme 6.62).

6.5 Conclusions

The fascinating architecture of oxindole-based spirocyclic alkaloids and the biological activity often associated to these structural motifs have been arising the interest of organic and medicinal chemists in the past two or three decades, and much work and ingenuity has been devoted to their synthesis, especially in non-racemic form. As a result, tremendous progress has been achieved in this area, and metal-catalyzed methods have been applied to a variety of strategic approaches. Remaining challenges in the synthesis of spirooxindoles and related compounds must address the issues of efficient catalytic methods to access all of the possible stereoisomers of a given structure, allowing the incorporation of different functional groups and of heterocyclic systems. Efforts in these directions will hopefully result in the opening of new avenues both in the synthesis of complex natural products and in the development of drug discovery programs.

Acknowledgements

The author acknowledges the financial support from FEDER, Ministerio de Ciencia, Innovación y Universidades, Agencia Estatal de Investigación (Spain), Proyectos CTQ2017-87864-C2-1-P and PID2020-116846GB-C21

References

1 For reviews, see among others:a)Marti, C. and Carreira, E.M. (**2003**). *Eur. J. Org. Chem.*: 2209–2219. b) Galliford, C.V. and Scheidt, K.A. (**2007**). *Angew. Chem. Int. Ed. 46*: 8748–8758. c) Yang, J., Wearing, X.Z., Quesne, P.W.L. et al. (**2008**). *J. Nat. Prod. 71*: 1431–1440. d) Trost, B.M. and Brennan, M.K. (**2009**). *Synthesis* (18): 3003–3025. e) Zhou, F., Liu, Y.-L., and Zhou, J. (**2010**). *Adv. Synth. Catal. 352*: 1381–1407. f) Rios, R. (**2012**). *Chem. Soc. Rev. 41*: 1060–1074.

2 a) Fensome, A., Adams, W.R., Adams, A.L. et al. (**2008**). *J. Med. Chem. 51*: 1861–1873. b) Tokunaga, T., Hume, W.E., Umezome, T. et al. (**2001**). *J. Med. Chem. 44*: 4641–4649. c) Badillo, J.J., Hanhan, N.V., and Franz, A.K. (**2010**). *Curr. Opin. Drug Discov. Devel. 13*: 758–776. d) Cao, J., Dong, S., Jiang, D. et al. (**2017**). *J. Org. Chem. 82*: 4186–4193.

3 a) Ball-Jones, N.R., Badillo, J.J., and Franz, A.K. (**2012**). *Org. Biomol. Chem. 10*: 5165–5181. b) Singh, G.S. and Desta, Z.Y. (**2012**). *Chem. Rev. 112*: 6104–6155.For a recent general review on the enantioselective synthesis of spiro compounds, see:c)Ding, A., Meazza, M., Guo, H. et al. (**2018**). *Chem. Soc. Rev. 47*: 5946–5996.

4 a) Overman, L.E. and Rosen, M.D. (**2000**). *Angew. Chem. Int. Ed. 39*: 4596–4599. b) Overman, L.E. and Rosen, M.D. (**2010**). *Tetrahedron 66*: 6514–6525.

5 a) Ashimori, A. and Overman, L.E. (**1992**). *J. Org. Chem. 57*: 4571–4572. b) Ashimori, A., Matsuura, T., Overman, L.E., and Poon, D.J. (**1993**). *J. Org. Chem. 58*: 6949–6951. c) Ashimori, A., Bachand, B., Overman, L.E., and Poon, D.J. (**1998**). *J. Am. Chem. Soc. 120*: 6477–6487.For studies on the diastereoselection in the formation of spirocyclic oxindoles by intramolecular Heck reaction, see:d)Overman, L.E. and Watson, D.A. (**2006**). *J. Org. Chem. 71*: 2587–2599. e) Overman, L.E. and Watson, D.A. (**2006**). *J. Org. Chem. 71*: 2600–2608. f) For an approach to the synthesis of the spirooxindole alkaloid (+)-gelsemine featuring an intramolecular non-reductive Heck reaction, see:Beyersbergen van Henegouwen, W.G., Fieseler, R.M., Rutjes, F.P.J.T., and Hiemstra, H. (**1999**). *Angew. Chem. Int. Ed. 38*: 2214–2217.

6 Kamisaki, H., Yasui, Y., and Takemoto, Y. (**2009**). *Tetrahedron Lett. 50*: 2589–2592.

7 Kamisaki, H., Nanjo, T., Tsukano, C., and Takemoto, Y. (**2011**). *Chem. Eur. J. 17*: 626–633.

8 Jaegli, S., Erb, W., Retailleau, P. et al. (**2010**). *Chem. Eur. J. 16*: 5863–5867.

9 a) Diethelm, S. and Carreira, E.M. (**2013**). *J. Am. Chem. Soc. 135*: 8500–8503.For a recent report on the intramolecular reductive Heck reaction of C2-substituted indoles, leading to 3,2′-spiropyrrolidine oxindoles, see:b)Liu, R.R., Xu, Y., Liang, R.X. et al. (**2017**). *Org. Biomol. Chem. 15*: 2711–2715.

10 Chen, P., Chen, Z.-C., Li, Y. et al. (**2019**). *Angew. Chem. Int. Ed. 58*: 4036–4040.

11 Hu, H., Teng, F., Liu, J. et al. (**2019**). *Angew. Chem. Int. Ed. 58*: 9225–9229.

12 Ladd, C.L., Roman, D.S., and Charette, A.B. (**2013**). *Org. Lett. 15*: 1350–1353.

13 Katayama, S. and Nishino, H. (**2019**). *Synthesis 51*: 3277–3286.

14 Alcaide, B., Almendros, P., and Rodríguez-Acebes, R. (**2006**). *J. Org. Chem. 71*: 2346–2351.

15 Trost, B.M. and Stiles, D.T. (**2007**). *Org. Lett. 9*: 2763–2766.

16 Trost, B.M. and Brennan, M.K. (**2006**). *Org. Lett. 8*: 2027–2030.

17 a) Aferweki, S., Ma, G., Ibrahem, I. et al. (**2015**). *ACS Catal. 5*: 1266–1272.For previous studies, see:b)Sun, W., Zhu, C., Wu, L. et al. (**2012**). *Chem. Eur. J. 18*: 13959–13963. c) Santoro, S., Deiana, L., Zhao, G.-L. et al. (**2014**). *ACS Catal. 4*: 4474–14484. d) Deiana, L., Ghisu, L., Aferweki, S. et al. (**2014**). *Adv. Synth. Catal. 356*: 2485–2492.

18 Zhang, Y.-Q., Yuan, Y.-A., Liu, Q.-S., and Xu, H. (**2013**). *Org. Lett. 15*: 3910–3913.

19 Crestini, C. and Saladino, R. (**1994**). *Synth. Commun. 24*: 2835–2841.

20 Meyers, C. and Carreira, E.M. (**2003**). *Angew. Chem. Int. Ed. 42*: 694–696.

21 Kumari, G., Nutan, Modi, M. et al. (**2011**). *Eur. J. Med. Chem. 46*: 1181–1188.

22 a) Awata, A. and Arai, T. (**2013**). *Synlett 24*: 29–32.For a subsequent Rh-catalyzed asymmetric cyclopropanation of *N*-Boc diazooxindoles with similar carboxylate

complexes, see:b)Chi, Y., Qiu, L., and Xu, X. (**2016**). *Org. Biomol. Chem. 14*: 10357–10361.

23 Cao, Z.-Y., Zhou, F., Yu, H.-Y., and Zhou, J. (**2013**). *Org. Lett. 15*: 42–45.

24 a) Hashimoto, S., Watanabe, N., Sato, T. et al. (**1993**). *Tetrahedron Lett.* 34: 5109–5112. b) Tsutsui, H., Abe, T., Nakamura, S. et al. (**2005**). *Chem. Pharm. Bull. 53*: 1366–1368.

25 Cao, Z.-Y., Wang, X., Tan, C. et al. (**2013**). *J. Am. Chem. Soc. 135*: 8197–8200.

26 a) Wang, L., Cao, W., Mei, H. et al. (**2018**). *Adv. Synth. Catal. 360*: 4089–4093.For a previous use of the same ligand in the asymmetric cyclopropanation of alkylidene oxindoles with phenyliodonium ylides, see:b)Guo, J., Liu, Y., Li, X. et al. (**2016**). *Chem. Sci. 7*: 2717–2721.For the Sc(III)-catalyzed asymmetric synthesis of chiral spirooxindole tetrahydroquinolines with the same ligand, see:c)Cao, W., Liu, X., Guo, J. et al. (**2015**). *Chem. Eur. J. 21*: 1632–1636.

27 Hao, X., Liu, X., Li, W. et al. (**2014**). *Org. Lett. 16*: 134–137.

28 Alper, P.B., Meyers, C., Lerchner, A. et al. (**1999**). *Angew. Chem. Int. Ed. 38*: 3186–3189.

29 Edmonson, D.E. and Danishefsky, S.J. (**1998**). *Angew. Chem. Int. Ed. 37*: 1138–1140.

30 He, F., Bo, Y., Altom, J.D., and Corey, E.J. (**1999**). *J. Am. Chem. Soc. 121*: 6771–6772.

31 Fischer, C., Meyers, C., and Carreira, E.M. (**2000**). *Helv. Chim. Acta 83*: 1175–1181.

32 a) Lerchner, A. and Carreira, E.M. (**2002**). *J. Am. Chem. Soc. 124*: 14286–14287. b) Lerchner, A. and Carreira, E.M. (**2006**). *Chem. Eur. J.* 12: 8208–8219.

33 Marti, C. and Carreira, E.M. (**2005**). *J. Am. Chem. Soc. 127*: 11505–11515.

34 a) Trost, B.M. and Chan, D.M.T. (**1979**). *J. Am. Chem. Soc. 101*: 6429–6432. b) Trost, B.M. and Chan, D.M.T. (**1979**). *J. Am. Chem. Soc. 101*: 6432–6433. c) Lautens, M., Klute, W., and Tam, W. (**1996**). *Chem. Rev. 96*: 49–92.

35 Trost, B.M., Mignani, S.M., and Nanninga, T.N. (**1986**). *J. Am. Chem. Soc. 108*: 6051–6053.

36 Trost, B.M., Cramer, N., and Bernsmann, H. (**2007**). *J. Am. Chem. Soc. 129*: 3086–3087.

37 Trost, B.M., Bringley, D.A., Zhang, T., and Cramer, N. (**2013**). *J. Am. Chem. Soc. 135*: 16720–16735.

38 Trost, B.M., Cramer, N., and Silverman, S.M. (**2007**). *J. Am. Chem. Soc. 129*: 12396–12397.

39 Antonchick, A.P., Gerding-Reimers, C., Catarinella, M. et al. (**2010**). *Nat. Chem. 2*: 735–740.

40 Chen, X.H., Wei, Q., Luo, S.W. et al. (**2009**). *J. Am. Chem. Soc. 131*: 13819–13825.

41 Liu, T.-L., Xue, Z.-Y., Tao, H.-Y., and Wang, C.-J. (**2011**). *Org. Biomol. Chem. 9*: 1980–1986.
42 Badillo, J.J., Arevalo, G.E., Fettinger, J.C., and Franz, A.K. (**2011**). *Org. Lett. 13*: 418–421.
43 Hanhan, N.V., Ball-Jones, N.R., Tran, N.T., and Franz, A.K. (**2012**). *Angew. Chem. Int. Ed. 51*: 982–992.
44 Badillo, J.J., Ribeiro, C.J.A., Olmstead, M.M., and Franz, A.K. (**2014**). *Org. Lett. 16*: 6270–6273.
45 Ball-Jones, N.R., Badillo, J.J., Tran, N.T., and Franz, A.K. (**2014**). *Angew. Chem. Int. Ed. 53*: 9462–9465.
46 Cobo, A.A., Armstrong, B.M., Fettinger, J.C., and Franz, A.K. (**2019**). *Org. Lett. 21*: 8196–8200.
47 a) Wang, G., Liu, X., Huang, T. et al. (**2013**). *Org. Lett. 15*: 76–79.For a subsequent use of Feng's chiral *N,N′*-dioxide ligand in the formal [3+2] annulation of alkylidene oxindoles with 1,4-dithiane-2,5-diol, see:b)Zhou, P., Cai, Y., Lin, L. et al. (**2015**). *Adv. Synth. Catal. 37*: 695–700.
48 Li, T.-R., Cheng, B.-Y., Fan, S.-Q. et al. (**2016**). *Chem. Eur. J. 22*: 6243–6247.
49 Mei, L., Wei, Y., Xu, Q., and Shi, M. (**2013**). *Organometallics 32*: 3544–3556.
50 Liu, Z., Feng, X., and Du, H. (**2012**). *Org. Lett. 14*: 3154–3157.
51 Chen, X., Chen, H., Ji, X. et al. (**2013**). *Org. Lett. 15*: 1846–1849
52 Trost, B.M. and Hirano, K. (**2012**). *Org. Lett. 14*: 2446–2449.
53 For a review on ProPhenol-catalyzed asymmetric additions, see:Trost, B.M. and Bartlett, M.J. (**2015**). *Acc. Chem. Res. 48*: 688–701.
54 Kato, S., Yoshino, T., Shibasaki, M. et al. (**2012**). *Angew. Chem. Int. Ed. 51*: 7007–7010.
55 Tan, F., Lu, L.-Q., Yang, Q.-Q. et al. (**2014**). *Chem. Eur. J. 20*: 3415–3420.
56 Xu, P.-W., Liu, J.-K., Shen, L. et al. (**2017**). *Nat. Commun.* 8: 1619. https://doi.org/10.1038/s41467-017-01451-1.
57 Meloche, J.L. and Ashfeld, B.L. (**2017**). *Angew. Chem. Int. Ed. 56*: 6604–6608.
58 Rodriguez, K.X., Kaltwasser, N., Toni, T.A., and Ashfeld, B.L. (**2017**). *Org. Lett. 19*: 2482–2485.
59 Rodriguez, K.X., Pilato, T.C., and Ashfeld, B.L. (**2018**). *Chem. Sci. 9*: 3221–3226.
60 Zheng, J., Lin, L., Fu, K. et al. (**2015**). *J. Org. Chem. 80*: 8836–8842.
61 Jayakumar, S., Louven, K., Strohmann, C., and Kumar, K. (**2017**). *Angew. Chem. Int. Ed. 56*: 15945–15949.
62 Hojo, D., Noguchi, K., Hirano, M., and Tanaka, K. (**2008**). *Angew. Chem. Int. Ed. 47*: 5820–5822.
63 Shintani, R., Hayashi, S., Murakami, M. et al. (**2009**). *Org. Lett. 11*: 3754–3756.
64 Zheng, H., Liu, X., Xu, C. et al. (**2015**). *Angew. Chem. Int. Ed. 54*: 10958–10962.

65 Malinakova, H.C. and Liebeskind, L.S. (**2000**). *Org. Lett. 2*: 4083–4086.
66 Kuang, Y., Lu, Y., Tang, Y. et al. (**2014**). *Org. Lett. 16*: 4244–4247.
67 Zhang, Y.-F., Yin, S.-J., Zhao, M. et al. (**2016**). *RSC Adv. 6*: 30683–30689.
68 Periyaraja, S., Shanmugam, P., and Mandal, A.B. (**2014**). *Eur. J. Org. Chem.* 2014: 954–965.
69 Whiting, M. and Fokin, V.V. (**2006**). *Angew. Chem. Int. Ed. 45*: 3157–3161.
70 Bae, I., Han, H., and Chang, S. (**2005**). *J. Am. Chem. Soc. 127*: 2038–2039.

7

Enantioselective Synthesis of Spiro Heterocycles

Michal Urban and Jan Veselý

Department of Organic Chemistry, Faculty of Science, Charles University, Prague, Czech Republic

7.1 Introduction

Spirocyclic framework is a common structural motif in many man-made and naturally occurring biologically active compounds, and it also represents valuable architecture successfully used in catalyst/ligand design and functional materials [1]. Thus, it is not surprising that there is a general interest in developing high-yielding and selective methods for their preparation. This chapter reviews catalytic synthetic methods allowing enantioselective construction of spiro heterocyclic compounds, excluding oxindole motif (for those compounds, please, see Chapters 5 and 6). This chapter is organized according to the kind of spirocyclic motif and present heterocycle. In the first part, enantioselective methods for the construction of spirocycles with one heteroatom are discussed, such as nitrogen-containing heterocycles (pyrrolidine, indolizine, and piperidine), oxygen- and sulfur-containing heterocycles (furanone, benzofuranone, pyrane, etc.). In the following part, synthetic strategies for the preparation of chiral spirocycles containing more than one heteroatom are presented, including pyrazolones, azlactones, barbituric acid derivatives, sultams, etc. Finally, catalytic methods for enantioselective preparation of other types of spiro heterocycles and bispiro compounds are mentioned. Each subchapter is organized according to the nature of the catalytic system used, starting with organometallic catalytic approaches, followed by organocatalysis and cooperative (synergistic) catalysis.

Spiro Compounds: Synthesis and Applications, First Edition. Edited by Ramon Rios Torres.

7.2 Enantioselective Synthesis of Spiro Heterocycle with One Nitrogen Atom

7.2.1 Enantioselective Formation of Spiropyrrolidines

7.2.1.1 Organometallic Approaches

In 2003, Hatano and Mikami reported the first transition metal-catalyzed approach for the synthesis of spiro compounds containing pyrrolidine ring **2**, **3** based on intramolecular ene-type cyclization of 1,6-enynes **1** (Scheme 7.1) [2]. The cyclization was efficiently catalyzed with Pd(II) complex and axially chiral N,P-ligand affording the final compounds in high yields with high enantioselectivity.

Soon after, the same group developed a highly enantioselective spirocyclization of 1,6-enynes **1** catalyzed by Rh(I) complex having (*S*,*S*)-skewphos ligand [3, 4]. While olefinic amide substrate gave the desired spiro amide-pyran in high yield and enantioselectivity (94% ee) without accompanying olefin migration, analogous olefinic ether substrates furnished spirocyclization product and also olefin-migration product; both were obtained with high enantioselectivity. Noteworthy, Hatano and Mikami also reported a highly enantioselective synthesis of spiroquinolines *via* ene-type cyclization of 1,7-enynes using Pd(II)/BINAP complex [5].

1,3-Dipolar cycloadditions can be also used for the construction of spirocycles containing pyrrolidine ring. That was demonstrated by Wang et al., who described Cu(I)-catalyzed 1,3-dipolar cycloaddition between azomethine ylides and cyclopropylidene acetates [6]. The presence of axially chiral biphenyl-based N,P-ligand exhibited excellent asymmetric induction affording functionalized 5-aza-spiro[2.4]heptanes in high yields with excellent diastereo- and enatioselectivity. Enantioselective 1,3-dipolar cycloaddition was also used for the construction of dispiropyrrolidine derivatives **5** from azomethine ylides **3** and α-alkylidne succinimides **4** by Yu, Deng et al. (Scheme 7.2) [7]. The combination of both Cu(I) and Cu(II) salts with newly designed 1,3-dihydroimidazolpyridine-based N,O-ligands (**II**) afforded dispiropyrrolidines in yields up to 90%, diastereoselectivities above 20 : 1, and enantioselectivites up to 97% ee. Recently, Li, Duan et al. reported another example of Cu(I)-catalyzed formation of dispiropyrrolidines *via* 1,3-dipolar cycloaddition [8], where new air-stable

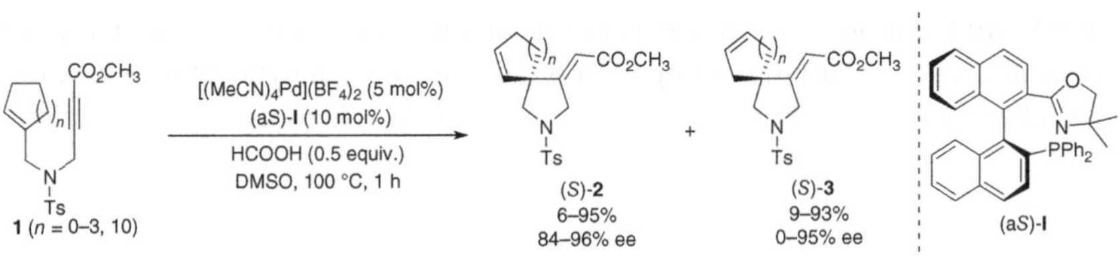

Scheme 7.1 Synthesis of spiropyrrolidines by Mikami. *Source:* Modified from Hatano and Mikami [2].

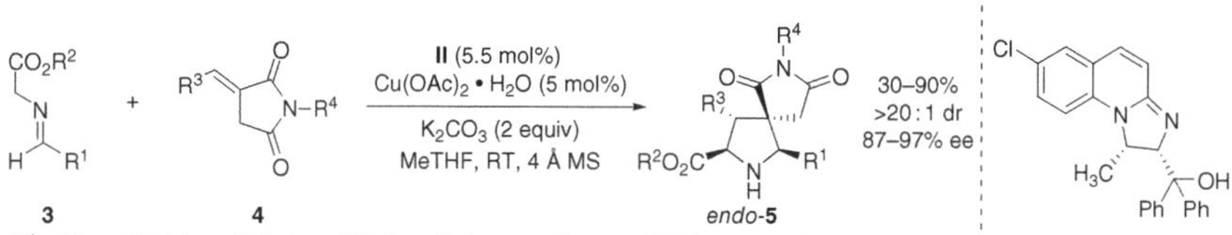

R^1 = Ph, *p*-ClC$_6$H$_4$, *m*-ClC$_6$H$_4$, *o*-ClC$_6$H$_4$, *p*-BrC$_6$H$_4$, *p*-FC$_6$H$_4$, *p*-CF$_3$C$_6$H$_4$, *p*-MeC$_6$H$_4$, *o*-MeC$_6$H$_4$, *p*-MeOC$_6$H$_4$, 2-Naphthyl, 2-Furyl, Cy
R^2 = Me, Et; **R**^{3} = Ph, *p*-BrC$_6$H$_4$, *p*-MeOC$_6$H$_4$, *p*-CF$_3$C$_6$H$_4$, *m*-BrC$_6$H$_4$, 2-Furyl, *n*-Pr, CH=CHPh;
R^4 = CO$_2$Me

Scheme 7.2 Asymmetric construction of a dispiropyrrolidine by Yu and Deng. *Source:* Based on Yang et al. [7].

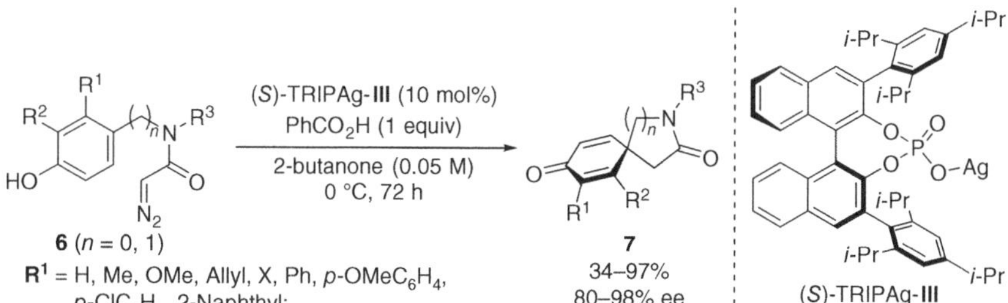

R^1 = H, Me, OMe, Allyl, X, Ph, *p*-OMeC$_6$H$_4$, *p*-ClC$_6$H$_4$, 2-Naphthyl;
R^2 - H, OMe;
R^3 = PMP, *t*-Bu, Allyl

Scheme 7.3 Silver phosphate-catalyzed asymmetric intramolecular dearomatization reported by Harada and Nemoto. *Sources:* Based on Bdiri et al. [9]; Nakayama et al. [10].

chiral 1-phosphanorbornadienes were used for diastereo- and enantiocontrol of the reaction.

Beside Cu-catalyzed methodologies, Ag(I) salts were also employed as efficient catalysts for the construction of chiral spiropyrrolidines (Scheme 7.3) [9, 10]. In 2017, Harada, Nemoto et al. reported an enantioselective intramolecular dearomatization of phenol-derived α-diazoacetamides **6** using chiral Ag(I) salt [6]. Among all catalysts screened, including Rh, Cu, Au, and Ag salts, TRIPAg was selected as the most efficient catalyst furnishing (in combination with benzoic acid) the desired spirocycles **7** with high level of chemoselectivity and enantiocontrol. Observed Ag-dependent chemoselectivity was explained by LUMO map analysis based on DFT calculations that showed Ag carbenoid more electrophilic and with longer distance than Rh carbenoid. Very recently, Li et al. reported an elegant Rh(III)-catalyzed [4+1] spiroannulation between nitrones **8** and *ortho*-quinone diazides **9** in the presence of AgF$_2$ as oxidant (Scheme 7.4) [11]. Regarding the scope of the reaction, excellent yields and enantioselectivities were observed for nitrones bearing EWG and EDG at phenyl ring, and for 7-substituted *ortho*-quinone diazides. Conversely, *para*-quinone diazide led to the formation of only racemic spiro product.

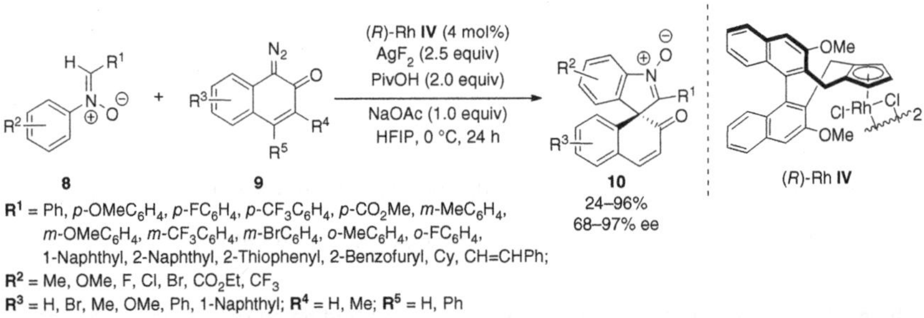

Scheme 7.4 Enantioselective [4+1] spiroannulation catalyzed by Rh. *Source:* Based on Kong et al. [11].

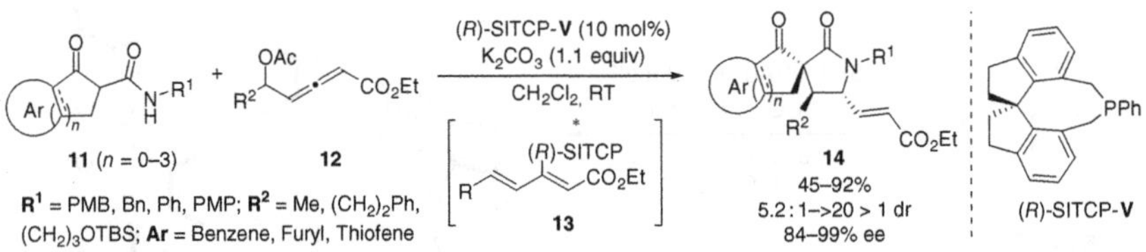

Scheme 7.5 Chiral phosphine-catalyzed synthesis of spiropyrrolidines reported by Zhou. *Source:* Based on Ni et al. [12].

7.2.1.2 Organocatalytic Approaches

The first organocatalytic synthesis of spiro compounds containing pyrrolidine ring was disclosed by Zhou et al. in 2017 using chiral phosphine-catalyzed [3+2] annulation of δ-acetoxy allenoates **12** with β-carbonyl amides **11** (Scheme 7.5) [12]. The stereochemistry of the reaction was controlled by an effective shielding of si-face in a key 2-phosphonium diene intermediate **13** by phenyl group present on (*R*)-SITCP catalyst **V**, allowing stereoselective 1,6-addition of β-carbonyl amide. The developed method features a wide substrate scope with respect to allenoate and β-carbonyl amides, affording the spirocyclic lactams in good yield with high diastereo- and enantioselectivity (up to 20 : 1 dr, 97% ee). Nearly simultaneously, Shi et al. reported a similar approach using PPh_3-catalyzed [3+2] spiroannulation, including their attempts for enantioselective variant [13]. Unfortunately, chiral BINAP and phosphine derived from phenylglycine afforded spirocycles with significantly lower yield and selectivity. Moderate level of enantioselectivity was also obtained in [3+2] spiroannulation between γ-phenyl allenoates and maleinimides catalyzed by chiral ferrocenylphosphines developed in Zhong's group [14]. Similarly, Waser et al. reported phase transfer-catalyzed [3+2] spiroannulation between trifluoropyruvate imines and benzylidene indandiones, but Cinchona alkaloid-based catalysts exhibited unsatisfying stereoselectivity in this process [15].

Recently, Zhao et al. reported another example of enantioselective formal [3+2] cycloaddition, employing isocyanoacetates **15** with *N*-itaconimides **16** (Scheme 7.6) [16]. The reaction was promoted by a dihydroquinine-derived squaramide catalyst **VI** affording succinimide-fused spiropyrrolidines in moderate to good yields with moderate diastereoselectivity and excellent enantioselectivity.

7.2.1.3 Cooperative-catalytic Approaches

An enantioselective synthesis of spiro compounds containing pyrrolidine ring can be also accomplished by combined catalysis, as demonstrated by Luo et al. in 2014 (Scheme 7.7) [17]. They reported a successful merger of photoredox catalysis with chiral primary amine catalysis in spirocyclization of β-ketoamides **18** with acylbromides **19**. The tandem α-photoalkylation/ketalization reaction is efficiently promoted by $Ru(bpy)_3Cl_2{\cdot}6H_2O$/diamine **VII** catalytic system rendering spiro-γ-lactams **20** as single diastereomers with high enantioselectivities. Another approach to enantioselective synthesis of spiropyrrolidines from *para*-quinone methanides and vinylaziridines was reported by Lin, Yao et al., using palladium and phosphine-thiourea cooperative catalysis [18]. Recently, Bernardi et al. disclosed cooperative palladium- and phosphoric acid-catalyzed [3+2] cycloaddition between vinylcyclopropanes and *N*-aryl imines; unfortunately, spiropyrrolidine was formed with a low level of enantiocontrol [19].

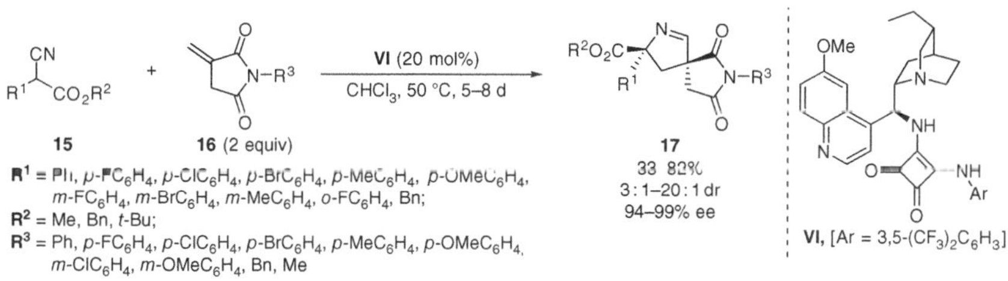

Scheme 7.6 Organocatalyzed asymmetric formal [3+2] cycloaddition of isocyanoacetates with *N*-itaconimides. *Source:* Based on Zhao et al. [16].

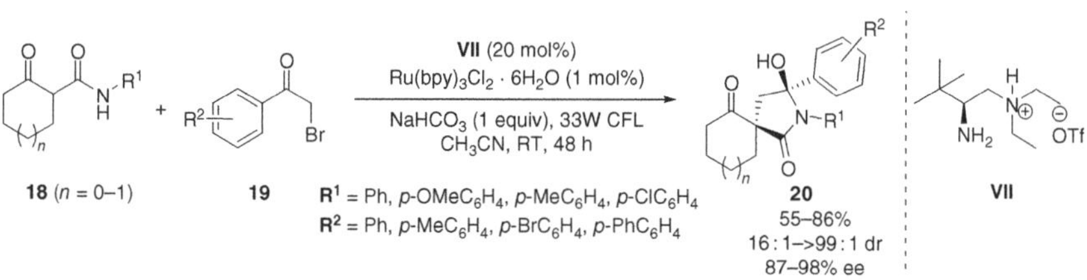

Scheme 7.7 The combination photoredox and primary amine catalysis used at the synthesis of spiropyrrolidine. *Source:* Based on Zhu et al. [17].

7.2.2 Enantioselective Formation of Spirocycles Containing Indoline and Indolenine Ring

7.2.2.1 Organometallic Approaches

The first highly enantioselective synthesis of spiroindolenines was developed by You et al. in 2010, using intramolecular Ir-catalyzed allylic alkylation of 3-indolyl allyl carbonates **21** (Scheme 7.8) [20]. Variously substituted spiroindolenines **22** were obtained in good yields with high dr and ee in the presence of $[Ir(cod)Cl]_2$ and fosforamidite **L-I** as the ligand, except 2-substituted spiroindolenines that were produced with moderate diastereoselectivity. The concept of Ir-catalyzed asymmetric dearomatization was further expanded by the same group in enantioselective construction of spiroindolines [21], and asymmetric allylic dearomatization of indoles by a desymmetrization strategy [22].

You et al. also reported Pd-catalyzed dearomative arylation of 3-substituted indoles using $[Pd(C_3H_5)Cl]_2$ and chiral spirophosphoramidate ligands, but resulting spiroindolenine derivatives were obtained only in moderate enantiomeric purity [23]. Another enantioselective approach to spiroindolenines was described by He, Liu et al., who used $Pd(dba)_2$ and chiral phosphoramidate **L-II** for cycloaddition between vinyl cyclopropanes **24** and arenesulfonyl indoles **23** (Scheme 7.9) [24]. The reaction is initiated by the formation of zwitterionic π-allylpalladium species that deprotonates arenesulfonyl indole forming α,β-unsaturated imine. Next, the conjugate addition of 1,3-dipole to imine and

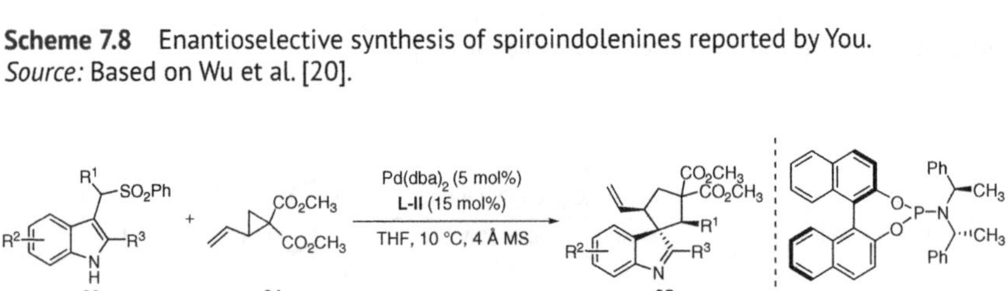

Scheme 7.8 Enantioselective synthesis of spiroindolenines reported by You. *Source:* Based on Wu et al. [20].

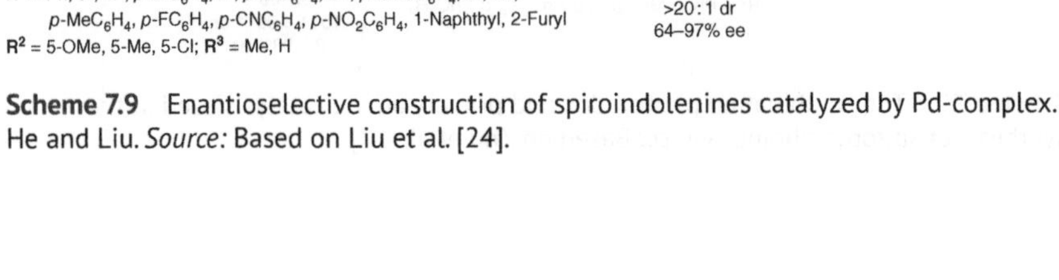

Scheme 7.9 Enantioselective construction of spiroindolenines catalyzed by Pd-complex. He and Liu. *Source:* Based on Liu et al. [24].

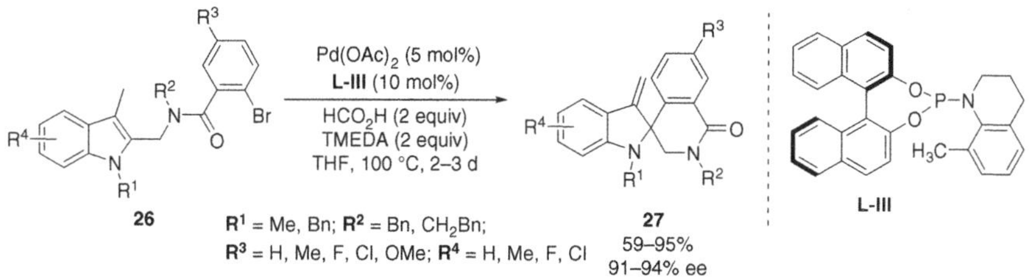

Scheme 7.10 Enantioselective synthesis of spiroindolines by intramolecular dearomative Heck reaction published by Jia. *Source:* Based on Li et al. [25].

subsequent intramolecular dearomatization furnishes spiroindolenines **25** as single diastereomers in good yields and high enantioselectivities. The main advantage of the intermolecular approach lies in the use of readily available starting materials.

In 2018, Jia et al. reported an enantioselective Pd-catalyzed intramolecular dearomative Heck reaction of 2-substituted indoles **26** using $Pd(OAc)_2$ and chiral BINOL- or H_8-BINOL-based phosphoramidates **L-III** (Scheme 7.10) [25]. The reaction renders the spiroindolines **27** in high yields with good to excellent enantioselectivities (up to 99% ee). In addition, the protocol was further expanded to other heterocycles, such as pyrrols, furans, and benzofuranes.

Au-catalyzed enantioselective synthesis of polycyclic spiroindolines from indoles substituted in C(3) position with propargylic alcohols was reported by Bandini et al. in 2012 [26]. The reaction in the presence of xylyl-BINAP ligand proceeds *via* hydroarylation/iminium trapping reaction sequence forming spirocycles with high stereoselectivity. Recently, a similar Au-catalyzed stereoselective approach to polycyclic spiroindolines was described by Ohno et al. [27], and the developed process was successfully applied for the synthesis of aspidosperma-type alkaloid, vindorosine.

Silver salts can also be used as catalysts for the enantioselective synthesis of spiroindolenines, as demonstrated by Taylor, Unsworth et al. [28]. They developed enantioselective dearomatizing spirocyclization of indoles bearing ynone moiety under catalysis with Ag salts of BINOL-based phosphoric acids, resulting in the formation of spiroindoline derivatives in good yields with moderate enantioselectivity.

7.2.2.2 Organocatalytic Approaches

In 2012, Wu, Cao, Zhao et al. developed the first enantioselective synthesis of highly substituted spiroindolenines using the organocatalytic approach. The Michael/Mannich cascade reaction between enals **29** and 2-methylindoles **28**, proceeding through an iminium intermediate, was efficiently catalyzed by secondary amine **VIII** and TFA, affording the tetracyclic spirocycles in high yields with moderate diastereoselectivity and high enantioselectivity [29]. Soon after, You et al. reported [(DHQD)$_2$PHAL]-catalyzed chlorinative dearomatization of indole-derived benzamides **31** with 1,3-dichloro-5,5-dimethylhydantoin **32**, giving chiral spiroindolines **33** in high yields with good to excellent enantioselectivity (Scheme 7.11) [30].

In 2017, the same group reported another organocatalytic approach for the preparation of chiral spiroindolenines [31]. In the presence of chiral phosphoric acid, indolyl enones undergo intramolecular dearomative Michael addition affording the desired spirocycles in good yields with high stereoselectivity (19 : 1 dr, 42–97% ee). A year later, You et al. developed organocatalytic dearomatization of indolyl dihydropyridines **36** using a chiral phosphoric acid **IX** and Hantzsch ester **37** (Scheme 7.12) [32]. The reaction sequence proceeds through acid-catalyzed enamine isomerization to iminium intermediate, followed by spirocyclization and subsequent transfer hydrogenation. The reaction proceeded under mild reaction conditions and gave spiroindolines as single diastereomers in high yields with high enantioselectivity.

N-Heterocyclic carbene catalysis can be also used for the enantioselective synthesis of spiroindolenines, as demonstrated by Studer et al. (Scheme 7.13) [33]. The reaction comprised 2-substituted indolyl benzaldehydes **39** in the presence of NHC precatalyst **X**, base (K$_2$CO$_3$), and bisquinone **40** as oxidant, and various spiroindolenines were attained in excellent yields and enantioselectivity.

In 2019, Liu, Chen, Ji et al. developed a cupreidine-catalyzed cyclization cascade reaction between 2-alkynyl-3,3-difluoro-*3H*-indoles **42** and 2-mercaptoimidazoles

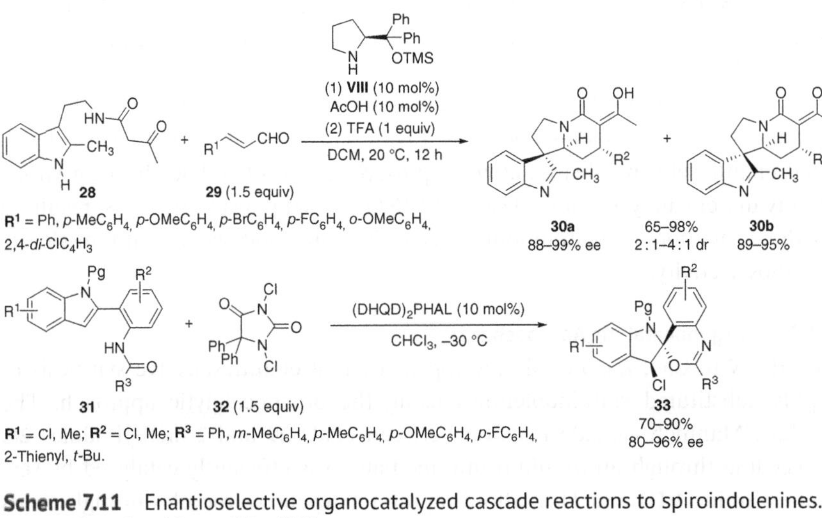

Scheme 7.11 Enantioselective organocatalyzed cascade reactions to spiroindolenines. *Source:* Based on Yin and You [30].

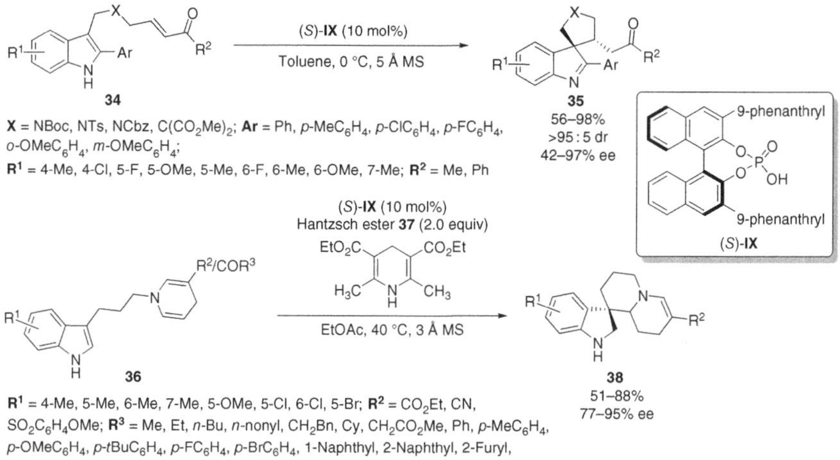

Scheme 7.12 Application of chiral phosphoric acid for the preparation of spiroindolines reported by You. *Source:* Based on Xia et al. [32].

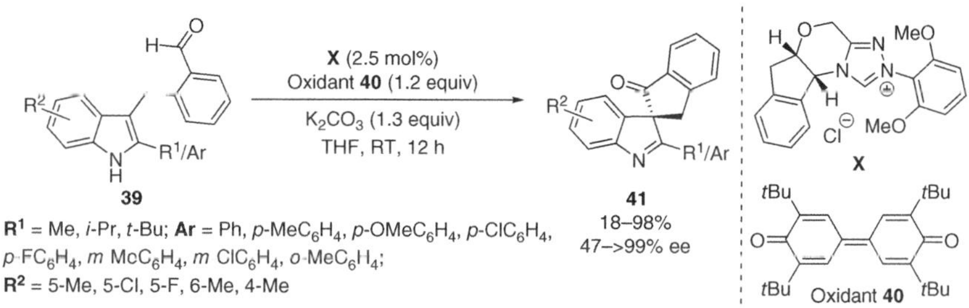

Scheme 7.13 Oxidative *N*-heterocyclic carbene catalyzed dearomatization of indoles to spirocyclic indolenines. *Source:* Based on Bera et al. [33].

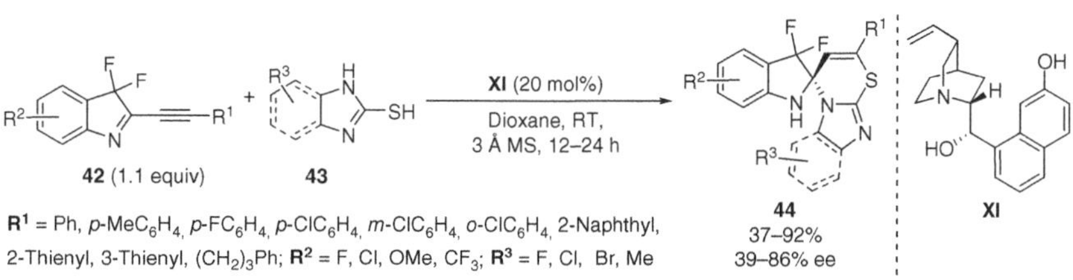

Scheme 7.14 Enantioselective access to gem-difluorinated spiroindolines. *Source:* Based on Luo et al. [34].

43 (Scheme 7.14) [34]. Generally, the chiral *gem*-difluorinated spiroindolines **44** were produced in moderate yields and enantioselectivities; unfortunately, substituted 2-mercaptoimidazoles led to the formation of mixture of two regioisomers with ratios ranging from 1 : 1 to 2 : 1.

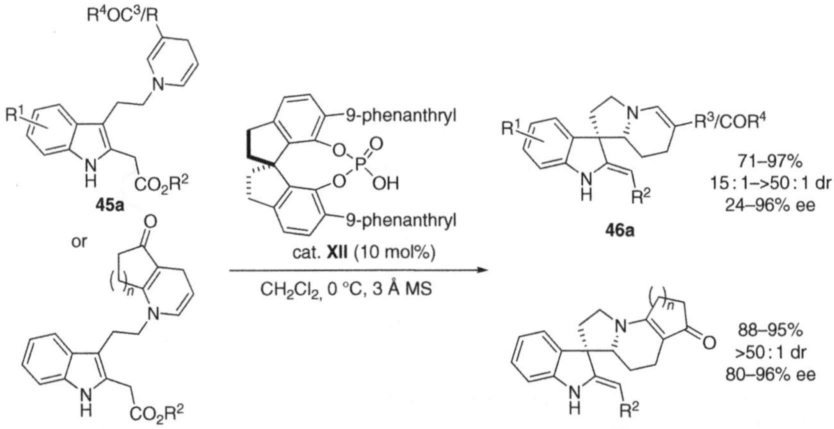

R[1] = 5-OMe, 5-Br, 6-OMe, 6-Cl, 4-OMe, 4-Cl, 7-Et; **R**[2] = Me, Et, *i*-Pr, *t*-Bu, Bn, Ph;
R[3] = CO_2Me, CO_2Ph, CN, SO_2Ph; **R**[4] = Me, Ph, *p*-ClC_6H_4, *m*-ClC_6H_4

Scheme 7.15 Enantioselective synthesis of spiroindolines via cascade reaction. *Source:* Based on Pan et al. [35].

Recently, Xia et al. reported another enantioselective synthesis of spiroindolines using chiral SPINOL-derived phosphoric acid **XII** (Scheme 7.15) [35]. In this protocol, the enamine izomerization/spirocyclization sequence is followed by the stabilization of the generated imine to the more stable enamine, which is promoted by the presence of a methylcarboxyl group on the C(2) of indole.

7.2.3 Enantioselective Formation of Spiropiperidines

7.2.3.1 Organometallic Approaches

In 2005, Romo et al. reported an enantioselective Cu-catalyzed Diels–Alder reaction between electron-deficient α-methylene lactam and electron-rich diene providing spirocyclic lactam, s key intermediate in the total synthesis of (−)-gymnodimine [36]. Evans' copper-bis(oxazoline) hexafluoroantimonate complex promoted the reaction with high diastereo-and enantiocontrol (dr > 19 : 1, 95% ee). Ten years later, Cai et al. developed a highly enantioselective synthesis of spirobilactams **48** through Cu-catalyzed intramolecular double *N*-arylation and phase separation (Scheme 7.16) [37]. Spirocyclization was efficiently promoted by CuI in the presence of chiral 1,2-diamine-derived ligand **L-IV** in acetonitrile and enantioenrichment of products was increased due to an amplification of solution ee through *in situ* precipitation of the racemates.

Prior to this work, in 2009, Sasai et al. reported an enantioselective synthesis of analogous spirobilactams [38], through intramolecular double *N*-arylation

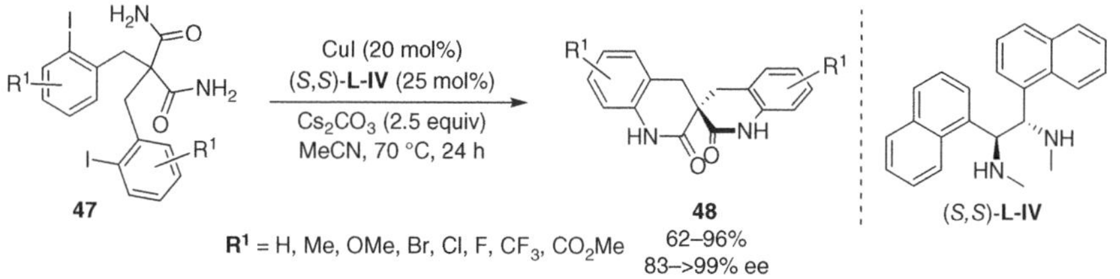

Scheme 7.16 Enantioselective synthesis of spirobilactams reported by Cai. *Source:* Based on Liu et al. [37].

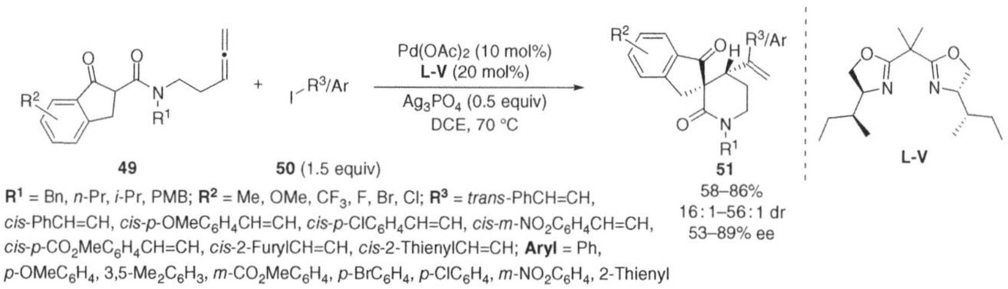

Scheme 7.17 Enantio- and diastereoselective Pd-catalyzed carbocyclization cascade reaction. *Source:* Based on Li et al. [39].

catalyzed by $Pd(OAc)_2$ in the presence of (*S*)-BINAP. In their work, excellent yields but only moderate enantiocontrol of the reaction was observed. Several years later, enantio- and diastereoselective Pd-catalyzed carbocyclization cascade between allene-linked ketoamides **49** and aryl and vinyl iodides **50** was disclosed by Dixon et al. (Scheme 7.17) [39]. The reaction, affording final spirolactams through allyl intermediate formation followed by intramolecular alkylation, is efficiently catalyzed with $Pd(OAc)_2$ in the presence of chiral bis(oxazoline) ligand **L-V** and Ag_3PO_4 enhancing the level of asymmetric induction. Very recently, enantioselective preparation of tetracyclic spiropiperines using Pd-catalyzed arylative dearomatization of *para*-aminophenols **52** was also realized by You et al. (Scheme 7.18) [40]. The reaction is mediated with Pd-catalyst generated from $[Pd(C_3H_5)Cl]_2$ and chiral TADDOL-derived phosphoramidite ligand **L-VI** affording tetracyclic spiroamines **53** in moderate yields (up to 73%) and excellent enantioselectivity (up to 97% ee).

In 2011, the same group reported the first enantioselective Ir-mediated intramolecular allylic dearomatization of phenols **54** (Scheme 7.19) [41]. $[Ir(cod)Cl]_2$ and BINOL-derived fosforamidites **L-VII** were employed to achieve chiral spiropiperidines and spiropyrrolidines in good yields and excellent enantioselectivities. Noteworthy, the concept of Ir-catalyzed asymmetric allylic dearomatization was also expanded to 2-substituted pyrroles [42].

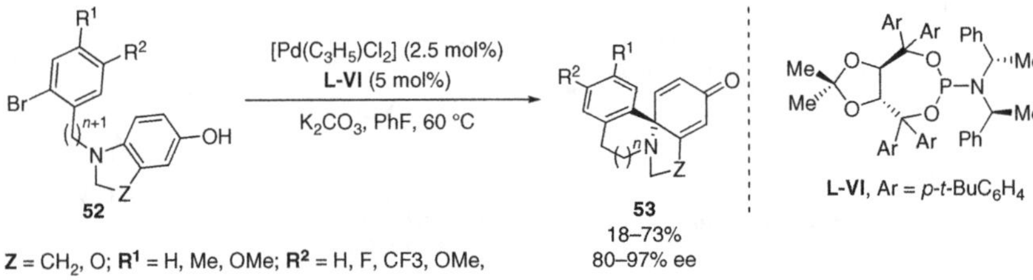

Scheme 7.18 Pd-catalyzed asymmetric intramolecular arylative dearomatization of para-aminophenols. *Source:* Based on Xu et al. [40].

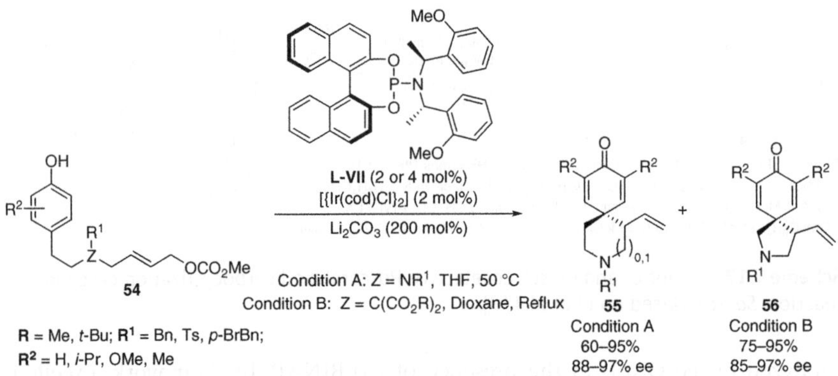

Scheme 7.19 Iridium-catalyzed intramolecular asymmetric allylic dearomatization of phenols reported by You. *Source:* Based on Wu et al. [41].

7.2.3.2 Organocatalytic Approaches

In 2014, Bonne and Rodriguez reported an enantioselective organocatalytic formal [3+3] spiroannulation of cyclic β-ketoamides **57** with α,β-unsaturated acyl cyanides **58** affording spiroimides **59** in good yields and high stereoselectivity (up to 10 : 1 dr and 92% ee, Scheme 7.20) [43]. The reaction proceeding through Michael addition/acylation reaction sequence is efficiently catalyzed by bifunctional tertiary amine-thiourea **XIII**.

Three years later, Peng, Han et al. [44], Huang et al. [45], and Meazza, Veselý, Rios et al. [46] independently reported other enantioselective protocols for the synthesis of enantiomerically enriched spiropiperidines. The developed methods are based on secondary amine-catalyzed enantioselective [3+3] spiroannulation of cyclic β-ketoamides with enals, proceeding *via* a Michael addition/hemiaminal annulation reaction sequence as a key step. In addition, Huang's one-pot multicomponent protocol involves the formation of cyclic β-ketoamides using heat-assisted Wolff rearrangement. Whereas the first two spiroannulations are mediated by chiral imidazolidinones, the latter transformation

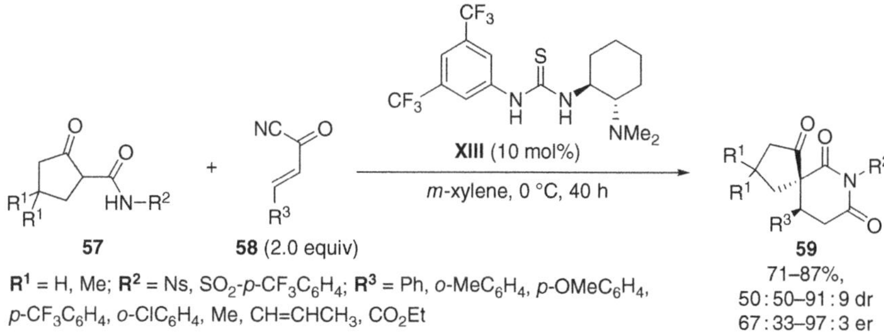

Scheme 7.20 Enantioselective organocatalyzed formal [3+3] spiroannulation reported by Bonne and Rodriguez. *Source:* Based on Goudedranche et al. [43].

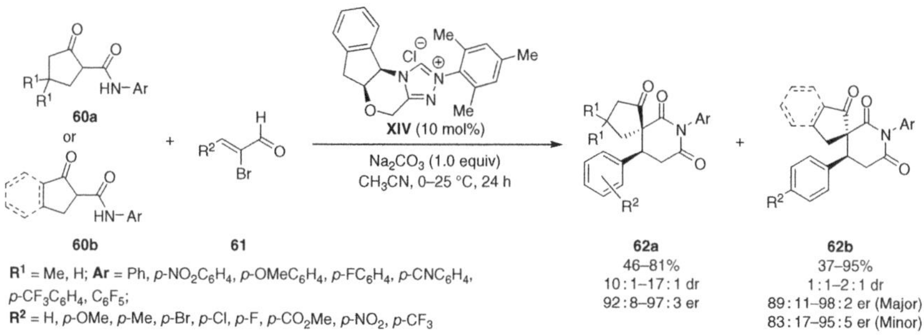

Scheme 7.21 NHC-catalyzed enantioselective synthesis of spiropiperidinones reported by Biju. *Source:* Based on Mondal et al. [47].

employed prolinol-based secondary amines. All three methods afforded optically active spiroamides in good yields and excellent enantioselectivities.

Another organocatalytic approach for the enantioselective synthesis of spiropiperidones was developed by Biju et al., in 2018 (Scheme 7.21) [47]. They described a novel *N*-heterocyclic carbene-catalyzed [3+3] spiroannulation of 2-bromoenals **61** and β-ketoamides **60a,b** affording functionalized spiropiperidinones **62a,b** in good yields with high stereoselectivities (up to 17 : 1 dr and 94% ee). A combination of chiral triazolinium salt **XIV** with Na_2CO_3 displayed the best catalytic activity in acetonitrile. Soon after, Xu et al. reported similar [3+3] spiroannulation using 2-bromoenals and β-ketothioamides [48]. A broader survey of triazolinium salts revealed precatalyst **XIV** in combination with DBU as the most active, affording functionalized spiropiperidinones in high yields and stereoselectivities (up to >20 : 1 dr, 99% ee).

The same year, an enantioselective dearomative formal [4+2] cycloaddition reaction between pyridinium salts **63** and enones **64** was reported by Liang and

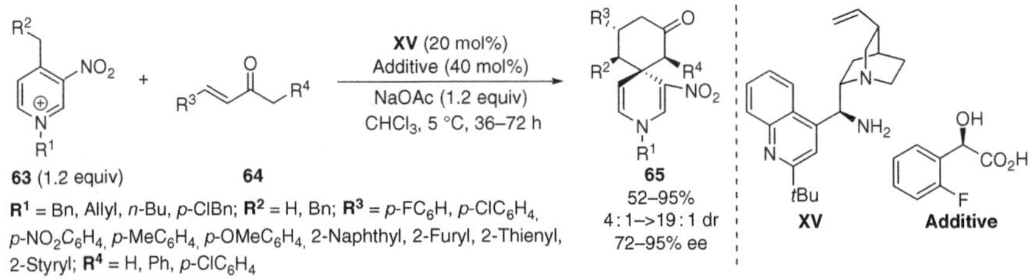

Scheme 7.22 Asymmetric dearomative formal [4+2] cycloadditions published by Liang and Chen. *Source:* Based on Yan et al. [49].

Chen (Scheme 7.22) [49]. The reaction proceeding *via* iminium/enamine catalysis in the presence of Cinchona-derived primary amine **XV** and 2-fluoromandelic acid produced final spirocyclic compounds in good to excellent stereoselectivity.

7.3 Enantioselective Formation of Spirocycles Containing Heterocycle with One Oxygen Atom

7.3.1 Enantioselective Formation of Spirocycles Containing Dihydrofuranone and Tetrahydrofuran Ring

7.3.1.1 Organometallic Approaches

In 2003, Mikami et al. reported the first example of enantioselective ene-type spirocyclization of allyl propargyl ethers **66** catalyzed by $[(CH_3CN]_4Pd](BF_4)_2$ with BINOL-based P,N-ligand bearing oxazoline motif **I** (Scheme 7.23) [50]. Two regioisomeric spirotetrahydrofurans **67a,b** containing 5- to 15-membered rings were obtained in high yields with good to high enantioselectivity.

Enantioselective preparation of spirotetrahydrofurans using Pd-catalyzed formal [3+2] cycloaddition between *para*-quinone methanides **68** and vinyl epoxides **69** was disclosed by Zhao et al. (Scheme 7.24) [51]. The reaction was efficiently catalyzed with only 1 mol% of $Pd_2(dba)_3$ and P,N-ligand bearing oxazoline **L-VIII** affording spirotetrahydrofuranes **70** in high yields with high diastereo- and

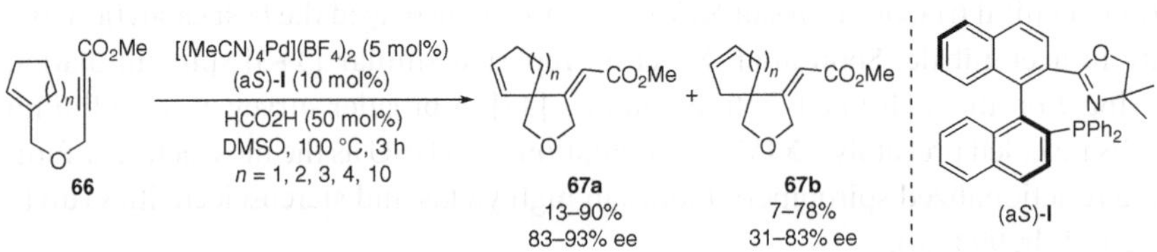

Scheme 7.23 Enantioselective ene-type spirocyclization developed by Mikami. *Source:* Based on Hatano and Mikami [50].

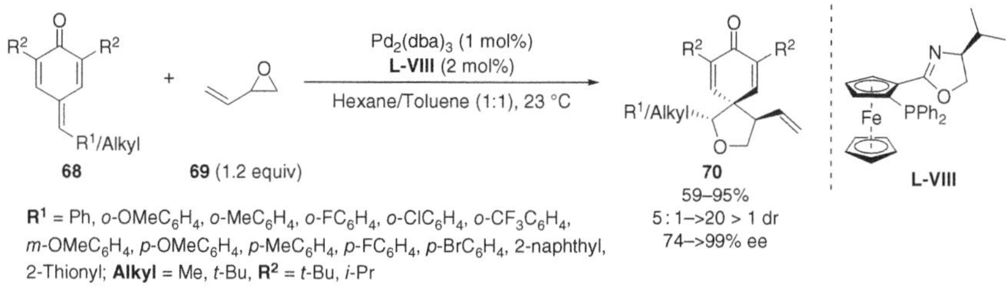

Scheme 7.24 Enantioselective 1,6-conjugate addition/annulation of para-quinone methides with vinyl epoxides reported by Zhao. *Source:* Based on Ma et al. [51].

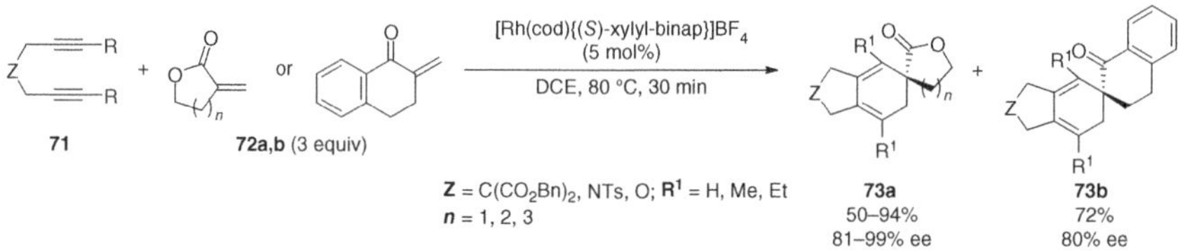

Scheme 7.25 Enantioselective synthesis of spirodihydrofuranes developed by Shibata. *Source:* Based on Tsuchikama et al. [52].

enantioselectivity (up to >20 : 1 dr and 99% ee). Noteworthy, the reaction is highly stereospecific to the solvent system used; a mixture of hexane and toluene (1 : 1) was found as optimal. Rh-catalyzed methods used for the stereoselective preparation of spirocycles containing dihydrofuranone ring are represented by the work of Shibata et al., who constructed chiral spirodihydrofurans **73a,b** *via* [2+2+2] cycloaddition of diynes **71** and methylene 5- to 7-membered lactones **72a,b** (Scheme 7.25) [52]. Final spirocycles were obtained in high yields and high enantioselectivity upon catalysis with [Rh(cod)(*S*)-xylyl-BINAP)BF_4 in short reaction times in DCE at elevated temperature.

To date, several Cu-catalyzed methods were also developed. In 2012, Gade et al. reported enantioselective alkylation of β-keto esters **74a,b** using *in situ* generated alkyl iodides **77** from alcohols, and including enantiocontrol of the alkylation process by Cu(II) complex bearing chiral pincer boxmi ligand **L-XI** (Scheme 7.26) [53]. As a part of the work, one-pot enantioselective synthesis of spiro- and bispirolactones **76**, **78** based on enantioselective alkylation with subsequent $BF_3{\cdot}Et_2O$-promoted cyclization was developed. Later, Chemler et al. announced an interesting preparation of spirocyclic ethers *via* enantioselective Cu-catalyzed carboetherification of alkenols **79** (Scheme 7.27) [54]. The reaction proceeds through enantioselective *cis*-oxycupration in the presence of $Cu(OTf)_2$ and (*S*,*S*)-*t*Bu-Box **L-X**, homolysis of formed C–Cu bond producing radical intermediate **81**, followed by *endo* or *exo* cyclization with subsequent oxidation or reduction. While

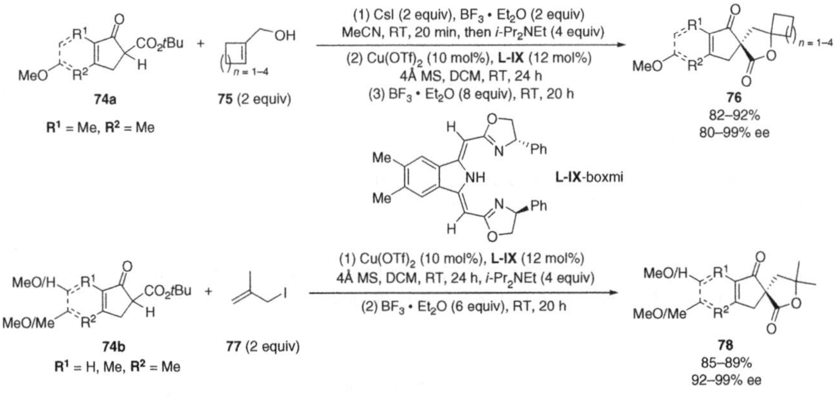

Scheme 7.26 The enantioselective synthesis of spirolactones reported by Gade. *Source:* Based on Deng et al. [53].

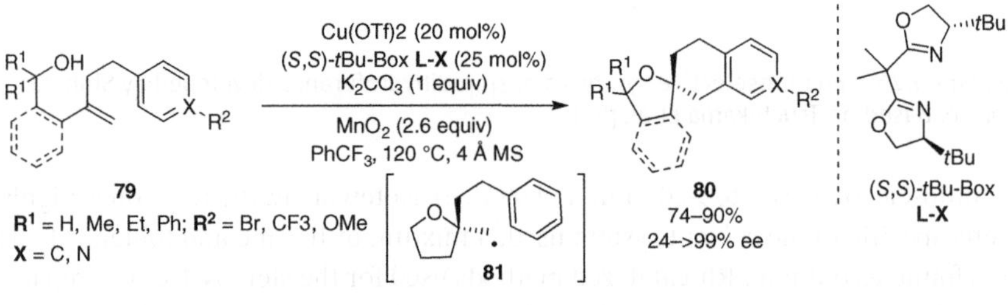

Scheme 7.27 Asymmetric copper-catalyzed preparation of spirotetrahydropyranes. *Source:* Based on Karyakarte et al. [54].

aryl-substituted alkenols favor *endo*-cyclization giving spiro-[4,5]-products, alkenyl- and alkynyl-substituted alkenols favor *exo*-cyclization providing spiro-[4,4]-products. Both types of spirotetrahydrofurans were obtained in high yields and with an excellent level of enantiomeric purity.

In 2013, an enantioselective desymmetrization of prochiral spiro cyclohexadienone lactones was reported by Wang et al. [55] 1,3-Dipolar cycloaddition between lactones and iminoesters was efficiently catalyzed by AgOAc and (*S*)-TF-BiphamPhos ligand, yielding the corresponding spirodihydrofuranones as single diastereomers in good yields (73–89%) with excellent enantioselectivity (up to 99% ee).

Recently, in 2019, Hua, Chang, Wang et al. reported an enantioselective Zn-catalyzed tandem Michael/transesterification reaction between α-hydroxy indanones and α,β-unsaturated esters. Final spirodihydrofuranones were formed with an excellent level of diastereo- and enantiocontrol (up to >20 : 1 dr and >99% ee) under the catalysis with dinuclear zinc complex formed from $ZnEt_2$ and AzePhenol ligand [56].

7.3.1.2 Organocatalytic Approaches

In 2008, Kita et al. reported the first enantioselective dearomatization of phenols **82** using chiral hypervalent iodine (III) reagent proceeding through oxidative spirolactonization reaction (Scheme 7.28) [57]. In addition, a catalytic version of spirolactonization was developed under catalysis with a chiral spirobiindane-based iodoarene (**XVII**) in the presence of *m*CPBA as oxidant, where final *ortho*-spirolactones were obtained in good yields and moderate enantioselectivity (65–69%). Preparation of a series of chiral spirobiindane-based iodoarenes, their evaluation for the oxidative spirolactonization, and theoretical studies revealed that the enantiocontrol is improved by the presence of the *ortho* substituents [58]. Thus, spirobiindane catalyst **XVII** efficiently catalyzed the enantioselective dearomatization of phenols furnishing *ortho*-spirolactones with enantioselectivity up to 92% ee.

In 2010, Ishihara et al. developed another highly effective precatalyst for the enantioselective Kita's oxidative spirolactonization. They employed a conformationally flexible iodoarene derived from lactate in combination with *m*CPBA, and obtained *ortho*-spirolactones with excellent enantioselectivity (up to 92% ee) [59]. The same group further studied the design of chiral organoiodine precatalyst and chiral organoiodine derived from 2-aminoalcohol **XVIII** was successfully used for enantioselective dearomatization of less reactive substrates, such as 1- and 2-naphthols (Scheme 7.29) [60]. Noteworthy, the use of alcohol additives (HFIP or EtOH) was found to be crucial for the induction of high enantioselectivity. Later, the same group expanded this approach also to other substrates, such as *ortho*- and *para*-hydroquinone derivatives, affording chiral spiroketals [61].

Besides the above mentioned, other chiral iodoarenes were prepared and successfully employed in oxidative Kita's spirolactonization in the presence of *m*CPBA. In 2015, Ibrahim et al. reported the use of *anti*-dimethanoantracene-based iodoaren in spirolactonization of 1-naphthols, affording final spirodihydrofuranones with a medium level of enantiocontrol [62]. Soon after, Pericàas, Nachtsheim et al. [63] developed a modular synthesis of triazole-based chiral

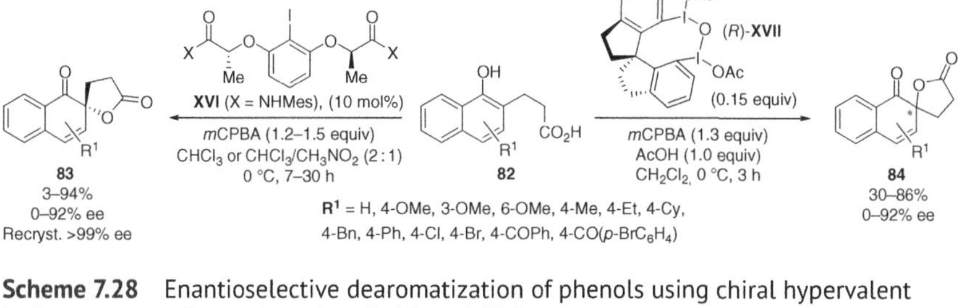

Scheme 7.28 Enantioselective dearomatization of phenols using chiral hypervalent iodine reagents. *Source:* Based on Dohi et al. [57].

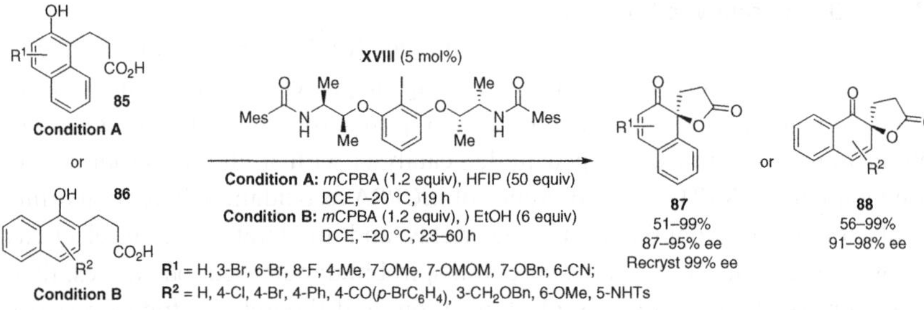

Scheme 7.29 Chiral hypervalent organoiodine-catalyzed enantioselective oxidative spirolactonization of naphthol derivatives. *Source:* Based on Uyanik et al. [60].

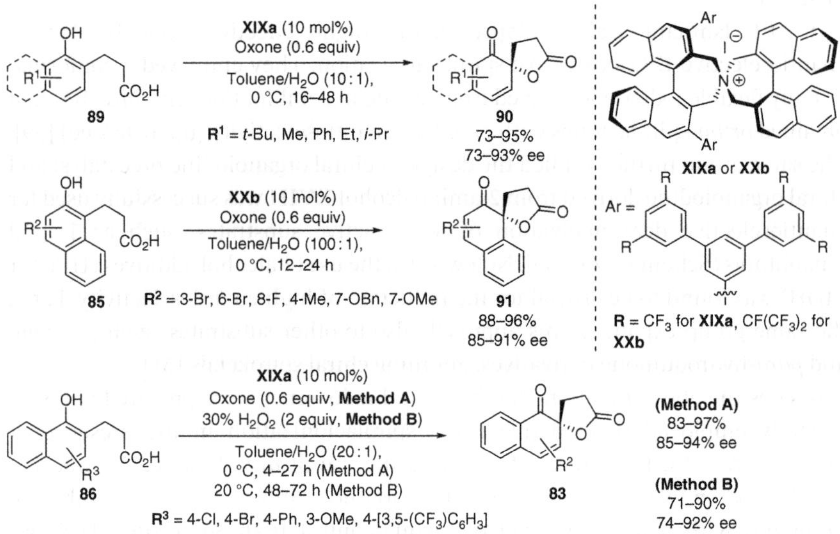

Scheme 7.30 Enantioselective oxidative dearomatization of arenols using chiral quaternary ammonium iodide. *Source:* Based on Uyanik et al. [65].

iodoarenes and Kita et al. [64] reported the synthesis of chiral atropoisomeric biaryl iodides. Both precatalysts were evaluated in Kita's spirolactonization of 1-naphthols; unfortunately, the corresponding spirodihydrofuranones were obtained with only moderate to a high level of enantiomeric purity. Recently, Ishihara et al. developed enantioselective oxidative dearomatization of arenols using chiral quaternary ammonium iodide catalyst **XIXa** in combination with an environmentally benign oxidant, oxone (Scheme 7.30) [65]. A wide variety of arenols including phenols, 1- and 2-naphthols can be readily converted to the

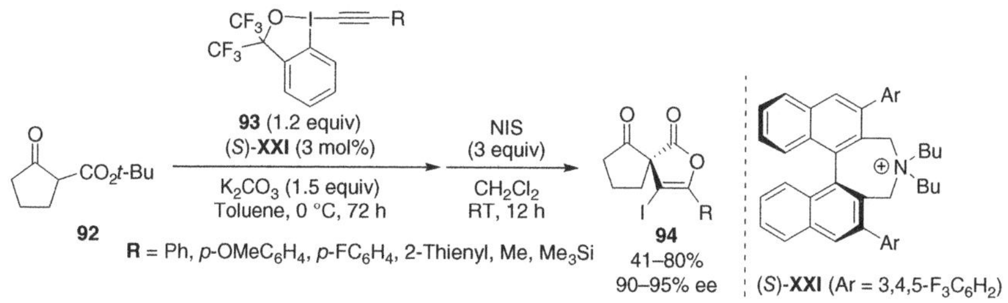

Scheme 7.31 Asymmetric synthesis of spiro-2(*3H*)-furanones *via* phase-transfer-catalyzed alkynylation. *Source:* Based on Wu et al. [67].

corresponding spirolactones in high yields (71–97%) and high enantioselectivity (73–94% ee). Based on control experiments and Raman analysis, I^+ species such as hypoiodous acid and $[SO_4I]^-$ are key active species in a proposed reaction mechanism.

Recently, Suzuki et al. reported an elegant application of enantioselective oxidative dearomatization strategy using chiral hypervalent iodine compounds in total synthesis of (−)-maldoxine [66].

An efficient enantioselective synthesis of spirofuranones was also achieved by Maruoka et al., in 2014 (Scheme 7.31) [67]. They developed an enantioselective synthesis of spiro-2(*3H*)-furanones via enantioselective alkynylation/electrophilic cyclization reaction sequence between β-keto esters **92** and hypervalent iodine reagents **93**, using chiral quaternary ammonium salts **XXI** as phase-transfer catalysts.

The same year, Albrecht et al. reported another enantioselective approach to spirofuranones, trienamine-mediated [4+2] cycloaddition between 3-alkylidene-furan-2(*3H*)-ones and dienals catalyzed by secondary amine **VIII** [68]. The final spirocycles were obtained in good yields with moderate diastereoselectivity and excellent enantioselectivity. BINOL-based phosphoric acids were also used as organocatalysts for the construction of spirotetrahydrofurans, as demonstrated by Jiang, Yin et al. [69]. They developed an asymmetric Pinacol-type rearrangement of furanyl cyclobutanols producing final spirotetrahydrofurans in high yields but with only moderate enantioselectivity (up to 73% ee). Recently, Ghorai et al. reported an interesting enantioselective desymmetrization reaction of keto-bischalcones with malononitrile, resulting in the formation of spirotetrahydrofurans in good yields, high diastereo- and excellent enantioselectivity (up to 32 : 1 dr and 99% ee) [70]. The reaction proceeding through Michael addition/1,2-addition/oxa-Michael addition cascade process is efficiently catalyzed by a quninine-derived amino-squaramide bifunctional catalyst.

7.3.2 Enantioselective Formation of Spirocycles Containing Benzofuranone and Benzodihydrofuran Ring

7.3.2.1 Organometallic Approaches

In 2007, Trost et al. reported a highly enantioselective Pd-catalyzed construction of spirocyclic oxindoles based on [3+2] cycloaddition between Pd-trimethylenemethane complex and 3-alkylidene oxindoles **95** (Scheme 7.32) [71]. The reaction was efficiently catalyzed by $Pd(dba)_2$ in the presence of BINOL-based phosphoramidates **L-XI** and one example of an enantioselective synthesis of spirobenzofuranone as a part of substrate scope was described. Pd catalysis in cooperation with Lewis acid catalysis was also used for diastereoselective synthesis of spirobenzofuranones reported by Zhao et al. [72] They developed Pd-Ti relay catalysis between enones and vinylethylene carbonates and attempts of enantioselective [4+2] cycloaddition were performed. Unfortunately, only a medium level of enantiocontrol was reached in the presence of TADDOL as a chiral ligand.

In 2014, Katsuki et al. reported an enantioselective Fe-catalyzed spirocyclization between naphthols **98** and phenols **99**, involving oxidative *ortho*-quinone methanide formation, Michael addition, and dearomatization reaction sequence (Scheme 7.33) [73]. A chiral Fe(salene) complex combined with air as oxidant and elevated temperature was used to achieve spirobenzodihydrofurans in high yields and enantioselectivities. Later, Katsuki, Uchida et al. further expanded this

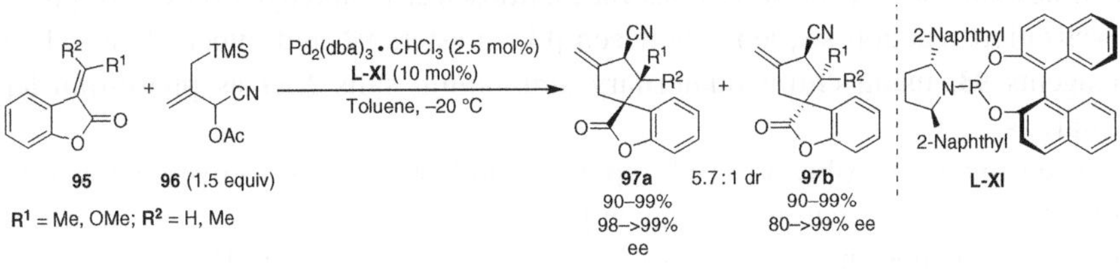

Scheme 7.32 Enantioselective construction of spirocyclic benzofuranone cyclopentanes by palladium-catalyzed trimethylenemethane-[3+2]-cycloaddition. *Source:* Based on Trost et al. [71].

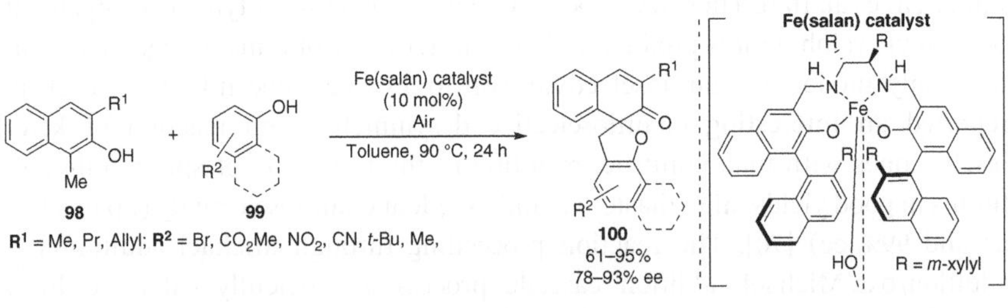

Scheme 7.33 Asymmetric tandem synthesis of spirocyclic (2H)-dihydrobenzofurans. *Source:* Based on Oguma and Katsuki [73].

concept to methylenebisarenols, prepared from 2-naphthols and salicyl alcohols under MW irradiation [74, 75].

Catalysis with chiral Sc(III)-phosphine oxide complex to construct chiral spirobenzodihydrofurans from 2-benzofuranylmethyl *N*-tosylcarbamates **101** was firstly used by Shi et al. (Scheme 7.34) [76]. The reaction can be performed in gram scale, and this method allows the synthesis of a wide variety of optically active oxazolidinone-fused spirobenzodihydrofurans **102** in high yields with excellent enantioselectivity (up to 97% ee). In 2018, Shao and He reported an enantioselective Ag-catalyzed [3+2] cycloaddition between isocyanoacetates and aurone analogs, affording chiral pyrroline-fused spirobenzofuran-3-ones in excellent yields, diastereo- and enantioselectivities. The reaction was catalyzed by the chiral Ag complex generated from Ag_2O and chiral Cinchona alkaloid-derived phosphine.

Recently, Tu, Wang et al. reported an efficient method to form chiral spirobenzofurans **105** through Cu-catalyzed [3+2] cyclization/semipinacol rearrangement sequence (Scheme 7.35) [77]. A catalytic system composed of $Cu(ClO_4)_2{\cdot}6H_2O$ and BOX ligand **L-XIII** efficiently produced key intermediates **106** from substituted benzoquinones **104** and allylic alcohols **103**. Subsequent treatment of intermediate **106** with $SnCl_4$ followed by reduction with $NaBH_4$ to avoid hemiketal formation provided final spirocycles in good yields with excellent diastereo and enantioselectivities (up to 67% yield, >20 : 1 dr, 92% ee) (Scheme 7.36).

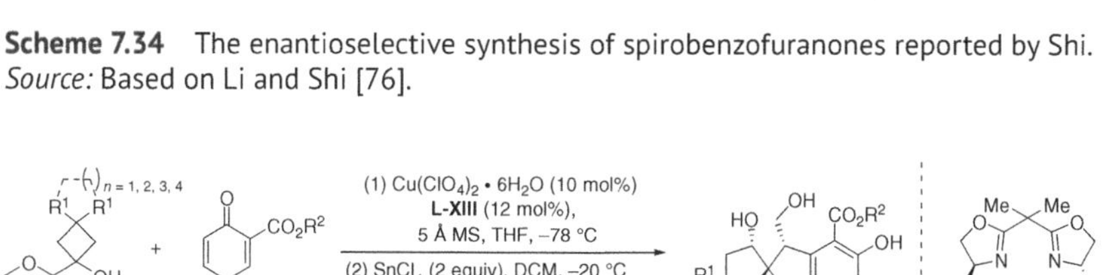

Scheme 7.34 The enantioselective synthesis of spirobenzofuranones reported by Shi. *Source:* Based on Li and Shi [76].

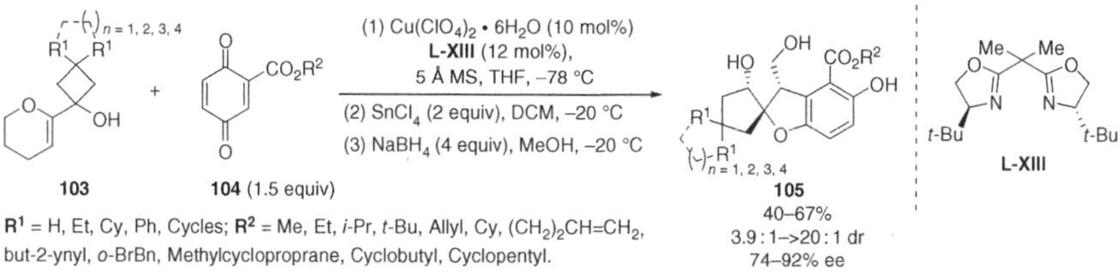

Scheme 7.35 The catalytic asymmetric one-pot [3+2] cyclization/semipinacol rearrangement sequence introduced by Tu and Wang. *Source:* Based on Liu et al. [77].

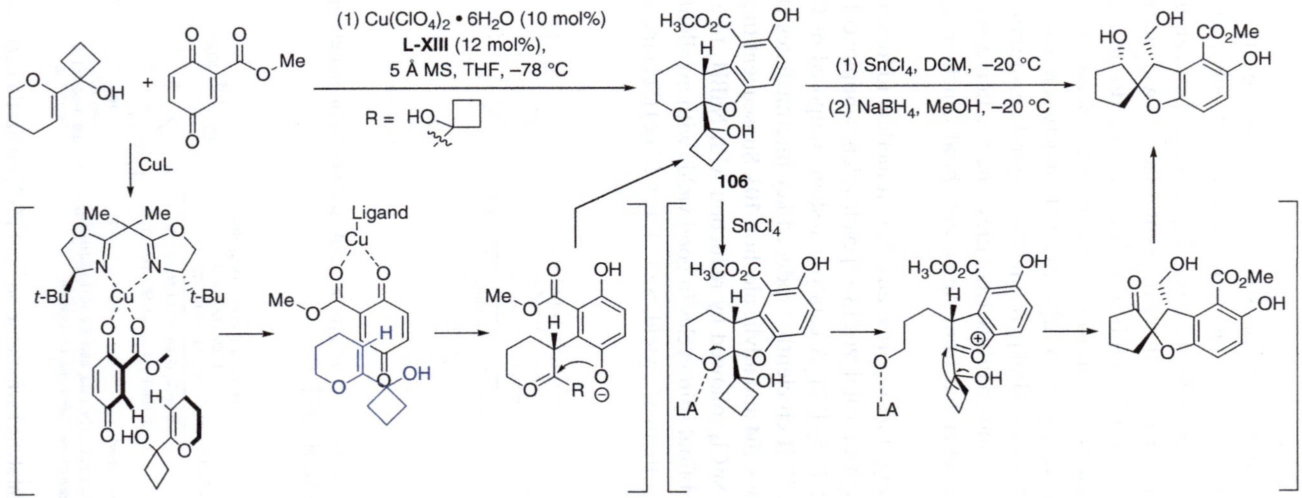

Scheme 7.36 Reaction mechanism.

7.3.2.2 Organocatalytic Approaches

In 2010, Rios et al. reported a highly enantioselective synthesis of various chiral spiroheterocycles from corresponding heterocycles and enals using secondary amine catalysis (Scheme 7.37) [78]. The cascade reaction between **107a,b** and **108** proceeds under secondary amine **VIII**/benzoic acid catalysis *via* double Michael addition followed by aldolization/dehydration sequence. Preparation of two regioisomeric spirobenzofurans from the corresponding benzofuranones and enal was accomplished with high enantioselectivity (up to 99% ee).

In 2011, Melchiorre et al. reported three organocatalytic cascade reactions allowing enantioselective preparation of spirobenzofuranones based on secondary and primary amine catalysis (Scheme 7.38) [79]. The first cascade reaction is based on an approach described by Rios, but broader substrate scope was surveyed. The second approach uses three-component cascade reaction comprising benzofuranone **95** together with aldehyde **109** and enal **29**, that upon enamine–iminium–enamine activation with secondary amine **VIII** and *ortho*-fluorobenzoic acid furnishes spirocycles **110** in good yields with excellent diastereo- and enantioselectivity. The last developed cascade reaction between **95** and **64** is based on enamine–iminium activation sequence under primary amine catalysis with 9-amino(9-deoxy)-*epi*-hydroquinine **XXII** and *ortho*-fluorobenzoic acid as cocatalyst.

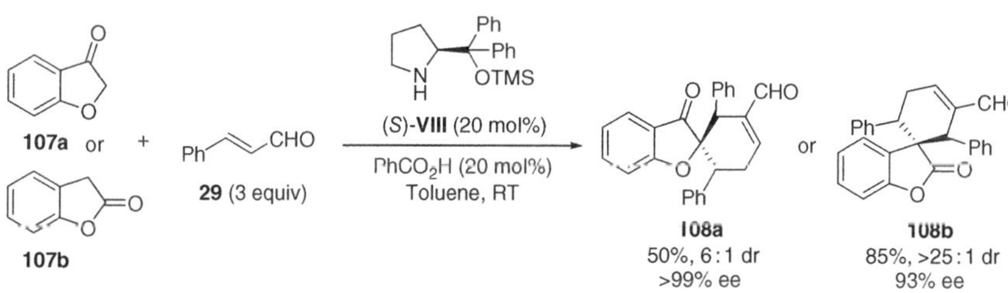

Scheme 7.37 Organocatalytic synthesis of spirobenzofuranones via cascade Michael/Michael/aldol reaction. *Source:* Based on Companyó et al. [78].

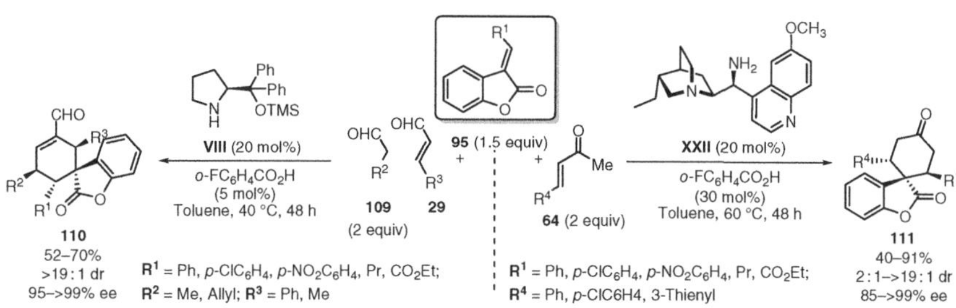

Scheme 7.38 Two approaches to chiral spirocyclic benzofuranones by using aminocatalysis. *Source:* Based on Cassani et al. [79].

Later, the same group expanded secondary amine-catalyzed cascade reactions with three-component process between benzofuranone, propanal, and 2,4-dienal proceeding *via* Michael/1,6-addition/vinylogous aldol sequence to give spirobenzofuran with high stereoselectivity [80].

In 2012, Xu and Wang reported an enantioselective double Michael reaction of benzofuranones with dienones **113** efficiently catalyzed by Cinchona-based primary amine **XXIII** in combination with BINOL-derived phosphoric acid **XXV** (Scheme 7.39) [82]. Final spirocycles **111** were reached in good yields and enantioselectivities except enones bearing strong electron-withdrawing or electron-donating substituents of phenyl ring.

Later, the same authors reported another approach to spirobenzofuranones, an enantioselective Michael/aldol/dehydration sequence between 3-acyl substituted benzofuranones **114** and enones **64** (Scheme 7.39) [81]. High yields of **115** and highly efficient stereocontrol of the reaction was observed under catalysis with primary amine **XXVI** and 3,5-dinitrobenzoic acid.

An attractive strategy to construct chiral spirobenzofuranes from benzofuran derivatives **107b** and dienones **113** was also developed by Cheng et al. (Scheme 7.40) [83]. An enantioselective double Michael addition reaction was efficiently catalyzed by tertiary amine-thiourea Cinchona alkaloid **XXVII** and spiroproducts **111** were obtained in high yields with high stereoselectivity (up to 99% yield, 19 : 1 dr, 92% ee). Later, the same group developed an elegant primary

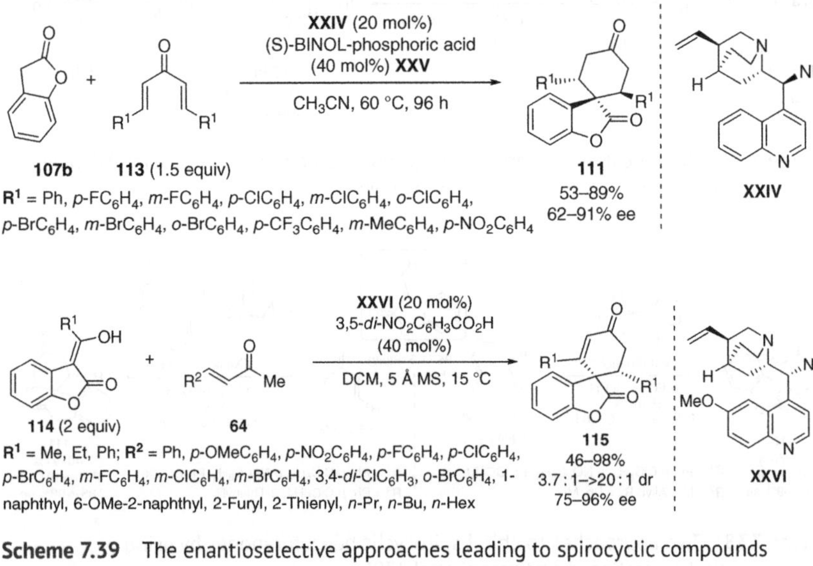

Scheme 7.39 The enantioselective approaches leading to spirocyclic compounds introduced by Xu and Wang. *Source:* Based on Qi et al. [81].

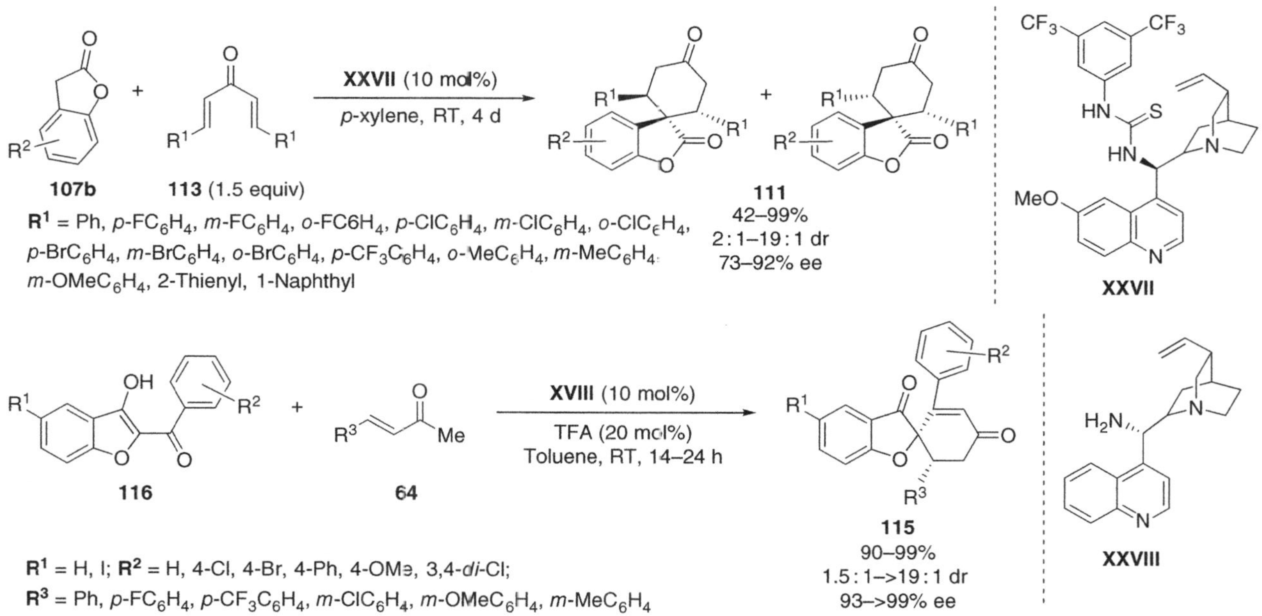

Scheme 7.40 Organocatalytic asymmetric double Michael addition and Michael/aldol reaction cascade providing spirobenzofuranones. *Source:* Based on Li et al. [33].

amine-catalyzed Michael/aldol cascade reaction between 2-substituted benzofuran-3-ones **116** and enones **64** affording chiral spirobenzofuranones **115** in excellent yields, diastereo- and enantioselectivities [84]. The same year, Chen et al. has developed diastereoselective synthesis of functionalized spirobenzofuranones via Michael/aldol reaction sequence; unfortunately, only a medium level of enantiocontrol was observed under catalysis with quinine-derived thiourea catalyst [85].

In 2013, Barbas et al. reported phosphine-catalyzed [3+2] cycloaddition between alkylidene benzofuranones **95** and MBH carbonates **117** providing chiral spirobenzofuranones in good yields with good stereoselectivity (up to 95% yield, 19 : 1 dr, 97% ee) (Scheme 7.41) [86]. The reaction is efficiently catalyzed by C_2-symmetrical diphosphine **XXIX** that activates MBH carbonate via the formation of a key intermediate, zwitterion **118**.

A different strategy to construct chiral spirobenzofuranones with high enantiomeric purity was disclosed by Li, Cheng et al., in 2014 (Scheme 7.42) [87]. They reported secondary amine-catalyzed [4+2] cycloaddition between electron-deficient alkylidene benzofuranones **95** and activated 2,4-dienals **112** with amine **VIIIa** through trienamine activation strategy. The same year, Jorgensen et al. reported another example of amine-catalyzed [4+2] cycloaddition between alkylidene benzofurane **95** and 2-(cyclohepta-1,3,5-trien-1-yl)acetaldehyde **121** providing spirobenzofurane in good yield with high diastereoselectivity, but only moderate enantioselectivity (68% ee) (Scheme 7.42) [88]. Similar example with acyclic trienal was also reported by Chen et al. [89]. Both methods are based on the generation of a key tetraenamine intermediate by condensation of trienals with secondary amine catalyst. A year later, Jørgensen et al. reported an elegant activation of cyclopropyl aldehydes through secondary amine catalysis [90]. As a part of the substrate scope of the developed cyclopropane ring-opening/[2+2] cycloaddition cascade sequence, one example of enantioselective synthesis of spirobenzofurane derivative was also achieved.

Soon after, Albrecht et al. developed enantioselective thio-Michael/aldol reaction sequence between α-arylidene benzofuran-3(*2H*)-ones **123** and 2-thiacetaldehyde

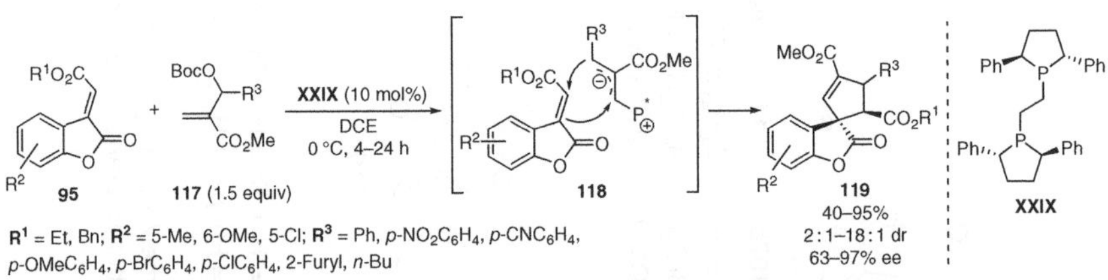

Scheme 7.41 Asymmetric construction of spirocyclopentenebenzofuranone core structures via phosphine-catalyzed [3+2] cycloaddition reaction reported by Barbas III. *Source:* Based on Albertshofer et al. [86].

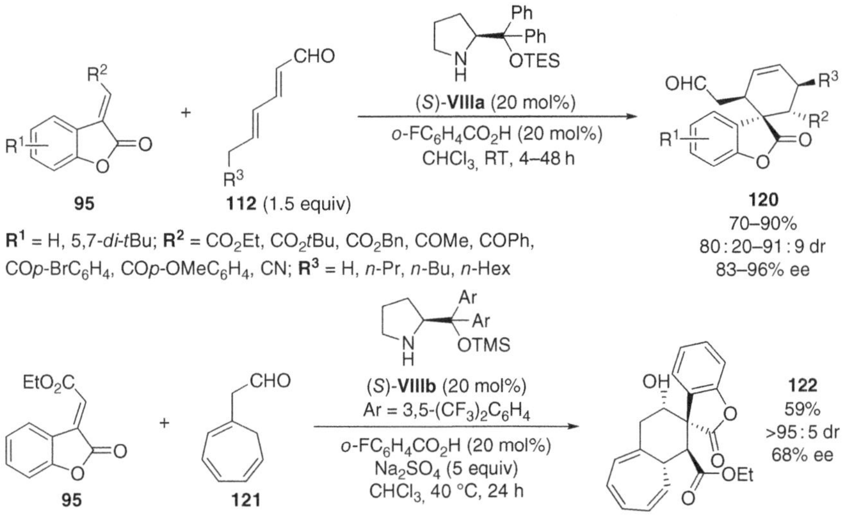

Scheme 7.42 Application of trienamine catalysis for the construction of spirobenzofuranones. *Source:* Li et al. [87]; Based on Stiller et al. [88].

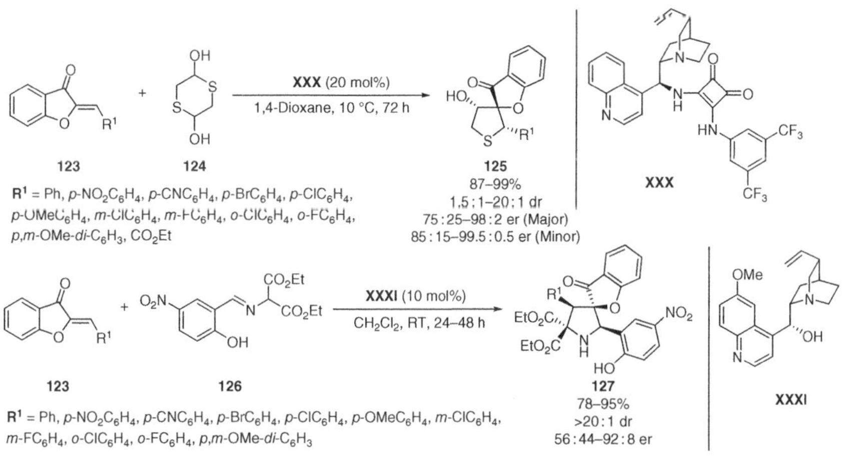

Scheme 7.43 Asymmetric synthesis of pyrrolidine- and tetrahydrothiophene-fused spirobenzofuran-3(2*H*)-ones. *Sources:* Adapted from Kowalczyk et al. [91]; Kowalczyk et al. [92].

generated *in situ* from 1,4-dithiane-2,5-diol **124**, affording chiral tetrahydrothiophene-fused spirobenzofurans under catalysis with chiral bifunctional tertiary amine-squaramide **XXX** with high efficiency (Scheme 7.43) [91]. The same group also reported an enantioselective approach to pyrrolidine-fused spirobenzofurans **127** based on [3+2] cycloaddition between α-arylidene benzofuran-3(2*H*)-ones **123** and imines **126** derived from salicylaldehyde and aminomalonates catalyzed by quinine **XXXI** (Scheme 7.43) [92]. Both reactions afforded the corresponding spirocyclic products in high to excellent yields, moderate to high diastereoselectivity, and good to excellent enantioselectivity. Later, chiral thiourea-quaternary ammonium salt-catalyzed [3+2] cycloaddition between α-arylidene benzofuran-3(2*H*)-ones and imino esters was described by Zhao et al. [93].

In 2015, Wei, Zhou, Shi et al. reported an enantioselective [3+2] cycloaddition between alkylidene benzofuranones and allenoates [94]. The reaction is efficiently catalyzed by phosphine catalyst (*R*)-SITCP affording two regioisomeric spirobenzofurans in good yields with high stereoselectivity. The regioselectivity of the reaction is controlled by the nature of the allenoate; simple allenoates exhibit opposite regioselectivity to γ-substituted allenoates. Two years later, Lan, Ullah, Lu et al. reported another phosphine-catalyzed enantioselective synthesis of spirobenzofurans [95]. Interestingly, the regioselectivity of the reported [3+2] cycloaddition between aurones and allenoates is controlled by the diastereomeric nature of the phosphine catalyst. In addition, the observed regioselectivity in both phosphine-catalyzed methods were also supported by DFT studies.

An elegant chlorinative dearomatization of benzofuran derivatives **128** giving chiral spirobenzofurans **129**, **130** in high yields and excellent enantioselectivity was developed by You et al. (Scheme 7.44) [96]. *N*-Chlorophthalimide was used as the chlorinating reagent, and hydroquinidine 1,4-phthalazinediyl diether [$(DHQD)_2$PHAL] promoted the reaction with high efficiency. Later, the same group reported an analogous protocol using *N*-bromoacetamide as the bromination reagent [97].

Other synthetic approaches to chiral spirobenzofurans were disclosed by Glorius et al., who used *N*-heterocyclic carbenes as catalysts in [3+2] annulation reaction and dearomatizing hydroacylation. Annulation reaction of α,β-unsaturated aldehydes **29** with aurone **123** in the presence of NHC precatalyst **XIV** and DBU furnished spirobenzofurane with moderate yield and selectivity (Scheme 7.45) [98]. Later, the same group developed enantioselective

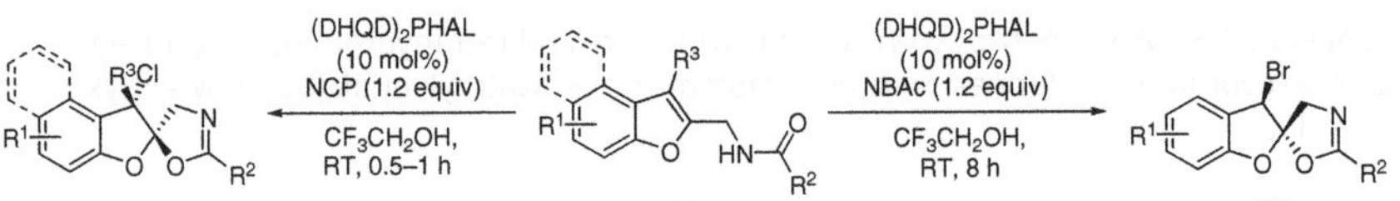

129
76–99%
94–99% ee

128
R[1] = H, Ph, Cl, F; **R**[2] = Ph, *m*-Tol, *p*-MeC_6H_4, *p*-*t*BuC_6H_4, *p*-$OMeC_6H_4$, *p*-FC_6H_4, *p*-BrC_6H_4, *m*-BrC_6H_4, 2-Thienyl, CH_2=CHPh; **R**[3] = H, Me

130
29–99%
21–98% ee

Scheme 7.44 Catalytic asymmetric brominative or chlorinative dearomatization reaction of benzofuranes. *Source:* Based on Liang et al. [96].

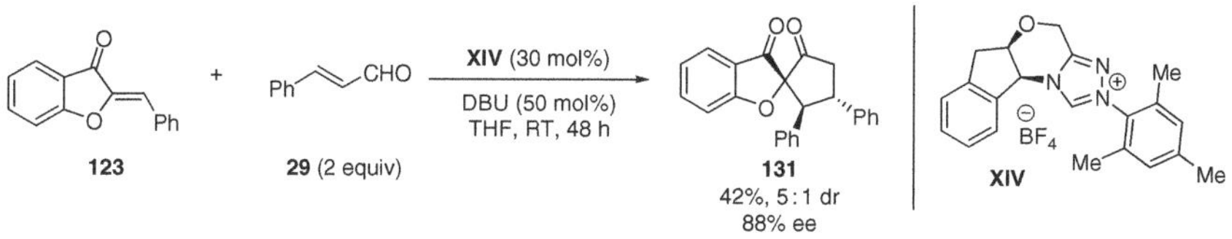

Scheme 7.45 The synthesis of spiro-heterocycles by the NHC-catalyzed annulation reaction. *Source:* Based on Guo et al. [98].

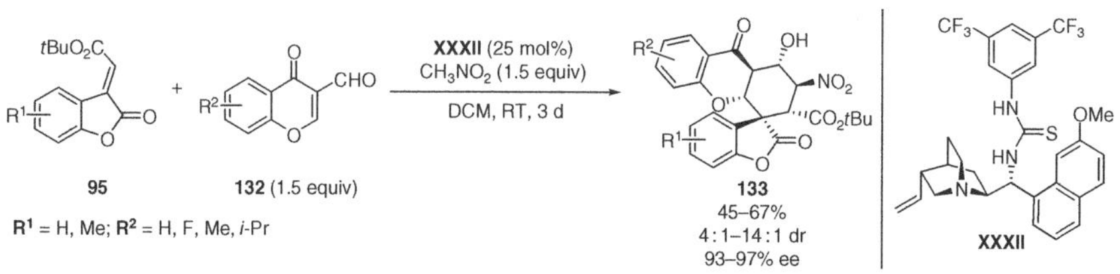

Scheme 7.46 The enantioselective formal [2+1+3] annulation forming spirocyclic benzofuranones. *Source:* Based on Chang et al. [100].

dearomatizing hydroacylation of benzofuran derivatives. The reaction was efficiently catalyzed by chiral NHC generated from the corresponding salt with K_2CO_3, affording final spirobenzofurans in good yields with excellent enantioselectivity (up to 97% ee) [99].

Recently, Liu et al. reported the construction of chiral benzofuranones through a formal [2+1+3] annulation strategy (Scheme 7.46) [100]. The reaction of alkylidene benzofuranones **95** with nitromethane and 3-formyl chromone **132** proceeds *via* Michael/Michael/Henry reaction sequence in the presence of quinine-derived thiourea catalyst **XXXII**. The final spirobenzofuranones were obtained with up to 67% yield, good diastereoselectivity (up to 14 : 1 dr), and excellent enantioselectivity (up to 67% ee). Very recently, Jiang and Wang developed enantioselective synthesis of spirodihydrobenzofurans through bifunctional phosphonium salt-catalyzed [4+1] cyclization of *ortho*-quinone methides with α-bromoketones [101].

7.3.2.3 Cooperative-catalytic Approaches

An enantioselective synthesis of spiro compounds containing benzofuranone ring can also be accomplished by combined catalysis, as demonstrated by Rios et al. in 2016 (Scheme 7.47) [102]. Initially, they developed enantioselective one-pot cascade reaction between 1,3-indanedione-derived vinylcyclopropanes and enals **29** catalyzed by the combination of $Pd_2(dba)_3$ complex and chiral secondary amine **VIII**. The developed method was further expanded to other

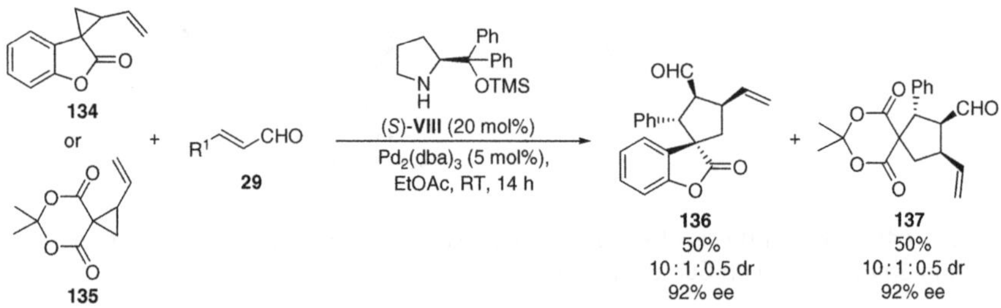

Scheme 7.47 Enantioselective ring expansion of vinyl cyclopropanes combining four catalytic cycles for the synthesis of spirocyclopentanes. *Source:* Based on Meazza and Rios [102].

vinylcyclopropanes, and chiral spirooxindole, spirobenozfuranone, and spirocyclic Meldrum's acid derivative were successfully prepared.

7.3.3 Enantioselective Formation of Spirocycles Containing Dihydrocoumarin, Dihydro- and Tetrahydropyran Ring

7.3.3.1 Organometallic Approaches

In 2002, Hoveyda et al. reported the first example of catalytic enantioselective synthesis of spirocyclic compounds, including spirodihydropyrans, by double olefin metathesis (Scheme 7.48) [103]. The reaction of cyclopentyl trienes **138** is catalyzed by *in situ*-generated BINOL-derived Mo catalysts **XXXIII** in THF, which has a positive effect on the enantioselectivity of the reaction. Final spirodihydropyrans **139** were prepared in high yields (45–90%) with moderate to high enantioselectivity (62–88% ee).

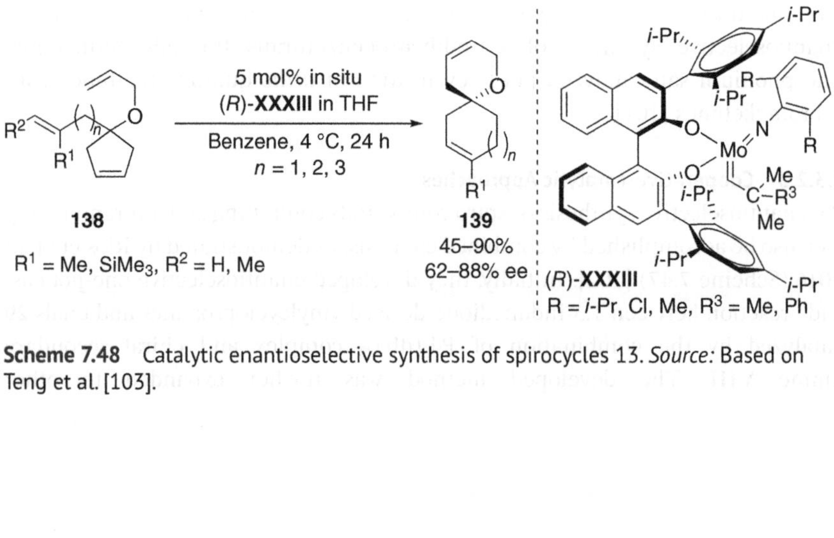

Scheme 7.48 Catalytic enantioselective synthesis of spirocycles 13. *Source:* Based on Teng et al. [103].

An attractive strategy to construct chiral spirobenzopyranones was reported by Tanaka et al. in 2008 (Scheme 7.49) [104]. They developed enantioselective [4+2] annulation reaction of 2-alkynylbenzaldehydes **140** with various electron-deficient carbonyls **141** catalyzed by a Rh(I) complex **L-XIV**. The reaction, proceeding through a key five-membered rhodacycle **143**, afforded highly enantioenriched spirocyclic benzopyranones in high yields (up to 87% yield and 98% ee). Later, Lam et al. and Ashfeld et al. reported different Rh-catalyzed methods allowing the construction of chiral spirocycles containing pyranon and benzopyranone ring, respectively. Lam's approach involves catalysis with chiral cyclopentadienyl Rh(III) complex **XXXIV** that efficiently promotes oxidative [3+2] spiroannulation of α-aryl cyclic 1,3-dicarbonyls **144** with alkynes **145** (Scheme 7.50) [105]. Ashfeld's team developed enantioselective formal [4+1] cycloaddition between diazo compounds **147** and vinyl ketenes **150** generated in

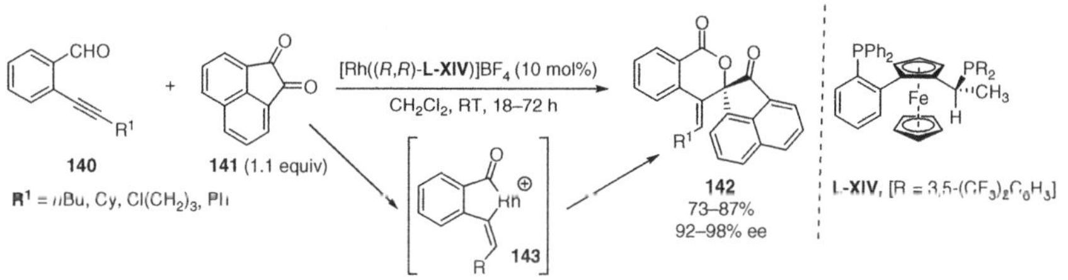

Scheme 7.49 Enantioselective synthesis of spirocyclic benzopyranones by rhodium-catalyzed intermolecular [4+2] annulation. *Source:* Based on Hojo et al. [104].

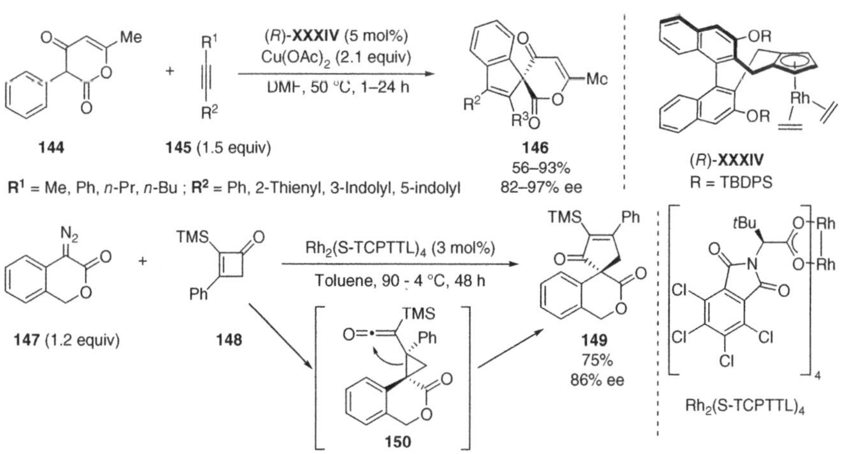

Scheme 7.50 Enantioselective organometallic approach to chiral spirobenzopyranones **149**. *Sources:* Based on Reddy Chidipudi et al. [105]; Rodriguez et al. [106].

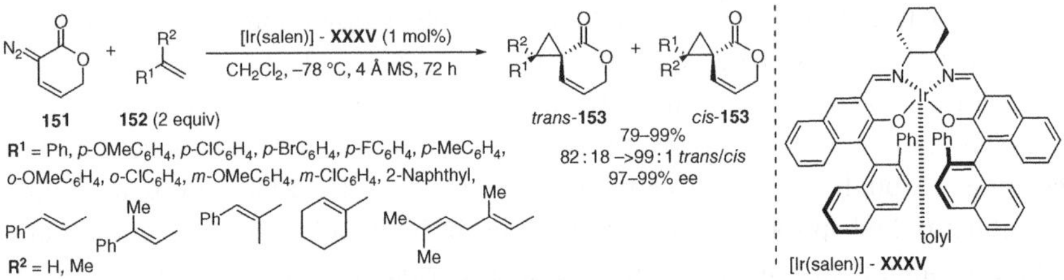

Scheme 7.51 Enantioselective synthesis of a-spirocyclopropyl lactones. *Source:* Based on Ichinose et al. [107].

situ from cyclobutenone **148** in the presence of $Rh_2(TCPTTL)_4$ (Scheme 7.50) [106]. Both methods afforded the corresponding spirocycles in good yields and high enantioselectivity, 97% ee, and 86% ee, respectively.

In 2009, Katsuki et al. reported enantioselective Ir-catalyzed cyclopropanation of aryl- and alkenyl-substituted olefins with vinyl diazolactone **151** (Scheme 7.51) [107]. The reaction was efficiently catalyzed with chiral Ir-salen complex **XXXV** furnishing spirodihydropyranones in high yields with excellent diastereo- and enantioselectivity (up to 99% yield, >99 : 1 dr, 99% ee).

7.3.3.2 Organocatalytic Approaches

In 2007, Miller et al. reported a single example of construction of chiral spirobenzopyranone *via* chiral phosphine-catalyzed [3+2] cycloaddition of allenoate and enone; unfortunately, with a low level of enantiomeric purity (65% ee) [108]. The first highly efficient organocatalytic system convenient for the construction of chiral spirodihydropyrans was developed by Ma et al. (Scheme 7.52) [110]. The

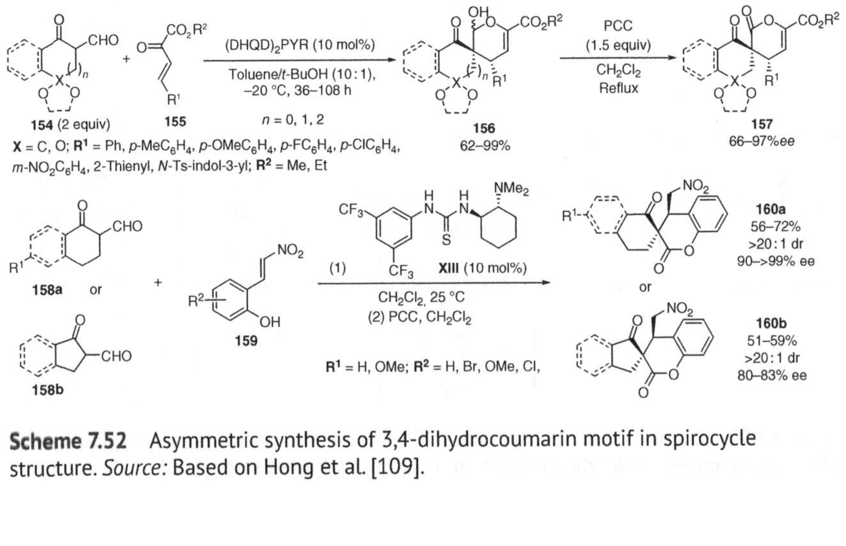

Scheme 7.52 Asymmetric synthesis of 3,4-dihydrocoumarin motif in spirocycle structure. *Source:* Based on Hong et al. [109].

Michael/hemiacetalization reaction sequence between cyclic β-oxo aldehydes **154** and aromatic β,γ-unsaturated α-keto esters **155** was catalyzed with $(DHQD)_2PYR$, resulting in the formation of spirodihydropyrans in good yields with high stereoselectivity (up to 97% ee). Soon after, Hong et al. developed a similar Michael/acetalization/oxidation reaction sequence between 2-hydroxynitrostyrenes **159** and 2-oxocyclohexanecarbaldehydes **158** providing chiral spirobenzopyranones **160** (Scheme 7.52) [109]. Hong's reaction was catalyzed by bifunctional amine-thiourea Takemoto catalyst **XIII** to form the corresponding cycloadducts, and after *in situ* oxidation with PCC final spirocycles were obtained with up to 72% yield, >20 : 1 dr, and >99% ee.

A year later, enantioselective synthesis of highly substituted chiral spirodihydrocoumarines from 2-(2-nitrovinyl)phenols **159** and keto-esters **161** via Michael/lactonization reaction sequence was reported by Ramachary et al. (Scheme 7.53) [111]. Michael addition step was efficiently catalyzed by quinine-based bifunctional amine-thiourea **XXXVI**, and lactonization proceeded in the presence of *p*-TSA at elevated temperature. The final spirocycles were obtained in high yields and high selectivity.

In 2013, Peng, Huang, Han et al. reported an enantioselective organocatalytic Morita–Baylis–Hillman/Michael/acetalization/oxidation reaction cascade of ninhydrin **163** with nitrostyrenes **164** and aldehydes **109** catalyzed by a combination of DMAP and chiral secondary amine **VIII** (Scheme 7.54) [112]. The spirocyclic

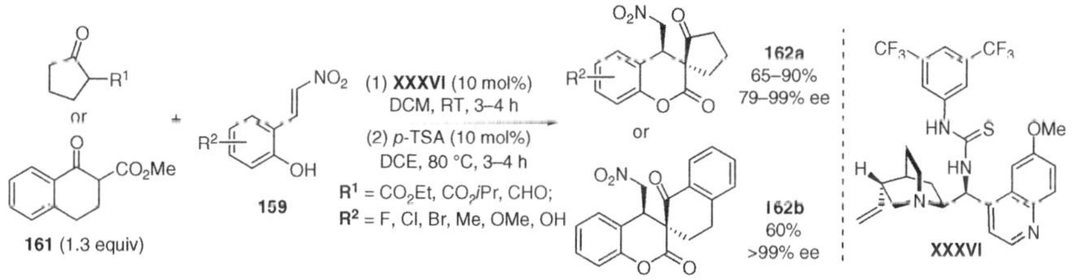

Scheme 7.53 Synthesis of highly substituted chiral spirodihydrocoumarins. *Source:* Based on Ramachary et al. [111].

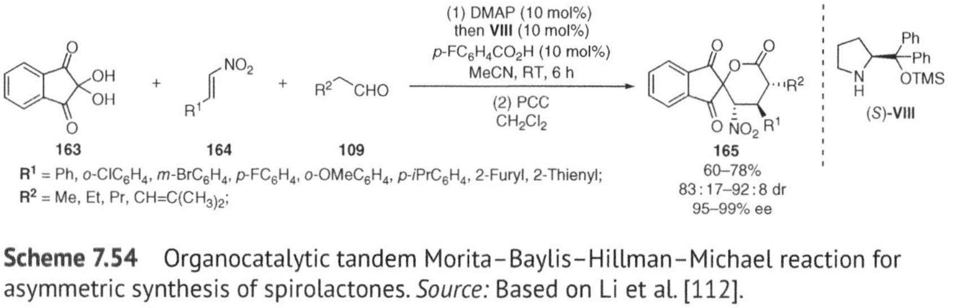

Scheme 7.54 Organocatalytic tandem Morita–Baylis–Hillman–Michael reaction for asymmetric synthesis of spirolactones. *Source:* Based on Li et al. [112].

tetrahydropyranones were obtained in high yields with good diastereoselectivity and excellent enantioselectivity. Later, Peng, Huang et al. also reported a variant of the developed method based on secondary amine-catalyzed Michael/Henry/acetalization reaction sequence [113].

In 2015, Du et al. reported enantioselective cyclopropanation of arylidenechromanones with bromonitromethane [114]. The Michael/alkylation cascade reaction is catalyzed with hydroquinine-derived squaramide, and produces the desired spirobenzopyranones in moderate to good yields (54–90%) and with excellent diastereo- (up to >99 : 1 dr) and enantioselectivities (up to >99% ee).

A phase-transfer-catalyzed dearomative spirolactonization of propanoic acid-tethered naphthols under activation with bis(trifluoracetoxyiodo)benzene was investigated by Piexoto, Quideau et al. [115]. Although chiral Cinchona alkaloid-derived quaternary ammonium iodides exhibited asymmetric inductions in the spirolactonization process, the final spiro compounds were obtained in moderate yields with low enantioselectivity.

In 2018, Li et al. developed an enantioselective 1,6-addition/acetalization cascade reaction of 1-oxotetraline-2-carbaldehydes **167** and *ortho*-hydroxyphenyl-substituted *para*-quinone methides **166** followed by oxidation with Dess–Martin periodinane (Scheme 7.55) [116]. This strategy provided chiral spirodihydrocoumarins **168** in high yields with excellent diastereo- and enantioselectivity, when bifunctional tertiary amine-squaramide catalyst **XXXVII** was employed. Later, Jiang and Wang developed an enantioselective synthesis of similar spirodihydrobenzofurans through bifunctional phosphonium salt-catalyzed [4+1] cyclization of ortho/quinone methides with α-bromoketones.

Recently, enantioselective synthesis of chiral spirodihydropyrans from cyanoketones and cyclic 2,4-dienones was reported by Pan et al. [117]. Cinchona alkaloid-derived bifunctional primary amine in combination with *N*-Boc-*tert*-leucine was found to be the most effective for the developed 1,6-addition/oxa-Michael reaction sequence and provided the spirocycles in good yields with high enantioselectivity but low diastereocontrol. The same year, Wang, Liu et al. reported organocatalytic cascade reaction between 2-hydroxycinnamaldehydes **169** and

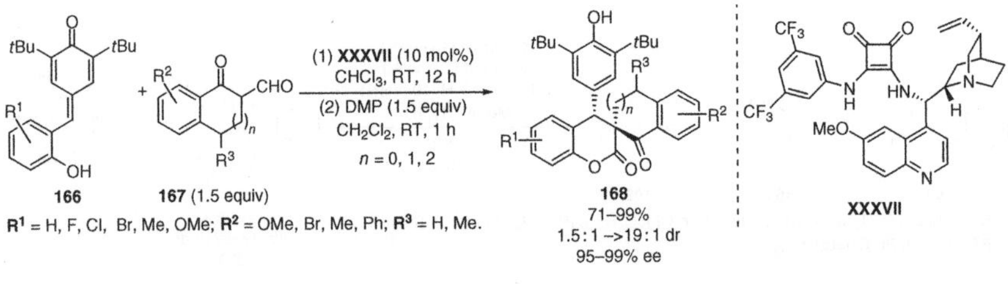

Scheme 7.55 A chiral squaramide-catalyzed approach constructing spiro-3,4-dihydrocoumarin structures. *Source:* Zhang et al. [116].

cyclic β-oxo aldehydes **167** catalyzed by a combination of bifunctional tertiary amine-thiourea **XXXIX**, secondary amine **XXXVIII**, and benzoic acid (Scheme 7.56) [118]. The developed cascade reaction, proceeding through a quinary catalyst–substrate complex as proposed, provided spiro-bridged products as single diastereomers with excellent enantioselectivity (up to 79% yield and >99% ee).

Beside the abovementioned methods, Alexakis et al. (Scheme 7.57) [119] and Luo et al. (Scheme 7.58) [120] reported single examples of synthesis of chiral

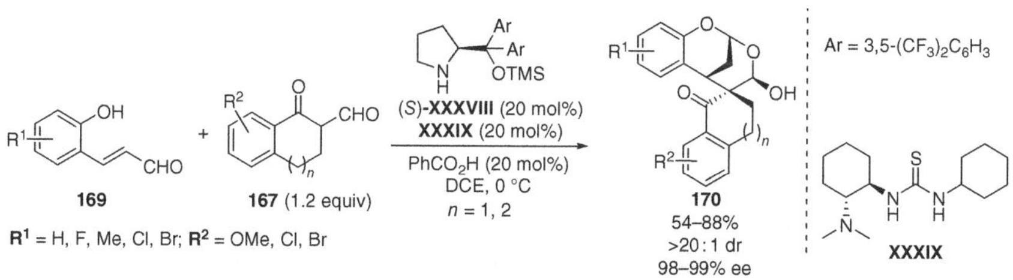

Scheme 7.56 Construction of spiro-bridged polyheterocyclic compounds. *Source*: Based on Wang et al. [118].

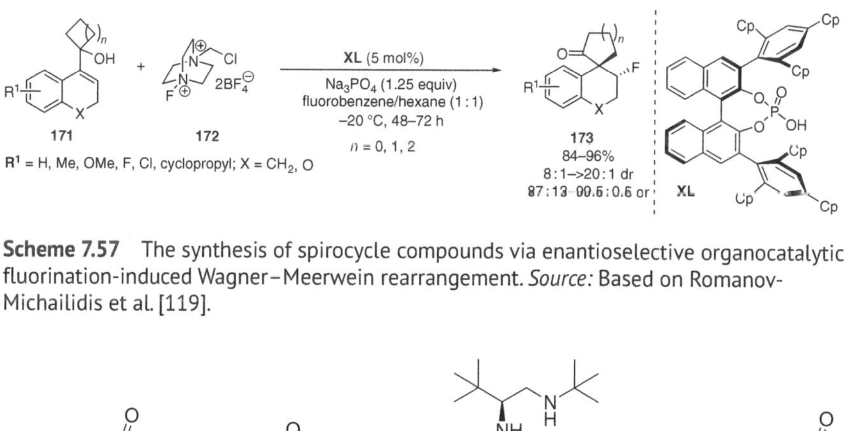

Scheme 7.57 The synthesis of spirocycle compounds via enantioselective organocatalytic fluorination-induced Wagner–Meerwein rearrangement. *Source:* Based on Romanov-Michailidis et al. [119].

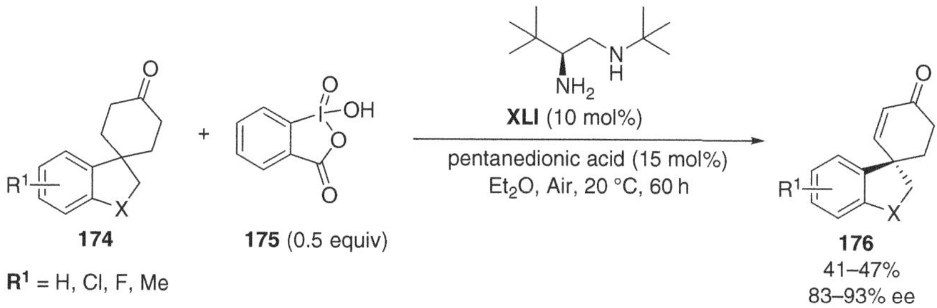

Scheme 7.58 Catalytic desymmetrizing dehydrogenation of 4-substituted cyclohexanones through enamine oxidation. *Source:* Based on Zhu et al. [120].

spirodihydropyrans using organocatalytic approach; the first work describes BINOL-derived phosphoric acid-catalyzed fluorination/semipinacol rearrangement cascade, the second is based on primary amine-catalyzed desymmetrizing dehydrogenation process.

7.4 Enantioselective Formation of Sulfur-containing Spirocycles

In the previous chapters, synthetic approaches providing spiro compounds containing nitrogen and oxygen atom were presented. In the last decade, chemists have also turned their attention to the preparation of chiral sulfur-containing spirocyclic compounds. If we focus on spiro compounds containing only one sulfur heteroatom, then benzothiophen-2-one is the most popular substrate.

Among catalytic approaches, asymmetric organocatalysis predominates in the preparation of chiral sulfur-containing spiro compounds. In 2011, Wang et al. performed an enantioselective Michael/cyclization reaction sequence of α-alkylidene benzothiophen-2-one **176** with α-isothiocyanato imide **177** using chiral thiourea catalysis **XLII** (Scheme 7.59) [121]. The reaction afforded the corresponding spirocyclic product in 85% yield with excellent selectivity (6.4 : 1 dr, >99%, >99% ee).

Similarly, Du et al. applied non-covalent H-bonding catalysis to prepare chiral sulfur-containing spirocycles, in 2014. In this work, they described sulfa-Michael/aldol cascade reaction between benzylidenchroman-4-ones **179** and 1,4-dithian-2,5-diol **124** efficiently catalyzed by bifunctional amine-squaramide **XLIII**. Chiral chromanone-fused spirotetrahydrothiophenes containing three new stereocenters were obtained in high yields and enantiomeric excesses (Scheme 7.60) [122]. A year later and in 2017, Du et al. prepared spirothiochromanes with excellent yields and enantioselectivity by the same enantioselective organocatalytic cascade reaction catalyzed by chiral bifunctional tertiary amine-squaramide **XLIV** (Scheme 7.60) [123].

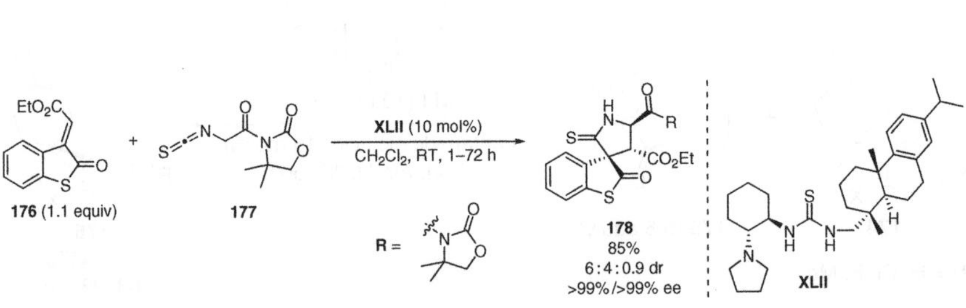

Scheme 7.59 Enantioselective synthesis of spirocyclic compounds containing benzothiophen-2-one. *Source:* Based on Cao et al. [121].

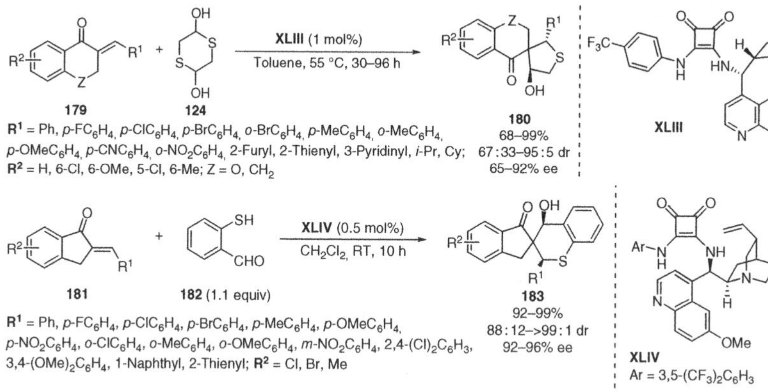

Scheme 7.60 Asymmetric synthesis of spirocyclic tetrahydrothiophene-chromanone derivatives and spirothiochromanones. *Sources:* Modified from Zhao et al. [122]; Adapted from Zhao and Du [123a]; Zhao and Du [123b].

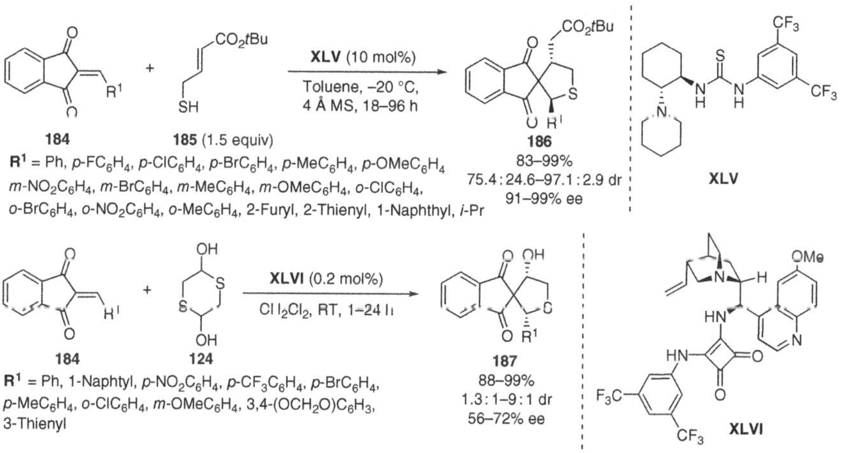

Scheme 7.61 Enantioselective synthesis of spiro compounds containing tetrahydrothiophene ring. *Source:* Modified from Duan et al. [124]; Based on Grošelj et al. [125].

Another way to prepare spiro compounds containing tetrahydrothiophene ring was introduced by Li et al. in 2015. Spiro compounds can be prepared by the enantioselective organocatalytic sulfa-Michael/Michael domino reaction between 2-arylidene-1,3-indanedione **184** and 4-thio-2-butenoate **185** (Scheme 7.61) [124]. The chiral bifunctional thiourea **XLV** proved to be the best organocatalyst in terms of yields and reaction selectivity. Soon after, Enders et al. prepared similar compounds, but he used 1,4-dithian-2,5-diol **124** in a squaramide-catalyzed sulfa-Michael/aldol domino reaction with 2-arylidene-1,3-indanedione **184** (Scheme 7.61) [126]. Noteworthy, prior to Enders's work, Wang et al. reported enantioselective synthesis of chromanone-fused spirothiochromans *via* sulfa-Michael/aldol cascade reaction promoted by originally developed bifunctional indane catalyst [127].

Chiral spirodihydrothiophenes can also be prepared by an organocatalytic formal [3+2] cyclization reaction published by Xie et al. in 2015 (Scheme 7.62) [128]. The reaction between thioamides **188** and nitrostyrenes **164** was efficiently promoted by a chiral Takemoto thiourea catalyst **XIII**. In 2017, Yan et al. used a chiral BINOL-derived organocatalyst **XLVII** in the presence of KF to catalyze organocatalytic cascade sulfa-Michael/aldol reaction. The final spirotetrahydrothiophenes **187** were prepared from 1,4-dithian-2,5-diol **124** and benzocyclic ketones **181** in good yields (up to 99%) with excellent selectivity (up to >20 : 1 dr, up to 99% ee) (Scheme 7.63) [129]. Later, similar sulfa-Michael/aldol cascade reaction between 1,4-dithian-2,5-diol and arylidene pyrrolinones was reported by Ričko et al., giving chiral pyrrolo-fused spirotetrahydrothiophenes [125].

Enantioselective synthesis of spiro compounds containing sulfur atom can also be performed by aminocatalysis. In 2013, Veselý et al. introduced the organocatalytic cascade reaction of enals **29** with benzothiophen-2-one **190** catalyzed by a chiral secondary pyrrolidine catalyst **VIII** (Hayashi-type organocatalyst). The cascade Michael/Michael/aldol reaction gave sulfur-containing spiro

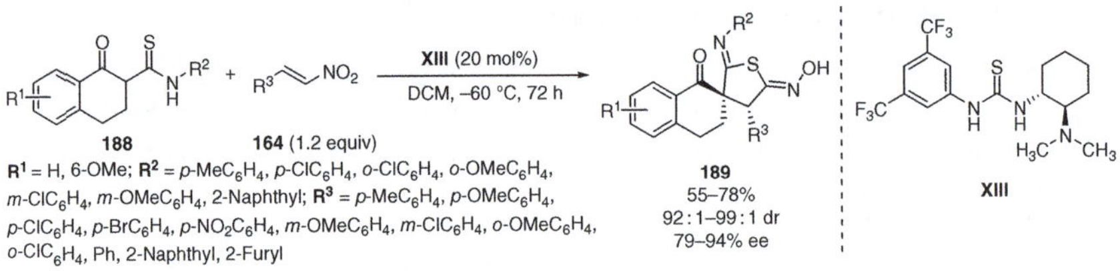

Scheme 7.62 Enantioselective construction of polyfunctionalized spiroannulated tetrahydrothiophenes via a formal thio [3+2] cyclization. *Source:* Based on Zeng et al. [128].

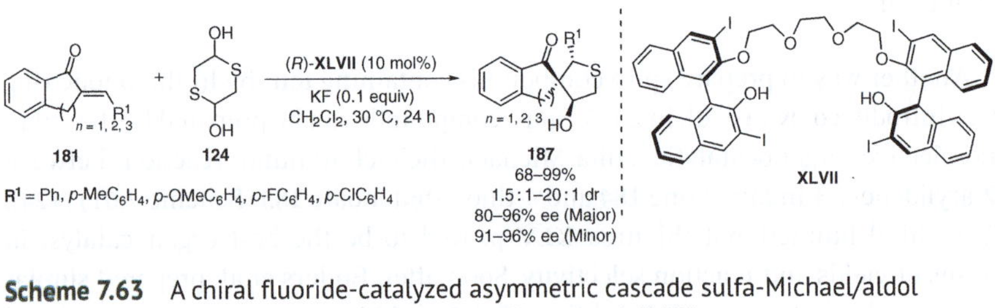

Scheme 7.63 A chiral fluoride-catalyzed asymmetric cascade sulfa-Michael/aldol condensation reaction. *Source:* Modified from Duan et al. [129].

compounds **191** in good yields (up to 77%) with high stereoselectivity (>20 : 1 dr, 99% ee). Extension of the substrates to dienones **113** in reaction with benzothiophen-2-one **190** also led to the preparation of spirocyclic compounds. However, a chiral primary amine was used as an organocatalyst in this Michael/Michael cascade reaction (Scheme 7.64) [131]. Subsequently, in 2014, Jørgensen and Albrecht performed a formal [4+2] cycloaddition between olefinic benzothiophen-3-one **192** and 2-(hydroxyalkyl) cyclopent-2-ene **193**. Even in this reaction, a chiral primary amine derived from quinine proved to be suitable for the preparation of a spirocycle (Scheme 7.65) [132]. Recently, the alkylidene benzothiophen-3-ones **192** were also used by Veselý's group in the organocatalytic Michael/Michael cascade reaction with enones **64** [130]. The cinchonidine-derived chiral primary amine used as the organocatalyst provided spirocyclic benzothiophen-3-ones in good yields (up to 96%) and excellent stereoselectivity (14 : 2 : 1 dr, up to 99% ee).

In 2016, Ye et al. applied chiral primary amine **XLVIII** catalysis in an asymmetric 1,6-addition/1,4-addition cascade reaction between indoline-2-thiones **195** and cyclic dienones **196**. The cascade reaction provided sulfur-containing spiro

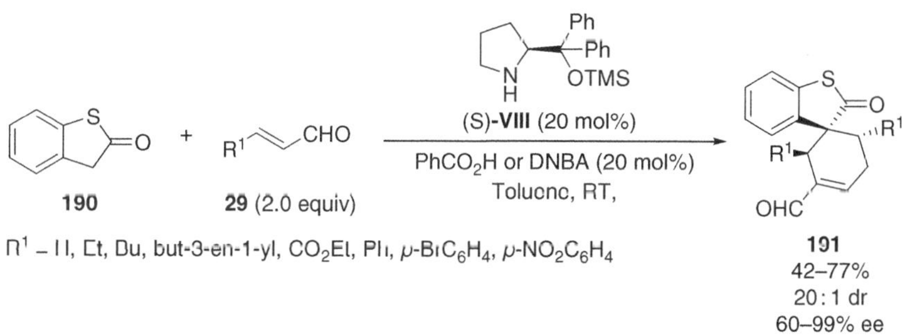

Scheme 7.64 Enantioselective synthesis of spirocyclic compounds containing benzothiophen-2-one reported by Veselý. *Source:* Based on Geánt et al. [131].

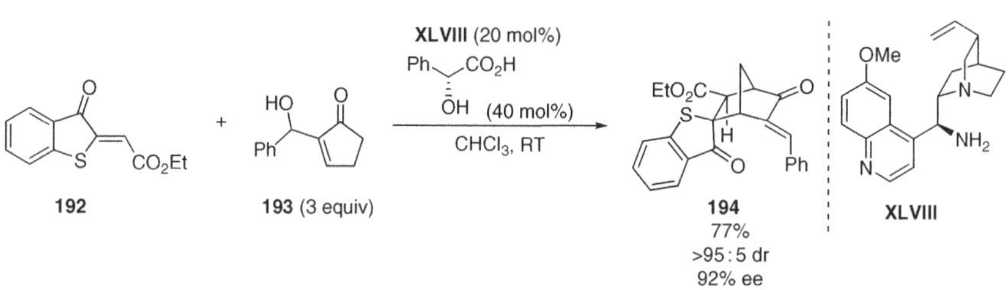

Scheme 7.65 Enantioselective synthesis of spirocyclic compounds containing benzothiophen-3-one. *Source:* Based on Stiller et al. [132].

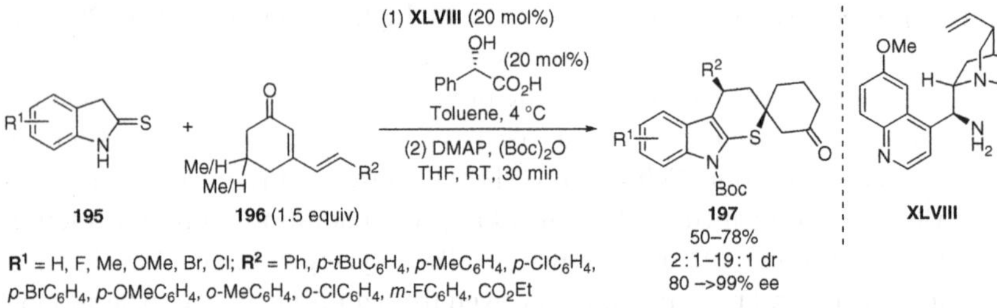

Scheme 7.66 The enantioselective approach provides chiral spiro[thiopyranoindolecyclohexanone] scaffold. *Source:* Based on Sun et al. [133].

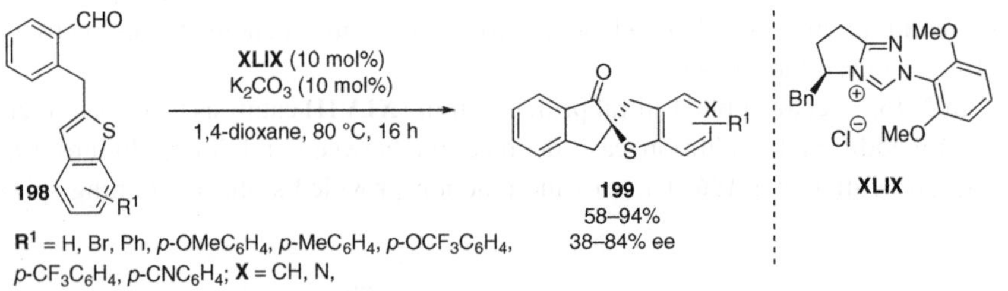

Scheme 7.67 NHC-catalyzed enantioselective dearomatizing hydroacylation of benzothiophenes for the synthesis of spirocycles. *Source:* Modified from Janssen-Müller et al. [99].

compounds with an attached indole ring in high optical purity and yields (Scheme 7.66) [133]. NHC catalysis has also found application in the preparation of sulfur-containing spiro compounds. Glorius and Neugebauer performed enantioselective dearomatic hydroacylation of benzothiophenones catalyzed by NHCs generated from triazolinium salts in the presence of K_2CO_3. The desired spiro compounds with a benzothiophene ring were obtained in good yields and enantiomeric excesses (Scheme 7.67) [99].

7.5 Enantioselective Formation of Spirocycles Containing Heterocycle with More Than One Heteroatom

7.5.1 Enantioselective Formation of Spirocycles Containing Pyrazolone Ring

7.5.1.1 Organometallic Approaches

Enantioselective synthesis of chiral spiropyrazolones through transition-metal-catalyzed methods is limited to Rh- and Sc-based catalysis. In 2017, You et al.

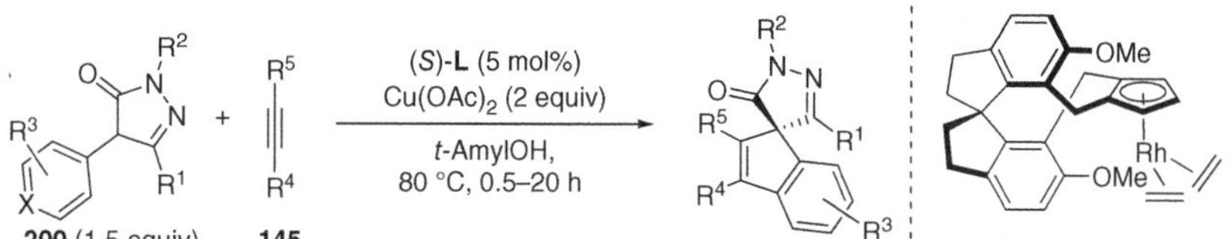

200 (1.5 equiv) 145 201 (S)-L

R¹ = Me, Et, CO_2Et, PMB; **R²** = Cy, tBu, Bn; **R³** = m-CNC_6H_4, m-MeC_6H_4, p-$OMeC_6H_4$, p-BrC_6H_4, p-FC_6H_4, p-CNC_6H_4, 2-Thienyl; **X** = CH, N; **R⁴** = **R⁵** = p-MeC_6H_4, p-$OMeC_6H_4$, p-FC_6H_4, p-BrC_6H_4, p-$CF_3C_6H_4$, nBu, Ph, Me, Et,Cyclopropyl, CO_2Et; **R⁵** = 2-Thienyl

19–99%
4.3 : 1–>19 : 1 rr
82–98% ee

Scheme 7.68 Asymmetric synthesis of spiropyrazolones by complex of rhodium-catalyst. *Source:* Based on Zheng et al. [134].

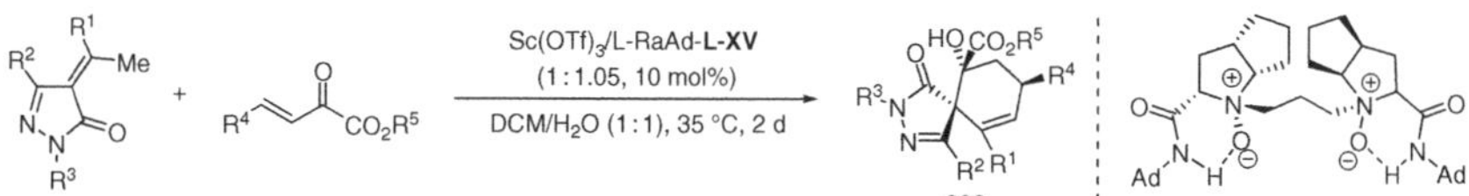

202 155 203 L-XV

R¹ = Ph, p-FC_6H_4, p-BrC_6H_4, p-ClC_6H_4, m-ClC_6H_4, o-ClC_6H_4, p-MeC_6H_4, m-MeC_6H_4, o-MeC_6H_4, p-$OMeC_6H_4$, m-$OMeC_6H_4$, o-$OMeC_6H_4$, piperonyl;
R² = Ph, Me, Ph; **R³** = Ph, tBu; **R⁴** = Ph, p-FC_6H_4, p-BrC_6H_4, p-ClC_6H_4, m-ClC_6H_4, o-ClC_6H_4, p-MeC_6H_4, m-MeC_6H_4, p-$OMeC_6H_4$, m-$OMeC_6H_4$, 2-Thienyl, 2-Naphthyl, Cy; **R⁵** = iPr, Cp, Me

60–99%
59 : 41–86 : 14 dr
58–94% ee

Scheme 7.69 Enantioselective vinylogous Michael/aldol reaction to synthesize spirocyclohexene pyrazolones. *Source:* Xu et al. [136].

reported the first enantioselective synthesis of spiropyrazolones based on Rh-catalyzed C–H functionalization/annulation reaction sequence (Scheme 7.68) [134]. The use of a Rh complex with chiral 1,1-spirobiindane-based Cp ligand **L** and $Cu(OAc)_2$ as oxidant enabled the synthesis of various spiropyrazolones **201** in up to 99% yield and 98% ee from readily available 4-aryl substituted pyrazolones **200** and alkynes **145**.

Two years later, Antonchick, Waldmann et al. reported another Rh(III)-catalyzed method allowing the synthesis of chiral spiropyrazolones in high yields and with high enantioselectivity [135]. In this protocol, α-arylidene pyrazolones **202** react with alkynes **145** through a formal sp3 C–H activation/annulation reaction sequence under mild conditions in the presence of Rh complex and $Cu(OAc)_2$. Soon after, in 2019, Feng et al. reported enantioselective vinylogous Michael/aldol cascade reaction between α-arylidene pyrazolones **202** and β,γ-unsaturated α-ketoesters **155** catalyzed by a chiral *N,N*-dioxide-Sc(III) complex (Scheme 7.69) [136]. The reaction provided the final spiropyrazolones in high yields with good diastereo- and enantioselectivity (up to 99% yield, 6 : 1 dr, and 94% ee), when aqueous media was used.

7.5.1.2 Organocatalytic Approaches

In 2011, Rios et al. reported the first enantioselective organocatalytic synthesis of spiropyrazolones from pyrazolones **204** and enals **29** using a combination of secondary amine catalyst **VIII** and benzoic acid (Scheme 7.70) [138]. The reaction proceeding via double Michael/aldol reaction cascade furnished the final spirocycles in good to excellent yields with excellent stereoselectivity (up to 84% yield, >25 : 1 dr, >99% ee). This concept was also used for the preparation of other types of chiral spiro heterocycles [78]. Soon after, the same group disclosed another highly enantioselective approach to chiral spiropyrazolones based on secondary amine-catalyzed double Michael/aldol reaction cascade of pyrazolones with enolizable aldehydes and enals [137].

In 2012, an asymmetric cascade [5+1] double Michael addition between pyrazolone **204** and dienones **113** was reported by Wang et al. (Scheme 7.70) [139]. When a combination of Cinchona-based chiral primary amine **XLVIII** and *N*-Boc-D-phenylglycine as the catalytic system was used, the final spiro products were obtained in moderate to high yields and with good to excellent diastereo- and enantioselectivity. However, the diastereoselectivity declined significantly in case of unsymmetrically substituted dienones. Thus, the same group developed an alternative organocatalytic method allowing the construction of unsymmetrically substituted spiropyrazolones **206** based on [4+2] double Michael addition between α-arylidene pyrazolones **207** and enones **64** catalyzed by a combination of a Cinchona-based primary amine **LI** and *ortho*-fluorobenzoic acid (Scheme 7.71) [140]. Independently, the same reaction was described by R. Wang et al. using a combination of primary amine and benzoic acid [141]. At the same time, Jiang, Wang et al. developed highly stereoselective synthesis of spiropyrazolone derivatives through Michael/cyclization reaction sequence of isothiocyanato imides and esters with α-arylidene pyrazolones **207** [142].

Soon after, Enders et al. disclosed an enantioselective synthesis of spiropyrazolones *via* sequential organocatalytic Michael/Michael/aldol of acetoacetates **155**, nitrostyrenes **164** and α-arylidene pyrazolones **207** (Scheme 7.72) [143]. This

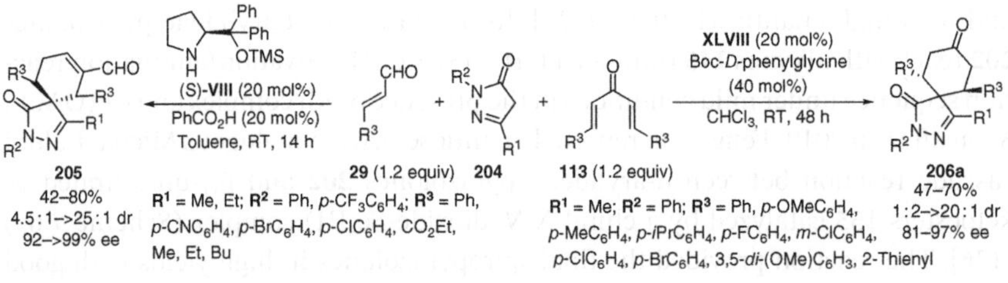

Scheme 7.70 Highly stereoselective synthesis of spiropyrazolones reported by Rios and Wang. *Source:* Based on Zea et al. [137].

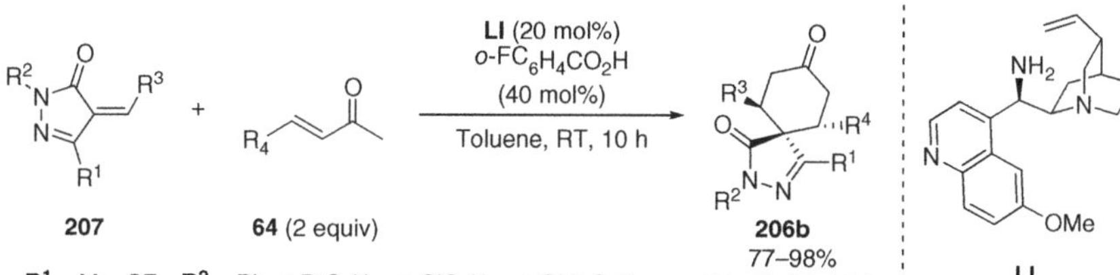

R^1 = Me, CF_3; R^2 = Ph, p-BrC_6H_4, m-ClC_6H_4, p-$OMeC_6H_4$, p-MeC_6H_4, ; R^3 = Ph, p-$OMeC_6H_4$, p-MeC_6H_4, p-FC_6H_4, m-ClC_6H_4, p-ClC_6H_4, p-BrC_6H_4, 3,5-di-(OMe)C_6H_3, m-MeC_6H_4, Cy; R^4 = o-$OMeC_6H_4$, p-MeC_6H_4, p-iPrC_6H_4, p-FC_6H_4, o-FC_6H_4, m-ClC_6H_4, p-$CF_3C_6H_4$, p-ClC_6H_4, p-BrC_6H_4, 2,4-di-(OMe)C_6H_3, 1-Naphthyl, 2-Thienyl

Scheme 7.71 The synthesis of unsymmetrical diaryl-substituted spirocyclohexanone pyrazolones through a cascade [4+2] double Michael addition. *Source:* Based on Zhang et al. [140].

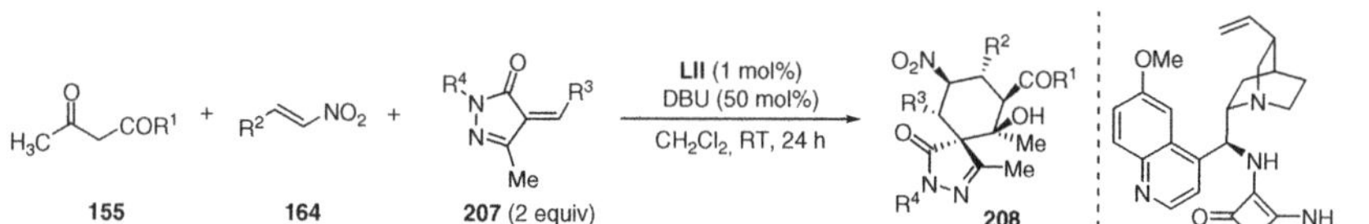

R^1 = OEt, OMe, Me; R^2 = Ph, p-FC_6H_4, p-ClC_6H_4, p-MeC_6H_4, p-$OMeC_6H_4$, 2-Furyl, 2-Thienyl; R^3 = Ph, p-$CF_3C_6H_4$, p-ClC_6H_4, p-BrC_6H_4, p-MeC_6H_4, p-$OMeC_6H_4$, o-MeC_6H_4, m-ClC_6H_4, m-$OMeC_6H_4$, 2-Thienyl; R^4 = Ph, o-ClC_6H_4

Scheme 7.72 The synthesis of spiropyrazolones via organocatalytic Michael/Michael/1,2-addition reaction. *Source:* Based on Chauhan et al. [143].

one-pot cascade sequence was promoted by a bifunctional tertiary amine-squaramide catalyst **LII** and DBU, producing the final spiropyrazolones bearing six stereocenters in good yields and excellent stereoselectivities (up to >25 : 1 dr and 97–99% ee). The same year, Xie et al. reported a similar multicomponent reaction sequence using acetoacetates, α-arylidene pyrazolones, and enals under catalysis with bifunctional tertiary amine-thiourea catalyst and piperidine as additive [144]. The final spirocycles were obtained as single diastereomers with good yields and high enantioselectivities.

Another organocatalytic three-component Michael/Michael/aldol reaction cascade allowing the formation of chiral spiropyrazolones and other spirocycles was reported by He, Wang, Peng et al. (Scheme 7.73) [145]. In this reaction, γ-nitro aldehydes **209**, generated from Michael addition of aldehydes **109** to nitrostyrenes **164** under catalysis with secondary amine **VIII**, undergoes Michael/aldol cascade reaction with α-arylidene pyrazolones **207** in the presence of **209** and K_2CO_3 as a base. Spiropyrazolones bearing six stereocenters were obtained in moderate yields with excellent stereoselectivities (up to 11 : 1 dr and 95% ee). A similar reaction was reported by Zhou et al. under catalysis with bifunctional primary

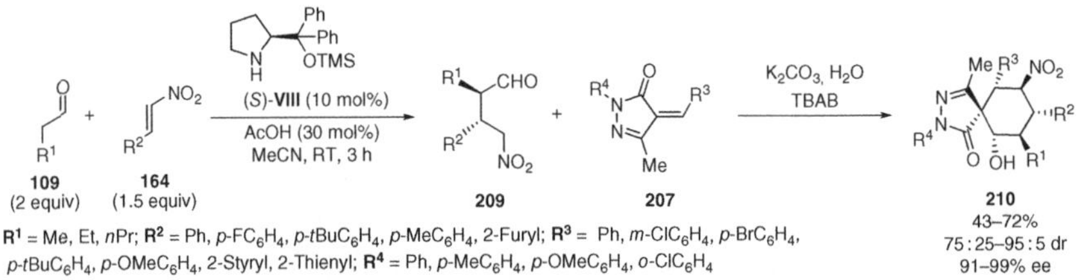

Scheme 7.73 Three-component Michael/Michael/aldol reaction cascade *reported by He, Wang, and Peng. Source:* Based on Han et al. [145].

amine-thiophosphinamide catalyst, resulting in the formation of spiropyrazolones with even higher stereoselectivity (up to >20 : 1 dr and >99% ee) [146]. In addition, the concept of secondary amine-catalyzed three-component Michael/Michael/aldol reaction was later applied for the preparation of novel chiral spirocyclic pyrazolone-fused ferrocenes by Han, He et al. [147]. In 2015, Rios et al. reported another stereoselective synthesis of spiropyrazolones from α-arylidene pyrazolones and glutaraldehyde using secondary amine catalysis. Unfortunately, Michael/aldol cascade reaction proceeded with low enantiocontrol [148].

In 2014, Du et al. developed stereoselective synthesis of tetrahydroquinoline-fused spiropyrazolones through aza-Michael/Michael reaction cascade reaction of 2-tosylaminoenones to α-arylidene pyrazolones [149]. Later, the same group reported extension of the reported procedure to aliphatic tosylaminomethyl enones and enoates [150]. Reaction cascade was catalyzed by bifunctional tertiary amine-squaramide and spirocyclic products were obtained in excellent yields and diastereoselectivities with moderate to high enantioselectivity. Similarly, organocatalytic oxa-Michael/Michael reaction cascade can be used for the construction of spiropyrazolone derivatives. Miao et al. reported highly enantioselective synthesis of chromane-fused spiropyrazolones from (*E*)-2-(2-nitrovinyl)phenols and α-arylidene pyrazolones [151]. The reaction proceeding *via* an oxa-Michael/Michael cascade was efficiently catalyzed by a carbohydrate-based bifunctional tertiary amine-thiourea catalyst. Later, Pan et al. reported enantioselective synthesis of tetrahydrofuran-fused spiropyrazolones from γ-hydroxyenones and α-arylidene pyrazolones promoted by quinine-derived squaramide catalyst [152]. This reaction furnished final spiropyrazolones in good yields with excellent diastereo- and enantioselectivity.

An attractive strategy to construct chiral spiropyrazolones was reported by Lu et al. (Scheme 7.74) [153]. An enantioselective [4+1] annulation reaction between pyrazolones **204** and 2-acetoxymethyl allenoate **211** was catalyzed with L-threonine-derived phosphine **LIII**, affording highly optically active spiropyrazolones **212** in yields 57–88% (up to 92% ee). Soon after, Guo et al. used bifunctional thiourea-based phosphine to catalyze [4+2] cycloaddition between α-arylidene

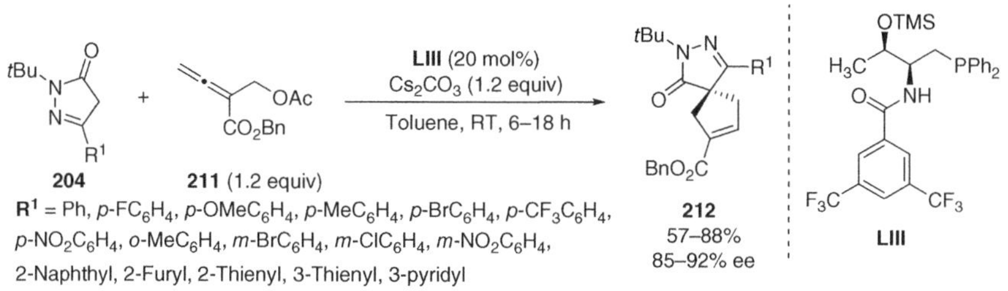

Scheme 7.74 Asymmetric synthesis of spiropyrazolones through phosphine-catalyzed [4+1] annulation. *Source:* Based on Han et al. [153].

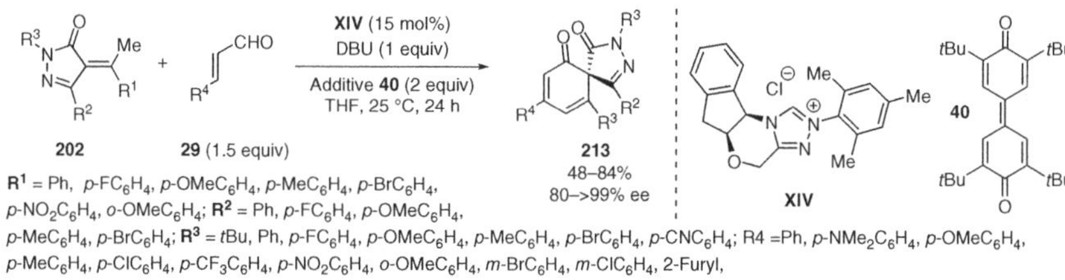

Scheme 7.75 Asymmetric, three-component, and one-pot synthesis of spiropyrazolones. *Source:* Based on Yetra et al. [156].

pyrazolones **207** and allenoates **211** [154]. The developed reaction afforded chiral spiropyrazolones in good yields with excellent enantioselectivity but only moderate diastereoselectivity (up to 4 : 1 dr). Zhong's attempt to prepare chiral spiropyrazolones via [3+2] annulation of α-arylidene pyrazolones **207** and γ-substituted allenoates under catalysis with bifunctional chiral ferrocenylphosphine-amide resulted with low efficiency [155].

In 2016, Biju et al. developed the first NHC-catalyzed [3+3] annulation reaction between enals **29** and α-arylidene pyrazolones **202** bearing a γ-hydrogen (Scheme 7.75) [156]. The reaction is initiated with 1,2-addition of carbene generated from chiral triazolium salt **XIV**, and followed by the formation of Breslow intermediate **A** *via* proton transfer. Oxidation of **A** with quinone **40** generates the key chiral α,β-unsaturated acyl azolium **B**, which is attacked by **C** *via* vinylogous Michael addition forming enolate **D**. Next, the intramolecular proton transfer generates acyl azolium **E**, which through an intramolecular alkylation affords spiropyrazolone **213** and regenerates the NHC catalyst (Scheme 7.76). Spiropyrazolones **213** were obtained in good yields and excellent enantioselectivities (50–84% yield and 94–99% ee). Also Enders et al. reported NHC-catalyzed annulation reactions allowing the formation of chiral spiropyrazolones. Firstly, they developed a one-pot enantioselective synthesis of spiropyrazolones through base-catalyzed aldol condensation of pyrazolone **204** and enal **29**, followed by

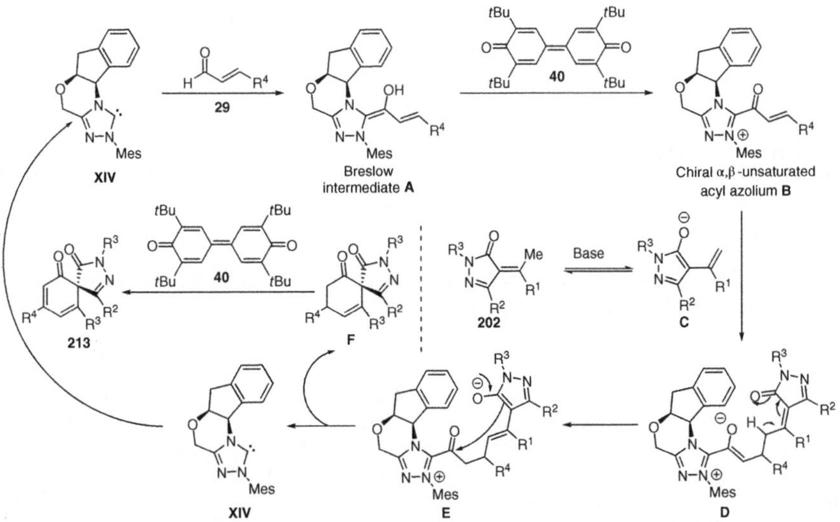

Scheme 7.76 Mechanism of asymmetric, three-component, one-pot synthesis of spiropyrazolones.

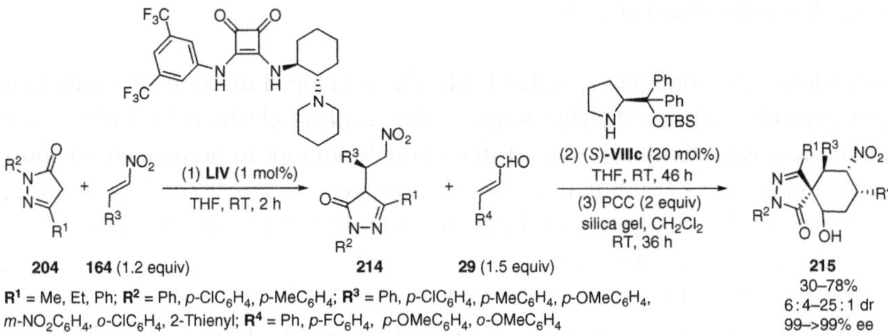

Scheme 7.77 Diastereo- and enantioselective construction of cyclohexanone-fused spirospyrazolones. *Source:* Based on Li et al. [159].

[3+2] annulation of formed α-arylidene pyrazolones **207** with a second equivalent of enal under catalysis with NHC, generated from precatalyst **XIV** [157]. The spiropyrazolones were synthesized in good yields and good to excellent stereoselectivities (up to 79% yield, >20 : 1 dr, and 96% ee). Later, they extended the NHC-catalyzed [3+2] annulation strategy by employing α-arylidene pyrazolones **207** and enals [158].

Another approach to enantioselective construction of spiropyrazolones through Michael/Michael/aldol cascade reaction followed by sequential oxidation with PCC was developed by Du et al. (Scheme 7.77) [159]. The cascade reaction of pyrazolones **204** with nitrostyrenes **164** and enals **29** was promoted by the combination of a bifunctional tertiary amine-squaramide catalyst **LIV** and secondary amine **VIIIc**, affording the spiro compounds bearing four stereocenters **215** in good yields with moderate to good diastereoselectivity and excellent enantioselectivity (up to >99% ee). Almost simultaneously, Chen et al. developed a sequential Michael/aldol reaction between pyrazolones **204** and (*E*)-5-nitro-6-aryl-hex-5-en-2-ones [160]. The desired spiropyrazolones were obtained in good yields with good to excellent stereoselectivity (up to >20 : 1 dr, 94% ee) under catalysis with a bifunctional quinine-derived squaramide catalyst **LVI** and DIPEA.

In 2017, Guo et al. developed an elegant Lewis base-catalyzed enantioselective [3+3] annulation reaction between MBH carbonates **117** and α-arylidene pyrazolones bearing a γ-hydrogen **202**, giving spiropyrazolones in high yields with high diastereoselectivity and excellent enantioselectivity (59–93% yield, up to >20 : 1 dr, 89–96% ee) (Scheme 7.78) [161]. The reaction was efficiently catalyzed by commercially available hydroquinidine **LV** in DCM and lowered temperature. The same α-arylidene pyrazolones **202** combined with (*E*)-nitro allylic acetates **217** were used by Xu et al. in another enantioselective [3+3] annulation reaction, affording chiral spiropyrazolones in high yields with excellent stereoselectivity (Scheme 7.79) [162]. The reaction was catalyzed with bifunctional tertiary amine-squaramide **LVI** responsible for synergistic activation of both substrates.

In 2017, Biju et al. reported an enantioselective secondary amine-catalyzed [3+3] annulation reaction between α-arylidene pyrazolones **202** and enals **29** (Scheme 7.80) [163]. The key step of the reaction involves a vinylogous Michael addition of dienolate generated from **202** to iminium formed via condensation of **29** with secondary amine **VIIIb**, with subsequent intramolecular aldol reaction. Soon after, the same reaction was reported by Han, Li et al., who used slightly different catalytic system, secondary amine **VIIIc** in combination with *ortho*-nitrobenzoic

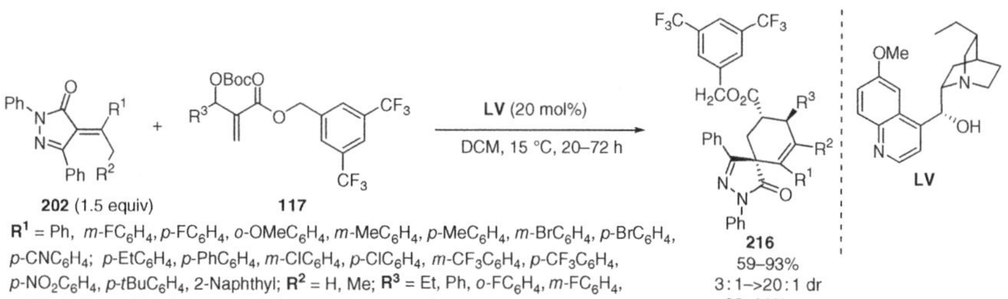

202 (1.5 equiv) **117** **216**

R[1] = Ph, *m*-FC_6H_4, *p*-FC_6H_4, *o*-$OMeC_6H_4$, *m*-MeC_6H_4, *p*-MeC_6H_4, *m*-BrC_6H_4, *p*-BrC_6H_4, *p*-CNC_6H_4; *p*-EtC_6H_4, *p*-PhC_6H_4, *m*-ClC_6H_4, *p*-ClC_6H_4, *m*-$CF_3C_6H_4$, *p*-$CF_3C_6H_4$, *p*-$NO_2C_6H_4$, *p*-*t*BuC_6H_4, 2-Naphthyl; **R**[2] = H, Me; **R**[3] = Et, Ph, *o*-FC_6H_4, *m*-FC_6H_4, *o*-$OMeC_6H_4$, *m*-$OMeC_6H_4$, *m*-MeC_6H_4, *m*-BrC_6H_4, *o*-BrC_6H_4, *p*-$OMeC_6H_4$; *p*-ClC_6H_4, *m*-ClC_6H_4, 2-Naphthyl, 2-Thienyl

Scheme 7.78 Lewis-base-catalyzed asymmetric [3+3] annulation reaction of Morita–Baylis–Hillman carbonates with alkylidene pyrazolones. *Source:* Based on Yang et al. [161].

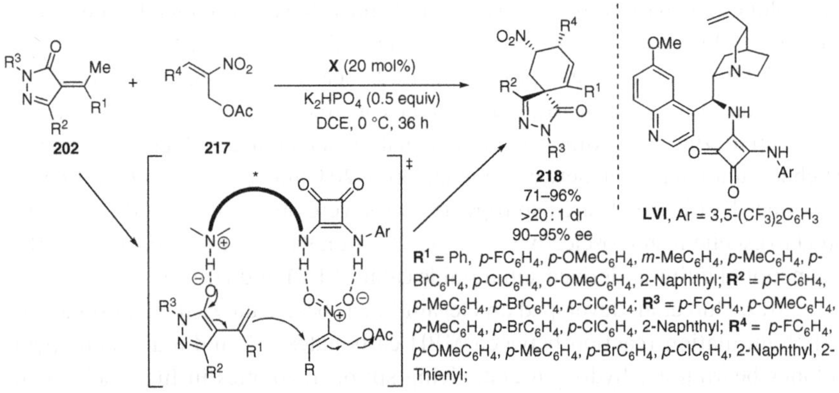

Scheme 7.79 Asymmetric synthesis of spirocyclohexene pyrazolones via formal [3+3] cycloaddition. *Source:* Based on Liu et al. [162].

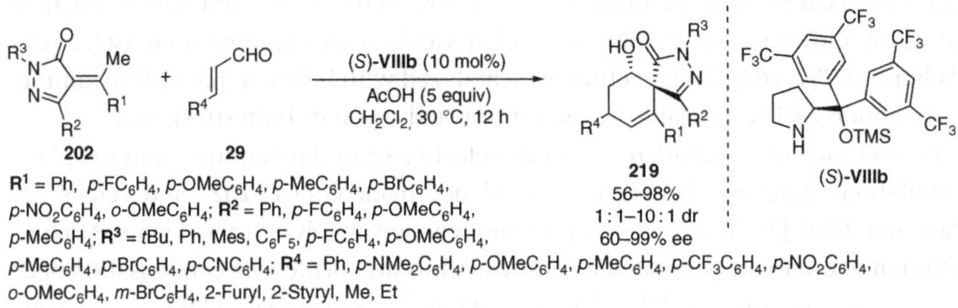

Scheme 7.80 Enantioselective synthesis of pyrazolone-fused spirocyclohexenols. *Source:* Mondal et al. [163].

acid [164]. Both reported methods provided the final spiropyrazolones in excellent yields with good diastereoselectivity and excellent enantioselectivity.

An enantioselective Michael/I_2-mediated cyclization strategy toward chiral spiropyrazolones was disclosed by Xia, Xu et al. [165, 166]. Bifunctional cinchonidine-derived squaramide **LVI** was used to catalyze enantiodetermining Michael addition step between 1,3-diketones or hydroxycoumarins and α-arylidene pyrazolones **207**, and a combination of I_2 and Na_2CO_3 mediated the following cyclization step with high efficiency. Later, the same authors reported an efficient synthesis of chiral dihydrobenzofuran-fused spiropyrazolones with high level of diastereo- and enantioselectivity (up to >99 : 1 dr and 99% ee) through one-pot Michael/iodization/nucleophilic substitution reaction sequence, starting from 2-hydroxynitroolefins **159** and pyrazolones **204** [167]. The reaction proceeds in the presence of bifunctional tertiary amine-squaramide catalyst, a catalytic amount of molecular iodine, and an excess of 30% aqueous solution of hydrogen peroxide. Based on additional experiments, the authors suggest that the organocatalyst is involved not only in Michael addition step but also in the iodine-mediated step.

In 2017, Lattanzi et al. reported diastereodivergent and enantioselective approach to oxirane-fused spiropyrazolones **220** via nucleophilic epoxidation of α-arylidene pyrazolones **207** (Scheme 7.81) [168]. The reported protocol enables efficient preparation of *trans*- and *cis*-spiropyrazolone epoxides **220** in high enantiocontrol by using readily available amine-thioureas. The same group reported an asymmetric sulfa-Michael/Michael reaction cascade between α-arylidene pyrazolones and 4-mercapto-2-butenoates, affording chiral tetrahydrothiophene-fused spiropyrazolones bearing three consecutive stereocenters [169]. Later, Lattanzi et al. developed an enantioselective vinylogous Michael/cyclization cascade reaction between α,α-dicyanoalkylidenes and α-arylidene pyrazolones **207** [170]. The process was catalyzed with commercially available Takemoto's catalyst **XIII** giving diastereoisomeric spiropyrazolones in high yields with high to excellent enantioselectivity. Soon after, Wang's team reported an enantioselective [3+3] cycloaddition between α-arylidene pyrazolones **202** and benzylidenemalononitriles promoted by a chiral bifunctional tertiary amine-squaramide **LVI**, giving the final spiropyrazolones in good yields with moderate stereoselectivity [171]. Recently, Han et al. reported an alternative three-component annulation of α-arylidene pyrazolones **202**, proceeding through benzylidene malonates generated in situ from malononitrile and benzaldehydes, and catalyzed by chiral Cinchona alkaloids [172].

In 2018, Jiang et al. reported the enantioselective synthesis of tetrahydrofuran-fused spiropyrazolones **223** through Michael addition of pyrazolones **204** to

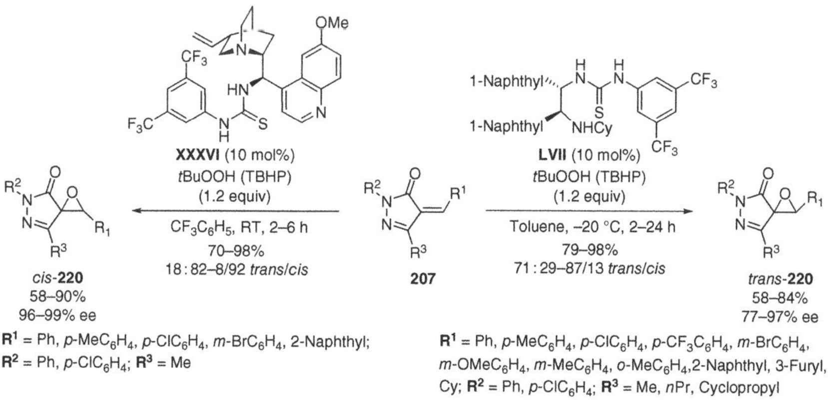

Scheme 7.81 Diastereodivergent and enantioselective access to spiroepoxides. *Source:* Based on Meninno et al. [168].

ortho-quinone methides, generated *in situ* from 2-(1-tosylalkyl)phenols **222**, followed by chlorination/cyclization cascade (Scheme 7.82) [173]. The Michael addition step was efficiently catalyzed by bifunctional quinine-derived squaramide **LVI**. According to the optimization of the subsequent chlorination/cyclization step, bifunctional cyclohexane-1,2-diamine-based tertiary amine-squaramide **LVIII** was found to be essential for high diastereocontrol of the process. Interestingly, an attempt to perform both steps as one-pot procedure led to the formation of spirocyclic products with significantly lowered diastereoselectivity. Recently, Wang, Xu et al. reported [4+1] annulation of *in situ* generated *ortho*-quinone methides and 4-halopyrazolones, affording final benzofuran-fused spiropyrazolones in high yields with high to excellent enantioselectivity [174].

An attractive strategy to construct chiral spiropyrazolones through NHC-catalyzed [4+2] annulation of α-arylidene pyrazolones **207** and γ-chloroenals **224** was developed by Yang, Zhong et al. (Scheme 7.83) [175]. The reaction proceeds via a vinyl enolate intermediate in the absence of any oxidant, affording spiropyrazolones in good yields with high diastereo- and enantioselectivity (up to 86% yield,

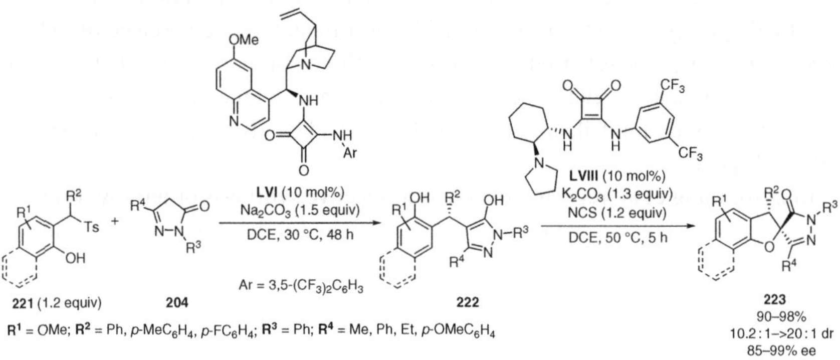

Scheme 7.82 Synthesis of chiral pyrazolone and spiropyrazolone derivatives through squaramide-catalyzed reaction. *Source:* Based on Zhou et al. [173].

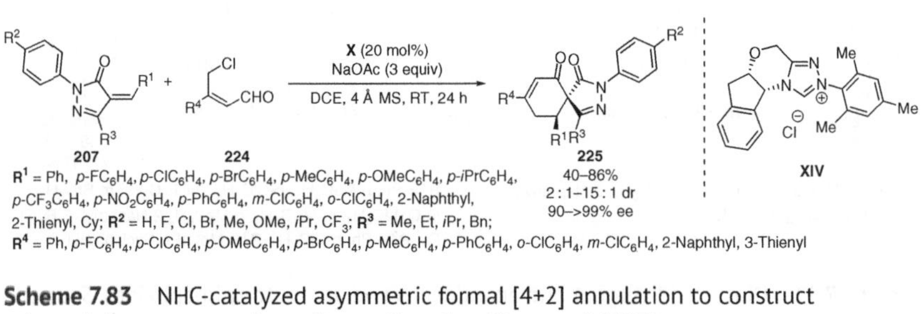

Scheme 7.83 NHC-catalyzed asymmetric formal [4+2] annulation to construct spirocyclohexane pyrazolone. *Source:* Based on Zhao et al. [175].

15 : 1 dr, >99% ee). Recently, the annulation process was studied in detail using DFT calculations, attributing the importance of AcOH as a proton shuttle in the formation of Breslow intermediate, as well as the cumulative effect of several week interactions in stereoselectivity-determining steps [176]. A secondary amine-catalyzed [4+2] annulation between α,β,γ,δ-unsaturated pyrazolones **226** and aldehydes **109** was reported by Han et al., and used for the enantioselective synthesis of multifunctional spiropyrazolones; several of them exhibiting potent toxicity against a panel of cancer cell lines [177]. Another enantioselective approach to spiropyrazolones, developed by Xu et al., is based on one-pot multicomponent Michael addition/chlorination/cyclization sequence using bifunctional tertiary amine-squaramide catalyst **LVI** [178]. Based on that approach, two series of highly enantiomerically enriched spirocycles were prepared, including spirocyclic cyclopropane-fused pyrazolones and dihydrobenzofuranone-fused pyrazolones. Very recently, the same group reported enantioselective [4+1] annulation of pyrazolones **204** with imine intermediate generated from γ,δ-unsaturated β-ketoesters **227** and nitrosobenzene, affording spiropyrazolones in high yields and excellent stereoselectivity (up to 85% yield, >20 : 1 dr, 96% ee) [179].

7.5.1.3 Cooperative-catalytic Approaches

An enantioselective synthesis of spiropyrazolones can also be accomplished by combined catalysis, as demonstrated by Schoenebeck, Enders et al. (Scheme 7.84) [180]. They developed a successful merger of squaramide-catalysis with Ag-catalysis in Michael/Conia-ene reaction sequence of pyrazolones **204** with alkyne-tethered nitroalkenes **228**, affording the desired spiropyrazolones in high yields and stereoselectivities. The reaction proceeds through the formation of a ternary complex in which nitroalkene and pyrazolone are synergistically activated by H-bonding with squaramide **LIX**.

In 2018, Roithová, Veselý, Rios et al. reported a new protocol for preparing chiral spiropyrazolones based on the ring-opening/ring-closing reaction sequence [181]. The reaction of readily available bicyclic pyrazolone derivatives **230** and enals **29** is initiated with their activation by $Pd_2(dba)_3$ and secondary

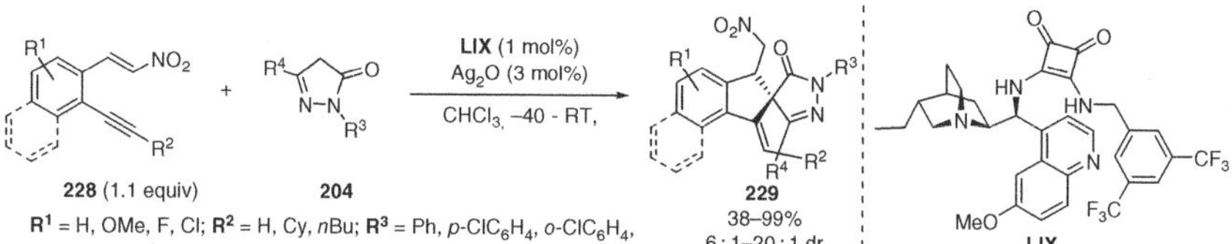

228 (1.1 equiv) **204** **229** 38–99% 6 : 1–20 : 1 dr 82–99% ee **LIX**

R¹ = H, OMe, F, Cl; **R**² = H, Cy, *n*Bu; **R**³ = Ph, *p*-ClC_6H_4, *o*-ClC_6H_4, *p*-MeC_6H_4; **R**⁴ = Me, Ph, Et, *i*Pr, CF_3

Scheme 7.84 Asymmetric synthesis of spiropyrazolones by sequential organo- and silver catalysis. *Source:* Based on Hack et al. [180].

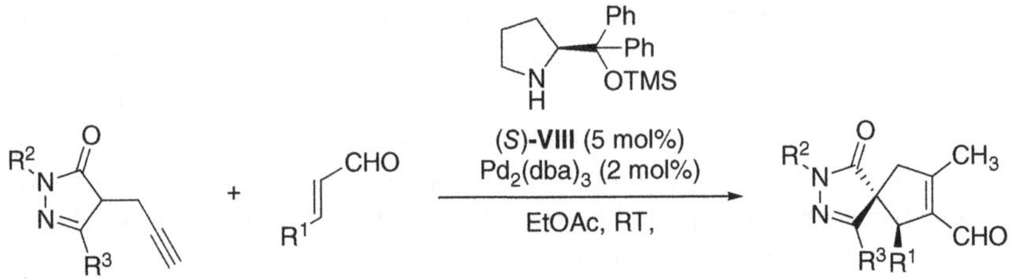

231 (1.5 equiv) **29** **232** 37–95% 1.1 : 1–5.6 : 1 dr 55–96% ee major 64–99% ee minor

R¹ = Ph, p-BrC$_6$H$_4$, p-ClC$_6$H$_4$, p-CNC$_6$H$_4$, p-MeC$_6$H$_4$, p-OMeC$_6$H$_4$, p-CF$_3$C$_6$H$_4$, p-NO$_2$C$_6$H$_4$, p-PhC$_6$H$_4$, m-MeC$_6$H$_4$, o-MeC$_6$H$_4$, 2-Naphthyl, 2-Thienyl, 2-Furyl, nPr, CO$_2$Et, But-3-en-1-yl;
R² = Ph, p-ClC$_6$H$_4$, p-MeC$_6$H$_4$, p-OMeC$_6$H$_4$, p-CF$_3$C$_6$H$_4$, p-NO$_2$C$_6$H$_4$, tBu; **R³** = Me, Et, Ph

Scheme 7.85 Michael/Conia-ene cascade reaction under synergistic catalysis providing spiropyrazolones. *Source:* Based on Putatunda et al. [182].

amine **VIII** to yield zwitterionic Pd-species (ring/opening step) and iminium intermediate. The formed Pd-intermediate as a nucleophile attacks intermolecularly iminium to generate enamine intermediate, which intramolecularly reacts with allylic cation in the following step. After hydrolysis and release of Pd catalyst, the final spiropyrazolones are achieved in good to high yields, moderate to high diastereoselectivity, and excellent enantioselectivity (up to 98% yield, 20 : 1 dr, >99% ee). Inspired by the previous works, Herrera, Rios, Veselý et al., expanded the use of enantioselective Michael/Conia-ene reaction cascade to the formation of spiropyrazolones (Scheme 7.85) [182]. They used alkyne-tethered pyrazolones **231** and enals **29** in combination with synergistic catalysis of Pd complex and secondary amine, giving the final spirocycles with high efficiency and stereocontrol. Interestingly, the reaction was catalyzed with a low catalyst loading (2 and 5 mol%), and the dual role of aryl substituents on organocatalyst was elucidated using computation studies. The same year, Singh et al. reported the synthesis of spiropyrazolones from α-arylidene pyrazolones and nitroalkynes using sequential bifunctional amine-thiourea catalysis and Ag-catalysis, unfortunately, with low enantiocontrol [183].

7.5.2 Chapter 7.5.2 – Enantioselective Formation of Spirocycles Containing Barbituric Acid and Hydantoin

The preparation of chiral spirocyclic compounds containing barbituric acid or hydantoin can be accomplished by enantioselective approaches based on organometallic catalysis and organocatalysis. In 2018, Zhao et al. reported the enantioselective [4+2] cycloaddition of alkylidene barbiturates **234** with vinyl benzoxazinanones **233** catalyzed by a chiral palladium-ligand complex. The spirocyclic barbiturates were obtained in isolated yields of up to 96% and with excellent diastero- and enantioselectivity (>99 : 1 dr, 97% ee) (Scheme 7.86) [184].

In the context of organocatalytic approaches leading to heterocyclic spiro compounds, the reactivity of the hydantoin derivatives was investigated firstly. In 2011, Fu et al. applied a new chiral phosphine catalyst **LX** in enantioselective [3+2] cycloaddition of methylene hydantoins **236** with allenolates **237a** and allenamides. The authors observed the formation of two possible regioisomers in a ratio ranging from 8 : 1 to 50 : 1. Major regioisomers were prepared in yields up to 98% with high stereoselectivity (>20 : 1 dr, up to 98% ee) (Scheme 7.87) [185]. Five years later, Guo et al. synthesized chiral cyclopentene-fused spirobarbiturates through enantioselective organocatalytic [3+2] annulation of MBH

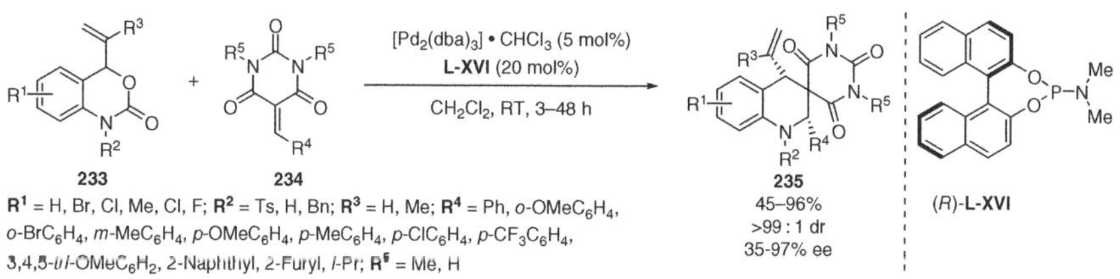

Scheme 7.86 Diastereoselective and enantioselective synthesis of barbiturate-fused spirotetrahydroquinolines. Zhao. *Source:* Based on Zhao et al. [184].

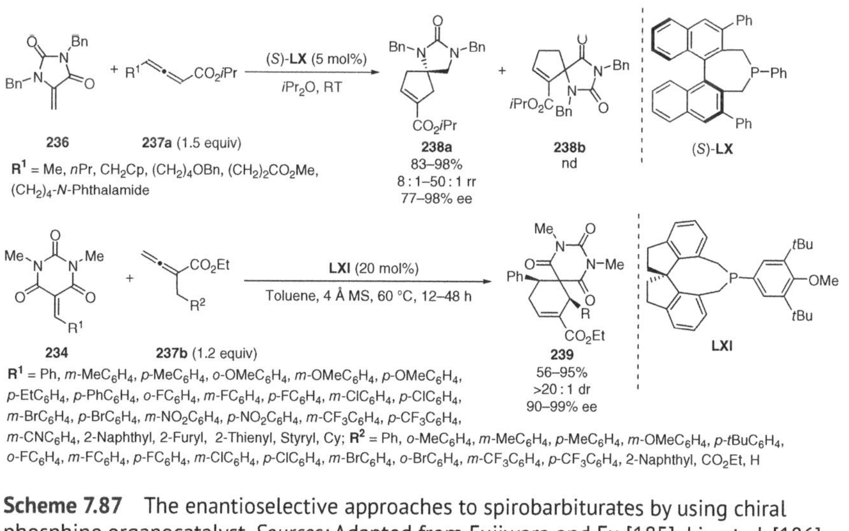

Scheme 7.87 The enantioselective approaches to spirobarbiturates by using chiral phosphine organocatalyst. *Sources:* Adapted from Fujiwara and Fu [185]; Liu et al. [186].

carbonates **117** with alkylidene barbiturates **234**. The enantioselective transformation was efficiently catalyzed by a chiral bifunctional amidophosphine, affording final spirobarbiturates in moderate to high yields and stereoselectivities [187]. Soon after, Guo et al. further expanded the area of catalytic methods allowing the formation of chiral spirobarbiturates and reported the enantioselective synthesis of cyclohexene-fused spirobarbiturates; a phosphine-catalyzed [4+2] cycloaddition of alkylidene barbiturate **234** derivatives with allenolates **237b**. In this work, good yields (56–95%), excellent diastereoselectivities (>20 : 1 dr), and enantiomeric excesses (90–99% ee) of final spiro heterocycles were achieved (Scheme 7.87) [186]. In 2018, Guo et al. developed enantioselective [8+2] annulation of heptafulvene barbiturate derivatives with allenolates promoted by chiral Kwon's *endo* phosphine catalyst. The target spirocyclic compounds containing barbituric acid were obtained in good yields (up to 67%) and with high enantioselectivity (up to 97% ee) [188].

In 2015, He, Wang, Peng et al. applied catalysis with chiral secondary amines for the preparation of chiral spirobarbiturates from readily available aldehydes, nitroalkenes, and alkylidene barbiturates [145]. The reported tandem Michael/Michael/aldol reaction furnished spirocyclic compounds derived from barbituric acid in good yields (up to 62%) and with high stereoselectivity (up to 19 : 1 dr and 98% ee). Later, Albrecht et al. published a sequential reaction involving Michael addition of α-isothiocyanates **240** to alkylidene barbiturates **234** followed by cyclization *via* nucleophilic addition to an isothiocyanate group. The organocatalytic transformation was catalyzed by a chiral bifunctional squaramide derived from Cinchona alkaloids **LVI**. Target spirobarbiturates containing a pyrrolidine heterocycle were prepared in high yields and with excellent stereocontrol (Scheme 7.88) [189].

In 2016, Michelet, Ratovelomanana-Vidal et al. developed an elegant synthetic strategy toward chiral cyclopentanes based on the synergistic merger of secondary amine catalysis with Pd(0)-catalysis. As a part of the substrate survey, an efficient synthesis chiral spirobarbiturate was also accomplished [190].

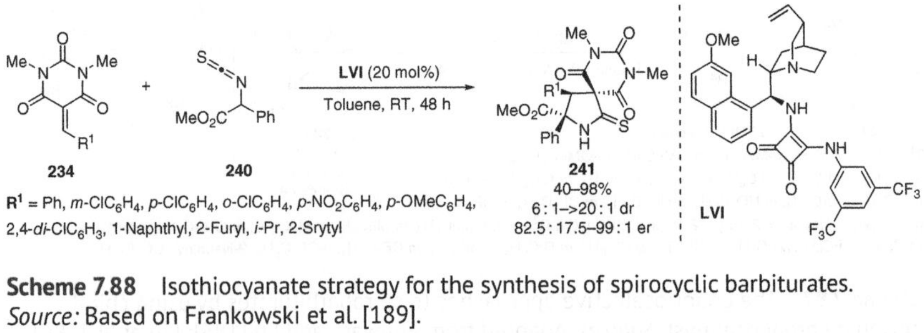

Scheme 7.88 Isothiocyanate strategy for the synthesis of spirocyclic barbiturates. *Source:* Based on Frankowski et al. [189].

7.5.3 Chapter 7.5.3 – Enantioselective Formation of Spirocycles Containing Azlactones

7.5.3.1 Organometallic Approaches

In 2011, Trost et al. introduced the enantioselective formal [3+2] cycloaddition of vinyl cyclopropanes **242** with alkylidene azlactones **243** to form cyclopentane-fused spiroazlactones. The authors used a complex of $[Pd_2dba_3] \cdot CHCl_3$ combined with Trost's ligand **L-XVIII** to catalyze the cycloaddition process. The spirocyclic azlactones **244** were obtained in excellent yields (up to 87%), with diastereoselectivity from 3 : 1 to >19 : 1 and with enantioselectivity up to 98% ee (Scheme 7.89) [192]. Later, the same group expanded this concept and enantioselective synthesis of spirocyclic Meldrum's acid derivatives was accomplished [193]. Another approach to spiroazlactones may be the catalytic enantioselective double Michael addition of dienones **113** to azlactones developed by Peters et al. in 2013. In this case, a chiral ferrocene bispalladacycle was used to give azlactone spirocycles in high yields and selectivity (Scheme 7.89) [191].

7.5.3.2 Organocatalytic Approaches

In the case of enantioselective organocatalysis without the participation of metals, phosphine catalysis was mainly used in the preparation of spiro compounds containing azlactone. In 2012, Jørgensen et al. reported a chiral phosphine (*S*)-BINEPINE-catalyzed [3+2] cycloaddition of allenoates **237c** with alkylidene azlactones **243**. In the reaction, both regioisomeric spirocycles were formed in a ratio ranging from 3.5 : 1 to 9 : 1, isolated in good yields (up to 83%) with high stereoselectivity (up to 94% ee) (Scheme 7.90) [194].The azlactone ring in the formed spiroazlactones was subsequently opened and various amino acid derivatives **248** were obtained. Nearly simultaneously, Shi et al. reported an analogous [3+2] annulation reaction catalyzed by chiral phosphine (*R*)-SITCP, affording only one of the two regioisomeric spiroazlactones in high yields (up to 96%) and excellent stereoselectivity (up to >19 : 1 dr, up to 99% ee) (Scheme 7.90) [195].

7.5.3.3 Cooperative-catalytic Approaches

An interesting approach to prepare spirocyclic azlactones is the combination of palladium-catalyzed cycloisomerization of enynamides **249** followed by trienamine catalysis. This concept was published in 2016 by Anderson et al. using alkylidene azlactones **243** to form spirocyclic azlactones **251** in yields up to 50%, with excellent diastereoselectivity (>20 : 1 dr) and enantioselectivity (>99% ee) (Scheme 7.91) [196]. Recently, Veselý et al. reported an enantioselective synthesis of spiroazlactones **253** through cyclization of enals **29** with racemic vinylcyclopropane azlactones **252** (Scheme 7.92) [197]. The reaction catalyzed by the

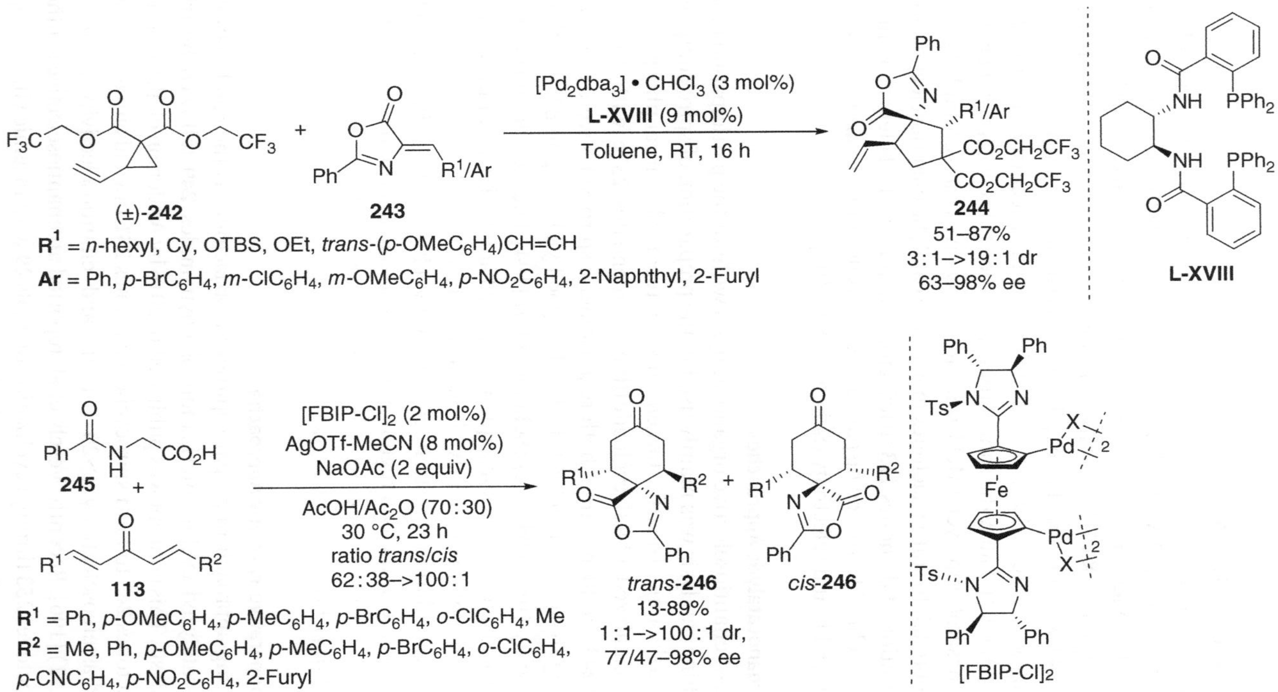

Scheme 7.89 Palladium-catalyzed enantioselective synthesis of spirocyclic azlactones published by Trost and Peters. *Source:* Based on Weber et al. [191].

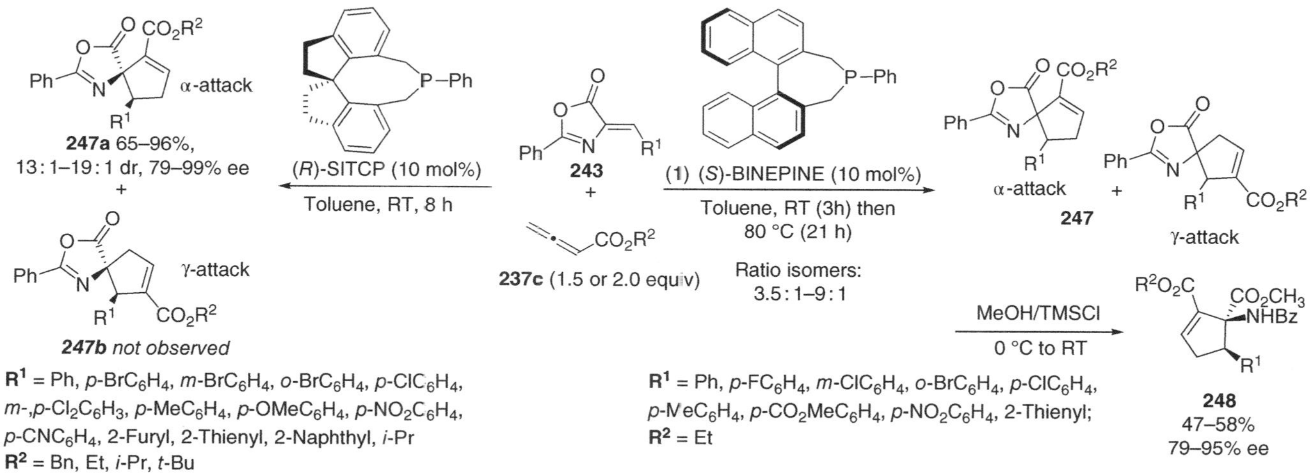

Scheme 7.90 Application of chiral phosphine in organocatalytic synthesis of spiroazlactone. *Sources:* Adapted from Steurer et al. [194]; Wang et al. [195].

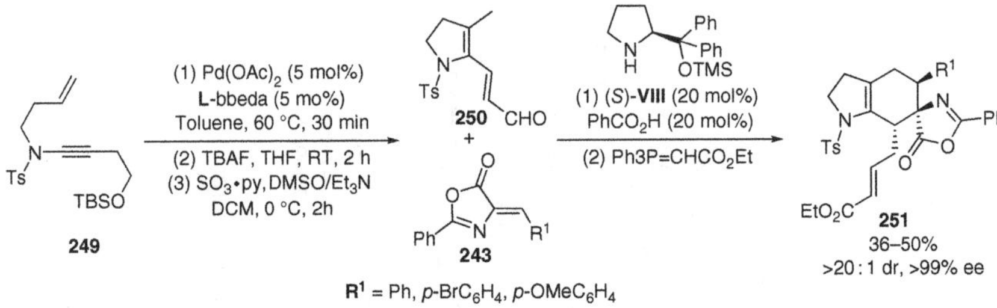

Scheme 7.91 Anderson's synthesis of spirocyclic azlactones. *Source:* Based on Chintalapudi et al. [196].

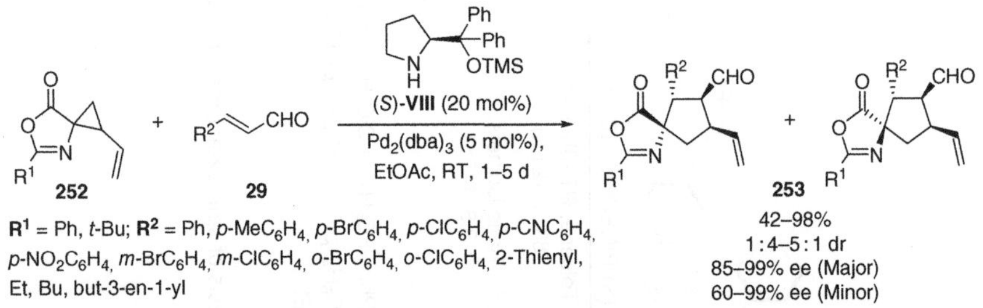

Scheme 7.92 Formal [3+2] cycloaddition of vinylcyclopropane azlactones to enals using synergistic catalysis reported by Veselý. Source: Based on Kamlar et al. [197].

combination of $Pd_2(dba)_3$ and chiral secondary amine **VIII** gave the spirocyclic products **253** in good yields with moderate diastereoselectivity and excellent enantioselectivity.

7.5.4 Chapter 7.5.4 – Enantioselective Formation of Spirocycles Containing Rhodanines and Sultams

In this part, we will focus on enantioselective approaches in the synthesis of spiro compounds containing rhodanine, sultam, and related types of sulfur-containing heterocycles.

7.5.4.1 Organometallic Approaches

In 2016, Cramer et al. reported a chiral Cp^x-Rh(III)-catalyzed [3+2] annulation of *N*-sulfonyl ketimines **254** with alkynes **145** in the presence of *N*-protected L-phenylalanine and $AgNTf_2$. The desired spirocycles were prepared in high yields (up to 99%) and with high enantioselectivity (up to 94% ee) (Scheme 7.93) [198]. Two years later, He and Wang performed an asymmetric Mannich reaction of α-diazocarbonyl compounds **257** with *N*-sulfonyl ketimines **254** catalyzed by Au(I)

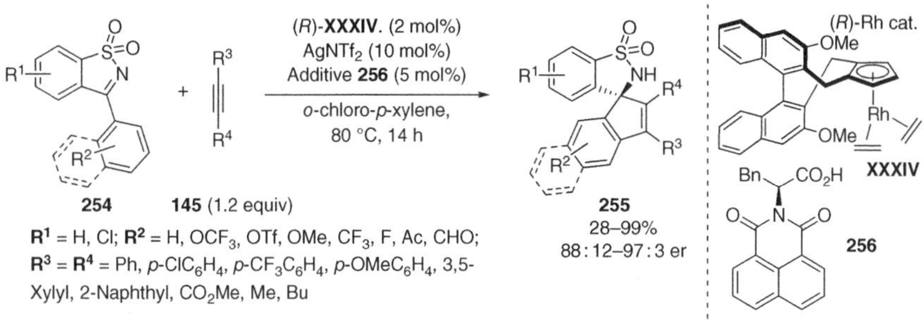

Scheme 7.93 Enantioselective access to spirocyclic sultams by chiral Cp^x–rhodium(III)-catalyzed annulations. *Source:* Based on Pham and Cramer [198].

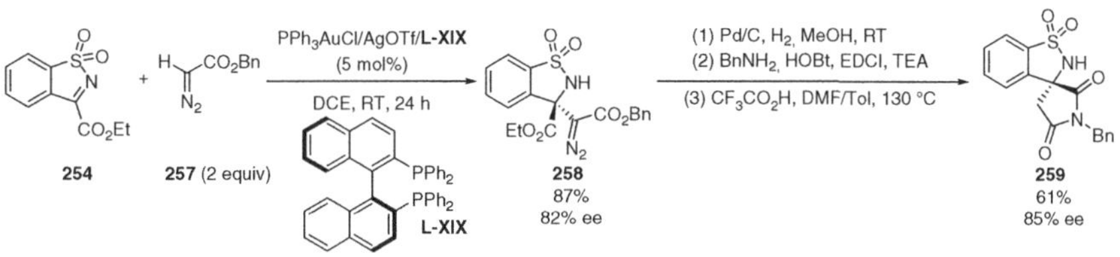

Scheme 7.94 The synthesis of chiral spirosultams by multi sequential reaction. *Source:* Based on Zhou et al. [199].

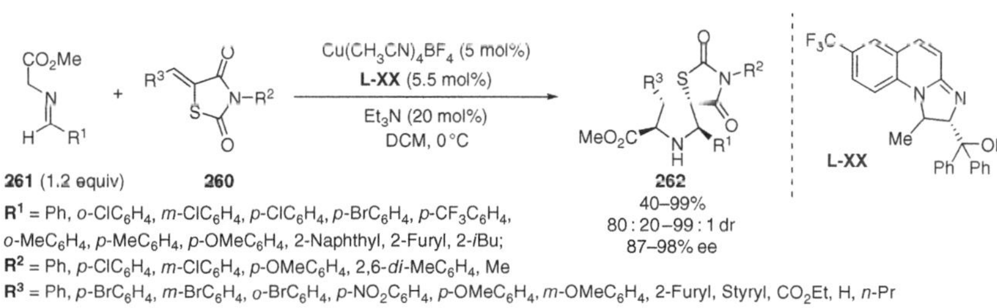

Scheme 7.95 Asymmetric construction of spirocyclic pyrrolidine-thiazolidinediones. *Source:* Yang et al. [200].

complexes. The obtained chiral Mannich adducts **258** were subsequently transformed in three steps to a spiro compound **259** containing a sultam moiety (Scheme 7.94) [199].

Enantioselective synthesis of spirothiazolidinediones was firstly reported by Yu, Deng et al. in 2015 (Scheme 7.95) [200]. They developed a highly enantioselective 1,3-dipolar cycloaddition of 5-alkylidene thiazolidine-2,4-diones **260** with azomethine ylides **261** catalyzed by [$Cu(CH_3CN)_4$]BF_4 in the presence of chiral

N,O-ligands. This method yielded structurally interesting pyrrolidine-fused spirothiazolidinediones **262** in good yields (up to 99%) and with excellent stereoselectivity (up to 99 : 1 dr, up to 98% ee).

7.5.4.2 Organocatalytic Approaches

In 2014, Chen, Gao et al. used the catalysis with chiral amines **LI** derived from Cinchona alkaloids for the formal [5+3] cycloaddition between β-substituted cyclopentenones **263** and sultam derivatives **264**. The Michael/Mannich cascade reaction proceeding through dienamine/dienamine catalytic pathway furnished spirocyclic sultams in high yields with excellent enantioselectivity (97–99% ee) (Scheme 7.96) [201]. A year later, Chen et al. reported a formal [3+3] cycloaddition between sultam derivatives and aliphatic ketones promoted by Cinchona-derived primary amine and involving enamine/enamine activation strategy [202].

In 2015, Ye et al. continued in this area and used unsaturated ketones in enantioselective formal [4+2] cycloaddition with cyclic sultams catalyzed by chiral primary amines [203]. In the same year, Zhou, Wang et al. applied hydrogen bond catalysis in this area. They used sultam derivatives **264** with indolin-2-ones **195** in an enantioselective formal [3+3] cycloaddition catalyzed by chiral bifunctional thiourea **XXXVI**. Optically active spirocyclic compounds containing sultam and thiopyranoindole rings were prepared in high yields (up to 99%) and with excellent diastereo- and enantioselectivities (up to >20 : 1 dr, up to 99% ee) (Scheme 7.97) [204].

Other representatives of sulfur heterocycles are rhodanines and thiazoles. In 2012, Ye et al. reported the first organocatalytic method for the preparation of spirocyclic compounds containing rhodanine moiety. The reaction between enones and alkylidene rhodanines was efficiently promoted by chiral *tert*-leucine **LXII**, and the corresponding spirocyclohexanone-rhodanines **269** were prepared in high yields (up to 95%) and with excellent selectivity (5 : 1 to >20 : 1 dr, 99% ee)

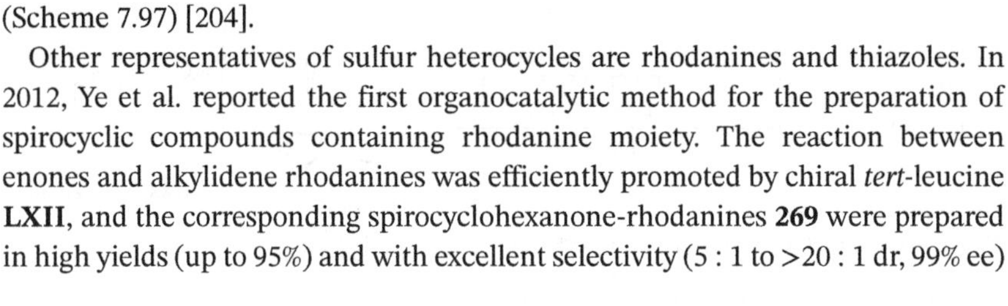

Scheme 7.96 The enantioselective cycloaddition of cyclic enones with sultams derivatives providing spirocyclic sultams. *Source:* Based on Yin et al. [201].

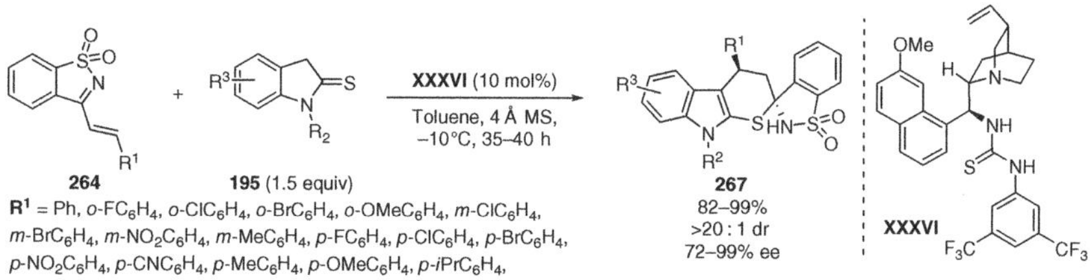

R^1 = Ph, o-FC_6H_4, o-ClC_6H_4, o-BrC_6H_4, o-$OMeC_6H_4$, m-ClC_6H_4, m-BrC_6H_4, m-$NO_2C_6H_4$, m-MeC_6H_4, p-FC_6H_4, p-ClC_6H_4, p-BrC_6H_4, p-$NO_2C_6H_4$, p-CNC_6H_4, p-MeC_6H_4, p-$OMeC_6H_4$, p-$iPrC_6H_4$, 2,4-$(Cl)_2C_6H_3$, 2,4-$(OMe)_2C_6H_3$, 1-Naphthyl, 2-Naphthyl, 2-Thienyl, 2-Styryl; R^2 = H, Et, Bn, Ph, Allyl, Me; R^3 = H, Br, F, I, Cl, Me

Scheme 7.97 Asymmetric construction of spiro thiopyrano-indole-benzoisothiazoles. *Source:* Chen et al. [204].

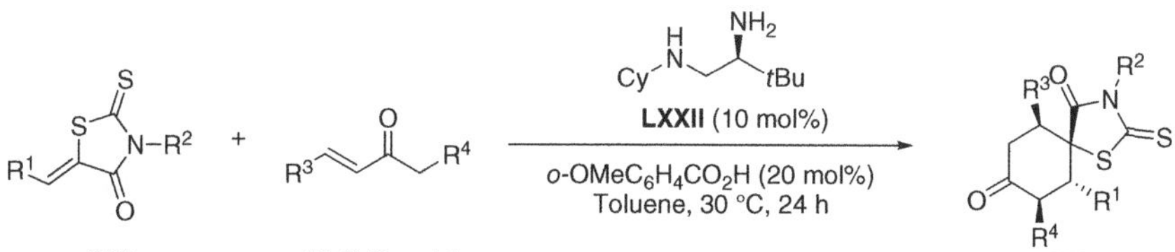

R^1 = CO_2Et, Ph, p-BrC_6H_4, p-FC_6H_4, p-$NO_2C_6H_4$, p-$OMeC_6H_4$; R^2 = H, Ph, iPr, Cy; R^3 = Ph, o-FC_6H_4, p-BrC_6H_4, o-$OMeC_6H_4$, m-MeC_6H_4, p-MeC_6H_4, 2-Thienyl, CO_2Et; R^4 = H, Me

35–95%
5 : 1–20 : 1 dr
87–99% ee

Scheme 7.98 Enantioselective synthesis of chiral spirocycles containing rhodanine ring. *Source:* Based on Wu et al. [205].

(Scheme 7.98) [205]. Spirocyclic compounds containing a thiazolidinedione were first prepared by Wang et al. in 2013. The preparation of these compounds was based on organocatalytic cascade reaction involving the Michael/Michael/aldol reaction sequence. The reaction between enals and thiazolidinedione or rhodanine derivatives was catalyzed by a chiral secondary amine derived from pyrrolidine. The obtained spirocyclic thiazolidinediones and spirocyclic rhodanines were prepared in good yields and with excellent stereoselectivity (up to >30 : 1 dr, up to 99% ee) [206].

In 2016, Jiang, Tan et al. used chiral bifunctional thioureas for enantioselective annulation between 5*H*-thiazol-4-one and *N*-itaconimides. In this reaction, spirocyclic piperidinones containing succinimide and a sulfur bridge were obtained with a high degree of enantioselectivity (up to 99% ee) and diastereoselectivity (>19 : 1 dr) [207]. Subsequently, in 2018, Du et al. introduced a bifunctional squaramide-catalyzed enantioselective cascade aza-Michael/Michael addition reaction between alkylidene thiazolidinone derivatives and 2-tosylaminochalcones **270**. Chiral spirothiazolidinones containing a tetrahydroquinoline ring **271** were synthesized in excellent yields (up to 99%) with excellent diastereoselectivity

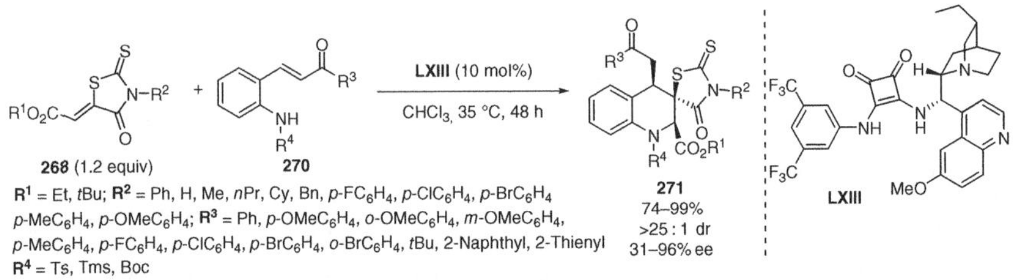

Scheme 7.99 Asymmetric synthesis of highly functionalized spirothiazolidinone tetrahydroquinolines. *Source:* Based on Song and Du [208].

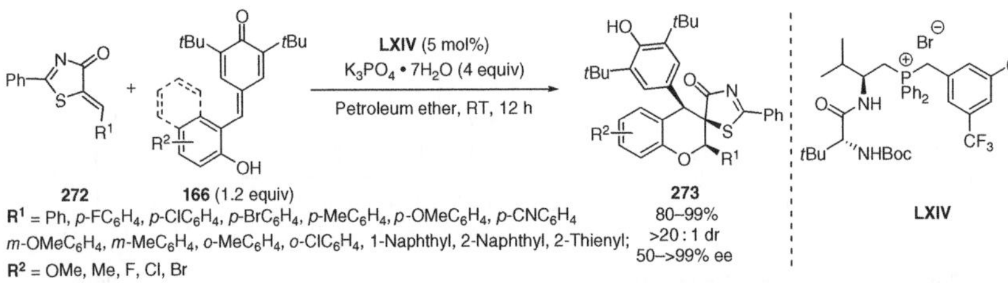

Scheme 7.100 Enantioselective construction of spiro chroman-thiazolones. *Source:* Based on Tan et al. [210].

(>25 : 1 dr) and enantioselectivity (up to 96% ee) (Scheme 7.99) [208]. The spirocyclic thiazolone-pyrrolidines were also prepared by Wang, Yan et al. in 2019 from an alkylidene thiazolone derivatives and CF_3-containing azomethine ylide catalyzed by β-isocupreidine. Enantioselective organocatalytic [3+2] cycloaddition gave products in high yields with high stereoselectivity with only 1 mol% of organocatalyst [209].

Recently, Wang et al. used chiral bifunctional phosphonium salts **LXIV** to catalyze [4+2] annulation between alkylidene thiazolones **272** and *ortho*-hydroxyphenyl-*para*-quinone methides **166**. The reported process gave chromane-fused spirothiazolones in excellent yields and stereoselectivities (>20 : 1 dr, up to >99.9% ee) (Scheme 7.100) [210]. The last example of spirocyclic compounds containing rhodanine and pyran moiety was announced by Xu et al. The developed organocatalytic [3+3] cycloaddition reaction between biselectrophilic rhodamine ketoesters and pyrazolones was efficiently catalyzed by a bifunctional Cinchona-derived squaramide [211].

7.5.5 Chapter 7.5.5 – Enantioselective Formation of Spiroketals (Spiroacetals)

Chiral spiroketals (spiroacetals) are common motifs in biologically active natural products, pharmaceuticals, and chiral ligands; thus, it is not surprising that several catalytic methods for their preparation were developed.

7.5.5.1 Organometallic Approaches

In 2012, Wang, Ding et al. reported the first enantioselective transition metal-catalyzed synthesis of aromatic spiroketals **275** through Ir-promoted hydrogenation of α,α′-bis(2-hydroxyarylidene) ketones **274** and spiroketalization of hydrogenation intermediates (Scheme 7.101) [212]. Final spiroketals were obtained in high yields and with excellent stereoselectivity (up to 95% yield, 49 : 1 dr, >99% ee) under catalysis with Ir(I) complex containing spiro[4.4]-1,6-nonadiene-based chiral phosphine-oxazoline ligand **LXV**, playing a dual catalytic role in the reaction. The authors improved the described protocol, and efficient preparation of aromatic spiroketals bearing electron-donating substituents was also accomplished [213]. At the same time, an example of spiroketal formation using Pd(II)-catalyzed cascade reaction was also reported by Wang, Shi et al. [214]. Later, inspired by Wang's and Ding's work, Dou et al. elaborated the synthesis of chiral spiroketals from chiral phenolic ketones bearing β,β′-stereogenic centers [215].

In 2017, Carreira et al. developed Ir-catalyzed stereoselective spiroketalization of racemic allylic carbonates **276** (Scheme 7.102) [216]. The reaction was efficiently catalyzed by $[Ir(cod)Cl]_2$ in the presence of chiral BINOL-derived fosforamidite **L-XXI**, affording final spiroketals in high yields and with high stereoselectivity (up to 99% yield, >25 : 1 dr, >98% ee). Soon after, Kang et al. developed an enantioselective synthesis of spiroketals using achiral Au(I) and

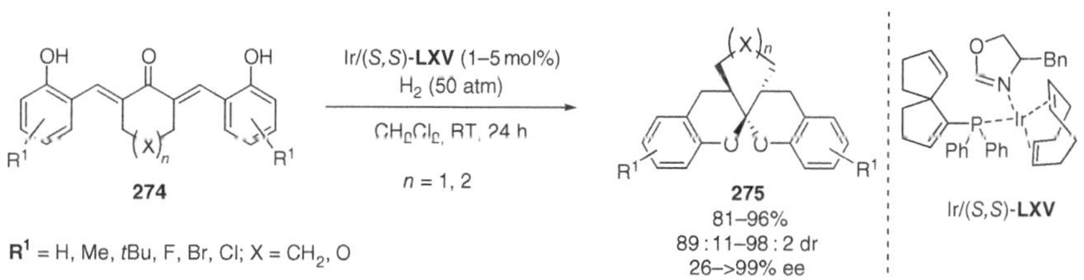

Scheme 7.101 Catalytic asymmetric synthesis of aromatic spiroketals. *Source:* Based on Wang et al. [212].

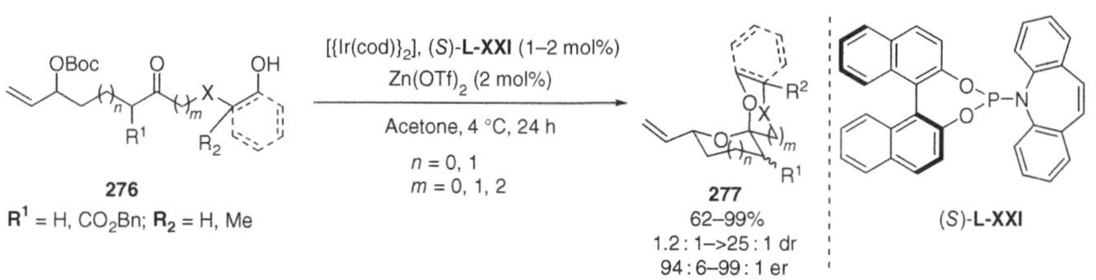

Scheme 7.102 Enantio- and diastereoselective spiroketalization catalyzed by chiral iridium complex. *Source:* Based on Hamilton et al. [216].

Rh(III) relay catalysis (Scheme 7.103) [217]. The reaction comprising alkynyl alcohols **278** and β,γ-unsaturated α-ketoesters **155** in the presence of AuCl·PPh_3 and chiral Rh(III) complex **LXVI** afforded the final spiroketals in high yields, good diastereoselectivities, and excellent enantioselectivity. Another highly efficient bimetallic catalytic system for the preparation of chiral spiroketals was introduced by Feng et al. [218] This approach was based on $AuCl_3$-catalyzed cycloisomerization of β-alkynyl ketones followed by enantioselective [4+2] cycloaddition with β,γ-unsaturated α-ketoesters **155** in the presence of chiral *N,N*-dioxide-Mg(II) complex.

7.5.5.2 Organocatalytic Approaches

An attractive strategy to construct chiral spiroketals through intramolecular nucleophilic addition of alcohol to oxocarbenium ion intermediate using chiral Brønsted acids was developed by List et al., followed by Nagorny et al. List's team designed and synthesized a new family of C_2-symmetric imidodiphosphoric acids, and they showed their potential in the enantioselective synthesis of various chiral spiroketals **281** from the corresponding enol ethers **280** (Scheme 7.104) [221].

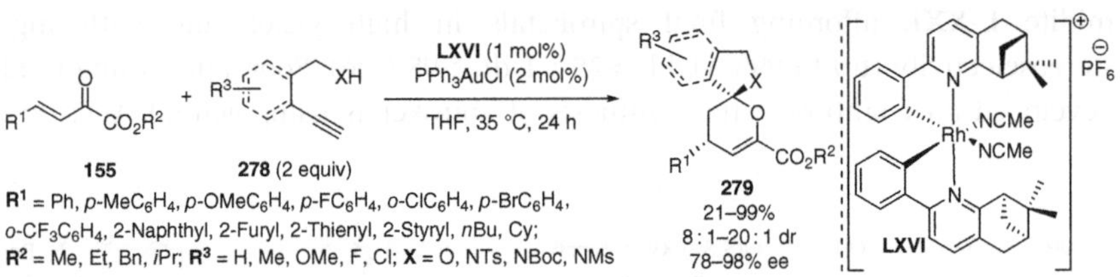

Scheme 7.103 Asymmetric synthesis of spiroketals and spiroaminals. *Source:* Based on Gong et al. [217].

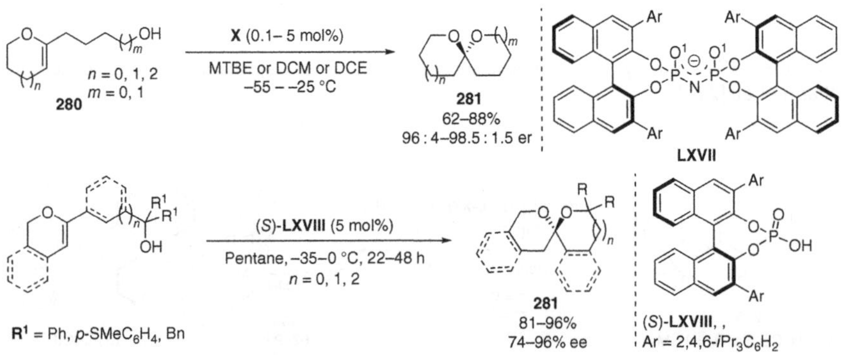

Scheme 7.104 Chiral phosphoric acid-catalyzed enantioselective and diastereoselective spiroketalizations. *Sources:* Adapted from Sun et al. [219]; Khomutnyk et al. [220].

Almost simultaneously, Nagorny et al. showed that also chiral BINOL-derived phosphoric **LXVIII** can efficiently promote enantioselective spiroketalization of various achiral and chiral cyclic enol ethers (Scheme 7.104) [219, 220]. Both methodologies afforded final spiroketals in high yields with high stereoselectivity.

Three years later, Asano, Matsubara et al. reported a synthesis of chiral spiroketals through an intramolecular hemiacetalization/oxa-Michael addition cascade [222]. Good to excellent yields and enantioselectivities of **283** were obtained with both electron-rich and electron-poor enones **282** in the presence of bifunctional Takemoto catalyst **XIII**. Other approaches to chiral spiroketals involving oxa-Michael addition as a part of the cascade process were developed by Ghorai et al. [223] The first cascade starting with alkoxyboronate spiro-ketalization step is catalyzed by chiral bifunctional amine-thiourea catalyst; the second approach is catalyzed by squaramide organocatalyst.

In 2016, Jiang, Xue, Li et al. developed the enantioselective synthesis of aromatic spiroketals with a high level of enantioselectivity (up to 98% ee) through organocatalytic bromination/nucleophilic substitution reaction sequence, starting from alcohols **284** (Scheme 7.105) [224]. The reaction proceeds in the presence of 1,3-dibromo-5,5-dimethylhydantoin and $(DHQD)_2N$, which is transformed to the chiral *N*-bromo quaternary ammonium salt intermediate playing a catalytic role in the process and also serving as an electrophilic source of bromine.

7.5.5.3 Cooperative-catalytic Approaches

From other catalytic approaches used to synthesize chiral spiroketals, methods based on combined catalysis are also important. In 2013, Gong et al. developed an elegant three-component Au(I)/BINOL-phosphoric acid relay catalyzed synthesis of chiral aromatic spiroketals from salicylaldehydes **286**, anilines **287**, and alkynols **278** (Scheme 7.106) [225]. The spiroketals were prepared in high yields with good to high selectivity. Nearly simultaneously, a similar procedure using Au(I)/BINOL-phosphoric acid relay catalysis for the stereoselective synthesis of spiroketals from glyoxylic acid, anilines, and alkynols was reported by Fañanás, Rodríguez

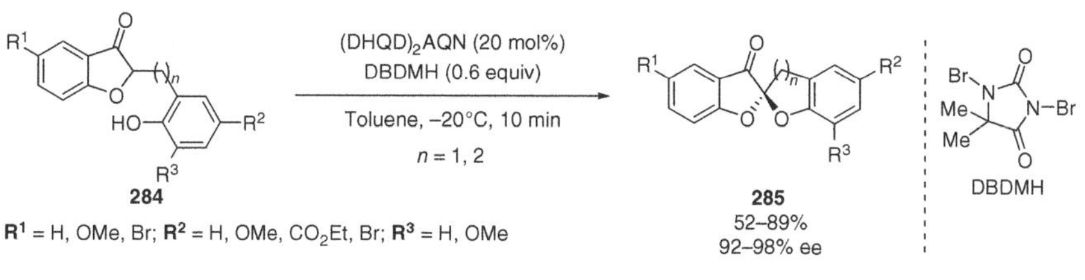

Scheme 7.105 Organohalogenite-catalyzed spiroketalization. *Source:* Based on Xue et al. [224].

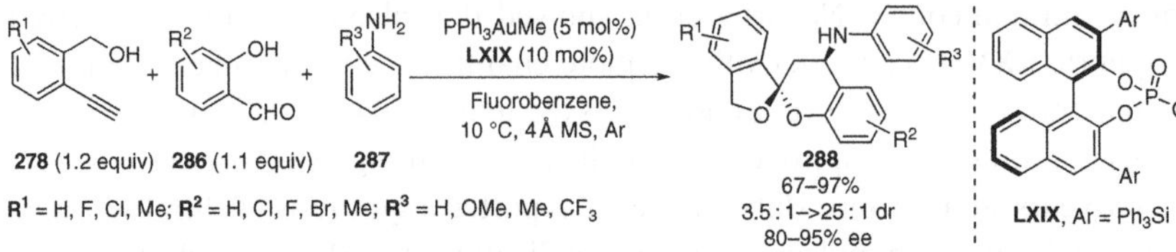

Scheme 7.106 Stereoselective three-component reaction of salicylaldehydes, anilines, and alkynols. *Source:* Modified from Wu et al. [225].

et al. [226]. Other attempts to prepare chiral spiroketals using Au(I) complexes in combination with chiral phosphoric acids were reported by Brimble et al. and Rexit et al. [227, 228].

7.6 Enantioselective Formation of Other Spiro Heterocyclic Systems

Beside the abovementioned, various catalytic methods were developed to access other spiro heterocyclic compounds; not only due to the occurrence of such systems in Nature or to explore the properties of newly designed materials but also to demonstrate the synthetic potential of newly developed methods.

7.6.1 Organometallic Approaches

Since Tamao's pioneering work [229] in the area of enantioselective synthesis of spiro compounds using catalytic methods based on transition metal complexes, a number of remarkable transition metal-catalyzed spirocyclization methods were developed. Most of the works were already mentioned in previous chapters; thus, here, we pay attention to catalytic methods allowing the formation of spirocycles containing less common heterocycles, and spirocycles with more heterocyclic motifs.

In 2016, Sansano, Cossío et al. developed a highly enantioselective synthesis of dihydrofuranone-fused spiropyrrolidines via 1,3-dipolar cycloaddition employing α-imino γ-lactones **289** as azomethine ylide precursors and nitroalkenes **164** as dipolarophiles (Scheme 7.107) [230]. The reaction was promoted by the complex formed from AgF and (*R,R*)-Me-DuPhos **L-XXXII**, acting as a bifunctional catalyst and providing the final spirocycles with high stereocontrol. Later, similar Ag(I)-catalyzed approach to this type of spirocycles was also reported by Wang et al. [231].

An attractive strategy to construct chiral spirocyclic β-lactams **293** in high yields with high stereoselectivity was reported by Enders et al. (Scheme 7.108) [232]. The

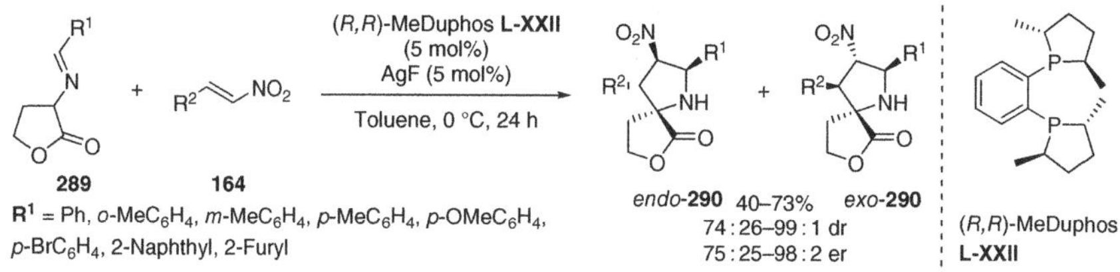

Scheme 7.107 The enantioselective synthesis spiropyrrolidines containing dihydrofuranone ring. *Source:* Based on Cayuelas et al. [230].

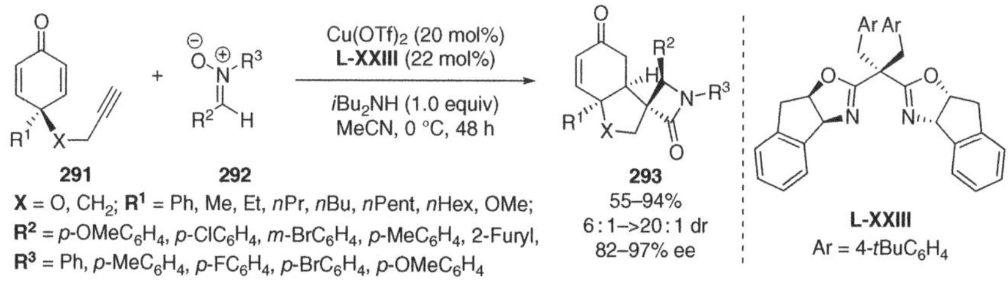

Scheme 7.108 Asymmetric synthesis of spirocyclic β-lactams through copper-catalyzed Kinugasa/Michael domino reactions. *Source:* Based on Shu et al. [232].

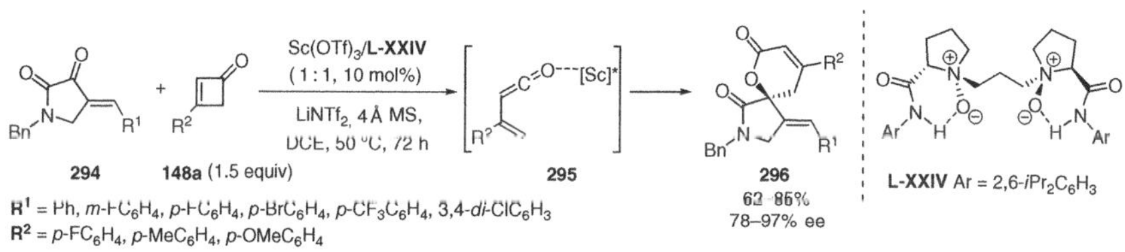

Scheme 7.109 Hetero-Diels–Alder process used for the synthesis of spiropyrrolidinone-dihydropyranone derivatives. *Source:* Based on Luo et al. [234].

process involving enantioselective Kinugasa/Michael cascade reaction between alkyne-tethered cyclohexadienones **291** and nitrones **292** was catalyzed by $Cu(OTf)_2$ in the presence of indane-BOX ligand **L-XXIII**. Soon after, Yang, Cai et al. reported Cu-catalyzed double *O*-arylation of propane-1,3-diol derivatives resulting in the formation of chiral *oxa*-spirocycles in good yields and enantioselectivities [233].

Recently, a highly enantioselective ring-opening/cycloaddition reaction cascade of cyclobutenones **148a** with (E)-dioxopyrrolidines **294** was disclosed by Dong and Feng (Scheme 7.109) [234]. The reaction, proceeding through vinylketene intermediate **295** generated from cyclobutenone under catalysis with chiral proline-derived *N,N*-dioxide/$Sc(OTf)_3$ complex, afforded final spirocyclic compounds in high yields with high enantioselectivity.

7.6.1.1 Organocatalytic Approaches

In 2011, Toste et al. reported an interesting example of enantioselective fluorocyclization of olefins **296** affording a wide range of fluorinated spiro-fused oxazolines **297** in high yields with excellent diastereo- and enantioselectivity (up to 95% yield, >20 : 1 dr, 97% ee) (Scheme 7.110) [235]. The developed chiral anion-based phase-transfer procedure, employing chiral cationic fluorinating agent generated from Selectfluor and BINOL-based phosphoric acid **LXX**, exhibits an improved tolerance toward sensitive functionalities and allows fluorocyclization even for less electron-rich olefins. Soon after, Voituriez and Marinetti reported the enantioselective synthesis of unprecedented heterocyclic spiranes and dispiranes through phosphine-catalyzed [3+2] annulation of allenoates and enones [236]. Spirocyclic products containing oxygen, nitrogen, and sulfur atom were obtained in good yields and excellent enantioselectivity under catalysis with chiral ferrocene-based phosphine.

In 2013, Chen et al. developed an elegant [4+2] cycloaddition of 2,4-dienals with α-methylene carbonyl compounds, generated in situ from stable methiodide salts of Mannich bases [237]. The reaction proceeding through trienamine activation with chiral secondary amine **VIII** afforded various heterocyclic spiranes with good to excellent diastereo- and enantioselectivity. Trienamine activation of 2,4-dienals **112** in combination with α,β,β-trisubstituted nitroolefins **298** was also employed by Jørgensen et al. to construct chiral oxetane-fused and azetidine-fused spirocyclohexenes **299** (Scheme 7.111) [238]. The developed process was

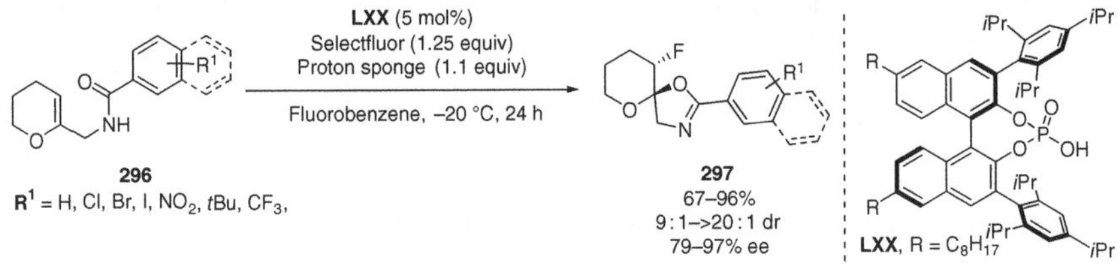

Scheme 7.110 The enantioselective fluorocyclization catalyzed by chiral BINOL-based phosphoric acid. *Source:* Based on Rauniyar et al. [235].

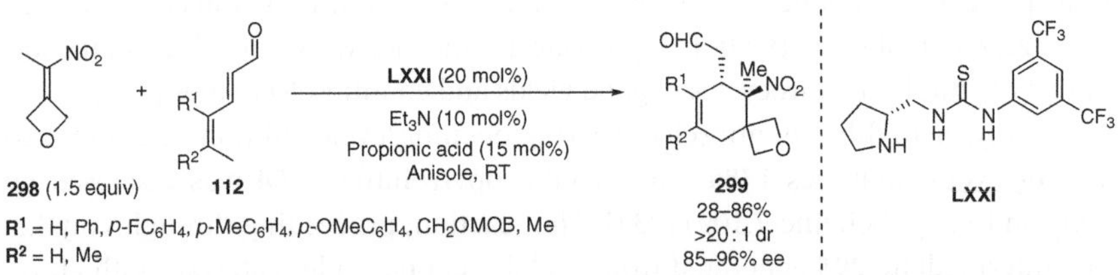

Scheme 7.111 Enantioselective cycloaddition to trisubstituted nitroolefins to create spirocyclohexene-oxetane scaffolds. *Source:* Based on Monleón et al. [238].

promoted by a bifunctional secondary amine-thiourea catalyst **LXXI** affording the spiro compounds in good yields and with excellent stereoselectivity (up to 82% yield, >20 : 1 dr, 98% ee).

Recently, Johnson et al. reported an interesting example of enantioselective oxidative dearomatization of 2-(hydroxymethyl)phenol dichloracetates with urea·H_2O_2 in the presence of Cinchona alkaloid-derived phase-transfer catalyst, resulting in the formation of chiral 1-oxaspiro[2.5]octadienones with high enantioselectivity [239].

7.7 Enantioselective Formation of Bispirocycles

The organic chemists have also studied catalytic approaches to synthesize chiral spirocyclic compounds containing more than one spiro atom; some of them led to structurally valuable substances.

In 2016, Yin et al. reported an enantioselective and diastereoselective approach to chiral bispirooxindoles based on Pd-catalyzed arylative dearomatization/ring expansion cascade of furfurylcyclobutanols **300** (Scheme 7.112) [240]. The reaction involves the formation of π-allyl palladium intermediate as a key intermediate. Final bispiro compounds **301** were prepared in yields up to 75% with moderate to high enantioselectivity (up to 83% ee).

Besides transition metal catalysis, asymmetric organocatalysis has also been used in the synthesis of bispirocompounds, for example, through organocatalytic 1,3-dipolar cycloaddition of alkylidene pyrazolones with α-imino γ-lactones. In this work, Cai et al. prepared optically active bispiro γ-butyrolactone-pyrrolidine-4,4′-pyrazolones containing two spiro centers in high yields (up to 93%) with diastereoselectivity >20 : 1 and enantiomeric excesses up to >99% ee. The process was efficiently promoted by a chiral tertiary amine-squaramide. In 2019, Ghorai et al. performed an enantioselective synthesis of bispiroketals **304** from *in situ*-generated cyclohexadienone alcohol derivatives **303** by a cascade

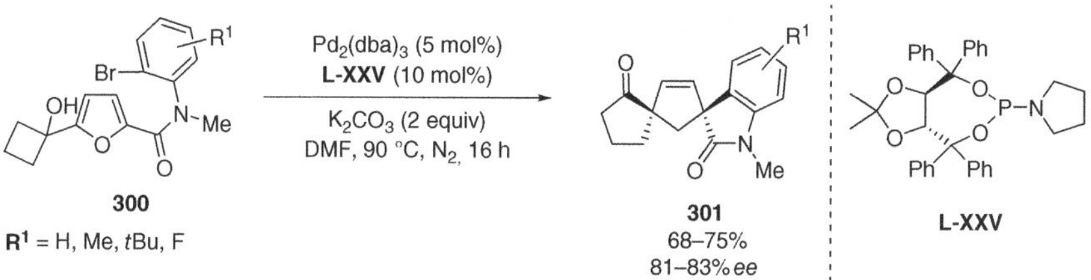

Scheme 7.112 Enantioselective access to dispirooxindoles from furfurylcyclobutanols. *Source:* Based on Liu et al. [240].

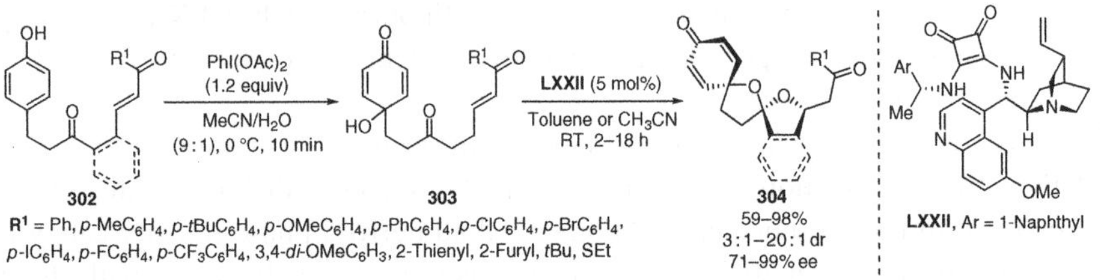

Scheme 7.113 The enantioselective synthesis of bis-spiroketals. *Source:* Based on Reddy et al. [241].

reaction including chiral bifunctional squaramide-catalyzed ketalization/oxa-Michael addition (Scheme 7.113) [241]. Again, high yields and excellent selectivity of bispiroketals were achieved.

Another type of bisspirocyclic compounds are those containing a pyrazolone and oxindole skeleton. These compounds were prepared by Zhou, Liu et al. by double Michael addition catalyzed by a chiral bifunctional thiourea. Chiral products were smoothly afforded in up to 87% yield, >20 : 1 dr, and >99% ee under mild conditions [242]. Subsequently, Wang et al. carried out an asymmetric construction of bispiro-cyclopropane-pyrazolones via a [2+1] cyclization reaction by dipeptide-based phosphonium salt catalysis. This method succeeded in preparing a large range of bisspiro heterocyclic compounds in high yields with excellent diastereo- and enatioselectivities [243]. Recently, bispiro compounds containing two benzofuranone units were prepared by Liu et al. This approach is based on a thiourea-catalyzed double Michael addition between chromone-oxindole synthons **304**, **306** and alkylidene benzofuranones **95**. The obtained spirocyclic compounds **305**, **307** were prepared in good yields (up to 93%) with high diastereoselectivities (up to >20 : 1 dr) and good enantioselectivities (up to >99% ee) (Scheme 7.114) [244].

7.8 Conclusion

This chapter illustrates the current evolution of enantioselective catalytic methods allowing an effective construction of chiral spiro heterocycles, including organometallic approaches, organocatalysis, and cooperative (synergistic) catalysis. As demonstrated in the previous pages, the remarkable achievements were made since the first enantioselective spirocyclization was disclosed. However, there is still enough room to further improve present catalytic approaches in terms of efficiency, substrate scope, catalyst loading, atom economy, etc. Moreover, detailed mechanism studies are highly desirable for better understanding the activation modes and further developing catalytic methodologies, allowing the construction of fascinating spirocyclic architectures.

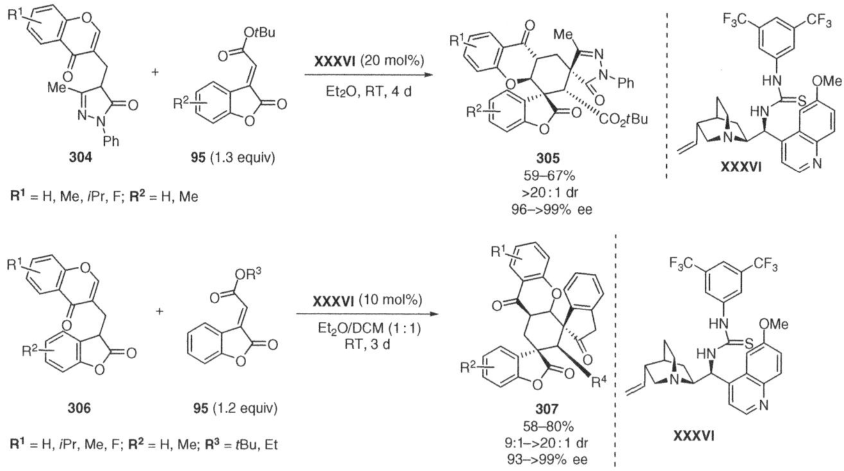

Scheme 7.114 Highly efficient enantioselective synthesis of bispiro benzofurans. *Source:* Based on Zhang et al. [244].

Acknowledgements

The authors gratefully acknowledge the Czech Science Foundation (No. 20-29336S) for financial support.

References

1 For selected recent reviews on the enantioselective preparation of spiro compounds, see:a)Zheng, C. and You, S.L. (**2020**). *Acc. Chem. Res. 53*: 974.b) Deng, Q. and Meng, X. (**2020**). *Chem. Asian J. 15*: 1225.c) Xu, P.-W., Yu, J.-S., Chen, C. et al. (**2019**). *ACS Catal. 9*: 1820.d) Ding, A., Meazza, M., Guo, H. et al. (**2018**). *Chem. Soc. Rev. 47*: 5946–5996.e) Bariwal, J., Voskressensky, L.G., and Van der Eycken, E.V. (**2018**). *Chem. Soc. Rev. 47*: 3831.f) Fang, X. and Wang, C.-J. (**2018**). *Org. Biomol. Chem. 16*: 2591.g) Griggs, S., Tape, D.T., and Clarke, P.A. (**2018**). *Org. Biomol. Chem. 16*: 6620.h) Xie, X., Huang, W., Peng, C., and Han, B. (**2018**). *Adv. Synth. Catal. 360*: 194.

2 Hatano, M. and Mikami, K. (**2003**). *Org. Biomol. Chem. 1*: 3871.

3 Mikami, K., Yusa, Y., Hatano, M. et al. (**2004**). *Chem. Commun.*: 98.

4 Mikami, K., Yusa, Y., Hatano, M., and Wakabayashi, K. (**2004**). *Tetrahedron 60*: 4475.

5 Hatano, M. and Mikami, K. (**2003**). *J. Am. Chem. Soc. 125* (16): 4704.

6 Liu, T.-L., He, Z.-L., Tao, H.-Y. et al. (**2011**). *Chem. Commun. 47*: 2616.

7 Yang, W.-L., Liu, Y.-Z., Luo, S. et al. (**2015**). *Chem. Commun. 51*: 9212.

8 Gan, Z., Zhi, M., Han, R. et al. (**2019**). *Org. Lett. 21*: 2782.
9 Bdiri, B., Li, C., and Zhou, Z.-M. (**2017**). *Tetrahedron Asymmetry 28*: 1044.
10 Nakayama, H., Harada, S., Kono, M., and Nemoto, T. (**2017**). *J. Am. Chem. Soc. 139*: 10188.
11 Kong, L., Han, X., Liu, S. et al. (**2020**). *Angew. Chem. Int. Ed. 59*: 7188.
12 Ni, C., Chen, J., Zhang, Y. et al. (**2017**). *Org. Lett. 19*: 3668.
13 Xing, J., Lei, Y., Gao, Y.-N., and Shi, M. (**2017**). *Org. Lett. 19*: 2382.
14 Luo, W., Hu, H., Nian, S. et al. (**2017**). *Org. Biomol. Chem. 15*: 7523.
15 Winter, M., Faust, K., Himmelsbach, M., and Waser, M. (**2019**). *Org. Biomol. Chem. 17*: 5731.
16 Zhao, M.-X., Liu, Q., Yu, K.-M. et al. (**2019**). *Org. Chem. Front. 6*: 3879.
17 Zhu, Y., Zhang, L., and Luo, S. (**2014**). *J. Am. Chem. Soc. 136*: 14642.
18 Yuan, Z., Wei, W., Lin, A., and Yao, H. (**2016**). *Org. Lett. 18*: 3370.
19 Corti, V., Marcantonio, E., Mamone, M. et al. (**2020**). *Catalysts 10*: 150.
20 Wu, Q.-F., He, H., Liu, W.-B., and You, S.-L. (**2010**). *J. Am. Chem. Soc. 132*: 11418.
21 Zhuo, C.-X., Zhou, Y., Cheng, Q. et al. (**2015**). *Angew. Chem. Int. Ed. 54*: 14146.
22 Wang, Y., Zheng, C., and You, S.-L. (**2017**). *Angew. Chem. Int. Ed. 56*: 15093.
23 Wu, K.-J., Dai, L.-X., and You, S.-L. (**2012**). *Org. Lett. 14* (14): 3772.
24 Liu, Z.-S., Li, W.-K., Kang, T.-R. et al. (**2015**). *Org. Lett. 17*: 150.
25 Li, X., Zhou, B., Yang, R.-Z. et al. (**2018**). *J. Am. Chem. Soc. 140*: 13945.
26 Cera, G., Chiarucci, M., Mazzanti, A. et al. (**2012**). *Org. Lett. 14* (5): 1350.
27 Matsuoka, J., Kumagai, H., Inuki, S. et al. (**2019**). *J. Org. Chem. 84*: 9358.
28 James, M.J., Cuthbertson, J.D., O'Brien, P. et al. (**2015**). *Angew. Chem. Int. Ed. 54*: 7640.
29 Wu, X., Liu, Q., Fang, H. et al. (**2012**). *Chem. Eur. J. 18*: 12196.
30 Yin, Q. and You, S.-L. (**2013**). *Org. Lett. 15*: 4266.
31 Zhou, Y., Xia, Z.-L., Gu, Q., and You, S.-L. (**2017**). *Org. Lett. 19*: 762.
32 Xia, Z.-L., Zheng, C., Wang, S.-G., and You, S.-L. (**2018**). *Angew. Chem. 130*: 2683.
33 Bera, S., Daniliuc, C.G., and Studer, A. (**2017**). *Angew. Chem. Int. Ed. 56*: 7402.
34 Luo, J., Fang, Y.-B., Mao, X.-Y. et al. (**2019**). *Adv. Synth. Catal. 361*: 1408.
35 Pan, Z., Liu, Y., Hu, F. et al. (**2020**). *Org. Lett. 22*: 1589.
36 Kong, K., Moussa, Z., and Romo, D. (**2005**). *Org. Lett. 7* (23): 5127.
37 Liu, J., Tian, Y., Shi, J. et al. (**2015**). *Angew. Chem. Int. Ed. 54*: 10917.
38 Takenaka, K., Itoh, N., and Sasai, H. (**2009**). *Org. Lett. 11* (7): 1483.
39 Li, M., Hawkins, A., Barber, D.M. et al. (**2013**). *Chem. Commun. 49*: 5265.
40 Xu, R.-Q., Yang, P., Zheng, C., and You, S.-L. (**2020**). *Chin. J. Chem. 38*: 683.
41 Wu, Q.-F., Liu, W.-B., Zhuo, C.-X. et al. (**2011**). *Angew. Chem. Int. Ed. 50*: 4455.
42 Zhuo, C.-X., Liu, W.-B., Wu, Q.-F., and You, S.-L. (**2012**). *Chem. Sci. 3*: 205.
43 Goudedranche, S., Bugaut, X., Constantieux, T. et al. (**2014**). *Chem. Eur. J. 20*: 410.
44 Li, X., Huang, W., Liu, Y.-Q. et al. (**2017**). *J. Org. Chem. 82*: 397.
45 Liu, Y., Ouyang, L., Tan, Y. et al. (**2017**). *Catalysts 7*: 46.

46 Zhang, K., Meazza, M., Dočekal, V. et al. (**2017**). *Eur. J. Org. Chem.*: 1749.
47 Mondal, S., Ghosh, A., Mukherjee, S., and Biju, A.T. (**2018**). *Org. Lett. 20*: 4499.
48 Lu, H., Tan, C.-Y., Zhang, H.-X. et al. (**2018**). *J. Org. Chem. 83*: 15245.
49 Yan, R.-J., Xiao, B.-X., Ouyang, Q. et al. (**2018**). *Org. Lett. 20*: 8000.
50 Hatano, M. and Mikami, K. (**2003**). *J. Mol. Catal. A Chem. 196*: 165.
51 Ma, C., Huang, Y., and Zhao, Y. (**2016**). *ACS Catal. 6*: 6408.
52 Tsuchikama, K., Kuwata, Y., and Shibata, T. (**2006**). *J. Am. Chem. Soc. 128*: 13686.
53 Deng, Q.-H., Wadepohl, H., and Gade, L.H. (**2012**). *J. Am. Chem. Soc. 134*: 2946.
54 Karyakarte, S.D., Um, C., Berhane, I.A., and Chemler, S.R. (**2018**). *Angew. Chem. Int. Ed. 57*: 12921.
55 Liu, K., Teng, H.-L., Yao, L. et al. (**2013**). *Org. Lett. 15* (9): 2250.
56 Liu, M.-M., Yang, X.-C., Hua, Y.-Z. et al. (**2019**). *Org. Lett. 21*: 7089.
57 Dohi, T., Maruyama, A., Takenaga, N. et al. (**2008**). *Angew. Chem. 120*: 3847.
58 Dohi, T., Takenaga, N., Nakae, T. et al. (**2013**). *J. Am. Chem. Soc. 135*: 4558.
59 Uyanik, M., Yasui, T., and Ishihara, K. (**2010**). *Angew. Chem. Int. Ed. 49*: 2175.
60 Uyanik, M., Yasui, T., and Ishihara, K. (**2017**). *J. Org. Chem. 82*: 11946.
61 Uyanik, M., Sasakura, N., Mizuno, M., and Ishihara, K. (**2017**). *ACS Catal. 7*: 872.
62 Murray, S.J. and Ibrahim, H. (**2015**). *Chem. Commun. 51*: 2376.
63 Hempel, C., Maichale-Mössmer, C., Pericàs, M.A., and Nachtsheim, B.J. (**2017**). *Adv. Synth. Catal. 359*: 2931.
64 Dohi, T., Sasa, H., Miyazaki, K. et al. (**2017**). *J. Org. Chem. 82*: 11954.
65 Uyanik, M., Kato, T., Sahara, N. et al. (**2019**). *ACS Catal. 9*: 11619.
66 Suzuki, T., Watanabe, S., Uyanik, M. et al. (**2018**). *Org. Lett. 20* (13): 3919.
67 Wu, X., Shirakawa, S., and Maruoka, K. (**2014**). *Org. Biomol. Chem. 12*: 5388.
68 Hejmanowska, J., Dzięgielewski, M., Kowalczyk, D., and Albrecht, Ł. (**2014**). *Synlett 25*: 2957.
69 Fei, C., Liu, J., Peng, H. et al. (**2018**). *Tetrahedron 74*: 6939.
70 Reddy, R.R., Gayen, P., Panda, S., and Ghorai, P. (**2019**). *Org. Lett. 21*: 5793.
71 Trost, B.M., Cramer, N., and Silverman, S.M. (**2007**). *J. Am. Chem. Soc. 129*: 12396.
72 Yang, L.-C., Tan, Z.-Y., Rong, Z.-Q. et al. (**2018**). *Angew. Chem. Int. Ed. 57*: 7860.
73 Oguma, T. and Katsuki, T. (**2014**). *Chem. Commun. 50*: 5053.
74 Kim, C., Oguma, T., Fujitomo, C. et al. (**2016**). *Chem. Lett. 45*: 1262.
75 Oguma, T., Doiuchi, D., Fujitomo, C. et al. (**2020**). *Asian J. Org. Chem. 9*: 404.
76 Li, Z. and Shi, Y. (**2015**). *Org. Lett. 17*: 5752.
77 Liu, L., Lei, L.-S., Zhan, Z.-S. et al. (**2019**). *Chem. Commun. 55*: 3789.
78 Companyó, X., Zea, A., Alba, A.-N.R. et al. (**2010**). *Chem. Commun. 46*: 6953.
79 Cassani, C., Tian, X., Escudero-Adán, E.C., and Melchiorre, P. (**2011**). *Chem. Commun. 47*: 233.
80 Chatterjee, I., Bastida, D., and Melchiorre, P. (**2013**). *Adv. Synth. Catal. 355*: 3124.
81 Qi, L.-W., Wang, L.-L., Peng, L. et al. (**2013**). *Tetrahedron 69*: 9303.
82 Luo, X., Wang, L., Peng, L. et al. (**2012**). *Chin. J. Chem. 30*: 2703.

83 Li, X., Yang, C., Jin, J.-L. et al. (**2013**). *Chem. Asian J. 8*: 997.
84 Dong, N., Li, X., Wang, F., and Cheng, J.-P. (**2013**). *Org. Lett. 15*: 4896.
85 Roy, S., Amireddy, M., and Chen, K. (**2013**). *Tetrahedron 69*: 8751.
86 Albertshofer, K., Tan, B., and Barbas, C.F. III (**2013**). *Org. Lett. 15*: 2958.
87 Li, X., Lin, M.-H., Han, Y. et al. (**2014**). *Org. Lett. 16*: 114.
88 Stiller, J., Poulsen, P.H., Cruz, D.C. et al. (**2014**). *Chem. Sci. 5*: 2052.
89 Zhou, Q.-Q., Xiao, Y.-C., Yuan, X., and Chen, Y.-C. (**2014**). *Asian J. Org. Chem. 3*: 545.
90 Halskov, K.S., Kniep, F., Lauridsen, V.H. et al. (**2015**). *J. Am. Chem. Soc. 137*: 1685.
91 Kowalczyk, D., Wojciechowski, J., and Albrecht, Ł. (**2016**). *Tetrahedron Lett. 57*: 2533.
92 Kowalczyk, D., Wojciechowski, J., and Albrecht, Ł. (**2017**). *Synthesis 49*: 880.
93 Du, T., Li, Z., Zheng, C. et al. (**2018**). *Tetrahedron 74*: 7485.
94 Wang, D., Wang, G.-P., Sun, Y.-L. et al. (**2015**). *Chem. Sci. 6*: 7319.
95 Ni, H., Yu, Z., Yao, W. et al. (**2017**). *Chem. Sci. 8*: 5699.
96 Liang, X.-W., Zheng, C., and You, S.-L. (**2016**). *Adv. Synth. Catal. 358*: 2066.
97 Liang, X.-W., Chen, X., Zhang, Z., and You, S.-L. (**2018**). *Chin. Chem. Lett. 29*: 1212.
98 Guo, C., Schedler, M., Daniliuc, C.G., and Glorius, F. (**2014**). *Angew. Chem. Int. Ed. 53*: 10232.
99 Janssen-Müller, D., Fleige, M., Schlüns, D. et al. (**2016**). *ACS Catal. 6*: 5735.
100 Chang, S.-Q., Zou, X., Gong, Y. et al. (**2019**). *Chem. Commun. 55*: 14003.
101 Zhang, S., Yu, X., Pan, J. et al. (**2019**). *Org. Chem. Front. 6*: 3799.
102 Meazza, M. and Rios, R. (**2016**). *Chem. Eur. J. 22*: 9923.
103 Teng, X., Cefalo, D.R., Schrock, R.R., and Hoveyda, A.H. (**2002**). *J. Am. Chem. Soc. 124*: 10779.
104 Hojo, D., Noguchi, K., Hirano, M., and Tanaka, K. (**2008**). *Angew. Chem. Int. Ed. 47*: 5820.
105 Chidipudi, S.R., Burns, D.J., Khan, I., and Lam, H.W. (**2015**). *Angew. Chem. Int. Ed. 54*: 13975.
106 Rodriguez, K.X., Pilato, T.C., and Ashfeld, B.L. (**2018**). *Chem. Sci. 9*: 3221.
107 Ichinose, M., Suematsu, H., and Katsuki, T. (**2009**). *Angew. Chem. Int. Ed. 48*: 3121.
108 Cowen, B.J. and Miller, S.J. (**2007**). *J. Am. Chem. Soc. 129*: 10988.
109 Hong, B.-C., Kotame, P., and Lee, G.-H. (**2011**). *Org. Lett. 13*: 5758.
110 Yao, W., Pan, L., Wu, Y., and Ma, C. (**2010**). *Org. Lett. 12*: 2422.
111 Ramachary, D., Madhavachary, R., and Prasad, M.S. (**2012**). *Org. Biomol. Chem. 10*: 5825.
112 Li, X., Yang, L., Peng, C. et al. (**2013**). *Chem. Commun. 49*: 8692.
113 Xie, X., Peng, C., Leng, H.-J. et al. (**2014**). *Synlett 25*: 143.

114 Zhao, B.-L. and Du, D.-M. (**2015**). *Eur. J. Org. Chem.*: 5350.

115 Antien, K., Viault, G., Pouységu, L. et al. (**2017**). *Tetrahedron 73*: 3684.

116 Zhang, Z.-P., Chen, L., Li, X., and Cheng, J.-P. (**2018**). *J. Org. Chem. 83*: 2714.

117 Mondal, B., Balha, M., and Pan, S.C. (**2019**). *Org. Biomol. Chem. 17*: 7849.

118 Wang, C., Chen, Y.-H., Wu, H.-C. et al. (**2019**). *Org. Lett. 21*: 6750.

119 Romanov-Michailidis, F., Guénée, L., and Alexakis, A. (**2013**). *Angew. Chem. 125*: 9436.

120 Zhu, L., Zhang, L., and Luo, S. (**2018**). *Angew. Chem. Int. Ed. 57*: 2253.

121 Cao, Y., Jiang, X., Liu, L. et al. (**2011**). *Angew. Chem. Int. Ed. 50*: 9124.

122 Zhao, B.-L., Liu, L., and Du, D.-M. (**2014**). *Eur. J. Org. Chem.*: 7850.

123 a) Zhao, B.-L. and Du, D.-M. (**2015**). *Asian. J. Org. Chem. 4*: 778.b) Zhao, B.-L. and Du, D.-M. (**2017**). *Org. Lett. 19*: 1036.

124 Duan, J., Cheng, J., Li, B. et al. (**2015**). *Eur. J. Org. Chem.*: 6130.

125 Grošelj, U., Ciber, L., Gnidovec, J. et al. (**2019**). *Adv. Synth. Catal.* 361: 5118.

126 Mahajan, S., Chauhan, P., Blümel, M. et al. (**2016**). *Synthesis 48*: 1131.

127 Gao, Y., Ren, Q., Wu, H. et al. (**2010**). *Chem. Commun.* ***46***: 9232.

128 Zeng, X.-M., Meng, C.-Y., Bao, J.-X. et al. (**2015**). *J. Org. Chem. 80*: 11521.

129 Duan, M., Liu, Y., Ao, J. et al. (**2017**). *Org. Lett. 19*: 2298.

130 Formánek, B., Tauchman, J., Císařová, I., and Veselý, J. (**2020**). *J. Org. Chem. 85*: 8510.

131 Géant, P.-Y., Urban, M., Remeš, M. et al. (**2013**). *Eur. J. Org. Chem.*: 7979.

132 Stiller, J., Kowalczyk, D., Jiang, H. et al. (**2014**). *Chem. Eur. J. 20*: 13108.

133 Sun, X., Fei, J., Zou, C. et al. (**2016**). *RSC Adv. 6*: 106676.

134 Zheng, J., Wang, S.-B., Zheng, C., and You, S.-L. (**2017**). *Angew. Chem. 129*: 4611.

135 Li, H., Gontla, R., Flegel, J. et al. (**2019**). *Angew. Chem. Int. Ed. 58*: 307.

136 Xu, J., Hu, L., Hu, H. et al. (**2019**). *Org. Lett. 21*: 1632.

137 Zea, A., Alba, A.-N.R., Mazzanti, A. et al. (**2011**). *Org. Biomol. Chem. 9*: 6519.

138 Alba, A.-N.R., Zea, A., Valero, G. et al. (**2011**). *Eur. J. Org. Chem.*: 1318.

139 Wu, B., Chen, J., Li, M.-Q. et al. (**2012**). *Eur. J. Org. Chem.*: 1318.

140 Zhang, J.-X., Li, N.-K., Liu, Z.-M. et al. (**2013**). *Adv. Synth. Catal. 355*: 797.

141 Liang, J., Chen, Q., Liu, L. et al. (**2013**). *Org. Biomol. Chem. 11*: 1441.

142 Liu, L., Zhong, Y., Zhang, P. et al. (**2012**). *J. Org. Chem. 77*: 10228.

143 Chauhan, P., Mahajan, S., Loh, C.C.J. et al. (**2014**). *Org. Lett. 16*: 2954.

144 Sun, P., Meng, C.-Y., Zhou, F. et al. (**2014**). *Tetrahedron 70*: 9330.

145 Han, B., Huang, W., Ren, W. et al. (**2015**). *Adv. Synth. Catal. 357*: 561.

146 Sun, J., Jiang, C., and Zhou, Z. (**2016**). *Eur. J. Org. Chem.*: 1165.

147 Zhang, Y., Wang, C., Huang, W. et al. (**2018**). *Org. Chem. Front. 5*: 2229.

148 Ceban, V., Olomola, T., Meazza, M., and Rios, R. (**2015**). *Molecules 20*: 8574.

149 Li, J.-H. and Du, D.-M. (**2014**). *Chem. Asian J. 9*: 3278.

150 Li, J.-H., Wen, H., Liu, L., and Du, D.-M. (**2016**). *Eur. J. Org. Chem.*: 2492.

151 Zheng, W., Zhang, J., Liu, S. et al. (**2015**). *RSC Adv. 5*: 91108.

152 Mondal, B., Maity, R., and Pan, S.C. (**2018**). *J. Org. Chem. 83*: 8645.
153 Han, X., Yao, W., Wang, T. et al. (**2014**). *Angew. Chem. Int. Ed. 53*: 5643.
154 Yang, W., Zhang, Y., Qiu, S. et al. (**2015**). *RSC Adv. 5*: 62343.
155 Luo, W., Shao, B., Li, J. et al. (**2019**). *Tetrahedron Lett. 60*: 151206.
156 Yetra, S.R., Mondal, S., Mukherjee, S. et al. (**2016**). *Angew. Chem. Int. Ed. 55*: 268.
157 Wang, L., Sun, L., Chauhan, P. et al. (**2016**). *Chem. Eur. J. 22*: 5123.
158 Li, S., Wang, L., Chauhan, P. et al. (**2017**). *Synthesis 49*: 1808.
159 Li, J.-H., Cui, Z.-H., and Du, D.-M. (**2016**). *Org. Chem. Front. 3*: 1087.
160 Amireddy, M. and Chen, K. (**2016**). *RSC Adv. 6*: 77474.
161 Yang, W., Sun, W., Zhang, C. et al. (**2017**). *ACS Catal. 7*: 3142.
162 Liu, J.-Y., Zhao, J., Zhang, J.-L., and Xu, P.-F. (**2017**). *Org. Lett. 19*: 1846.
163 Mondal, S., Mukherjee, S., Yetra, S.R. et al. (**2017**). *Org. Lett. 19*: 4367.
164 Leng, H.-J., Li, Q.-Z., Zeng, R. et al. (**2018**). *Adv. Synth. Catal. 360*: 229.
165 Zhang, X.-L., Tang, C.-K., Xia, A.-B. et al. (**2017**). *Eur. J. Org. Chem.*: 3152.
166 Xia, A.-B., Zhang, X.-L., Tang, C.-K. et al. (**2017**). *Org. Biomol. Chem. 15*: 5709.
167 Tang, C.-K., Zhou, Y.-Z., Xia, A.-B. et al. (**2018**). *Org. Lett. 20*: 5840.
168 Meninno, S., Roselli, A., Capobianco, A. et al. (**2017**). *Org. Lett. 19*: 5030.
169 Meninno, S., Overgaard, J., and Lattanzi, A. (**2017**). *Synthesis 49*: 1509.
170 Meninno, S., Mazzanti, A., and Lattanzi, A. (**2019**). *Adv. Synth. Catal. 361*: 79.
171 Sun, B.-B., Zhang, J.-Q., Chen, J.-B. et al. (**2019**). *Org. Chem. Front. 6*: 1842.
172 Ji, Y.-L., Li, H.-P., Ai, Y.-Y. et al. (**2019**). *Org. Biomol. Chem. 17*: 9217.
173 Zhou, J., Huang, W.-J., and Jiang, G.-F. (**2018**). *Org. Lett. 20*: 1158.
174 Chu, M.-M., Qi, S.-S., Wang, Y.-F. et al. (**2019**). *Org. Chem. Front. 6*: 1977.
175 Zhao, C., Shi, K., He, G. et al. (**2019**). *Org. Lett. 21*: 7943.
176 Li, Y. and Zhang, Z. (**2020**). *New J. Chem. 44*: 11643.
177 Li, X., Chen, F.-Y., Kang, J.-W. et al. (**2019**). *J. Org. Chem. 84*: 9138.
178 Lu, H., Zhang, H.-X., Tan, C.-Y. et al. (**2019**). *J. Org. Chem. 84*: 10292.
179 Tan, C.-Y., Lu, H., Zhang, J.-L. et al. (**2020**). *J. Org. Chem. 85*: 594.
180 Hack, D., Dürr, A.B., Deskers, K. et al. (**2016**). *Angew. Chem. 128*: 1829.
181 Meazza, M., Kamlar, M., Jašíková, L. et al. (**2018**). *Chem. Sci. 9*: 6368.
182 Putatunda, S., Alegre-Requena, J.V., Meazza, M. et al. (**2019**). *Chem. Sci. 10*: 4107.
183 Shukla, K., Shah, S., Rana, N.K., and Singh, V.K. (**2019**). *Tetrahedron Lett. 60*: 92.
184 Zhao, H.-W., Feng, N.-N., Guo, J.-M. et al. (**2018**). *J. Org. Chem. 83*: 9291.
185 Fujiwara, Y. and Fu, G.C. (**2011**). *J. Am. Chem. Soc. 133*: 12293.
186 Liu, H., Liu, Y., Yuan, C. et al. (**2016**). *Org. Lett. 18*: 1302.
187 Liu, Y., Yang, W., Wu, Y. et al. (**2016**). *Adv. Synth. Catal. 358*: 2867.
188 Gao, Z., Wang, C., Zhou, L. et al. (**2018**). *Org. Lett. 20*: 4302.
189 Frankowski, S., Gajda, T., and Albrecht, Ł. (**2018**). *Adv. Synth. Catal. 360*: 1822.
190 Laugeois, M., Ponra, S., Ratovelomanana-Vidal, V. et al. (**2016**). *Chem. Commun. 52*: 5332.
191 Weber, M., Frey, W., and Peters, R. (**2013**). *Chem. Eur. J. 19*: 8342.

192 Trost, B.M. and Morris, P.J. (**2011**). *Angew. Chem. Int. Ed. 50*: 6167.
193 Trost, B.M., Morris, P.J., and Sprague, S.J. (**2012**). *J. Am. Chem. Soc. 134*: 17823.
194 Steurer, M., Jensen, K.L., Worgull, D., and Jørgensen, K.A. (**2012**). *Chem. Eur. J. 18*: 76.
195 Wang, D., Wei, Y., and Shi, M. (**2012**). *Chem. Commun. 48*: 2764.
196 Chintalapudi, V., Galvin, E.A., Greenaway, R.L., and Anderson, E.A. (**2016**). *Chem. Commun. 52*: 693.
197 Kamlar, M., Franc, M., Císařová, I. et al. (**2019**). *Chem. Commun. 55*: 3829.
198 Pham, M.V. and Cramer, N. (**2016**). *Chem. Eur. J. 22*: 2270.
199 Zhou, M., Su, Q., Addepalli, Y. et al. (**2018**). *Org. Biomol. Chem. 16*: 2923.
200 Yang, W.-L., Tang, F.-F., He, F.-S. et al. (**2015**). *Org. Lett. 17*: 4822.
201 Yin, X., Zheng, Y., Feng, X. et al. (**2014**). *Angew. Chem. Int. Ed. 53*: 6245.
202 He, X.-L., Xiao, Y.-C., Du, W., and Chen, Y.-C. (**2015**). *Chem. Eur. J. 21*: 3443.
203 Fei, J., Qian, Q., Sun, X. et al. (**2015**). *Org. Lett. 17*: 5296.
204 Chen, X., Zhang, J.-Q., Yin, S.-J. et al. (**2015**). *Org. Lett. 17*: 4188.
205 Wu, W., Huang, H., Yuan, X. et al. (**2012**). *Chem. Commun. 48*: 9180.
206 Zhang, Y., Wang, S., Wu, S. et al. (**2013**). *ASC Comb. Sci. 15*: 298.
207 Qui, S., Tan, C.-H., and Jiang, Z. (**2016**). *Beilstein J. Org. Chem. 12*: 2293.
208 Song, Y.-X. and Du, D.-M. (**2018**). *Org. Biomol. Chem. 16*: 9390.
209 Li, B., Liu, J., Gao, F. et al. (**2019**). *Org. Biomol. Chem. 17*: 2892.
210 Tan, J.-P., Zhang, H., Jiang, Z. et al. (**2020**). *Adv. Synth. Catal. 362*: 1058.
211 Ji, D.-S., Luo, Y.-C., Hu, X.-Q., and Xu, P.-F. (**2020**). *Org. Lett. 22*: 1028.
212 Wang, X., Han, Z., Wang, Z., and Ding, K. (**2012**). *Angew. Chem. Int. Ed. 51*: 936.
213 Wang, X., Guo, P., Wang, X. et al. (**2013**). *Adv. Synth. Catal. 355*: 2900.
214 Wang, F., Chen, F., Qu, M. et al. (**2013**). *Chem. Commun. 49*: 3360.
215 Liu, N., Zhu, W., Yao, J. et al. (**2020**). *ACS Catal. 10*. 2596.
216 Hamilton, J.Y., Rössler, S.L., and Carreira, E.M. (**2017**). *J. Am. Chem. Soc. 139*: 8082.
217 Gong, J., Wan, Q., and Kang, Q. (**2018**). *Adv. Synth. Catal. 360*: 4031.
218 Ge, S., Cao, W., Kang, T. et al. (**2019**). *Angew. Chem. Int. Ed. 58*: 4017.
219 Sun, Z., Winschel, G.A., Borovika, A., and Nagorny, P. (**2012**). *J. Am. Chem. Soc. 134*: 8074.
220 Khomutnyk, Y.Y., Argüelles, A.J., Winschel, G.A. et al. (**2016**). *J. Am. Chem. Soc. 138*: 444.
221 Čorić, I. and List, B. (**2012**). *Nature 483*: 315.
222 Yoneda, N., Fukata, Y., Asano, K., and Matsubara, S. (**2015**). *Angew. Chem. Int. Ed. 127*: 15717.
223 Midya, A., Maity, S., and Ghorai, P. (**2017**). *Chem. Eur. J. 23*: 11216.
224 Xue, J., Zhang, H., Tian, T. et al. (**2016**). *Adv. Synth. Catal. 358*: 370.
225 Wu, H., He, Y.-P., and Gong, L.-Z. (**2013**). *Org. Lett. 15*: 460.
226 Cala, L., Mendoza, A., Fañanás, F.J., and Rodríguez, F. (**2013**). *Chem. Commun. 49*: 2715.

227 Quach, R., Furkert, D.P., and Brimble, M.A. (**2013**). *Tetrahedron Lett. 54*: 5865.
228 Rexit, A.A. and Mailikezati, M. (**2015**). *Tetrahedron Lett. 56*: 2651.
229 Tamao, K., Nakamura, K., Ishii, H. et al. (**1996**). *J. Am. Chem. Soc. 118*: 12469.
230 Cayuelas, A., Ortiz, R., Nájera, C. et al. (**2016**). *Org. Lett. 18*: 2926.
231 Hou, Q., You, Y., Song, X. et al. (**2020**). *Catalysts 10*: 28.
232 Shu, T., Zhao, L., Li, S. et al. (**2018**). *Angew. Chem. Int. Ed. 57*: 10985.
233 Yang, W., Wang, X., Jin, X. et al. (**2019**). *Adv. Synth. Catal. 361*: 562.
234 Luo, Y., Zhang, H., Wang, S. et al. (**2020**). *Org. Lett. 22*: 2645.
235 Rauniyar, V., Lackner, A.D., Hamilton, G.L., and Toste, F.D. (**2011**). *Science 334*: 1681.
236 Duvvuru, D., Pinto, N., Gomez, C. et al. (**2012**). *Adv. Synth. Catal.* **354**: 408.
237 Zhang, S.-J., Zhang, J., Zhou, Q.-Q. et al. (**2013**). *Org. Lett. 15*: 968.
238 Monleón, A., Glaus, F., Vergura, S., and Jørgensen, K.A. (**2016**). *Angew. Chem. Int. Ed. 55*: 2478.
239 McLaughlin, M.F., Massolo, E., Liu, S., and Johnson, J. (**2019**). *J. Am. Chem. Soc. 141*: 2645.
240 Liu, J., Peng, H., Lu, L. et al. (**2016**). *Org. Lett. 18*: 6440.
241 Reddy, R.R., Panda, S., and Ghorai, P. (**2019**). *J. Org. Chem. 84*: 5357.
242 Liu, X.-L., Zuo, X., Wang, J.-X. et al. (**2019**). *Org. Chem. Front. 6*: 1485.
243 Lu, D., Liu, X., Wu, J.-H. et al. (**2020**). *Adv. Synth. Catal. 362*: 1966.
244 Zhang, M., Wang, J.-X., Chang, S.-Q. et al. (**2020**). *Chin. Chem. Lett. 31*: 381.

8

Enantioselective Synthesis of all Carbon Spiro Compounds

Ramon Rios

Department of Organic Chemistry, University of Southampton, Southampton, UK

Spirocycles have become one of the more challenging goals for organic chemists, and at the same time a source of inspiration for the development of new reactions. The fascination for these compounds relies in the inherent difficulty for their synthesis. A spirocycle consists in two cycles joined by a single atom. Most of the times, this single atom is a carbon that contains four different substituents making the molecule chiral. This feature makes the synthesis of this compound even more difficult, thus not only we have to assemble a quaternary carbon but also we need to control the stereochemical approach.

The discovery of spirocycles can be traced back to the pioneering works of Von Baeyer in the late 1890s [1]. He proposed the name spirocyclane for bicyclic hydrocarbons having two rings with a common carbon atom (spiro carbon atom). In the last decade, spiro compounds have attracted a lot of interest in drug discovery, but despite their importance, spiro compounds are still under-represented in screening libraries and they present low diversity. This limited occurrence in modern drug discovery is not the result of low activities or inappropriate physicochemical properties, but clearly reflects the difficulty to synthesize these compounds. Clearly, the presence of spiro compounds in drug discovery scaffolds provides more vectors for functionalization (more 3D/sp3-rich structures) compared with the old school flat/aromatic scaffolds which are over-represented in classic drug discovery libraries.

Spiro Compounds: Synthesis and Applications, First Edition. Edited by Ramon Rios Torres.

Several efforts have been done in the organic community for the development of efficient asymmetric methodologies for their synthesis. In this chapter, we will give a detailed view of the more important reactions to obtain all-carbon spiro compounds. Several issues arise for their synthesis, like the functional group compatibility or the addition of reactive handles in the newly formed ring. This, together with the previously cited necessity to control the newly stereogenic center formed and the difficulties of the synthesis of quaternary centers, attracted the interests of synthetic chemists. With the advent of organocatalysis and the unstoppable advances in organometallic chemistry, Chemists overcome these issues and in the last decade, a plethora of new reactions for the synthesis of spiro compounds have been released.

In this chapter, we will focus on the all-carbon-spiro compounds avoiding overlapping with other chapters of this book. We will focus our efforts on the most recent enantioselective methodologies based on the strategy for the synthesis of the spiro compounds.

We divided the strategies for the synthesis of spiro compounds in six different approaches as shown in Figure 8.1: (a) alkylation strategies, (b) metal-catalyzed methods, (c) ring closure methods, (d) cycloaddition strategies, (e) cascade reactions strategies, and (f) rearrangement strategies (Figure 8.1).

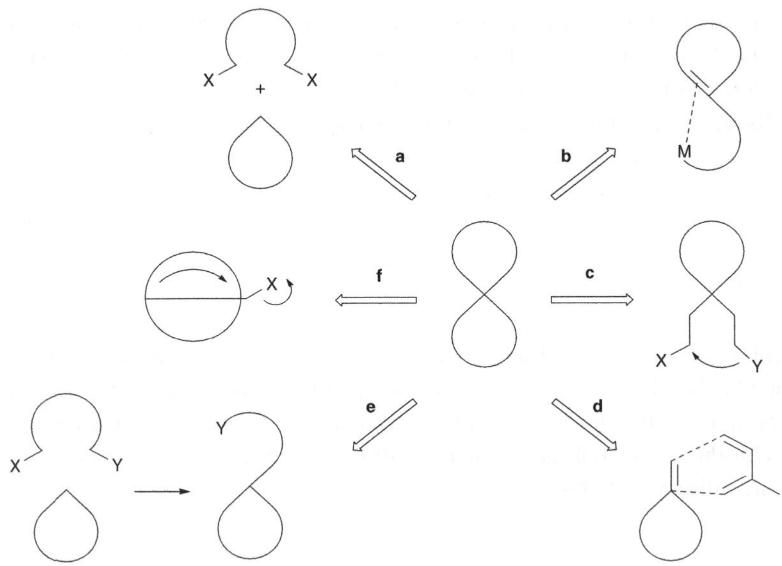

Figure 8.1 Common strategies for the synthesis of spirocycles: (a) alkylation methods, (b) metal catalyzed methods, (c) ring closure methods, (d) cycloaddition strategies, (e) cascade reactions strategies, (f) rearrangement strategies.

8.1 Enantioselective Synthesis of All-Carbon Spiro Compounds Based on Alkylation Methods

Alkylation methodologies are one of the most common resources to synthesize cyclic compounds by organic chemists. For the synthesis of spiro compounds easily, enolizable oxindoles, pyrazolones, or similar heterocycles have been used with amazing success as stated in other chapters. Herein, we will disclose several approaches based on organometallic alkylation methods for the enantioselective synthesis of all-carbon spirocycles.

In 2012, a stereoselective alkylation of cyclic β-ketoesters led to the synthesis of spirolactones or bispirolactones as reported by Deng [2]. The reaction is catalyzed by Cu(II) complex with chiral pincer ligands boxmi (**I**). Cyclic β-ketoesters (**1** or **4**) react with benzylic or allylic iodides (X), which are prepared "in situ" from the corresponding alcohols by means of CsI and BF_3. The chiral copper complex (**II**) promotes the alkylation between the cyclic β-ketoesters and the vinyl iodides and the following lactonization in a highly enantioselective fashion (ee up to 99%) and excellent yields (>90%) when bulky *tert*-butyl β-ketoesters are used. The scope of the reaction is limited to indanones and cyclopentanones, failing with six-membered rings and acyclic β-ketoesters (Scheme 8.1).

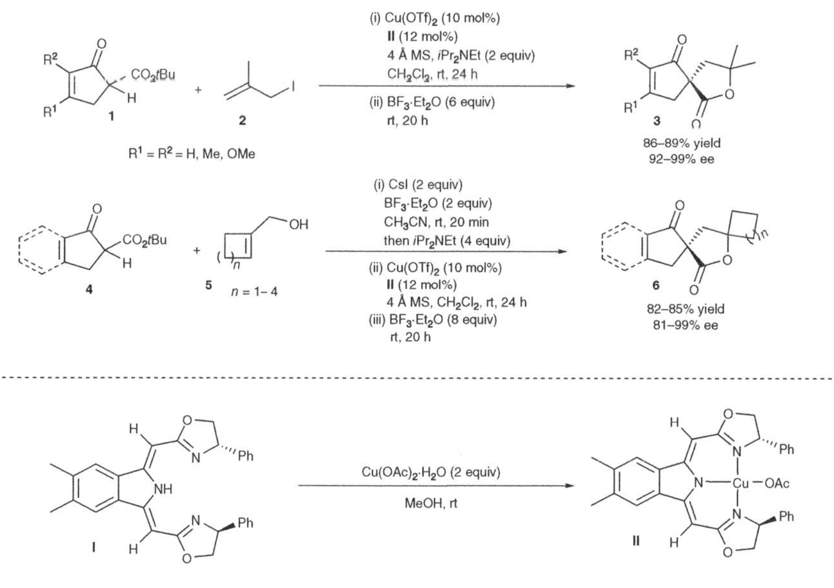

Scheme 8.1 Spirocyclizations reported by Gade. *Source:* Based on Deng et al. [2].

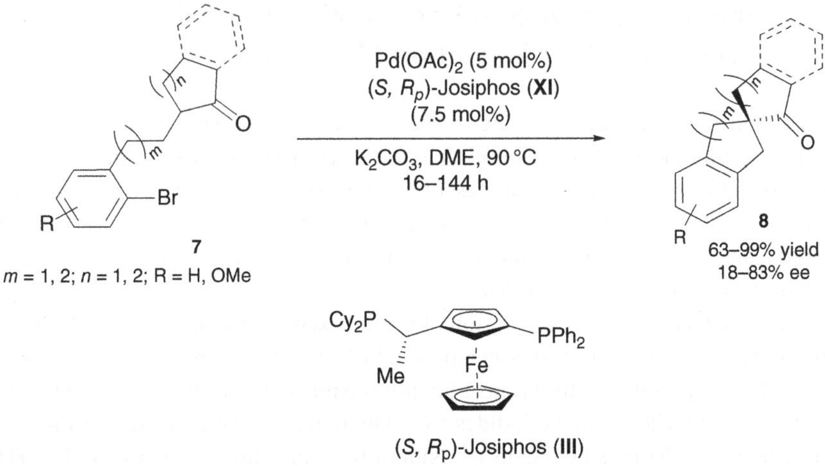

Scheme 8.2 Spirocyclization reported by Takisawa. *Source:* Based on Fan et al. [1].

Few years later, an intramolecular α-arylation of α-substituted cyclic ketones (**7**) was reported by Takisawa and Sasai for the synthesis of spirocyclic ketones (**8**) [1]. The reaction was catalyzed by Pd complexes using enantiopure Josiphos (**III**) as a ligand. The spirocyclic ketones (**8**) were synthesized in good to excellent yields (63–99%) but the enantioselectivity was highly dependent on the spirobicycle formed. The best enantioselectivities were obtained in the formation of spiro[4,4]nonanes, while spiro[5,5] or spiro [4,6] alkanones render moderate enantioselectivities. Another important limitation of the present methodology is the need of an *ortho*-OMe substituent in the bromo-phenyl ring (Scheme 8.2).

Smith and coworkers reported in 2016, a C-acylation approach for the synthesis of spirobisindanones (**10**) [3]. Spirobisindanones were obtained by taking advantage of the good leaving group located in the ester moiety (OC_6F_5) which is attacked intramolecularly by the enolate generated using phase transfer catalysts chiral ammonium salt (**IV**). As it is shown in Scheme 8.3, the spirobisindanones (**10**) were obtained in excellent yields (94–99%) and enantioselectivities(84–98%).

8.2 Enantioselective Synthesis of All-Carbon Spirocycles by Metal-catalyzed Methods

Metal-catalyzed reactions have been one of the classic approaches for the development of new scaffolds in synthetic chemistry. Herein, some examples of its applications for the synthesis of spiro compounds are highlighted.

Hashimoto, in 2001, reported the first enantioselective methodology for the synthesis of all-carbon spiro stereocenters [4]. They developed a double

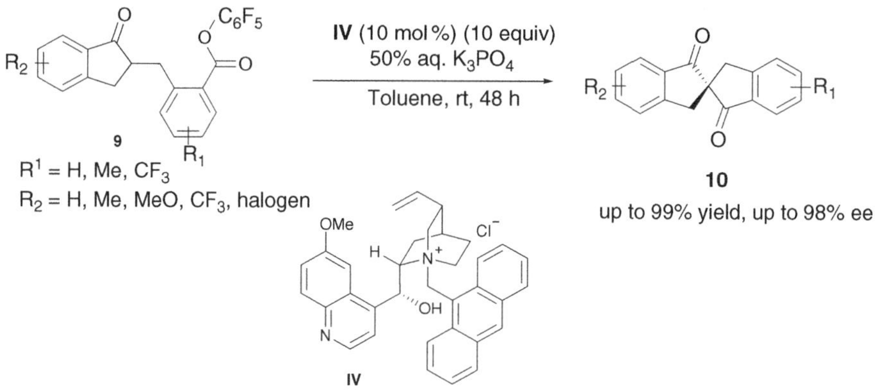

Scheme 8.3 Spirocyclization reported by Smith. *Source:* Based on Rahemtulla et al. [3].

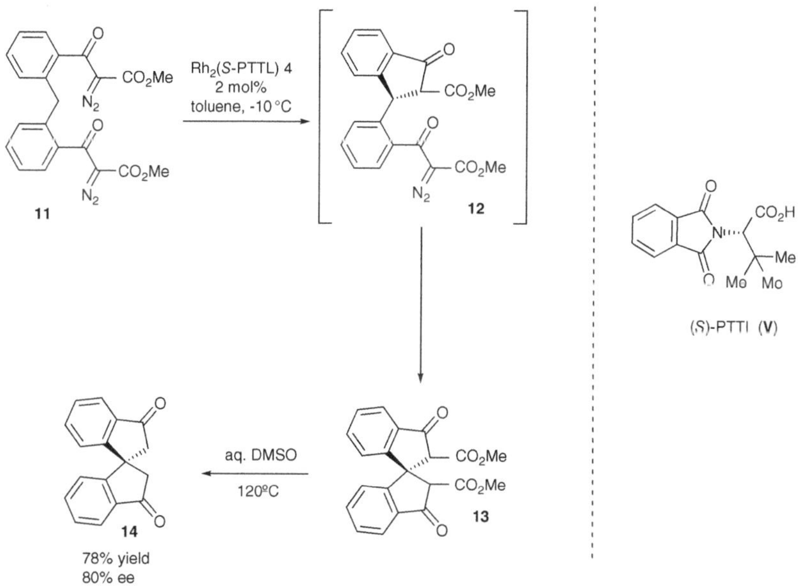

Scheme 8.4 Synthesis of spirocycles reported by Hashimoto et al. *Source:* Takahashi et al. [4].

intramolecular C–H insertion starting from of bis(α-diazo-β-keto ester) (**11**) which led to the synthesis of 1,1′-spirobi[indan-3,3′-dione] (**14**). The reaction affords excellent yields and enantioselectivities using chiral dirhodium (II) carboxylates derived from *N*-phthaloyl-(*S*)-*tert*-leucine (**V**) as catalysts (Scheme 8.4).

A highly enantioselective spirocyclization of 1,6-enynes (**15**) was developed by Mikami and coworkers in 2004 [5]. The key to the success for the reaction is the right choosing of the Rhodium ligand. Rh(I)/(*S*,*S*)-skewphos (**VI**) complex shows better catalytic behavior than other chiral diphosphines due to its dynamic (*tropos*) nature affording the spirofuranes (**23, 24**) in moderate yields and good enantioselectivities. Remarkably during the reaction, an olefin migration takes place, leading to a mixture of isomers. As it is shown in Scheme 8.5, the same authors expanded the scope of the reaction to the synthesis of spiropyrrolidines **26** with good yields and excellent enantioselectivities and without the olefin migration.

Soon after, Nicolaou applied a similar rhodium-catalyzed asymmetric enyne cyclization in his synthesis of the antibiotic (−)-platensimycin (**22**) [6]. In this case, the reaction rendered the spiro cycle in 91% yield and 95% ee (Scheme 8.6).

The reaction between aryl boron reagents (**28**), as surrogates of 1,2-dimetalloarenes, and 2-cycloaken-1-ones tethered at the 3-position with another

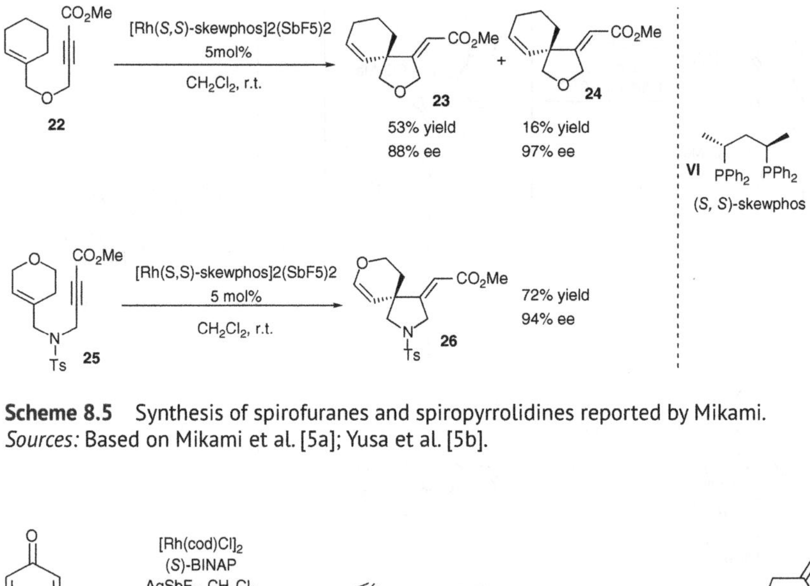

Scheme 8.5 Synthesis of spirofuranes and spiropyrrolidines reported by Mikami. *Sources:* Based on Mikami et al. [5a]; Yusa et al. [5b].

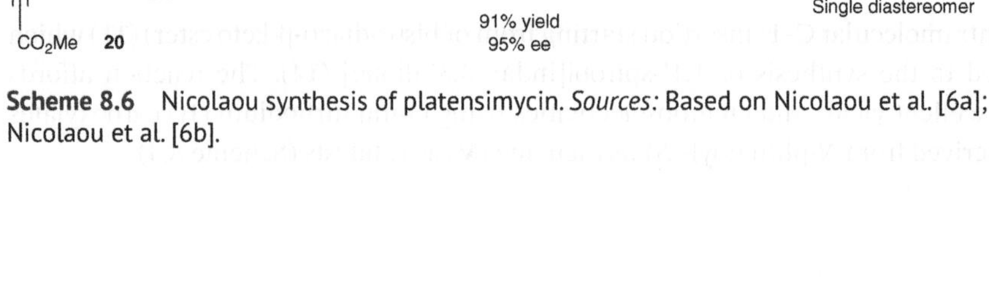

Scheme 8.6 Nicolaou synthesis of platensimycin. *Sources:* Based on Nicolaou et al. [6a]; Nicolaou et al. [6b].

electrophilic substituent such as an alkyne (**24**) leading to the synthesis of spirocycles **29** was reported by Hayashi and Shintani in 2010 [7]. The reported mechanism, depicted in Figure 8.2, starts with a transmetallation between the arylboron **28** and Rodhium catalyst to obtain the rhodium species **23** that undergoes insertion of the alkyne group of **24** to afford the alkenyl rhodium species **25**. Next, a 1,4-rhodium shift renders rhodium complex **26** which reacts with the enone in a Michael fashion to afford the oxa-p-allyl rhodium intermediate **27**; hydrolysis of this intermediate releases the spiro compound **29** along with a rhodium complex whose transmetallation with the Boro-aryl compound regenerates the phenyl Rhodium species **23**.

Remarkably, the use of phosphines as ligands gave low yields, being the best ligand chiral dienes such as Bn-bod (**VII**). With the optimized conditions, the reaction renders the spiro compounds (**29**) in good yields (up to 93%) and excellent enantioselectivities (up to 97%). The limitations of the present methodology are determined for the structure of the precursor needed to perform the metal-catalyzed cascade reaction (Scheme 8.7).

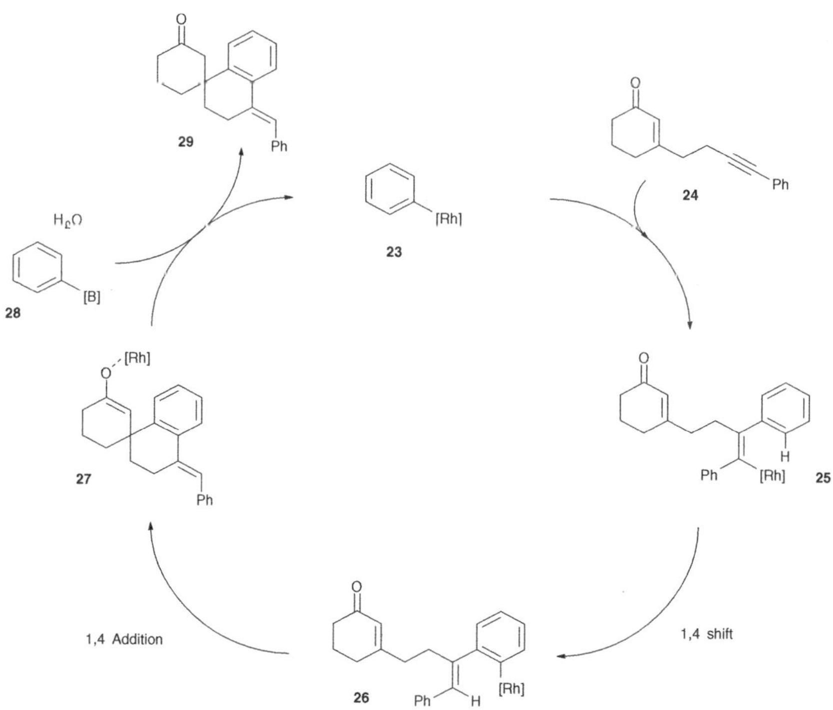

Figure 8.2 Proposed mechanism for the Rh-catalyzed synthesis of spirocycles reported by Shintani and Hayashi. *Source:* Based on Shintani et al. [7].

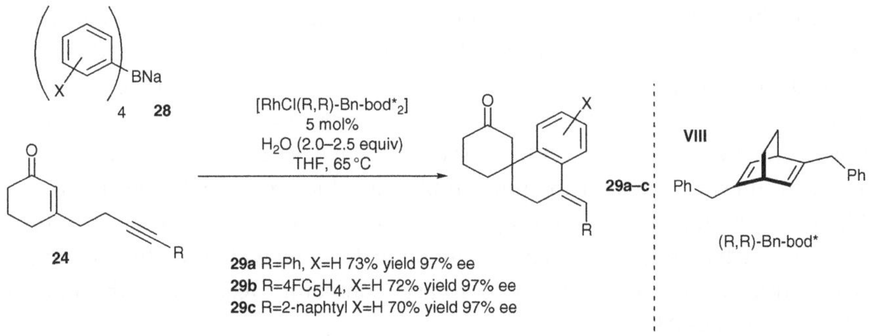

Scheme 8.7 Rh-catalyzed synthesis of spirocycles reported by Shintani and Hayashi. *Source:* Based on Shintani et al. [7].

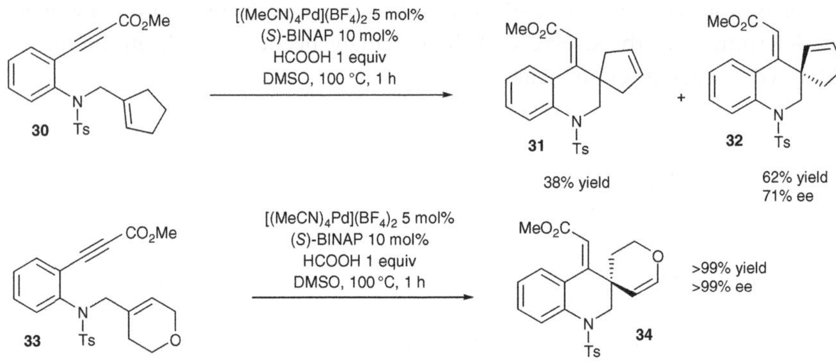

Scheme 8.8 First asymmetric palladium-catalyzed synthesis of spirocycles.

In 2001, Mikami and coworkers reported a similar synthesis but using Pd complex as catalysts this time [8]. Ene-type palladium-catalyzed cyclizations of 1,7-enynes (**30**) led to the formation of spiroquinolines (**31**, **32**) when cationic BINAP-Palladium (II) complex is used as catalysts. The ring size of the newly formed cyclic olefin has a big influence on the enantioselective outcome of the reaction. For example, five-membered rings give moderate enantioselectivities, along with olefin migration product (**32**), while the formation of six-membered rings render the final spiroquinolines (**34**) in excellent yields and enantioselectivities as illustrated in Scheme 8.8.

Soon after, the same strategy was used for the synthesis of chiral spyropyrrolidines (**36, 37**) [9]. In order to achieve high rates and enantioselectivities, PN ligand (**IX**) was used. This time, the size of the new cycle formed does not have a huge effect on the result of the reaction as can be seen in Scheme 8.9.

The same research group expanded the scope of this reaction using allyl propargyl ethers (**38**) [10]. Using the same ligand **IX**, as shown in Scheme 8.10, they

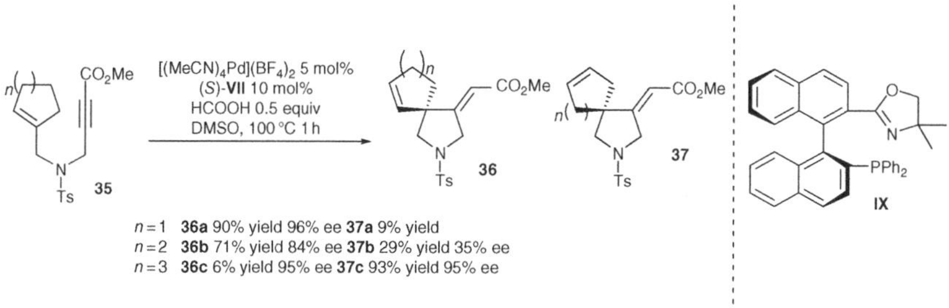

Scheme 8.9 Pd-catalyzed synthesis of spirocycles reported by Mikami. *Source:* Based on Hatano and Mikami [9].

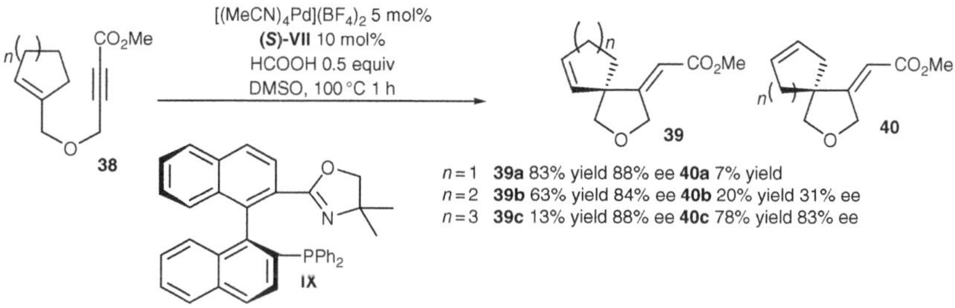

Scheme 8.10 Pd-catalyzed spirocyclization of allyl propargyl ethers.

obtained moderate results for the synthesis of spirofuran (**39**, **40**) derivatives. Two major issues were faced in this reaction: (i) low enantioselectivities were afforded and (ii) a mixture of regioisomers was obtained via olefin migration.

Silyloxy-1,6-enynes (**41**) were used as precursor for the synthesis of chiral spirocycles (**42**) by Toste and Corkey in 2007 [11]. Silyloxy-1,6-enynes (**41**) undergo intramolecular enantioselective cyclization catalyzed by Pd complexes (using as ligand diphosphine **X**) affording the spiro compounds (**42**) in good yields and good diastereoselectivities. The authors expand the scope of the reaction by using *O*-silylketene aminals achieving spirocyclic amides (**44**) with excellent results (Scheme 8.11).

A synthesis of spirobilactams was reported by Cai in 2015 [12]. The reaction consists in a copper-catalyzed intramolecular double *N*-arylation. 2,2-Bis(2-iodob enzyl)-malonamides **45** react intramolecularly with the aryl iodide, in the presence of Cu complex (starting from CuI and chiral ligand **XI**) and using $CsCO_3$ as a base. In the model reaction, the lactam was reduced to benzobispiramine to allow better HPLC measure. The product was obtained from the filtrate in yields of around 65–78% and ee up to 99% (due to enrichment in the precipitation process). Regarding the scope of the reaction, several substituents in the phenyl ring are well tolerated (Scheme 8.12).

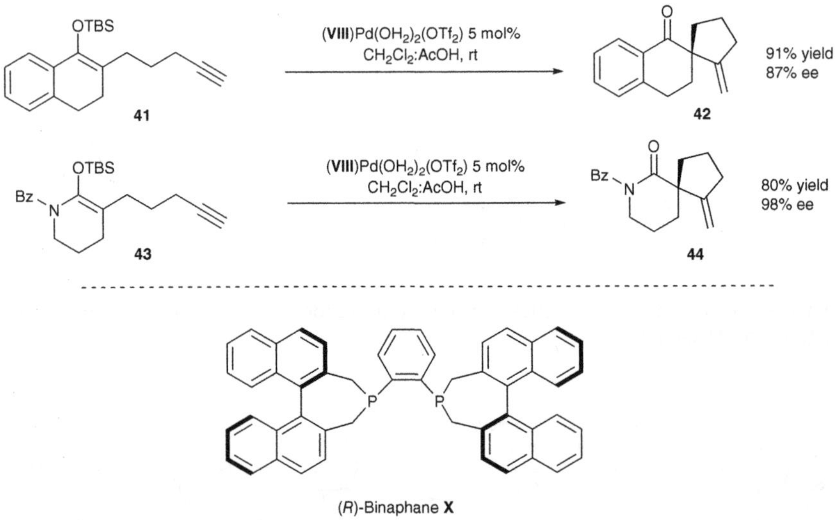

Scheme 8.11 Pd-catalyzed spirocyclization reported by Toste. *Source:* Based on Corkey and Toste [11].

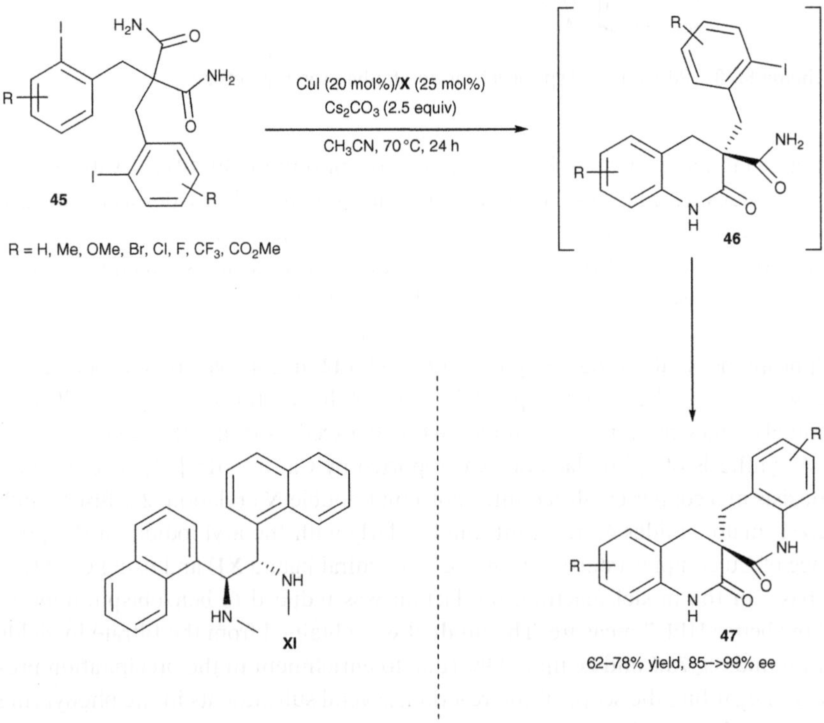

Scheme 8.12 Spirocyclization reported by Cai. *Source:* Based on Liu et al. [12].

A Pd-catalyzed spirocyclization was developed by Dixon and coworkers for the synthesis of heavily substituted spirolactams [13]. The reaction is mediated by silver phosphate and allenes bearing a ketoamide **48** react with aryl or vinyl iodides (**49**) under Pd catalysis to form first the π-allyl palladium complex **50** **t**hat undergoes intramolecular alkylation to form the ketoamide **51**. The final spirolactams **51** were obtained in good yields (61–82%), excellent diastereoselectivities (up to 42 : 1 dr), and good enantioselectivities (65–89%) with good group compatibility using as a chiral ligand bisoxazoline **XII** (Scheme 8.13).

Another methodology for the synthesis of spiro compounds was proposed by Luan: a palladium-catalyzed dynamic kinetic asymmetric reaction [14]. Conformationally, labile biaryls presenting axial chirality (due to the presence of an *ortho*-bromo substituent) **53** react with Pd(0)-NHC (**XIII**) chiral complexes to transform the axial chirality into central chirality. The Palladium (0)-NHC (**XIII**) complex is able to discriminate between the two chiral conformations of the biaryl achieving a dynamic kinetic resolution. Next, coordination of the alkyne **52** and subsequent insertion leads to spiro compound **69** in excellent yields and stereoselectivities (Scheme 8.14).

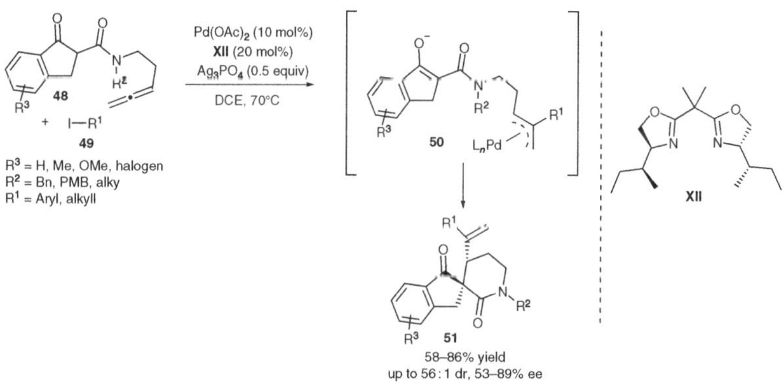

Scheme 8.13 Spirocyclization reported by Dixon. *Source:* Based on Li [13].

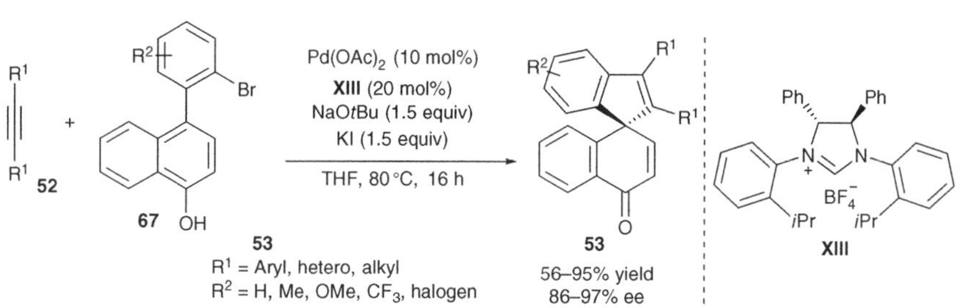

Scheme 8.14 Spirocyclization reported by Luan. *Source:* Based on Yang et al. [14].

8.3 Enantioselective Synthesis of All-Carbon Spirocycles by Cycloaddition Strategies

Cycloadditions have been a common approach for the synthesis of spiro compounds; herein, we report the use of [2+2+2], Diels–Alder, or 3+2 cycloadditions to synthesize spiro scaffolds with good selectivity degrees.

Shibata in 2006 reported a [2+2+2] cycloaddition for the synthesis of spiro compounds catalyzed by Rh(I) [15]. In this work, dienes (**55**) react with exo-unsaturated cyclic esters (**56**) to afford the spiro compounds (**57**) in a [2+2+2] cycloaddition fashion. Cationic Rh(I) complexes using as ligand (*S*)-xylyl-BINAP are able to catalyze the reaction affording spiro compounds **57** with moderate yields and high enantioselectivities (Scheme 8.15).

Unsymmetrical dienes (**58**) can be used with almost perfect regioselectivity control of the alkene and excellent enantioselectivities in the final spiro compounds **59** (Scheme 8.16).

Soon after, a similar methodology was developed by Tanaka [16]. In this work, a double intramolecular [2+2+2] cycloaddition takes place between two alkyne moieties and a nitrile as can be seen in Scheme 8.17. The reaction lead to the synthesis of C2-symmetric spiro bipyridine compounds in good to excellent yields and moderate enantioselectivities using as a catalyst[$Rh(cod)_2$]BF_4/Segphos. Remarkably, the scope of the reaction allows to synthesize not only spirobipyridines possessing

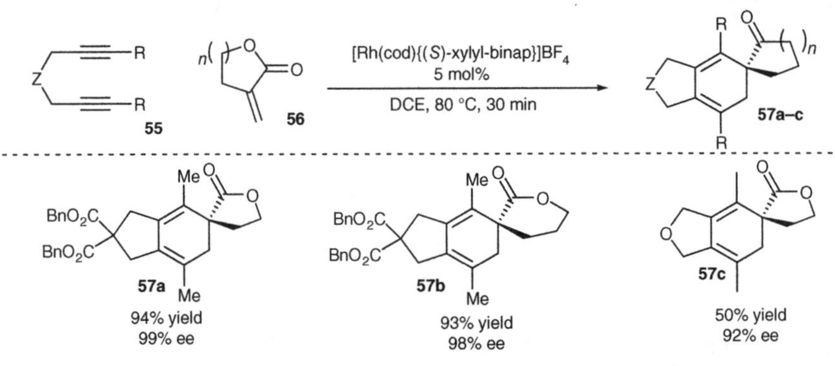

Scheme 8.15 [2+2+2] Cycloaddition reported by Shibata et al. *Source:* Based on Tsuchikama et al. [15].

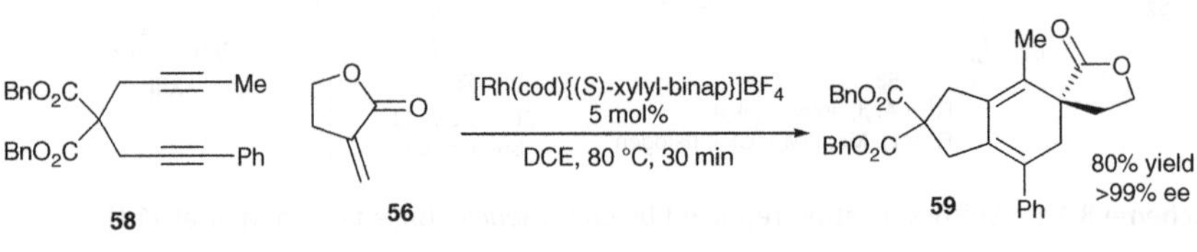

Scheme 8.16 [2+2+2] Cycloaddition with unsymmetrical dialkynes.

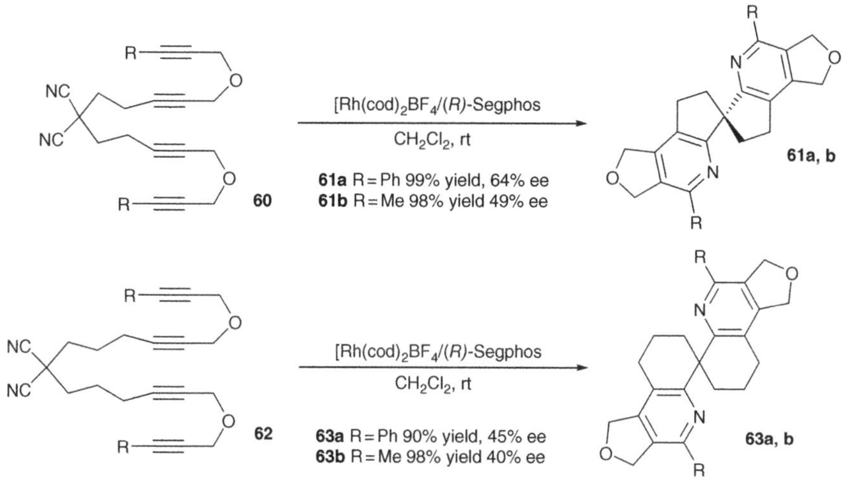

Scheme 8.17 Synthesis of spirocyclanes via a [2+2+2] cycloaddition reported by Tanaka. *Source:* Based on Wada et al. [16].

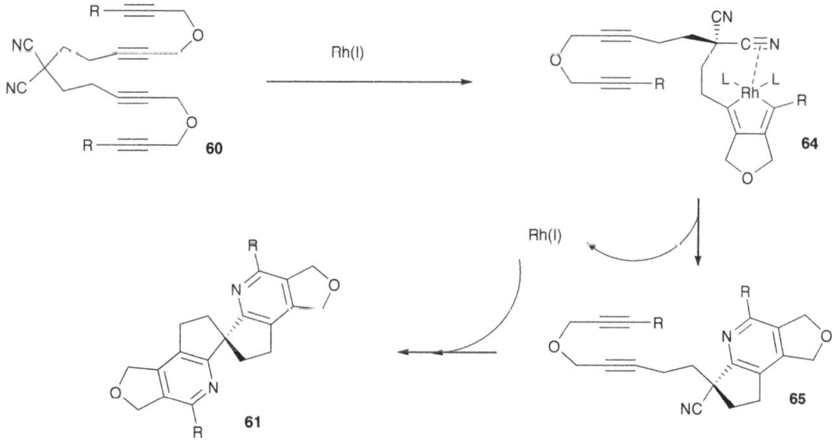

Figure 8.3 Proposed mechanism by Tanaka. *Source:* Based on Wada et al. [16].

a five-membered spiroskeleton (**61**) but also possessing a six-membered spiroskeleton (**63**) with similar results.

Tanaka proposed a plausible mechanism for the [2+2+2] cycloaddition as shown in Figure 8.3. First, the Rh coordinates with two enynes, followed by oxidative coupling and coordination of the nitrile moiety to form **64**. Next, insertion of the nitrile followed by the reductive elimination of Rh gives the pyridine compound **65** and regenerates the rhodium catalyst. Next, another [2+2+2] cycloaddition takes place to form the spiro compound **61**. As can be seen, the scope of the reaction is quite narrow due to the structural requirements for the double [2+2+2] cycloaddition and only moderate enantioselectivities were obtained.

The same research group reported few years later a Rh-catalyzed [2+2+2] cycloaddition of 1,6-enynes **66** with cyclopropylideneacetamides **67** to generate spirocyclohexenes **69** with excellent results (up to 77% yield and >99% ee) when H_8-BINAP (**XVI**) was used as a ligand [17, 18]. 1,6-Enynes **66** react with the Rh complex to generate rhodacyclopentane **68**, followed by insertion of the alkene that generates the rhodacycle. Next, reductive elimination afforded spiro compound **69** (Scheme 8.18).

A [4+2] Diels–Alder cycloaddition toward the synthesis of (−)-gymnodimine was reported by Romo in 2005 [19]. The key step for the synthesis of marine toxin (−)-gymnodimine is a copper-catalyzed Diels–Alder using bisoxazoline ligand **XVII** as a chiral inductor. As it is shown in Scheme 8.19, exo-methylene lactam **71** reacts with decorated diene **70** under the optimized conditions (CH_2Cl_2, rt) to

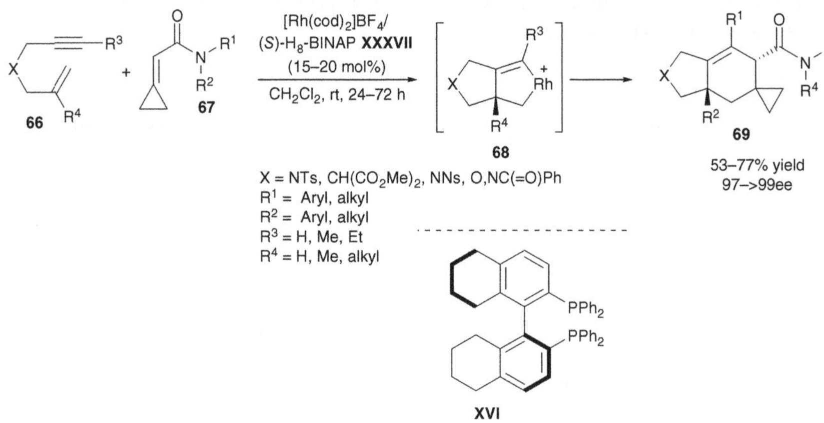

Scheme 8.18 [2+2+2] Spirocyclization reported by Tanaka. *Source:* Based on Wada et al. [16].

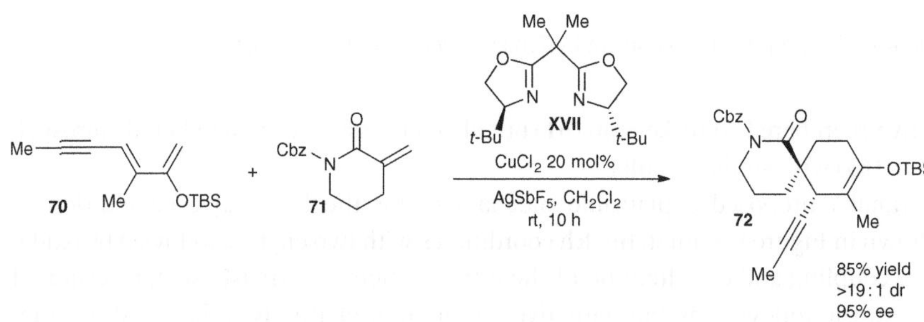

Scheme 8.19 Copper-catalyzed enantioselective Diels–Alder reaction. *Source:* Based on Romo et al. [19].

afford the corresponding spirolactam **72** in 85% yield and almost perfect stereoselectivity. An important remark of this reaction is that the corresponding (*Z*)-diene was totally inert under the reaction conditions probably caused by steric interactions between the copper catalyst and the adduct in the concerted process. This feature allows the obtention of high stereoselectivities.

A [3+2] cycloaddition catalyzed by copper was reported by Wang for the enantioselective synthesis of spiro compounds [20]. Iminoesters (**73**) react with ethylcyclopropylidene acetate (**74**) under copper catalysis using as chiral ligand (*S*)-TF-BiphamPhos. The reaction renders the final bioactive 5-aza-spiro[2,4]heptane derivatives (**75**) in excellent yields and enantioselectivities (Scheme 8.20).

A similar approach was reported by Deng for the synthesis of spirobispyrrolidine derivatives **77** [2]. The [3+2] cycloaddition between azomethine ylides (**73**) and α-alkylidene succinimides **46** is catalyzed by $Cu(OAc)_2$/*N*,*O*-ligand systems leading to *endo*-dispiropyrrolidines **47** in very good yields with diastereoselectivity >20 : 1 and up to 97% ee (Scheme 8.21).

The only limitation of the reaction relies on the presence of an electron-withdrawing group like CO_2Me or Cbz on the nitrogen of the succinimide, in order

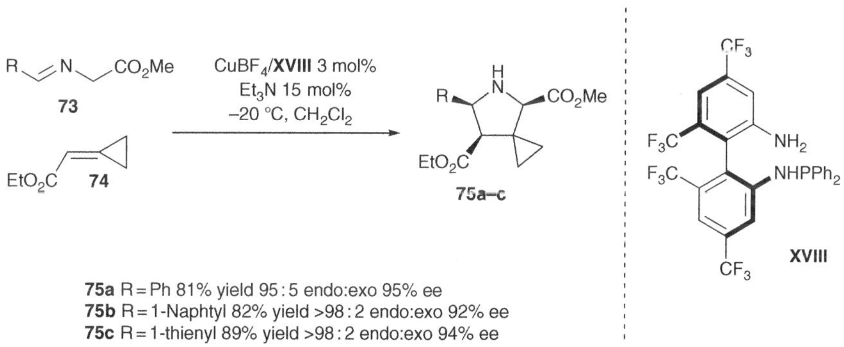

Scheme 8.20 [3+2] Cycloaddition reported by C.-J. Wang. *Source:* Based on Liu et al. [20].

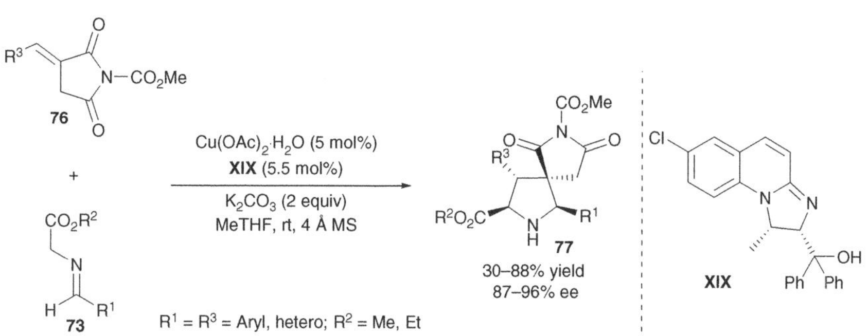

Scheme 8.21 Spirocyclization reported by Deng. *Source:* Based on Deng et al. [1].

to afford good results in terms of stereoselectivity. Among the chiral *N,O*-ligands tested, **XIX** gave the best yields and stereoselectivities. In the structure of the catalyst is crucial the methyl substituent in position 1 to obtain high enantioselectivities. The reaction works well with azomethine ylides **73** derived from glycine esters in which the substituents on the aromatic group R^1 can be indifferently neutral, electron-donating or electron-withdrawing, heteroarylic, and even sterically hindered.

A slightly different [3+2] cycloaddition catalyzed by palladium was reported by Trost for the synthesis of spiroazlactones (**80**) [21]. The reaction starts with the activation of vinyl cyclopropanes **79** under Pd catalysis to form the corresponding zwitterion which undergoes a formal [3+2] cyclization with the alkylidene azlactone **78**. The reaction achieves good to excellent enantioselectivities using chiral phosphine ligand **XX**. Interestingly, a key aspect to obtain reactivity was the type of substituents in the malonate moiety of the vinyl cyclopropane. The authors proposed that the use of trifluoroethylester increases the half-life of the intermediate dipole formed, without losing reactivity. The spiroazlactones **80** were obtained in good yields and good stereoselectivities (Scheme 8.22). Soon after, the same research group expanded the scope of the reaction to alkylidene Meldrum's acid derivatives (**81**) with good yields and stereoselectivities [23]. The main difference is the use of DPPBA biphosphine ligands (**XXI**) to achieve good enantioselectivities (Scheme 8.22, bottom). Using the same approach, Zhao and coworkers reported the synthesis of spiro compounds starting with *para*-quinonemethides and vinyl epoxides or vinyl cyclopropanes with excellent results few years later [22].

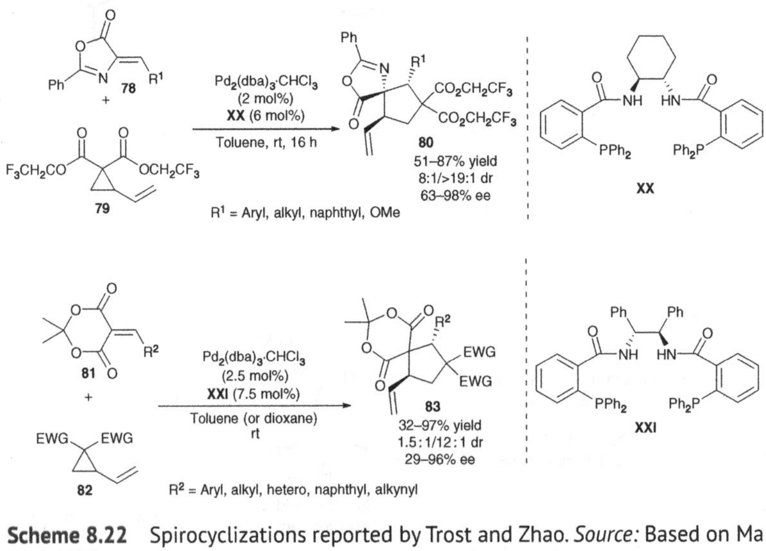

Scheme 8.22 Spirocyclizations reported by Trost and Zhao. *Source:* Based on Ma et al. [22].

Another [3+2] cycloaddition using vinylcylopropanes was developed by Liu in 2015. This time, Nitrostyrenes **84** react with vinyl cyclopropanes **85** derived from 1,3-indanone under Pd(0) catalysis, using diamine ligand **XXII** derived from trans-cyclohexyldiamine to produce spiro compounds **86** with good yields (75–90%), moderate to good diastereoselectivities (1.2 : 1 to 14 : 1), and excellent enantioselectivities (92–99% ee) (Schemes 8.19 and 8.23) [24].

A synergistic [3+2] cycloaddition for the synthesis of spiroindanones was reported almost at the same time by the research groups of Michelet [25] and Rios [26]. The reaction consists in a ring opening of vinyl cyclopropanes **85** *via* an allyl palladium complex to form the classical zwitterion and cycloaddition with enals **87**, activated by secondary amine **XXIII**. The reaction affords the final compounds **88** in good yields (76–99%) and diastereoselectivities (up to 12 : 1) and excellent enantioselectivities (87–99%) (Scheme 8.24).

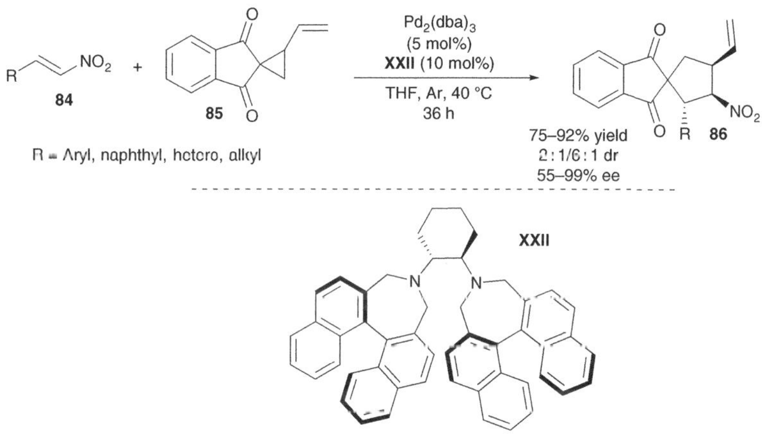

Scheme 8.23 Spirocyclization reported by Liu. *Source:* Based on Wei et al. [23].

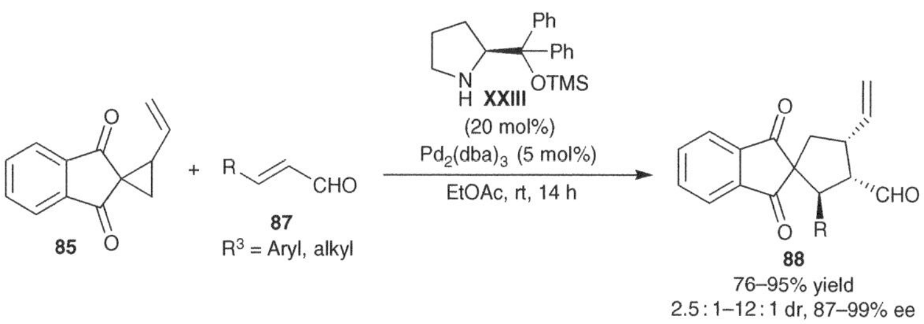

Scheme 8.24 Spirocyclization reported by Rios. *Source:* Based on Meazza and Rios [26].

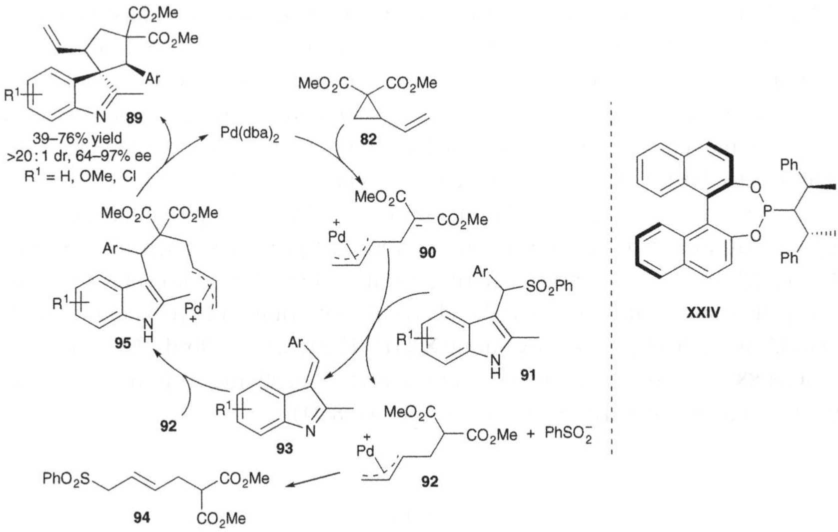

Scheme 8.25 Spirocyclization reported by Liu. *Source:* Based on Liu et al. [27].

A similar reaction was reported by Liu starting from arenesulfonyl indoles **91** and vinyl cyclopropanes [27]. As stated before, the Pd complex opens vinyl cyclopropane **82** to form the zwitterion **90**. On the other hand, arenesulfonyl indole **91** undergoes elimination to form conjugated compound **93**. Next, Michael addition of the zwitterion to **93** followed by intramolecular Tsuji–Trost allylation renders the final spiro compound in good yields (up to 74%) and stereoselectivities (up to 97% ee with total diastereoselectivity). The key of the reaction is the use of phosphoramidite XXIV as ligand which is the responsible for the high yields and stereoselectivities (Scheme 8.25).

8.4 Enantioselective Synthesis of All-Carbon Spirocycles by Radical Strategies

An elegant strategy for the synthesis of spiropyrrolidines was developed by Bach in 2005 [28]. The reaction started with compound **96** that undergoes an intramolecular cyclization driven by photo-induced electron transfer (PET). The catalysts employed present two key features: (i) the photocatalysts bind with the reactant by two hydrogen bonds from the bridgehead lactam and (ii) the catalyst contains a benzophenone unit responsible for the PET. Under optimized conditions, the intramolecular cyclization leads to the final spiropyrrolidizidine **97** with moderate yield (64%) and good enantioselectivities (70% ee; Scheme 8.26).

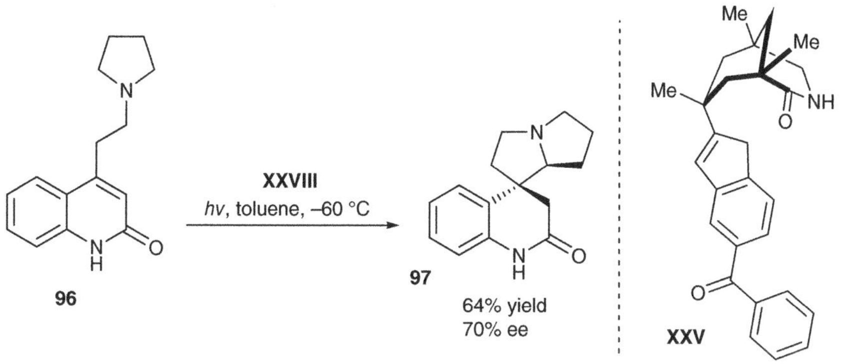

Scheme 8.26 Enantioselective PET reaction developed by Bach. *Source:* Based on Bauer et al. [28].

8.5 Enantioselective Synthesis of All-Carbon Spirocycles by Cascade Reactions

Cascade reactions have been one of the most used approaches for the synthesis of spiro compounds. With the advent of organocatalysis, the development of new enantioselective methodologies increased exponentially.

One of the first cascade reactions leading to spiro compounds was reported by Romo in 2008 [29]. In this work, an intramolecular cyclization takes place catalyzed by tetramisole catalyst **XXVI**. The key step is the enantioselective desymmetrization of the diketone leading to tricyclic compound **99**, that after treatment with $Zn(OTf)_2$ delivers the spirocycle **100**. The main limitation is the need to use stoichiometric amounts of the tetramisole catalyst to afford moderate yields and excellent enantioselectivities (Scheme 8.27).

Some years later, Wand and coworkers developed a Michael aldol cascade reaction for the synthesis of spiro compounds [30]. (*E*)-benzylidenechromanone derivatives **101** react with 2-mercapto benzaldehydes (**102**) via a sulfa-Michael

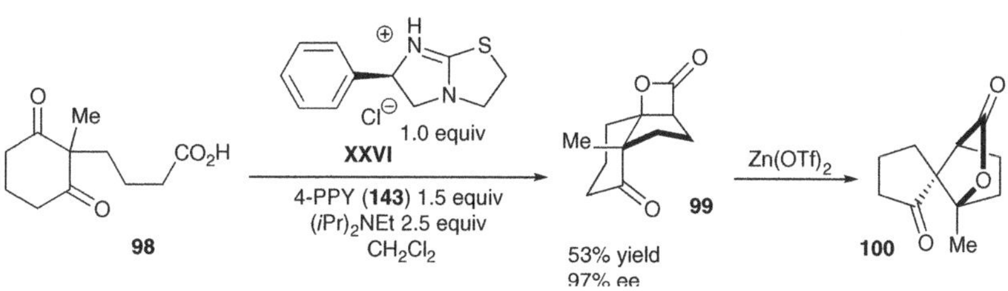

Scheme 8.27 Synthesis of spirocyclic β-lactones developed by Romo. *Source:* Based on Purohit et al. [29].

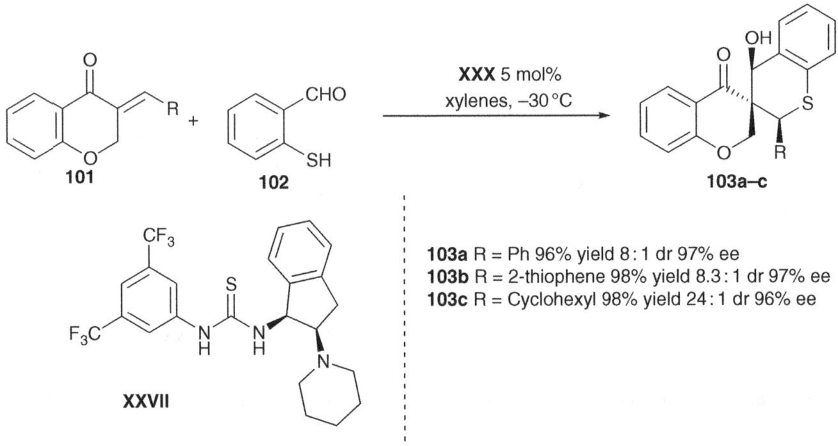

Scheme 8.28 Cascade reaction reported by Wang. *Source:* Based on Gao et al. [30].

reaction followed by intramolecular aldol reaction. As shown in Scheme 8.28, the reaction is catalyzed by bifunctional tertiary-amine/thiourea catalysts **XXVII** synthesizing the spirochromane derivatives **103** with excellent yields (up to 98%) and excellent stereoselectivities (up to 24 : 1 dr and 97% ee).

A cascade reaction between nitroalkenes and keto thioamides was reported by Xie and coworkers for the synthesis of spirodihydrothiophenes **105** [31]. This time, the reaction was catalyzed by Takemoto's catalyst (**XXVIII**). First, Ketothiamide (**104**) reacts with nitrostyrene (**84**) in a Michael fashion followed by nucleophilic attack of the sulfur atom to the α-carbon atom of the nitro group and dehydration. As it is shown in Scheme 8.29, the reaction furnishes the final spiro compounds **105** in moderate to good yields (55–78%), excellent diastereoselectivities (up to 99 : 1), and good enatioselectivities (82–94%).

With the advent of organocatalysis, numerous tandem reactions have been developed due to the easy stereoprediction and access of the catalysts. One organocatalytic example for the synthesis of spiro compounds based on trienamine catalysis was developed by Chen in 2013 [32]. In this work, dienal **107** reacts with the secondary amine **XXIX** to form the trienamine. Next, the trienamine goes through a formal Diels–Alder cycloaddition with the methyl iodide salt of the Mannich base **106** which first undergoes elimination to form the enone. The spiro compounds **108** were obtained in moderate to good yields (41–72%), excellent diastereoselectivities (up to 19 : 1 dr), and good enantioselectivities (83–96% ee). The reaction has a reasonable group compatibility and requires the use of a H bond cocatalysts (TADDOL, **XXX**) to achieve high stereoselectivities (Scheme 8.30).

Soon later, Albrech and coworkers reported a different trienamine-catalyzed synthesis of spiro compounds [33]. In this work, (*E*)-3-alkylidende-5-arylfuran-2(3H)-ones react with dienals under secondary amine catalysis to yield

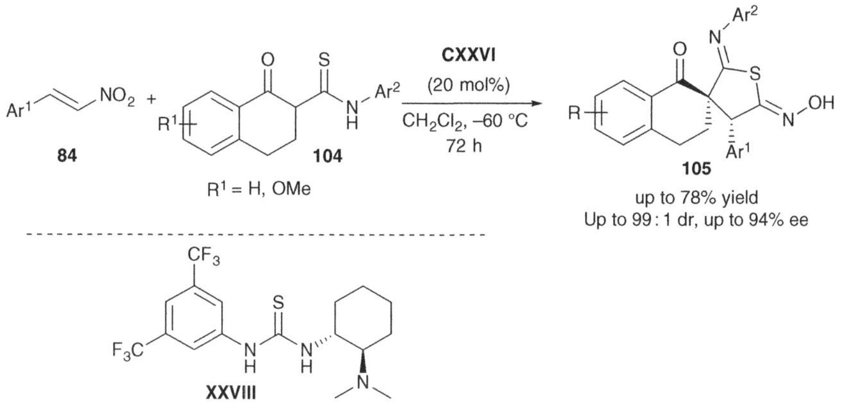

Scheme 8.29 Spirocyclization reported by Xie. *Source:* Based on Zeng et al. [31].

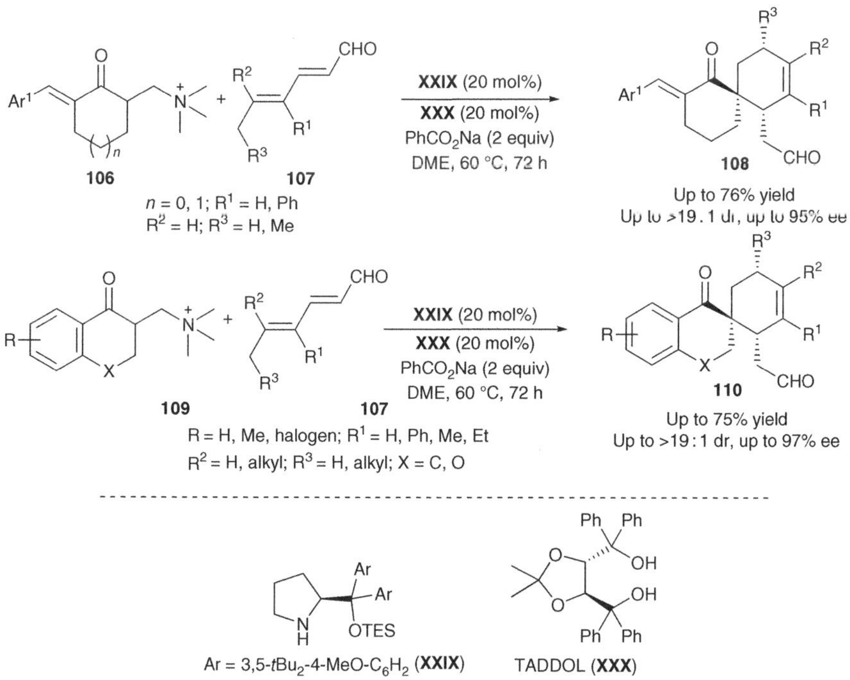

Scheme 8.30 Spirocyclization reported by Chen. *Source:* Based on Zhang et al. [32].

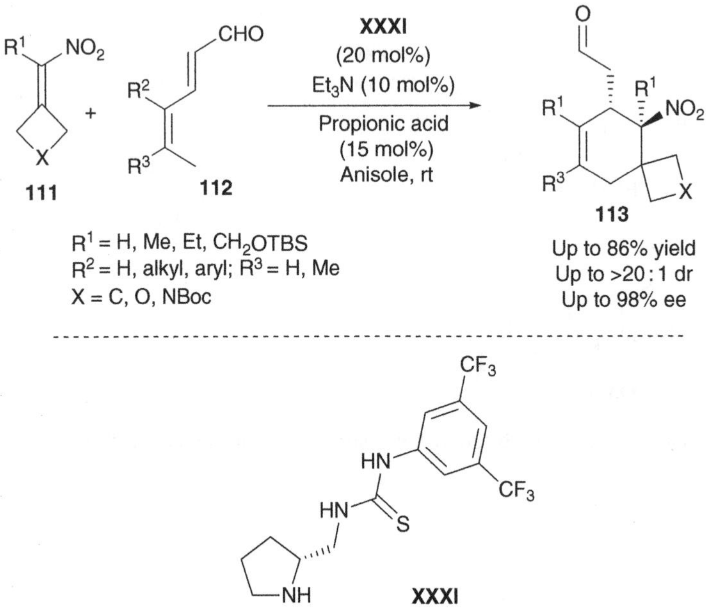

Scheme 8.31 Spirocyclization reported by Jørgensen. *Source:* Based on Donslund et al. [35].

the spirocyclic compounds with moderate yields (28–65%), diastereoselectivities (up to 3 : 1 dr), and excellent enantioselectivities (89–99% ee).

A similar approach was developed by Jørgensen dienal **112** under secondary amine catalysis to form the trienamine, which reacts in a Diels–Alder fashion with α,β,β-trisubstituted nitroolefins **111** [34]. A key aspect is the presence of a thiourea moiety in the catalyst. The hydrogen bond interaction between the thiourea and the nitrostyrene is fundamental to obtain good stereoselectivities. The reaction catalyzed by **XXXI** furnishes the final spiro compounds **113** with good yields (61–82%) and excellent stereoselectivities (up to 20 : 1 dr and 96% ee). The same research group developed another spiro reaction using trienamine catalysis with benzofulvenes achieving similar results (Scheme 8.31) [35].

Spirocyclopropanes have become one of the most coveted structures for organic chemists. For example, Benzofulvenes **114** react with bromomalonates **115** under phase-transfer catalysis affording the final cyclopropanes with excellent yields (88–98%), diastereoselectivities (up to 20 : 1 dr), and good enantioselectivities (68–98%) *via* a Michael/intramolecular alkylation sequence. The scope of the reaction is quite narrow as only bromomalonates gave excellent results; in the case of 2-bromoketoesters, lower yields and stereoselectivities were obtained (Scheme 8.32).

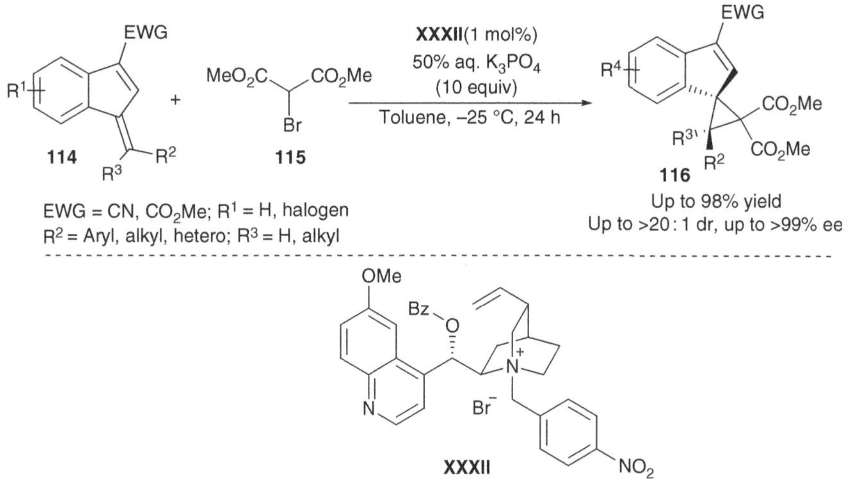

Scheme 8.32 Spirocyclopropanation of benzofulvenes. *Source:* Based on Donslund et al [35].

A similar work using as starting material 2-arylidene-1,3-indandiones (**117**) was reported by Lin [36]. In this work, 2-arylidene-1,3-indandiones react with 1-bromo-nitroalkanes **118** to afford spirocyclopropanes through a Michael/intramolecular alkylation pathway. The organocascade reaction was efficiently catalyzed by bifunctional tertiary amine/thiourea catalyst **XXXIII** obtaining the spirocyclopropanes with good yields (63–88%) and excellent stereoselectivities (up to 19 : 1 dr and 98% ee). The main limitation of the present reaction is that larger groups in substitute 1-bromo-1-nitromethane decrease the stereoselectivity of the organocascade. A similar approach was reported soon later by Du, using 3-arylenechroman-4-ones as starting material instead of indanones and affording the spironitrocyclopropanes with good yields and excellent stereoselectivities (Scheme 8.33) [37].

Spirolactams have attracted much attention recently, for example Bonne and Rodriguez reacted 1,3-cyclic ketoamides **120** with α,β-unsaturated acyl cyanides **121** under bifunctional tertiary amine/thiourea catalyst to form the spirolactam **122** (Scheme 8.34) [38]. The reaction starts with a Michael addition of 1,3-cyclicketoamide **120** to the α,β-unsaturated acyl cyanide followed by an intramolecular nucleophilic displacement of the cyanide by the amide moiety furnishing the spirolactam **122** with good yields (73–82%) and good stereoselectivities (up to 91 : 9 dr and 94% ee). Soon after, Vesely and Rios reported a similar reaction using cyclicketoamides and enals and catalyzed by secondary amine achieving the spirolactams with excellent yields and enantioselectivities [39].

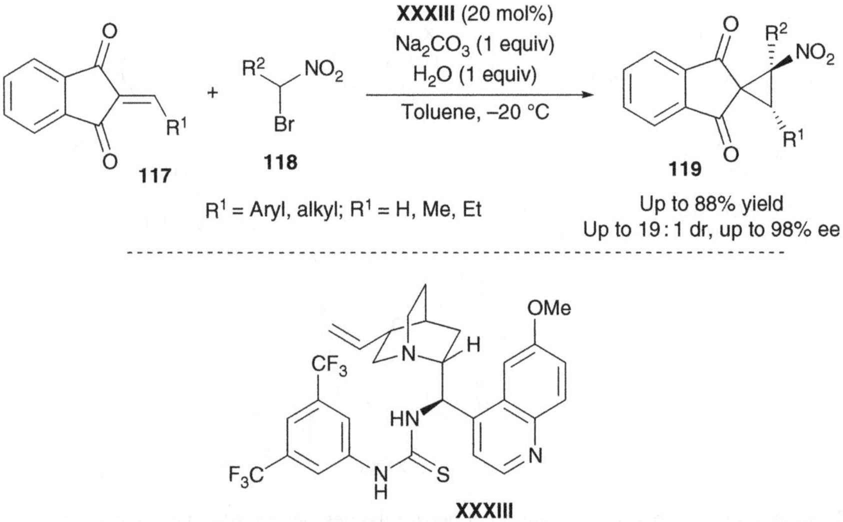

Scheme 8.33 Spirocyclization reported by Lin. *Source:* Based on Das et al. [36].

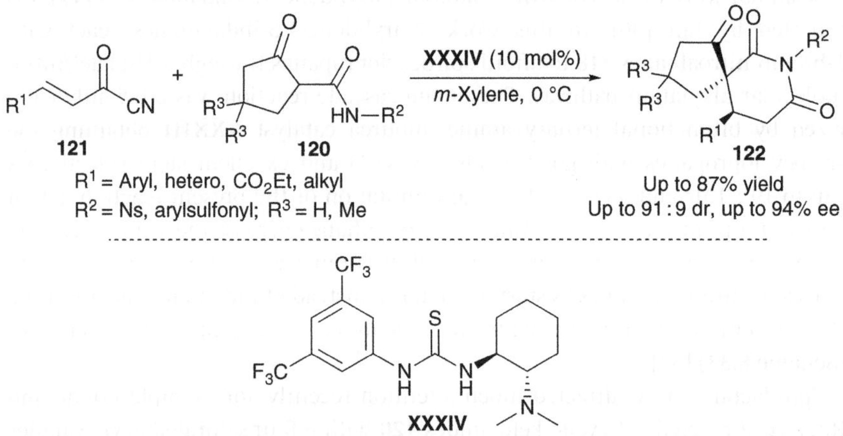

Scheme 8.34 Spirocyclization reported by Rodriguez and Bonne. *Source:* Based on Goudedranche et al. [38].

A synthesis of spiroindanones through a cascade nitro-Michael/aldol reaction was reported by Li in 2015 [40]. Nitroaldehydes **123** react with 2-Arylideneindane-1,3-diones (**117**) under bifunctional tertiary amine/squaramide catalysis to afford the spiroindanones with good yields and stereoselectivities (up

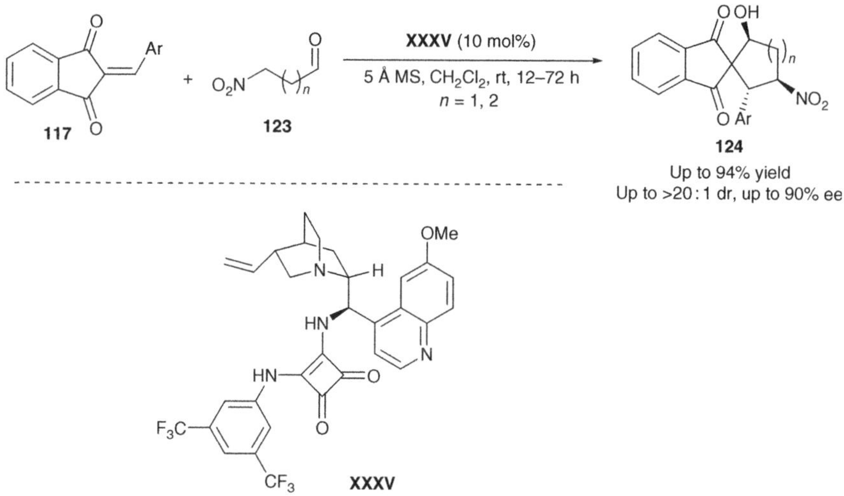

Scheme 8.35 Spirocyclization reported by Li. *Source:* Based on Duan et al. [40].

to 84% yield, 20 : 1 dr, and 88% ee) Scheme 8.35. The reaction starts with a nitro-Michael addition between **123** and 2-Arylideneindane-1,3-dione **117** followed by intramolecular aldol reaction between the "in situ" formed enolate and the aldehyde moiety of nitro alkene. The reaction affords [5,6,0]spirocycles and [5,5,0]spirocycles in good yields and stereoselectivities (up to 84% yield, 20 : 1 dr, and 88% ee). Soon after, the same research group reported a similar cascade reaction, but this time using mercapto-2-butenoates obtaining similar results [41].

8.6 Enantioselective Synthesis of All-Carbon Spirocycles by Rearrangement Strategies

Tu and coworkers reported a rearrangement strategy for the synthesis of spirocyclic compounds [42]. Hydroxyenones **125** undergo semipinacol-type 1,2-carbon migration catalyzed by a primary amine derived from cinchona alkaloid (**XXXVI**) through an iminium catalysis. The reaction affords the spirocyclic diketones **126** with good yields and excellent stereoselectivities. However, the scope of the reaction is rather limited because of the requirement of the presence of cyclobutanol moieties in the starting materials (Scheme 8.36).

A similar migratory ring expansion for the synthesis of spirocyclic rings was reported by Raney [43]. Indene rings bearing a cyclobutanol **127** first coordinate with a Pd/phosphoric acid complex to undergo a enantioselective allylic CH activation and, at the same time, a semipinacol ring expansion takes place to build the

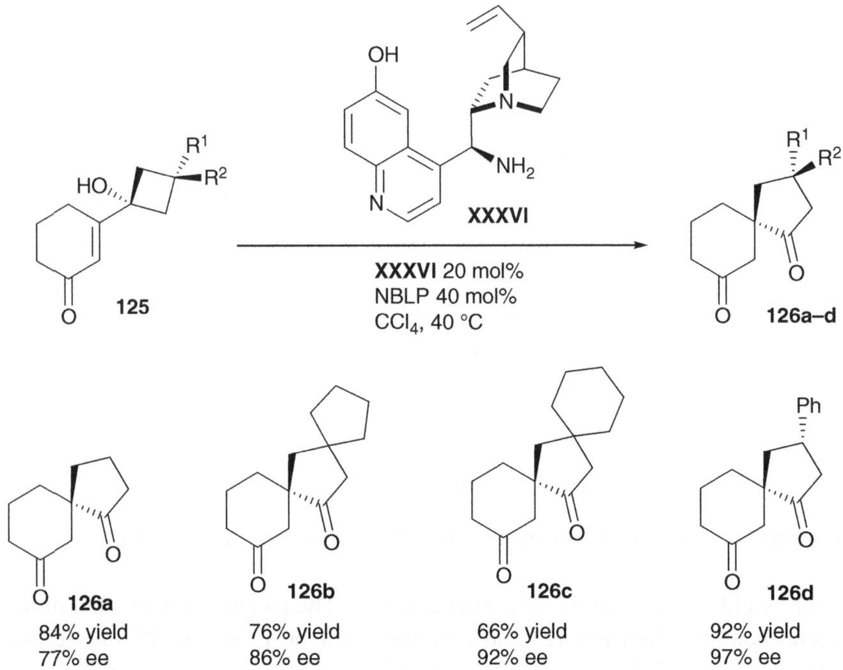

Scheme 8.36 α-ketol rearrangement reported by Tu. *Source:* Based on Zhang et al. [42].

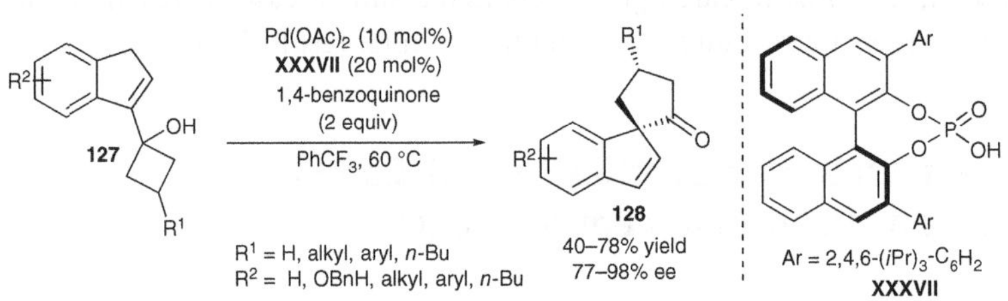

Scheme 8.37 Ring expansion reported by Takisawa. *Source:* Reported by Takisawa.

spiroindanones with moderate to good yields (40–78%) and good enatioselectivities (77–98% ee) (Scheme 8.37).

A redox-Mannich-pinacol rearrangement sequence for the synthesis of spirodiketones was reported by Zhang in 2013 [44]. In this work, aryl nitrones **129** first coordinate with Au-JohnPhos complexes and then undergo a pinacol rearrangement followed by intramolecular Mannich reaction (this later catalyzed by phosphoric acid **XXXVIII**) to afford spirocycle **130** with good to excellent yields (65–93%) and excellent enatioselectivities (90–99%) (Scheme 8.38).

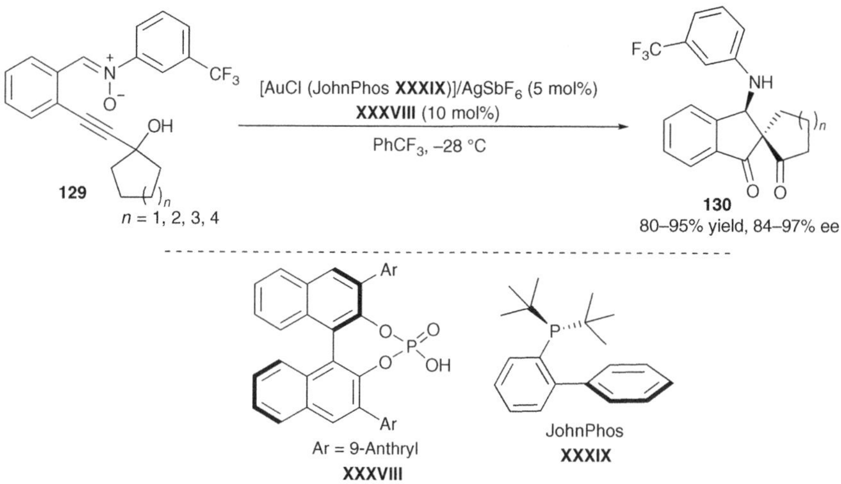

Scheme 8.38 Spirocyclization reported by Zhang. *Source:* Based on Qian and Zhang [44].

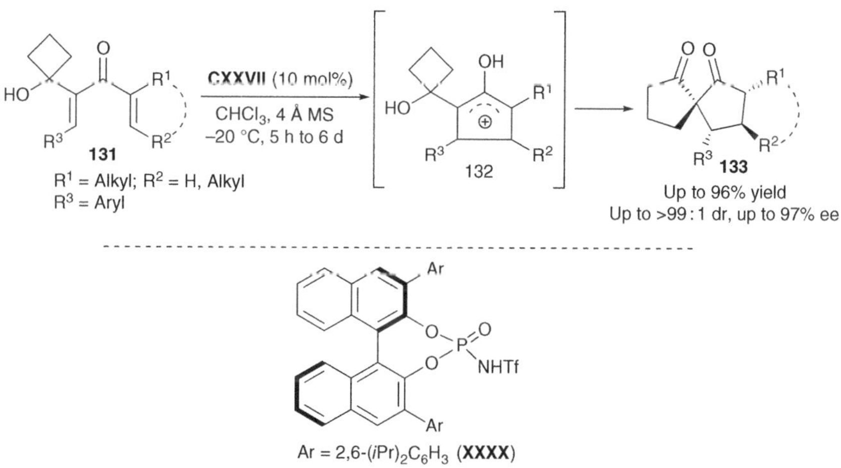

Scheme 8.39 Spirocyclization reported by Zhang. *Source:* Based on Qian and Zhang [44].

A tandem Nazarov/semipinacol rearrangement was reported by Zhang for the synthesis of chiral spiro[4.4]nonane-1,6-diones (**133**) in 2015 [45]. As shown in Scheme 8.39, 1,4-dien-3-ones decorated in the alpha position with a cyclobutanol motif **131** undergo Nazarov reaction under chiral acid catalysis to furnish an oxoallyl intermediate **132** which induces the semipinacol rearrangement with the cyclobutanol moiety affording the spirocycle **133** with good yields (72–92%) and excellent diastereo- and enantioselectivities (up to 99 : 1 dr and 97% ee).

8.7 Conclusions

In this chapter, the most interesting approaches for the enantioselective synthesis of spiro compounds have been discussed. From cycloadditions, to metal-catalyzed or organocascade reactions, organic chemists overcome the inherent difficulty to synthesize spiro compounds. The complexity and at the same time beauties of the molecules obtained are astonishing, and give us a clear vision how the art of organic synthesis flourished in recent years. However, despite these achievements, there are still enormous challenges for the future that organic chemists will address putting their efforts in this considerable task.

References

1 Fan, L., Takizawa, S., Takeuchi, Y. et al. (**2015**). *Org. Biomol. Chem. 13*: 4837–4840.
2 Deng, Q.-H., Wadepohl, H., and Gade, L.H. (**2012**). *J. Am. Chem. Soc. 134*: 2946–2949.
3 Rahemtulla, B.F., Clark, H.F., and Smith, M.D. (**2016**). *Angew. Chem. Int. Ed. 55*: 13180–13183.
4 Takahashi, T., Tsutsui, H., Tamura, M. et al. (**2001**). *Chem. Commun.*: 1604–1606.
5 (a) Mikami, K., Yusa, Y., Hatano, M. et al. (**2004**). *Chem. Commun.*: 98–100. (b) Mikami, K., Yusa, Y., Hatano, M. et al. (**2004**). *Tetrahedron 60*: 4475–4480.
6 (a) Nicolaou, K.C., Edmonds, D.J., Li, A., and Tria, S.S. (**2007**). *Angew. Chem. Int. Ed. 46*: 4712. (b) Nicolaou, K.C., Li, A., Ellery, S.P., and Edmonds, D.J. (**2009**). *Angew. Chem. Int. Ed.* 48: 6293.
7 Shintani, R., Isobe, S., Takeda, M., and Hayashi, T. (**2010**). *Angew. Chem. Int. Ed. 49*: 3795–3798.
8 Hatano, M., Terada, M., and Mikami, K. (**2001**). *Angew. Chem. Int. Ed. 40*: 249–253.
9 Hatano, M. and Mikami, K. (**2003**). *Org. Biomol. Chem. 1*: 3871–3873.
10 Hatano, M. and Mikami, K. (**2003**). *J. Mol. Catal. A 196*: 165–169.
11 Corkey, B.K. and Toste, F.D. (**2007**). *J. Am. Chem. Soc. 129*: 2764.
12 Liu, J., Tian, Y., Shi, J. et al. (**2015**). *Angew. Chem. Int. Ed. 54*: 10917–10920.
13 Li, M., Hawkins, A., Barber, D.M. et al. (**2013**). *Chem. Commun. 49*: 5265–5267.
14 Yang, L., Zheng, H., Luo, L. et al. (**2015**). *J. Am. Chem. Soc.* ***137***: 4876–4879.
15 Tsuchikama, K., Kuwata, Y., and Shibata, T. (**2006**). *J. Am. Chem. Soc. 128*: 13686–13687.
16 Wada, A., Noguchi, K., Hirano, M., and Tanaka, K. (**2007**). *Org. Lett. 9*: 1295–1298.
17 Yoshida, T., Tajima, Y., Kobayashi, M. et al. (**2015**). *Angew. Chem. Int. Ed. 54*: 8241–8244.

18 Yoshizaki, S., Nakamura, Y., Masutomi, K. et al. (**2016**). *Org. Lett. 18*: 388–391.

19 Romo, D., Moussa, Z., and Kong, K. (**2005**). *Org. Lett. 7*: 5127.

20 Liu, T.L., He, Z.L., Tao, H.Y. et al. (**2011**). *Chem. Commun. 47*: 2616.

21 Trost, B.M. and Morris, P.J. (**2011**). *Angew. Chem. Int. Ed. 50*: 6167–6170.

22 Ma, C., Huang, Y., and Zhao, Y. (**2016**). *ACS Catal. 6*: 6408–6412.

23 Trost, B.M., Morris, P.J., and Sprague, S.J. (**2012**). *J. Am. Chem. Soc. 134*: 17823–17831.

24 Wei, F., Ren, C.-L., Wang, D., and Liu, L. (**2015**). *Chem. Eur. J. 21*: 2335–2338.

25 Laugeois, M., Ponra, S., Ratovelomanana-Vidal, V. et al. (**2016**). *Chem. Commun. 52*: 5332–5335.

26 Meazza, M. and Rios, R. (**2016**). *Chem. Eur. J. 22*: 9923–9928.

27 Liu, Z.-S., Li, W.-K., Kang, T.-R. et al. (**2015**). *Org. Lett. 17*: 150–153.

28 Bauer, A., Westkaemper, F., Grimme, S., and Bach, T. (**2005**). *Nature 436*: 1139–1140.

29 Purohit, V.C., Matla, A.S., and Romo, D. (**2008**). *J. Am. Chem. Soc. 130*: 10478–10481.

30 Gao, Y., Ren, Q., Wu, H. et al. (**2010**). *Chem. Commun. 46*: 9232–9234.

31 Zeng, X.-M., Meng, C.-Y., Bao, J.-X. et al. (**2015**). *J. Org. Chem. 80*: 11521–11528.

32 Zhang, S.-J., Zhang, J., Zhou, Q.-Q. et al. (**2013**). *Org. Lett. 15*: 968–971.

33 Hejmanowska, J., Dziegielewski, M., Kowalczyk, D., and Albrecht, L. (**2014**). *Synlett 25*: 2957–2961.

34 Monleon, A., Glaus, F., Vergura, S., and Jorgensen, K.A. (**2016**). *Angew. Chem. Int. Ed. 55*: 2478–2482.

35 Donslund, B.S., Nielsen, R.P., Monsted, S.M.N., and Jorgensen, K.A. (**2016**). *Angew. Chem. Int. Ed. 55*: 11124–11128.

36 Das, U., Tsai, Y.-L., and Lin, W. (**2013**). *Org. Biomol. Chem. 11*: 44–47.

37 Zhao, B.-L. and Du, D.-M. (**2015**). *Eur. J. Org. Chem.*: 5350–5359.

38 Goudedranche, S., Bugaut, X., Constantieux, T. et al. (**2014**). *Chem. Eur. J. 20*: 410–415.

39 Zhang, K., Meazza, M., Docekal, V. et al. (**2017**). *Eur. J. Org. Chem.*: 1749–1756.

40 Duan, J., Cheng, J., and Li, P. (**2015**). *Org. Chem. Front. 2*: 1048–1052.

41 Duan, J., Cheng, J., Li, B. et al. (**2015**). *Eur. J. Org. Chem.*: 6130–6134.

42 Zhang, E., Fan, C.-A., Tu, Y.-Q. et al. (**2009**). *J. Am. Chem. Soc. 131*: 14626–14627.

43 Chai, Z. and Rainey, T.J. (**2012**). *J. Am. Chem. Soc. 134*: 3615–3618.

44 Qian, D. and Zhang, J. (**2013**). *Chem. Eur. J. 19*: 6984–6988.

45 Yang, B.-M., Cai, P.-J., Tu, Y.-Q. et al. (**2015**). *J. Am. Chem. Soc. 137*: 8344–8347.

9

Transition-Metal-Catalyzed Dearomative Spiroannulation Reactions

Lu Bai, Jingjing Liu and Xinjun Luan

Key Laboratory of Synthetic and Natural Functional Molecule of the Ministry of Education, College of Chemistry & Materials Science, Northwest University, Xi'an, China

9.1 Introduction: Discovery of Aromatic Compounds

Aromatic compounds, one of the most fundamental and abundant classes of organic compounds, are widely used across all areas of molecular sciences. Not surprisingly, the chemistry of these molecules is an exceptionally rich field and is distinctive due to their unique reactivity patterns and inherent stability. Michael Faraday first isolated the high-purity liquid benzene in 1825, while Eilhard Mitscherich obtained this substance and named it "benzin." Several years later, August Wilhelm von Hoffman and his students separated benzene from coal tar and subsequently he used the term "aromaticity" for describing the characteristics of this family of compounds in 1855. Since then, many chemists made the studies on an array of these substances. In 1865, the German chemist Kekulé predicted the structure-kekulé formula for benzene, and determined that the stability of such compounds was due to their high electron resonance energy. With the development of organic chemistry, chemists have realized that not only benzene has the aromatic property, other heterocyclic systems, neutral compounds, ions, and even inorganic compounds might have the similar property. In 1931, physical chemist Hüeckel came up with the famous Hüeckel rule and used it to determine the aromaticity of cyclic compounds. Since then, the conception of aromatic compounds has been significantly improved, and now unsaturated cyclic compounds with special stability are generally known as aromatic compounds.

9.2 Development of Dearomatization Reactions

Due to the importance of this variety of aromatic and heterocyclic aromatic compounds in modern synthetic chemistry, such compounds are widely used as raw materials or research objects in industrial production, pharmaceutical research,

Spiro Compounds: Synthesis and Applications, First Edition. Edited by Ramon Rios Torres.

and development of academic research and other fields. These compounds generally contain ring structures with conjugated double bonds, and each atom on the ring has unhybridized p orbitals, which form a continuous, overlapping, and parallel system. These characteristics make this kind of compounds extremely stable and the transformations of the derivative are mainly concentrated in nucleophilic and electrophilic substitution reactions, with the aromaticity reserved. In contrast, dearomatization reaction is to disrupt the relatively stable aromatic system, converting the compounds from the two-dimensional structure to three-dimensional structure. The specific feature of building quaternary carbon centers and interesting structures makes them straightforward to construct spiro or bridged compounds. The history of dearomative reactions can be dated back to the nineteenth century. Early studies on dearomatization reactions include photochemical processes, hydrogenation processes, oxidative processes, transition-metal-mediated processes, etc. In modern organic chemistry, transition-metal-catalyzed reaction is the most rapidly developing branch, which is widely used in organic synthesis, nano science, green chemistry, bioscience, and medicine field. Notably, transition-metal-catalyzed reactions could solve many chemical transformations that traditional organic chemistry cannot achieve. In recent years, given the high efficiency and economic irreplaceability of transition-metal-catalysis, the transition-metal-catalyzed dearomative reaction has developed rapidly and made great progress.

9.3 Development of Dearomative Spiroannulation Reactions

Spirocyclic frameworks are commonly occurring structural motifs in many bioactive compounds and have been used as core building blocks for a wide range of functional materials and chiral ligands. Thus, exploring an efficient method for accessing spirocyclic compounds has been of vital significance, as well as it is challenging but meaningful topic in organic synthesis. This chapter reviews transition-metal-catalyzed dearomative spiroannulation of aromatic compounds. This chapter is organized according to the kind of aromatic compounds, including phenols, indoles, furans, thiophenes, pyrroles, aryl diazo salt, arylamine, and so on.

9.4 Dearomative Spiroannulations of Phenols

Phenols are one kind of important chemical raw materials [1]; diverse transformations of the derivative have been reported. Many natural products and functional materials with biological activities can be prepared from them [2]. Meanwhile, due to their cheap and easy availability, chemists are constantly exploring and studying the chemical properties and applications of phenols. They were frequently used as oxygen nucleophiles to give allyl aryl ethers, while the Friedel–Crafts-type nucleophilic aromatic substitutions could not occur [3]. In recent years, a novel derivatization transition-metal-catalyzed dearomative spiroannulation has become a hot research topic and great progress has been made in obtaining a number of spirocyclic compounds with diverse skeletons [4]. This chapter focuses on two parts: dearomative spiroannulation without oxidants and dearomative spiroannulation with oxidants.

9.4.1 Dearomative Spiroannulation Without Oxidants

9.4.1.1 Dearomative Alkylation of Phenols

In the traditional chemical research process, phenols are usually used as nucleophiles to generate ethers [5]. In 2010, different from the previous work of phenols as oxygen nucleophiles, Hamada and coworkers found that *meta*-substituted phenols could produce Friedel-Crafts-type adducts **2a** and **2b** under the catalysis of Pd(0)(Scheme 9.1a). Inspired by this work, they investigated a novel dearomatization reaction: Pd-catalyzed intramolecular *ipso*-Friedel–Crafts allylic alkylation of phenols for constructing spiro[4.5]cyclohexadienones [6] (Scheme 9.1b). As shown in Scheme 9.2, the catalytic cycle is initiated with oxidative addition of allylic carbonates to the Pd(0) and generate a π-allyl-Pd intermediate. Subsequently, assisted by the endogenous methoxide anion, the *para*-position of phenol's nucleophilicity increases, and forms the carbon–carbon bond via transition states **TS-1** or **TS-2** to give the spirocyclic product. The observed diastereoselectivity could be explained by steric repulsion between the π-allyl unit and the *ortho*-substituent (R^1) in **TS-1**. Notably, when chiral ligand L1 was used for this transformation, 89% ee, 9.2 : 1 dr were obtained (Scheme 9.3). In 2012, they continued to explore

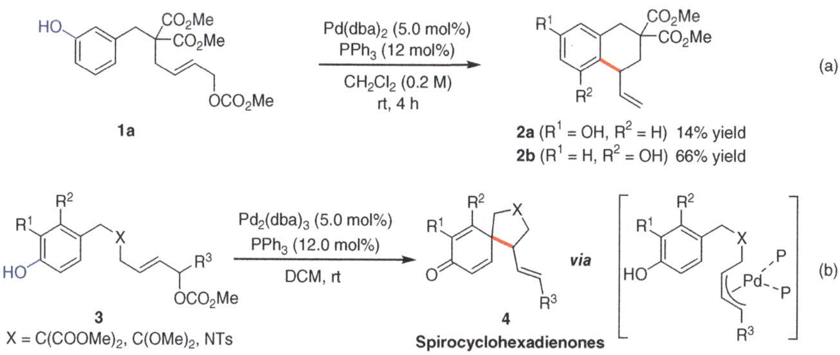

Scheme 9.1 Intramolecular dearomative reaction of *para*-substituted phenol.

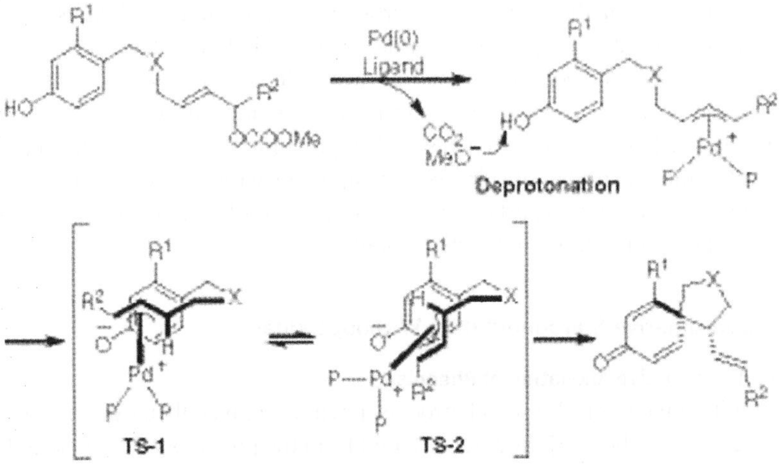

Scheme 9.2 Proposed dearomative reaction pathway.

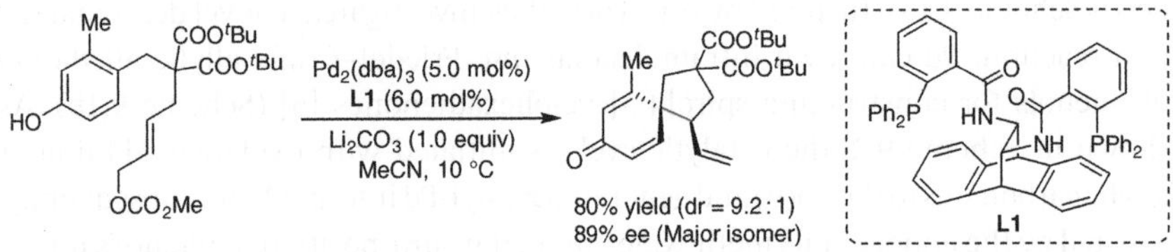

Scheme 9.3 Preliminary enantioselective construction of all-carbon spirocenter.

asymmetric transformation of this reaction and try a large number of chiral ligands, and then optimized the reaction conditions. Finally a novel catalytic asymmetric synthetic method was developed to afford spirocyclohexadienones with an all-carbon quaternary center [7]. The C-3 site is not only substituted by the methyl group, but also by the chloride or methoxy. Besides, 98% yield and 71% ee were obtained when used naphthols as starting material with the additive Li_2CO_3. Furthermore, the spiro[4.5]decane core architecture was found in alkaloid and terpene natural products, such as magellanine, lycojaponicumin B, and conidiogenone (Scheme 9.4). Donahue utilized the phenolic dearomatization strategy to construct the spiro[4.5] decane core and then the tricyclic cores were demonstrated utilizing a 5-exo-trig free radical cyclization of a primary carbon radical onto a cyclohexadienone generated with tri-*n*-butyl-germanium hydride [8].

The catalytic asymmetric reaction for constructing spirocyclic molecules is one of the most challenging topics in synthetic organic chemistry. In 2011, You and coworkers developed the first Ir-catalyzed intramolecular direct asymmetric

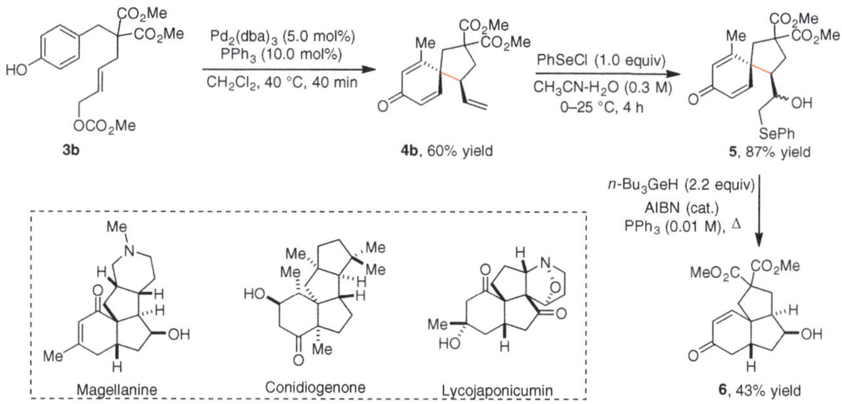

Scheme 9.4 Synthesis of spiro[4.5]decane core architecture via dearomative spiroannulation.

allylicdearomatization of phenols for constructing enantiopure spirocyclohexadienones [9], overcoming the possible *O*-allylation and other *C*-allylation byproducts. The reaction proceeds in the presence of 2 mol% $[Ir(cod)Cl]_2$, 4 mol% L2, and 2.0 equiv Li_2CO_3 in THF or Dioxane and the results are summarized in Scheme 9.5: not only the N-Ts linked five-member-ring formation but also six-member-ring construction were achieved in good to excellent yields and ee values. In addition, the reaction proceeded smoothly for the carbocyclic ring formation.

Soon after, You and coworkers turned their attention to *ortho*-substituted naphthols. In the presence of $[Ir(dbcot)Cl]_2$ and a new THQphos ligand L3, this reaction proceeds successfully in high chemo-, diastereo-, and enantioselectivity [10] (Scheme 9.6). In the reaction, not only β-naphthols behaved very well, α-naphthols which was a more challenging substrate could be tolerated with a moderate C/O alkylation ratio. Furthermore, a series of novel spiro[5.5] skeletons were synthesized which greatly enriched the spirocyclic skeleton database. Notably, the spirocyclic product could be transferred to ether **12** easily, and then complex alkaloids and terpenes could be synthesized.

In order to extend the method to more kinds of reactions and to obtain more helical skeleton compounds, Hamada and coworkers continued to explore the dearomatization of phenols linked with alkynes to investigate the feasibility [11] (Scheme 9.7). Firstly, the reaction proceeded rapidly through a similar process with the above reaction giving the initial product **14**. Then, a rearomatization-assisted oxidative of **14** to Pd(0) subsequently occurs and produce η^3-propargyl palladium(II) intermediate **15** through η^1-propargyl palladium(II). The above mechanism has been further proved by the direct transformation from compound **14** to **16**.

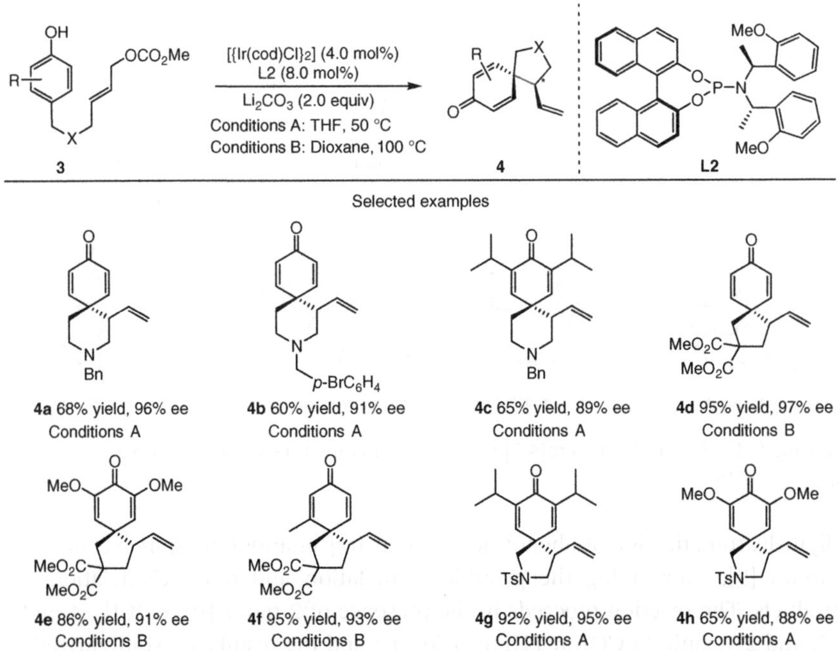

Scheme 9.5 Iridium-catalyzed asymmetric allylic dearomatization of phenols.

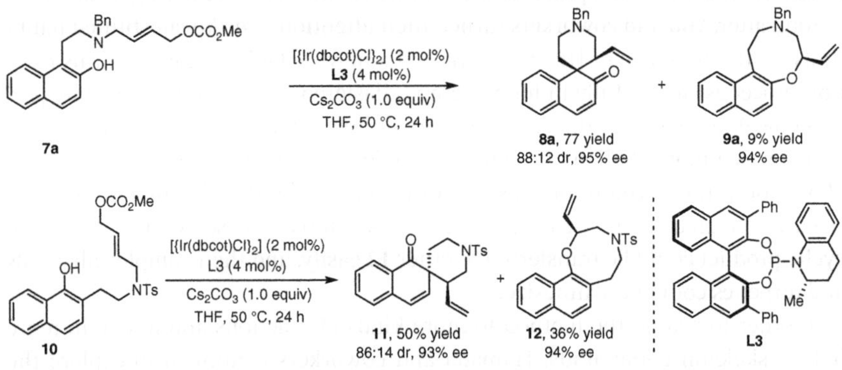

Scheme 9.6 C-alkylation for *ortho*-substituted naphthols.

To make the dearomatization more efficient, Hamada and coworkers also tried intermolecular reactions. They used phenols containing diene fragments and iodobenzene as substrates and the two components were connected via Heck reaction firstly, then intramolecular ipso-Friedel–Crafts allylic alkylation occurred [12]

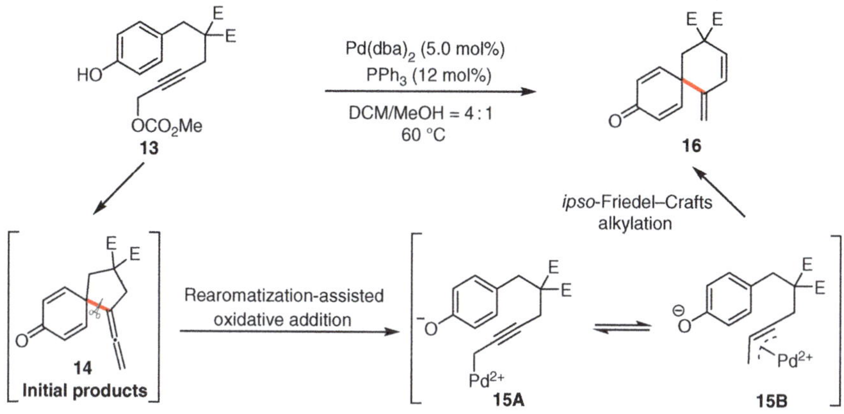

Scheme 9.7 Intramolecular dearomative reaction of alkyne-substituted phenol.

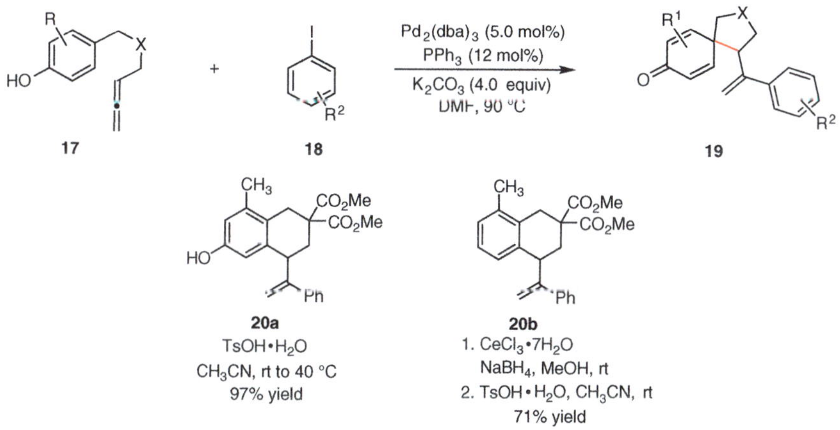

Scheme 9.8 Intermolecular dearomative spiroannulation of phenol and iodobenzene.

(Scheme 9.8). In the catalyst of 5 mol% palladium and 12 mol% PPh_3, a novel variety of spirocycles has been accessed in good to excellent yields which have been applicable to the synthesis of tetrahydronaphthalene derivatives.

9.4.1.2 Dearomative Arylation of Phenols

In 2011, Buchwald and coworkers developed palladium-catalyzed arylative dearomatization for the construction of a series of spirocyclic products in good to excellent yield [13]. As shown in Scheme 9.9, aryl phenols were reacted in Dioxane

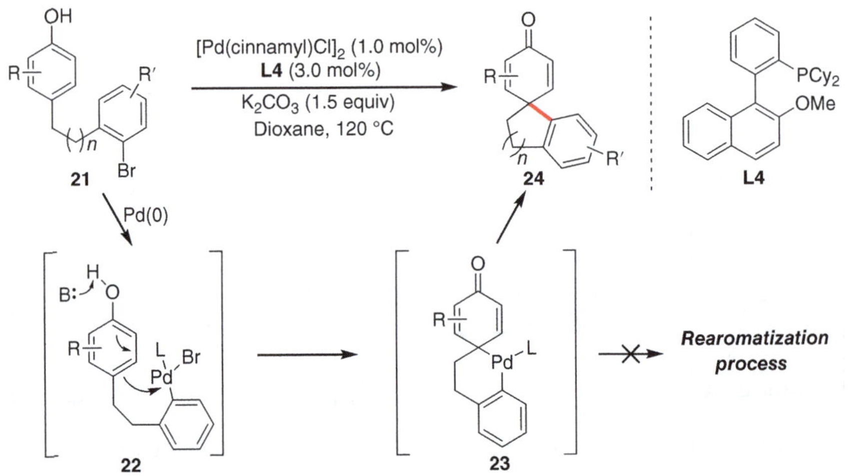

Scheme 9.9 Strategy for dearomative arylation of phenols.

(0.2M) at 120 °C in the presence of 1 mol% [Pd(cinnamyl)Cl]$_2$, 3 mol% **L4**, and 1.5 equiv base for 16 hours. The challenge of this reaction is to avoid the diaryl ether formation from a competitive intermolecular C–O cross-coupling reaction while the dearomatization of palladacycle **23** is favorable. In addition, when using methyl- or benzyl-protected derivatives of phenols, trace or no product was observed. Notably, deprotonation is required in order to produce nucleophile attack at the Pd(II) center. It is worth noting that the procedure offers possibility for enantioselectivity when using asymmetric ligands. A catalyst system with Kenphos L5 enabled the formation of **26a** in 91% yield and 65% ee. When L5 was replaced with L6, the yield and ee were further improved. In 2017, You and coworkers also reported arylative dearomatization using β-naphthols hindering the *O*-arylation by-products and a series of structurally diverse spirocarbocycles were obtained efficiently [14] (Scheme 9.10).

The erythrina alkaloids have curare-like and hypnotic activity, and many of them display biological activities (Figure 9.1). So, exploring methods for constructing this skeleton is very necessary. In 2014, You and coworkers carefully designed 5-hydroxyl indoline as reaction materials and developed an intramolecular dearomative arylation reaction, affording tetracyclic sterically congested spiroamines in excellent yields, which opens up a new pathway for the synthesis of alkaloids containing such skeletons (Scheme 9.11). They have successfully synthesized alkaloid (±)3-demethoxyerythratidinone which shows that this method has great potential application in the total synthesis. In addition, primary asymmetric results using chiral TADDOL-derived phosphoramidite **L8** provide good

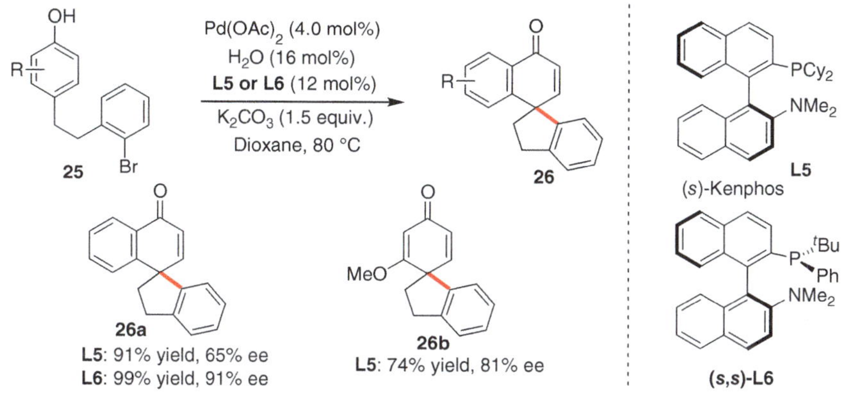

Scheme 9.10 Preliminary results of asymmetric dearomative arylation of phenols.

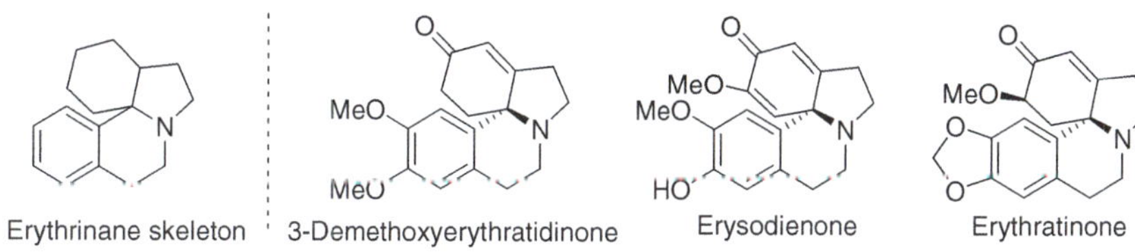

Figure 9.1 Natural erythrina alkaloids.

routes for asymmetric synthesis of natural products and biological molecules contained sprioamines skeletons (Scheme 9.12) [15].

In 2015, Luan and coworkers reported a microwave-assisted Pd(0)-catalyzed vinylative dearomatization reaction of β-naphthols with internal alkynes to access a class of valuable spirocyclic compounds [16] (Scheme 9.13). This work was initiated with substrate **29** as the coupling partner which might undergo a cyclization reaction with alkynes involving **30** as the key intermediate. In the beginning, the chemoselectivity for the prospective transformation was quite unpredictable, because potential side reactions usually occur quite easily to form dibenzofuran **33**. Suitable conditions were also discovered to form the favorable cyclization product between palladacycle **30** and alkyne which proceed much faster than the possible direct reductive elimination of **30**, thus providing spirocyclic **32** as the desired product with excellent chemoselectivity. Notably, this intriguing structural moiety is often present in natural products and functional materials.

Encouraged by the above work, Luan and coworkers embarked to develop a catalytic DYKAT of racemic biaryls to access a new type of enantioenriched compound

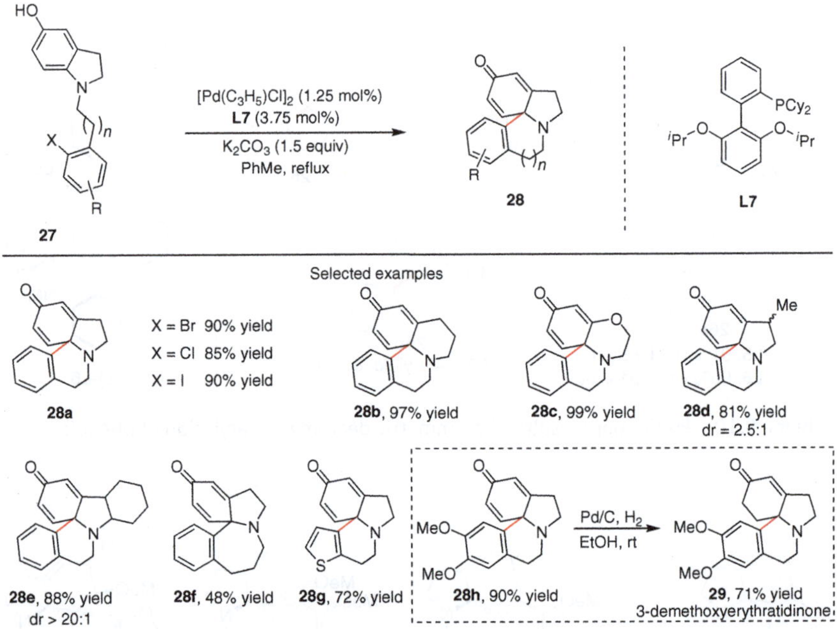

Scheme 9.11 Pd(0)-catalyzed intramolecular arylative coupling of 5-hydroxyl indolines.

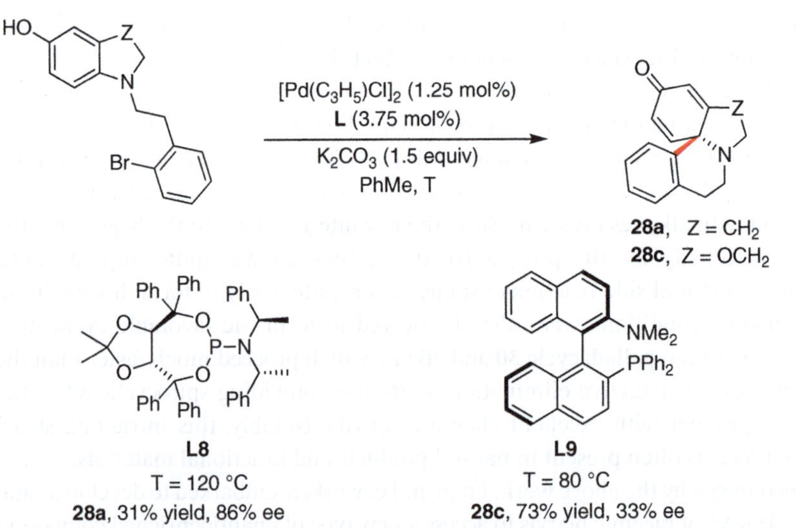

Scheme 9.12 Preliminary asymmetric studies for constructing erythrinane skeleton.

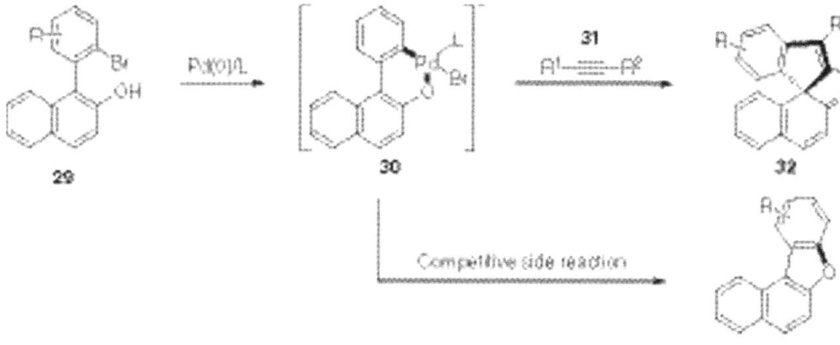

Scheme 9.13 Proposed one-step construction of spirocarbocycles and potential challenge. *Source:* Based on Zheng et al. [16].

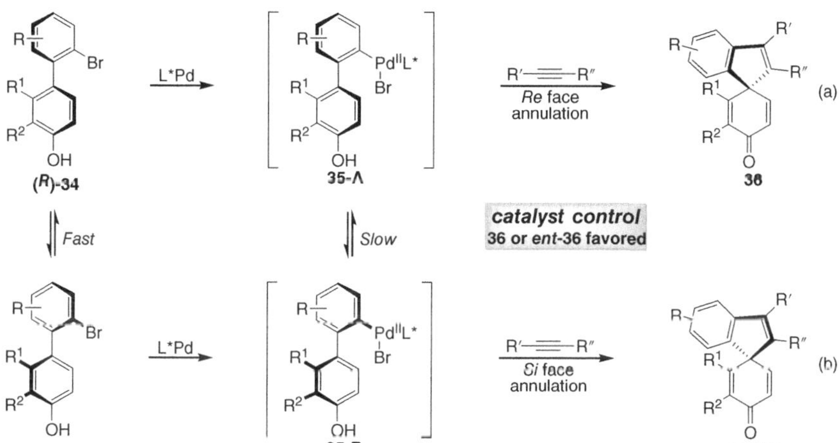

Scheme 9.14 A DYKAT of phenolic derivative via axial-to-central chirality transfer.

bearing a sp^3 stereocenter. The synthetic value of these racemic reactions has been reflected by the capacity of transferring one biaryl scaffold into another spirocyclic framework bearing an all-carbon quaternary stereocenter, which offers a potential opportunity for enabling the DYKAT of biaryls via axial-to-central chirality transfer. The phenolic derivative **34**, which interconverts rapidly between *(R)*-**34** and *(S)*-**34**, would be an ideal candidate for the anticipated DYKAT (Scheme 9.14). However, catalyst-controlled asymmetric dearomatization of phenolic derivatives being restricted from a single enantioface is very challenging, and to date, only a handful of leading example has been disclosed [17]. In this work, chiral Pd-NHC complexes were found to catalyze the dynamic kinetic asymmetric spiroannulation of 4-(2-bro moaryl)-naphthalen-1-ols (or 2′-bromo-[1,1′-biphenyl]-4-ols) with internal alkynes, affording a series of enantioenriched spirocyclic products bearing an all-carbon quaternary stereocenter in good yields (up to 95%) with excellent enantioselectivities (up to 97% ee) (Scheme 9.15).

Nevertheless, it is important to note that the coupling partners for dearomative [3+2] spiroannulations were limited to internal alkynes. From the viewpoint of synthetic and mechanistic relevance, this transformation called for exploring the performance of unsaturated partners other than alkynes. Soon after, a novel Pd(0)-catalyzed intermolecular carbocyclization of phenol-derived biaryls with 1,3-diynes through a sequence of oxidative addition to the C–I bond, regioselective olefin insertion, and allylative dearomatization has been implemented [18]. This method provides a broad range of attractive spirocyclic compounds bearing two contiguous tertiary/quaternary carbon centers in good yields with excellent chemoselectivity and regioselectivity. Moreover, preliminary results indicate that asymmetric control of this process is feasible with chiral ligands. Then, they extended the conjugated partners to diynes, and this potential [3+2] spiroannulation proceeded efficiently, involving a key step of regioselective allylative dearomatization of phenol by overcoming undesired β-hydride elimination which provided a variety of highly valuable spiro[cyclohexane-1,1′-indene]-2,5-dien-4-ones [19]. Preliminary asymmetric studies showed that high enantioselectivity could be realized by using a commercially available PHOX ligand (Scheme 9.16).

Transition-metal-catalyzed polycyclization reactions represent some of the most powerful and efficient approaches for creating highly functionalized, architecturally complex molecules from readily available precursors in one step. Luan and coworkers have developed a novel and efficient palladium(0)-catalyzed dearomative cyclization of bromophenols with tethered diynes for constructing new polycyclic architectures bearing an all-carbon quaternary center [20]. By using the *para*-bromophenols as the coupling partner for (1,*n*)-diynes ($n = 6$–9), the reaction undergoes a [2+2+1] spiroannulation, thus leading to a class of tricyclic molecules containing spirofused ring motifs. As shown in Scheme 9.17, the carbon tethered with the 1,6-diynes could also tolerate a dicyanide (**41d**) and diketone (**41e**). Moreover, the 1,7-diyne **41f** was also tolerated, thus giving rise to **42f** in 88% yield.

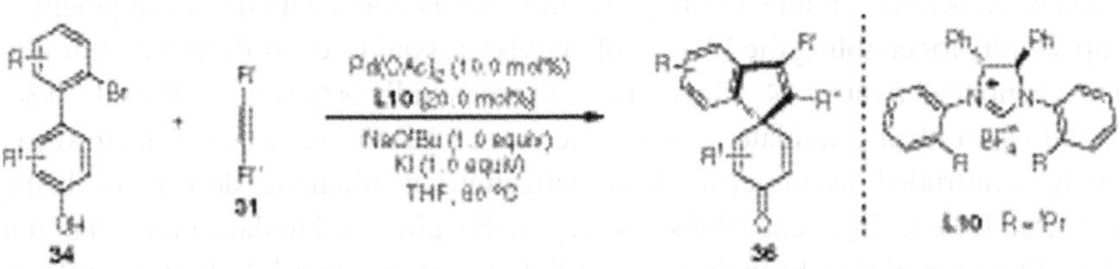

Scheme 9.15 Palladium-catalyzed dynamic kinetic asymmetric transformation.

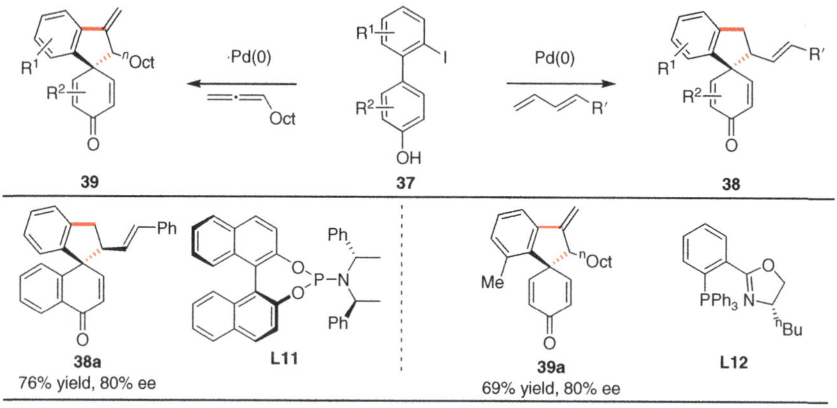

Scheme 9.16 [3+2] Spiroannulation of phenol-derived biaryls and diynes.

Notably, nitrogen- or oxygen-tethered (1,6)-diynes were well tolerated to provide the tricyclic compounds (**42g–h**) with a heterocyclic. Gratifyingly, the size of this five-membered heterocycle could be easily increased by lengthening the tether of diynes. Remarkably, a series of commercially available *para*-bromophenols were then tested and the results indicate that their corresponding reactions proceed smoothly to provide the anticipated products.

9.4.2 C–H Bond Activation/Dearomative Spiroannulation Reactions

Transition-metal-catalyzed C–H functionalization has unimpeachably emerged as a powerful tool to render shorter synthetic routes for complex molecules by eliminating the need for prior preparation of activated substrates [21]. In 2013, Luan and coworkers firstly launched a dual activation research program focusing on the simultaneous activation of C–H bond and dearomatization by Ru(II)-catalyzed oxidative annulation reaction of 1-aryl-2-naphthols with alkynes [22]. They carried out a series of deuterium-labeling experiments to shed more insights on the reaction mechanism. The results of the reaction with D_2O indicates that cycloruthenation is irreversible and the kinetic isotope effects observed in both intramolecular ($k_H/k_D = 3.8$) and intermolecular ($k_H/k_D = 5.0$) competition experiments suggest that the C–H bond cleavage is likely involved in the rate-limiting step and might follow the concerted metalation/deprotonation mechanism (Scheme 9.18). Based on the above results, a possible mechanism is proposed in Figure 9.2: this reaction proceeded via sequential deprotonation, cleavage of C–H bond and migratory insertion of the alkyne to form a rather strained eight-membered intermediate **47**. The most important step of this reaction should be the unique enolketo

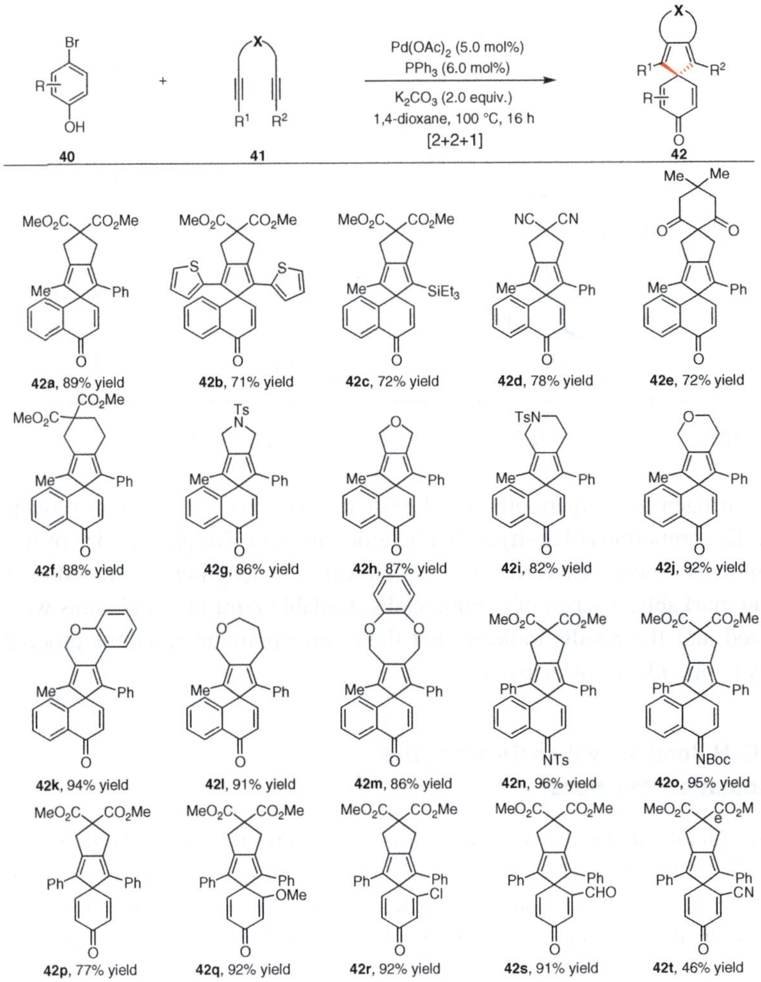

Scheme 9.17 Formal [2+2+1] cycloaddition routes to new tricycles.

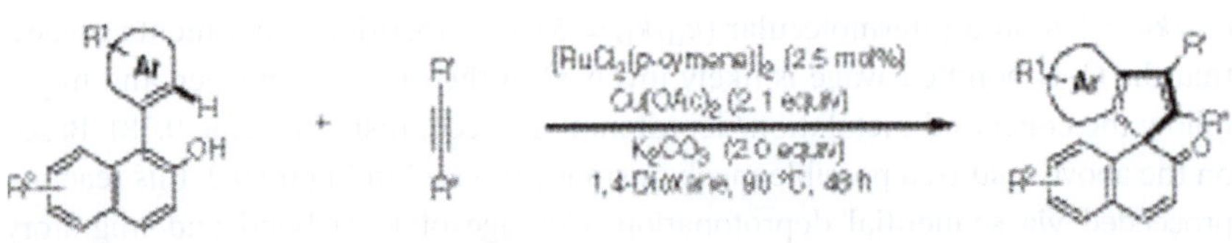

Scheme 9.18 Ru^{II}-catalyzed dearomative spiroannulation of naphthols via C(sp^2)-H activation.

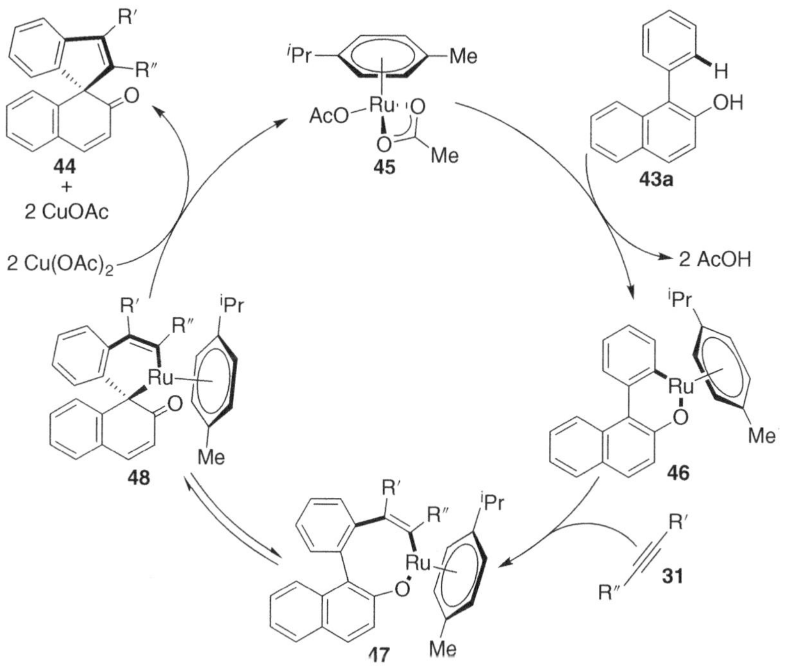

Figure 9.2 Proposed mechanism of C–H bond activation/dearomative reaction.

tautomerization of the phenol ring to deliver the key intermediate **48**, which is a C-bound Ru enolate; finally, reductive elimination takes place to generate a new class of spirocyclic compounds bearing the spiro[indene-1,1′-naphthalen] skeleton and regenerates Ru^{II} catalyst **45**.

The energy barrier for breaking the aromaticity of phenols is significantly higher than that of naphthol compounds [23]. In the previous study, the phenolic fragment was restricted on 2-naphthols, and the reactions with 2-phenylphenol could not produce the desired spirocyclic molecule in a synthetically useful yield (<5%) under the reported reaction conditions. Fortunately, when the solvent was replaced with ^{t}AmOH, they overcome the substrate limitation and realized the dearomatization of 2-arylphenols with alkynes acquiring a class of highly functionalized spirocyclic compounds (Scheme 9.19) [24]. Notably, 2-arylphenols or 2-arylnaphthols was pushed forward from 1-bromo-2-naphthol to **43** by the palladium(0). To develop a more economical reaction featuring the dual activation concept, Luan and coworkers proposed a concise route for the synthesis of spirocyclic compounds **51** from 2-naphthol via a one-step metal-catalyzed oxidative dearomatizing reaction (Scheme 9.20). The key point of this reaction is the replacement of intermediate **46** with a metallacycle **50**, which could be generated from 2-naphthol and one equivalent of alkyne by palladium(II). This process

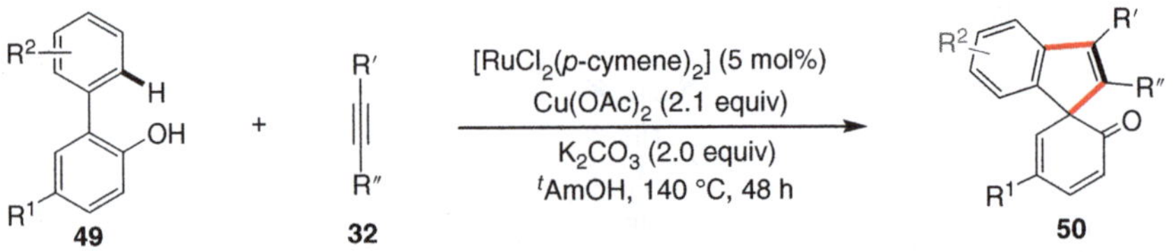

Scheme 9.19 Spiroannulation of 2-arylphenols via C–H activation/dearomative strategy.

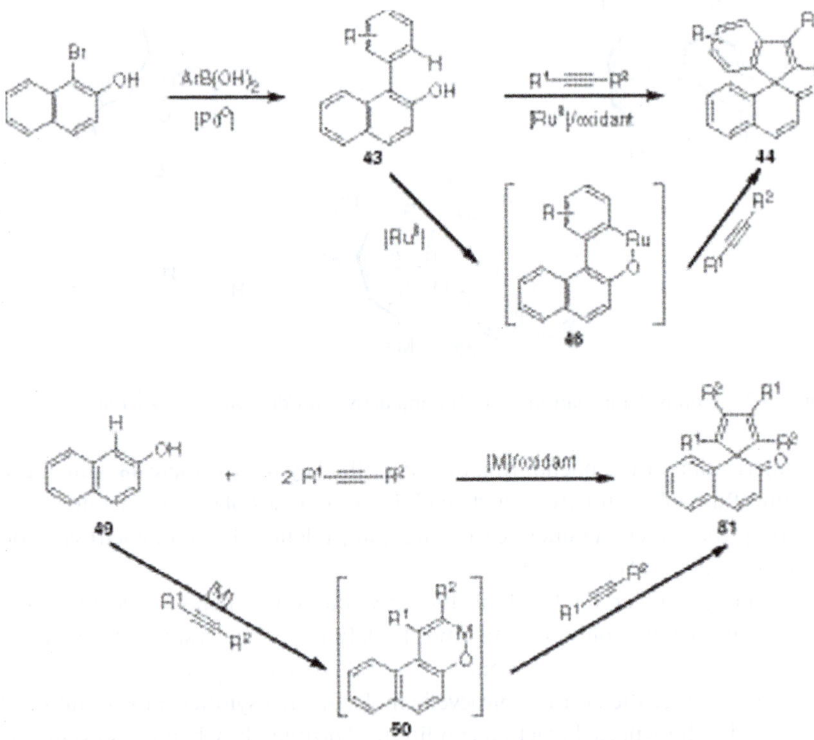

Scheme 9.20 Pd-catalyzed oxidative dearomatization of free naphthols.

relies on the C–H functionalization and naphthyl ring dearomatization at the 1-position of 2-naphthols to provide a variety of spirocyclic compounds. Using alkyl–aryl alkynes as the coupling partners led to regioisomeric mixtures in favor of the head-to-tail isomer bearing a quaternary carbon stereocenter (Scheme 9.21) [25], In this strategy, Wang and coworkers changed phenols to N-aryl ureas and the corresponding spirocyclic imine derivatives were obtained in good to excellent yields [26].

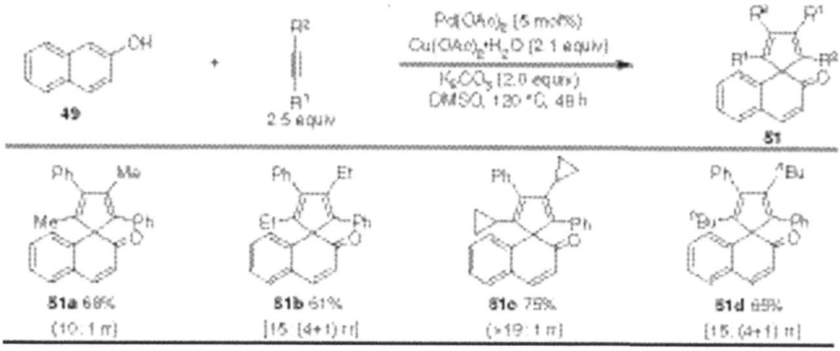

Scheme 9.21 [2+2+1] Spiroannulation of unsymmetrical alkynes.

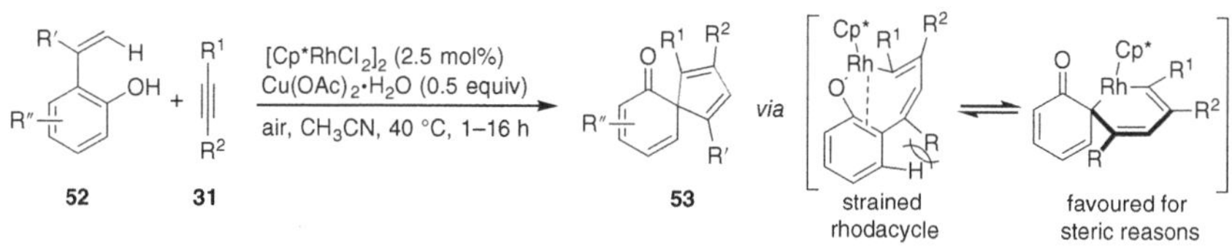

Scheme 9.22 Spirocyclopentadienes of 2-alkenylphenols and alkynes.

In an elegant extension of this methodology, Mascareñas and Gulías explored the activation of olefinic or aliphatic C–H bonds. Firstly, they used *ortho*-vinylphenols(R = H) reacting with alkynes in the presence of Rh(III) catalysts to give benzoxepine products [27]. Inspired by this work, they explored the performance of alkenylphenol derivatives equipped with a substituent at the internal position of the alkene which generated a steric clash that favors a reductive elimination, providing a versatile and efficient approach to highly appealing spirocyclic skeletons (Scheme 9.22) [28]. At the same time, Lam and coworkers also realized the dearomative spiroannulation of 2-alkenylphenols in this strategy [29]. In addition, enynes were also been tested, this reaction proceeds with alkynes in high chemoselectivity and delivered a series of spirocyclic compounds which could be further transformed (Scheme 9.23). In order to expand the practicability of this method, they then explored more challenging 1,3-diynes which might have occurred in Heck reaction [30]. Instead of using catalysis Rh or Ru, the reaction proceeds smoothly under Pd^{II} and $Cu(OAc)_2$ as the oxidant. The spirocyclic compounds were not also accessed by oxidative annulation of 2-aryl cyclic 1,3-dicarbonyl compounds but also by dearomatization of 1-aryl-2-naphthols. In 2015, You and coworkers firstly developed asymmetric annulative dearomatization reaction of β-naphthols with alkynes via C–H activation by chiral Rh catalyst which was developed by the Cramer Group. Various highly enantioenriched

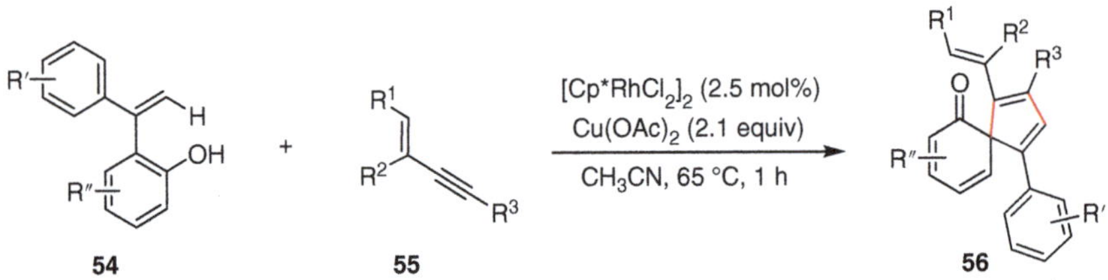

Scheme 9.23 Spirocyclopentadienes of 2-alkenylphenols and conjugated enynes.

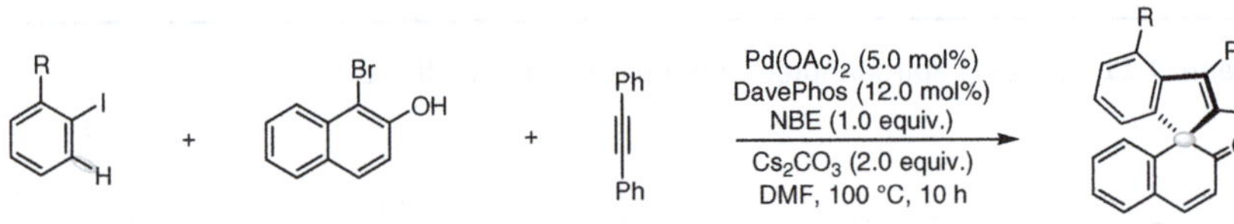

Scheme 9.24 Three-component dearomatizing [2+2+1] spiroannulation.

spirocyclic enones bearing an all-carbon quaternary stereogenic center were constructed in good to excellent yields and enantioselectivity [31].

The reported C–H activation/dearomative reactions are inevitably using high valence metal and oxidant. However, Catellani reaction could activate the C–H bond of aryl iodide with the assistance of NBE which was originally discovered by Catellani et al. [32]. Luan and coworkers developed a novel palladium(0)-catalyzed dearomatizing [2+2+1] spiroannulation of 1-bromo-2-naphthols with aryl iodides and alkynes for constructing spiro[indene-1,1′-naphthalen]-2′-ones [33]. This three-component cascade reaction was realized through consecutive Catellani-type C–H activation, unsymmetrical biaryl coupling, alkyne migratory insertion, and arene dearomatization (Scheme 9.24). Notably, the two-component reactions of **59** and 1-bromo-2-naphthol, which were heated in DMF at 130 °C for 10 hours, showed excellent performance when using 10.0 mol% of $Pd(OAc)_2$, 20.0 mol% of $P(o\text{-}Tol)_3$, 1.0 equivalent of NBE, and 2.0 equiv of Cs_2CO_3, affording the desired product **60a–e** in 52–68% yields (Scheme 9.25). Remarkably, the main polycyclic skeleton of dalesconols A and B could be rapidly assembled from two readily available starting materials in one step through the formation of three C–C bonds (Scheme 9.25). In this strategy, two readily available starting materials, phenol-based biaryls and bromoalkyl alkynes, were efficiently coupled by forming three C–C bonds, leading to a wide range of highly valuable polycyclic frameworks [34] (Scheme 9.26).

Above all, C–H activation/dearomative reactions are limited in C(sp^2)-H activation; the development of such processes involving the functionalization of alkane C(sp^3)-H bonds, which lack direct interaction with a metal center [35], is still a

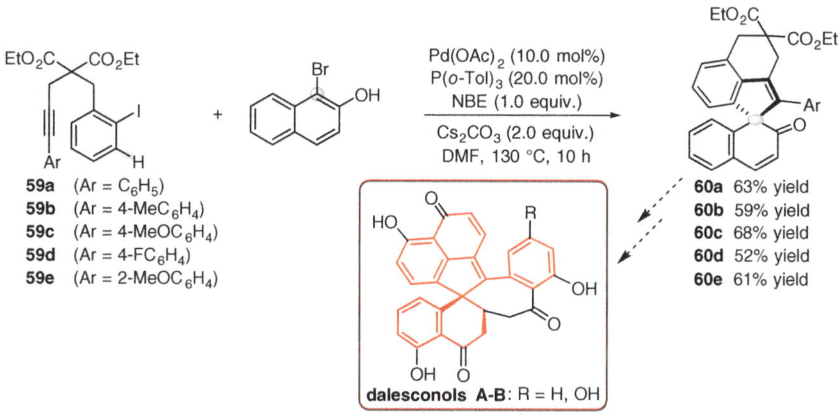

Scheme 9.25 Preliminary studies on the synthetic application of dearomatizing [2+2+1] spiroannulation.

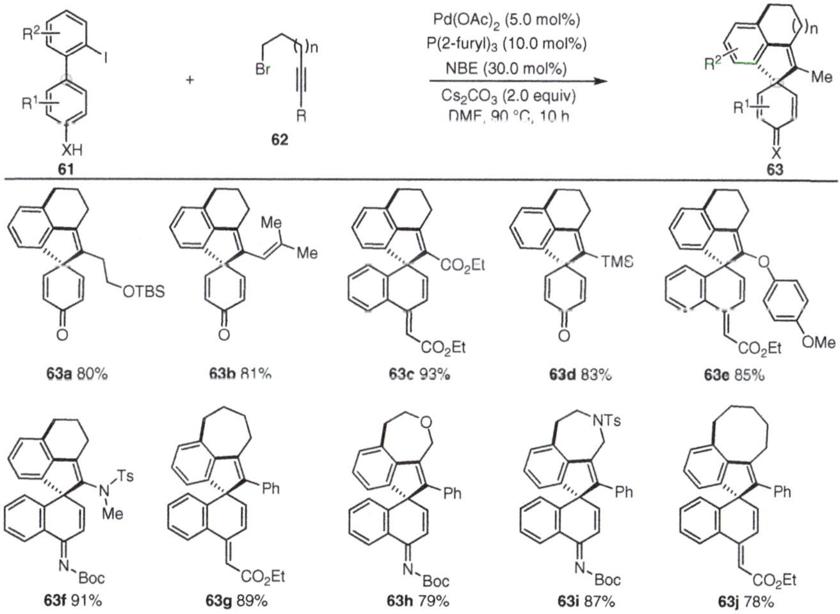

Scheme 9.26 Pd/NBE-catalyzed C–H activation/arene dearomatization reaction. *Source:* Nan et al. [34].

daunting challenge. In 2019, Luan and coworkers developed an unprecedented example of palladium-catalyzed [4+1] spiroannulation through a C(sp^3)-H activation/naphthol dearomatization domino process [36] (Scheme 9.27). The key to success of this proposal is the involvement of an electron-rich palladacyclic

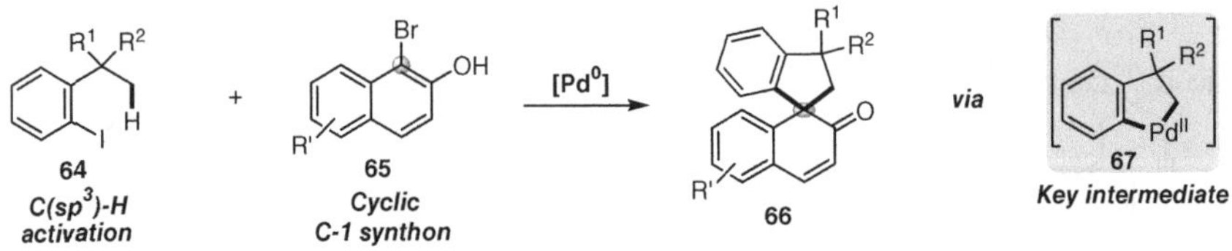

Scheme 9.27 [4+1] Spiroannulation by C(sp^3)-H activation and naphthol dearomatization.

intermediate **67** for enabling the cross-coupling of two electrophilic aryl halide components. This bimolecular domino reaction of two aryl halides was realized through a sequence of cyclometallation-facilitated C(sp^3)-H activation, biaryl cross-coupling, and naphthol dearomatization, thus rendering the rapid assembly of a new class of spirocyclic molecules in good yields with broad functional-group tolerance. In addition, the results of kinetic isotope effects (k_H/k_D = 5.8 and 5.7) suggest that primary methyl C(sp^3)-H bond cleavage was most likely involved in the rate-determining step.

9.5 Dearomative Spiroannulation of Indoles

Indoles are important heterocyclic compounds and the conjugated system of atomic electrons makes N and C3 more nucleophilic. In general, many functional reactions occur in indoles C3, such as aromatic electrophilic substitution reaction, Michael addition reaction, Mannich reaction, and so on. Over the years, researchers have found that many indolin heterocyclic compounds exhibit anticancer, anti-inflammatory, and blood-lowering properties. As a result, these compounds have potential medical and medicinal value. In view of the importance of indoles' aromatization as a synthesis step, a series of in-depth studies have been carried out in recent years and many ingenious strategies have been developed to achieve this goal. Among them, indoles C2 and C3-cyclocyclization is one of the favored methods. While achieving the goal of aromatization, the construction of complex multi-parallel ring framework can be realized, thus effectively shortening the reaction steps and reaction time. The spiroindolenine and spiroindoline units are privileged heterocyclic motifs that form the structural core for a large family of alkaloid natural products such as koumine, perophoramidine, and communesins.

9.5.1 Intramolecular Dearomative Allylic Alkylation

The key step for constructing spiroindolenine and spiroindoline from indoles is to construct a quaternary stereocenter via a dearomatization process. In 2010, You and coworkers firstly reported Ir-catalyzed highly enantioselective synthesis of spiroindolenines via indole's dearomative allylic alkylation [37]. Many chiral ligands have been tested, L14–L17 could catalyze the reaction in excellent ee but with moderate dr, and L18-L19 were not effective for the substrate, and finally L13 gave the best result in terms of yield, ee and dr ratios. The spiroindolenine derivatives were obtained in excellent yields with up to >99/1 dr and 97% ee (Scheme 9.28). In addition, they have tested Bandini's substrate by shortening one carbon of **68** in this catalyst system and tetrahydro-γ-carboline product was obtained. The experiment result explained that construction of spiroindolenine is likely caused by the favorable six-member ring product by reacting at the C-3 position over the seven-member ring product at the C-2 position. However, the protecting group on the linking N atom was limited to electron-donating groups. To construct more spiroindolenines, they transformed to ruthenium catalytic system and tremendous developments have been made. The reaction proceeded smoothly with a broad range of substrates including the ones not tolerated in the previous work, such as the allyl carbonates with electron-withdrawing protecting group [38].

Intrigued by the previous work, they tested carbon-tethered substrate to examine if five-membered spiroindolenine could be constructed. When the reaction was performed using 2 mol% Ir catalyst, 4 mol% of chiral phosphoramidite L20, and 2 equiv of Cs_2CO_3 in dioxane, five-membered spiroindolenine could be successfully synthesized. In addition, some products caused purification difficulties, so the corresponding products were subjected to $NaBH_3CN$ reduction, affording a series of indolines which contained a quaternary carbon center (Scheme 9.29). Interestingly, when the

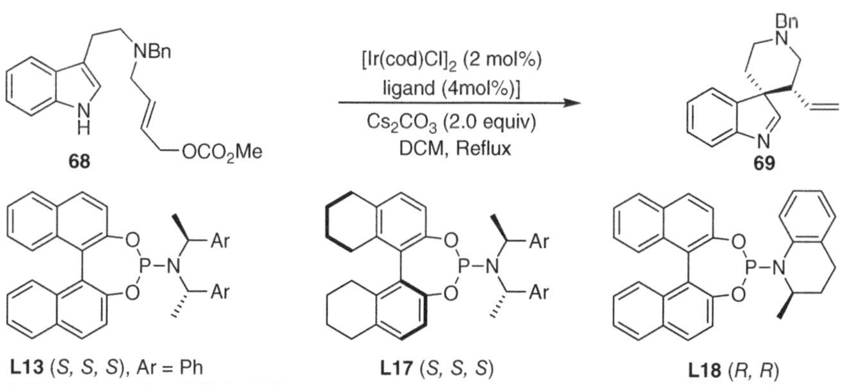

L13 (*S*, *S*, *S*), Ar = Ph
L14 (*S*, *S*, *S*), Ar = 2-MeO-C_6H_4
L15 (*S*, *S*, *S*), Ar = 2-naphthyl
L16 (*S*, *S*, *R*), Ar = Ph

L17 (*S*, *S*, *S*)

L18 (*R*, *R*)
L19 (*R*, *S*)

Scheme 9.28 Ir-catalyzed asymmetric allylic alkylation for constructing spiroindolenines.

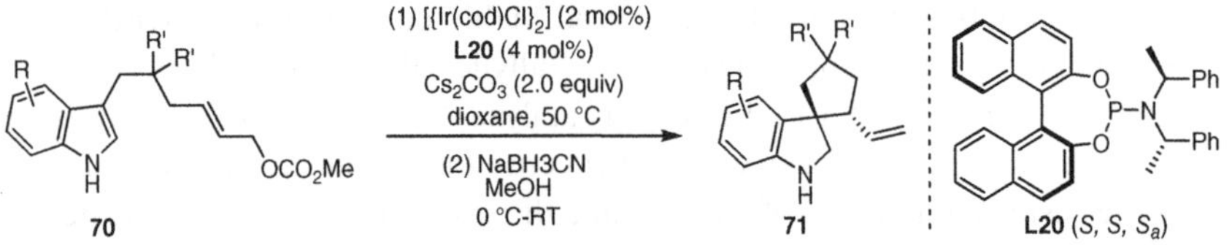

Scheme 9.29 Allylic dearomatization for the synthesis of spiro cyclopentene-1,3′-indoles.

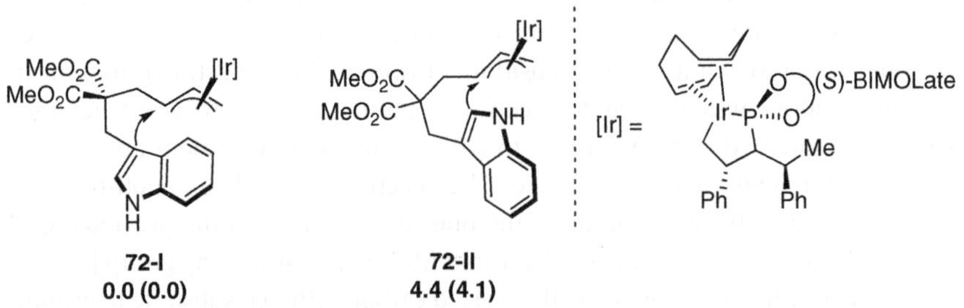

Figure 9.3 The energetic barrier of transition states.

spiroindolenine was treated with the catalyst of TsOH, the stereospecific migration could exactly occurr which led to enantioenriched 2,3,4,9-tetrahydro-1*H*-carbazoles. Finally, a DFT calculation was carried out to explain the above experiment result. As shown in Figure 9.3, C3 is the favored process through comparing energetic barrier between **72-I** and **72-II**. Furthermore, steric hindrance is also the influence factor, the transition states **72-I** which avoid steric congestion of one phenyl group of L20 could probably result in C3 attack [39]. In 2017, they developed efficient CADA reactions by applying desymmetrization strategy with the aid of Ir-catalyzed allylic dearomatization of indoles. The six-membered-ring spiroindolenines containing three contiguous stereogenic centers and an all-carbon quaternary center were obtained in excellent enatioselectivities [40].

9.5.2 Intramolecular Dearomative Arylation

Despite the significant importance of the target spirocycles, the 3-substitued indoles have been limited to the allylic alkylation reaction. You and coworkers reported the dearomative arylation of 3-substituted indole via palladium(0)-catalyzed cross-coupling reaction and various spiroindolenines derivatives have been obtained in good to excellent yields. In addition, asymmetric transformation is preliminarily realized with chiral ligands (Scheme 9.30) [41].

Though many reports about dearomatization of indoles have been reported, the aforementioned reactions mainly focused on dearomative functionalization

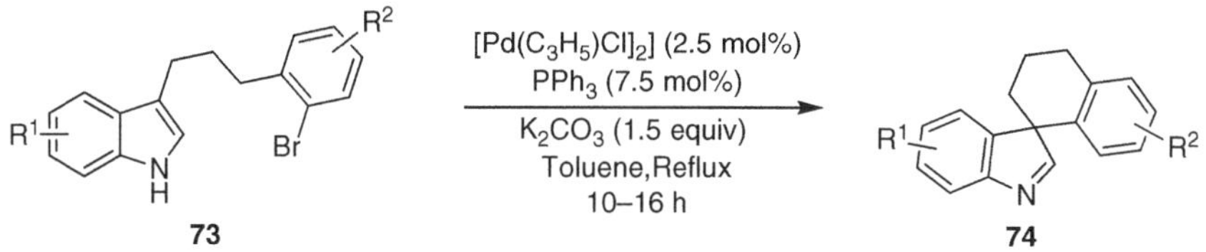

Scheme 9.30 Dearomative arylation of indoles for spiroindolenine derivatives.

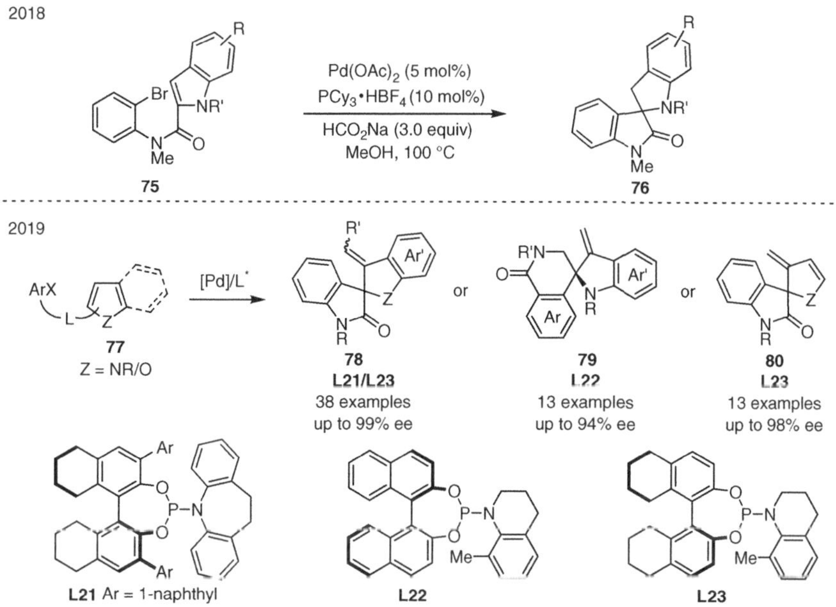

Scheme 9.31 Dearomatization of C2-substitued indoles for spirooxindoles.

of N- or C3-substituted indoles. In order to build more diverse skeletons of spirooxindoles, Jia and coworkers developed dearomative reductive-Heck reaction of C2-substituted indoles, which provides a reliable approach to unique 3,2′-spiropyrrolidine. Moreover, C2-substitued benzofuran was also tested in this reaction system, and the desired spiro-product was successfully acquired [42]. Soon after, with the help of new BINOL- or H8-BINOL-based chiral phosphoramidite ligands and a chiral diphosphine ligand, they realized the construction of chiral spiro and benzofused heterocycles bearing nitrogen-/oxygen-substituted quaternary carbon stereocenters (Scheme 9.31) [43]. They extended the method to indoles, pyrroles, benzofurans, and furans, showing a broad scope of the present protocol. More importantly, this method was applied to the synthesis of the core skeleton of phalarine efficiently (Scheme 9.32) [44].

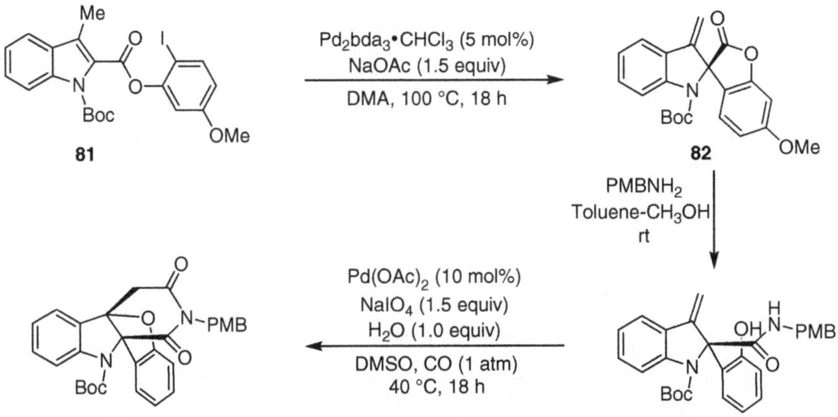

Scheme 9.32 Synthesis of the core structure of phalarine.

Though the former methods have constructed varieties of spiroindolenine, the disadvantage of this method is the difficult accessibility of halogenated indole precursors. In 2012, Van der Eycken reported the direct hydroarylation of the triple bond relying on a cationic gold catalysis [45]. In this catalyst system, four types of products have been accessed including spiro cyclized indoles **91** and **92** in low yields which might be considered as an indirect evidence for the mechanism reported by Echavarren previously [46]. The triple bond activated by the metal catalyst first reacts with the alkynes, the most reactive position of the indole core via exo or endo-mode resulting in the formation of spirofused intermediates **87** or **88**, respectively. Subsequent proton transfer in these intermediates could provide **91** and **92** with concomitant demetalation (Scheme 9.33). In 2015, Taylor and Unsworth developed a general, high-yield strategy for the conversion of indoles into complex, spirocyclic enones using low loading of simpler silver(I) or copper(II) salts [47]. Notably, asymmetric dearomatizing spirocyclization reactions can be achieved in high yield and good e.r. when treated with chiral phosphoric acids (Scheme 9.34). Soon after, they also briefly explored another strand of scaffold diversity starting from more functionalized ynones. Tetracyclic scaffolds could be formed with complete diastereoselectivity by a telescoped spirocyclization/nucleophilic addition sequence in one pot [48] (Scheme 9.35). To obtain a broad array of functionalized spirocyclic indole derivatives, 2-haloindole could be precursors for accessing indoleninyl halides. These products have been shown to react via three distinct modes: hydrolysis, nucleophilic substitution, and cross-coupling, further expanding their synthetic utility [49] (Scheme 9.36).

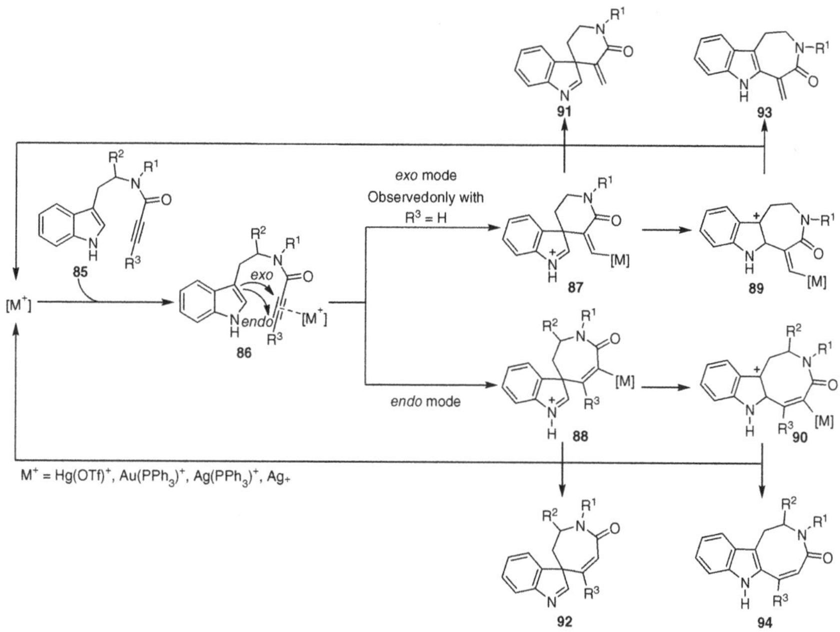

Scheme 9.33 Process of gold-catalyzed intramolecular alkyne hydroarylation.

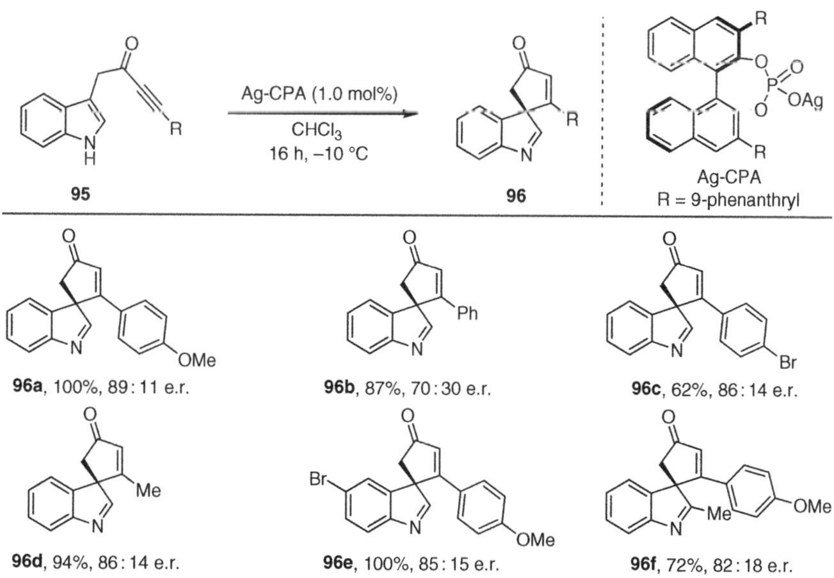

Scheme 9.34 Dearomatization of aromatic ynones for spirocyclic scaffolds.

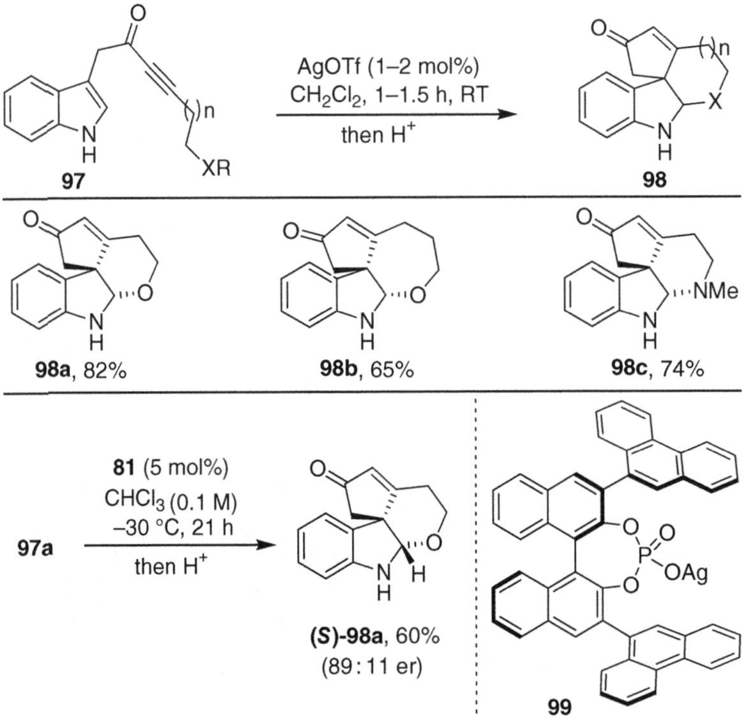

Scheme 9.35 One-pot spirocyclization/trapping to form tetracycles.

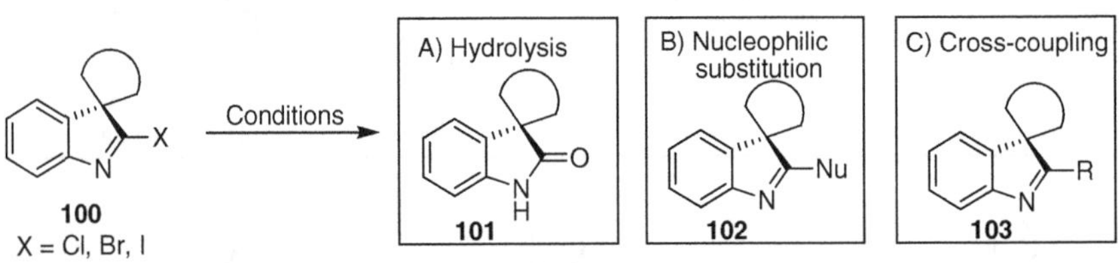

Scheme 9.36 Scaffolds for the synthesis of spirocyclic indole derivatives.

9.5.3 Intermolecular Dearomative Spiroannulation of Indoles

In the challenging area of C-3 spirocyclic indolines, successful approaches have taken advantage of the nucleophilicity of indoles and a preinstalled intramolecular electrophile; the relatively long synthetic routes for preparing these designed substrates has greatly restricted the development of this method. To build spirocyclic indolenine derivatives in a very convergent and direct manner, You and coworkers developed an intermolecular Pd-catalyzed dearomatization cascade sequence of a simple indole containing a tethered nucleophile with a bifunctional dual electrophile (Scheme 9.37). In the presence of Pd(0) catalyst, propargyl carbonates could be transformed to η^3-π-allenylpalladium intermediates, followed by cyclization giving spiroindolines [50].

Rawal and coworkers reported other indole-containing bis-nucleophiles. Tosyltryptamines were competent substrates for the spirocyclization and providing heterocyclic compounds containing nitrogen atom. The tether length of the linker attached to the amine could be extended and seven-membered spirocyclization could be accessed. Notably, the corresponding carboxylic acid was also examined for the propargylation reaction. When treated with the base NEt_3, the desired spirocyclized lactone was produced by forming a C–C bond to the central carbon of the propargyl unit unlike other examples [51] (Figure 9.4).

Allyl acetate is a considerably commercial starting material that can be prepared on an industrial scale [52]. Sheppard and coworkers reported the functionalized spirocyclic indolines by using this cheaper material and easily accessible indoles [53]. The reaction proceeded through two key procedures. The corresponding 3,3-diallylindolines were firstly accessed which was controlled by

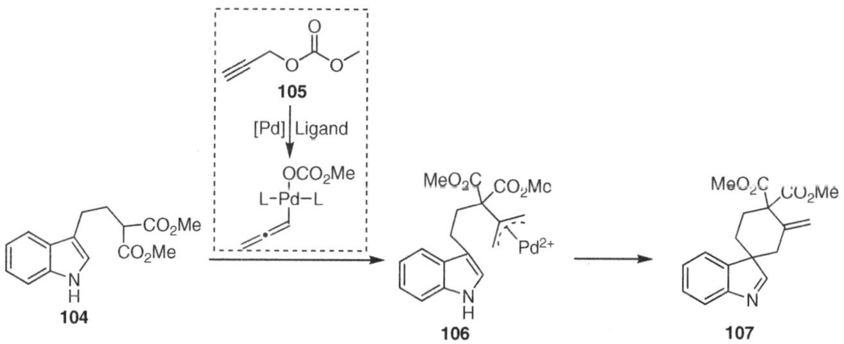

Scheme 9.37 Intermolecular spirocyclization of indoles and propargyl carbonate.

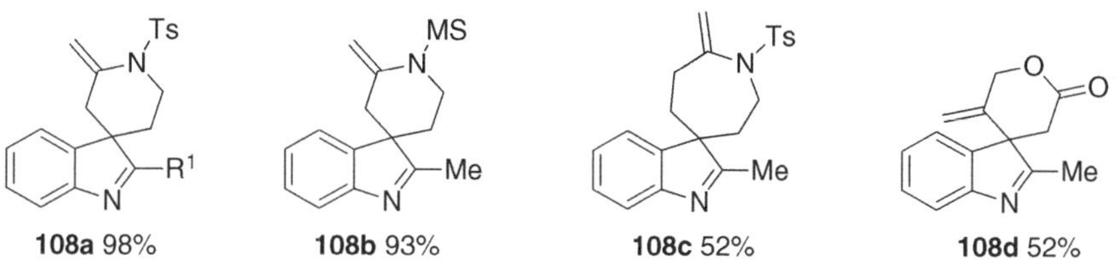

Figure 9.4 Novel spiroindolenine products.

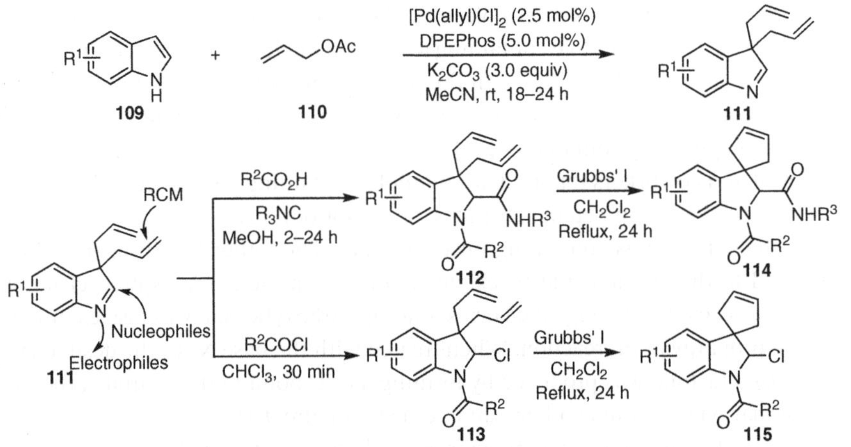

Scheme 9.38 Stepwise synthesis spirocyclic indolines of indoles with allyl acetate.

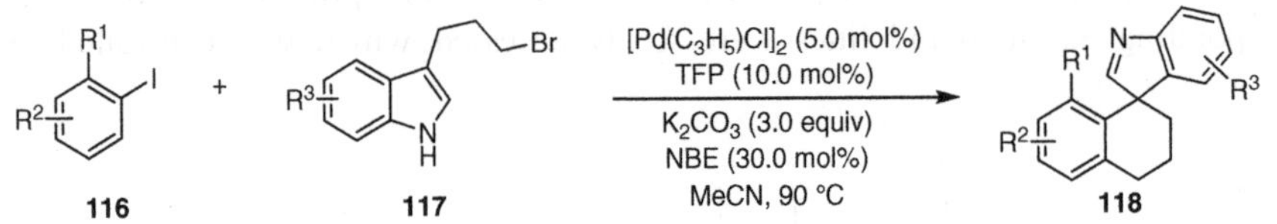

Scheme 9.39 Pd/NBE-catalyzed intermolecular dearomative spiroannulation of indoles.

DPEPhos avoiding other by-products. Then, using ring-closing metathesis, functionalized spirocyclic indoline scaffolds were isolated in good to excellent yield by the Grubbs' first-generation catalyst (Scheme 9.38).

In 2015, a new strategy that led to a series of spiroindolenine was first reported via a modified Catellani-type C–H functionalization by Liang and coworkers (Scheme 9.39) [54]. Based on the reactive mode of Catellani reaction, the catalytic cycle is shown in Figure 9.5. The reaction is initiated by NBE-assisted C–H palladation to form five-membered palladacycle **121**, which then reacts with bromopropylindole to generate ortho alkylated arene **123** from the reductive elimination of Pd(IV) species **122**. Subsequent elimination of NBE through β-carbon elimination then occurs to give the key intermediate **124**. Finally, with the assistance of base, dearomative reaction could take place, leading to a spirocyclic palladacycle **125**, which is finally converted into product **118a**.

In 2015, Liu and coworkers reported a palladium-catalyzed [3+2] cycloaddition allowing sulfonyl indoles and vinyl cyclopropane derivatives as the starting materials which provides five-membered spiroindolenines containing an all-carbon quaternary center and two tertiary stereocenters [55]. Generally, the vinyl

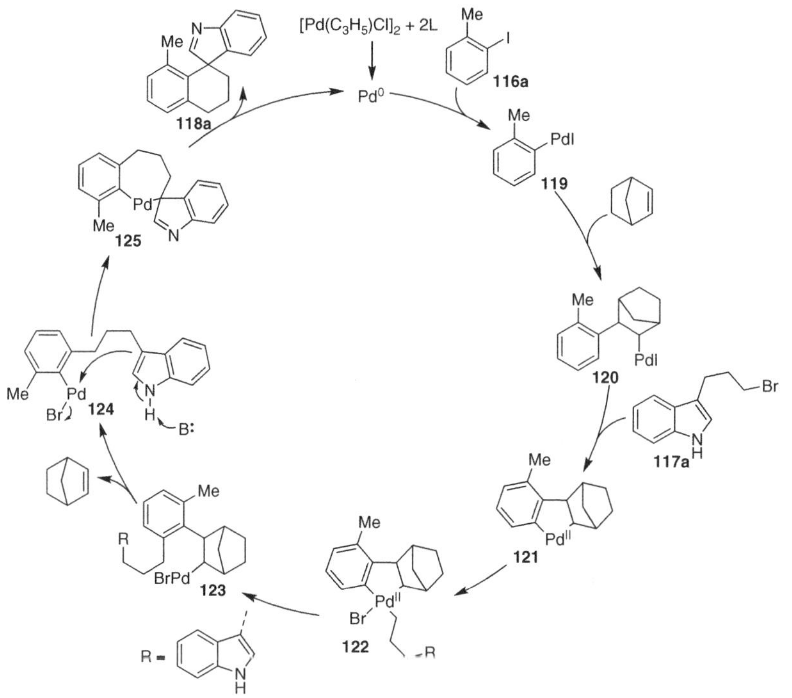

Figure 9.5 Proposed reaction mechanism for palladium-catalyzed dearomatization via a modified Catellani-type C–H functionalization.

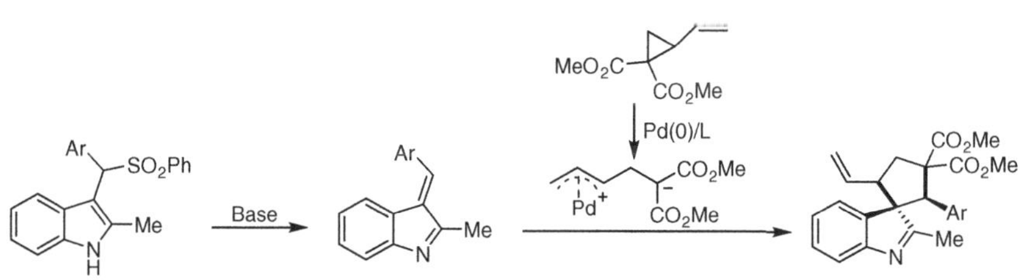

Scheme 9.40 Asymmetric cycloadditions of vinylcyclopropanes with indoles.

cyclopropane containing electron-withdrawing groups could be transformed to the amphiphilic ion in the presence of a palladium catalyst. The formed 1,3-dipoles could act as a transient base and deprive a proton from sulfonyl indole, and α,β-unsaturated imines could be formed. Then, the [3+2] cycloaddition of benzenesulfinic anion and π-allylpalladium complex could proceed to yield five-membered spiroindolenines in high yields and enantioselectivity (Scheme 9.40).

9.6 Dearomative Spiroannulation of Aromatic Heterocyclic Compounds

9.6.1 Dearomative Spiroannulation of Pyrroles

The pyrrole and pyrroline structural cores exist widely in bioactive compounds and pharmaceutical. Pyrroles have similar structural and electronic properties as indoles, which readily participate in selective Friedel–Crafts alkylation reactions as a consequence of the fact that the 3-position of indoles is much more reactive with electrophiles. However, the electrophilicity of 2- and 3-sites of pyrrole displayed similar reactivity. Moreover, alkylated pyrroles are more reactive than their parent pyrroles which probably makes the formation of mono-alkylated more difficult. In 2012, You and coworkers explored the reaction of allylic carbonate-tethered pyrrole **129**. In the presence of $[Ir(cod)Cl]_2$ and a BINOL-derived phosphoramidite ligand, a series of spiro-2*H*-pyrrole derivatives have been successfully acquired. Remarkably, several transformations of the spiro-2H-pyrrole products were also carried out, and novel spiro ring structure frameworks have been accessed in high yields as shown in Scheme 9.41 [56].

Soon after, they also developed the dearomative arylation of pyrroles. In order to optimize the reaction conditions, various palladium precursor, base, and ligands were tested. The desired product spironaphthalenepyrrole **132** could be obtained in 90% yield in the presence of 2 mol% of $[Pd(\eta^3\text{-}C_3H_5)Cl]_2$, 4 mol% of **L25**, and 1.5 equiv of K_2CO_3 in refluxed toluene (Scheme 9.42). The obtained spironaphthalene-1,2′-pyrrole derivatives are the structural cores of bioactive compounds and pharmaceuticals (Figure 9.6) [57].

Interestingly, when the tether in the pyrrole was shortened by one carbon, the reaction proposed to firstly proceed asymmetric allylic dearomatization affording

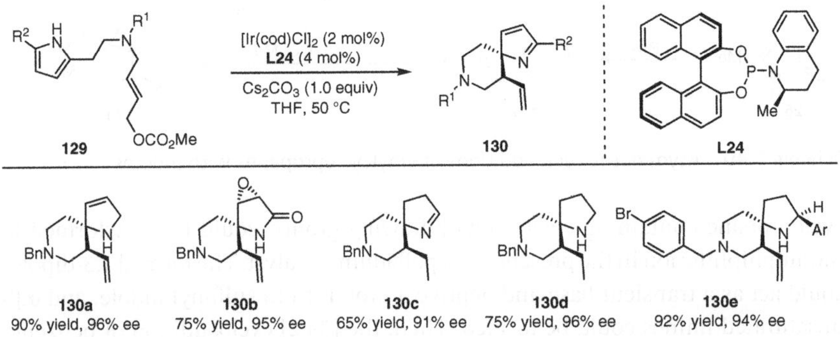

Scheme 9.41 Allylic dearomatization of pyrroles for six-membered spiro-2H-pyrroles.

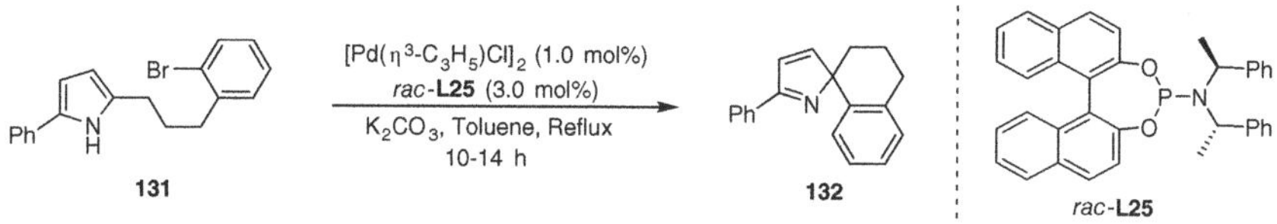

Scheme 9.42 Arylative dearomatization of pyrroles for six-membered spiro-2H-pyrroles.

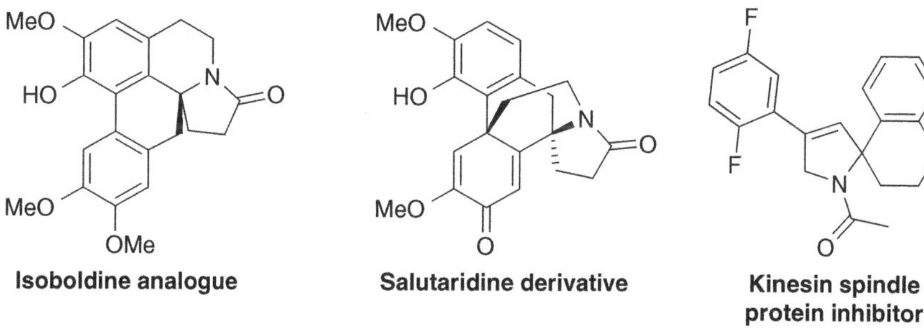

Figure 9.6 Selected bioactive compounds bearing spironaphthalene-pyrrolidine units.

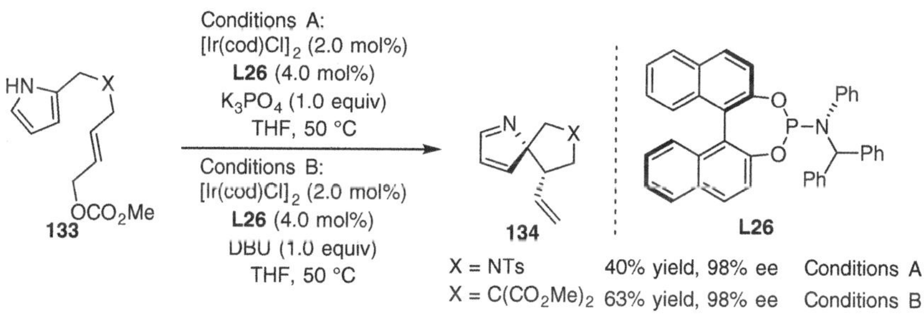

Scheme 9.43 Allylic dearomatization of pyrroles for five-membered spiro-2H-pyrroles.

the five-membered spiro-2*H*-pyrrole intermediate. However, the key intermediate was highly reactive and could not be isolated, which then underwent rapid *in situ* migration from C2 to C3 affording the bicyclic pyrrole compound. Therefore, You and coworkers imagined that the five-membered spiro-2*H*-pyrrole might be isolated as a stable product by turning the electronic nature of the linkage (Scheme 9.43). To decrease the migratory ability of the methylene portion, the substituent on the nitrogen was switched to an electron-withdrawing group, and fortunately spiro-2*H*-pyrrole was obtained with an excellent *ee* albeit with low yield. In addition, they also tested the substrate **133** bearing an all-carbon tethered with the allylic carbonate, and a series of five-membered spiro-2*H*-pyrrole have been accessed with high diastereoselectivity and enantioselectivity [58]. Notably,

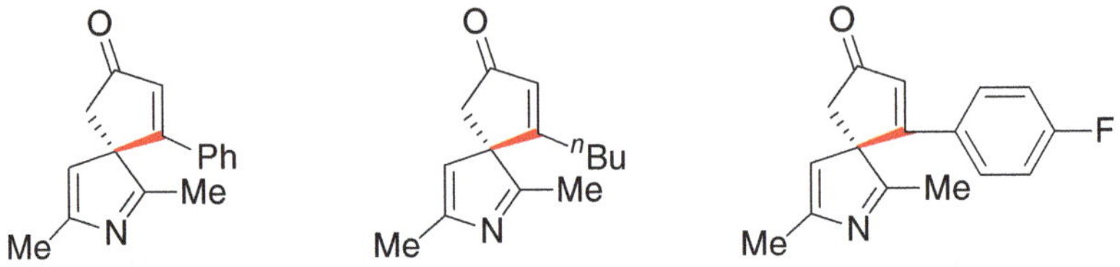

134a
20 min, RT
100% yield

134b
10 min, RT
100% yield

134c
15 min, RT
100% yield

Figure 9.7 Quantitative formation of spirocycles from 3-pyrroles.

with $AgNO_3$-SiO_2 promotion, the quantitative formation of spirocycles **134a–c** were reported by Taylor and Unsworth, given the rarity of dearomatized products derived from 3-pyrroles [59] (Figure 9.7).

9.6.2 Dearomative Spiroannulation of Furans

Furan derivatives are a kind of heterocyclic aromatic compounds and available from a renewable feedstock such as biomass. Owing to the low aromaticity of furan rings, the chemical property is similar to alkenes, 1,3-diynes, and enol ethers. Yin and coworkers reported a palladium-catalyzed cross-coupling of furfural tosylthydrazones with aryl halides to provide spiroacetal enol ethers. The reaction was started with oxidative addition of aryl halides to Pd(0) and affording intermediate **136**, then **137** was reacted with diazo compound **136** to afford Pd carbene complex **138**. The key intermediate arylmethylenepalladium halide **138** was successfully accessed. The η^3-furylmethylenepalladium halide **139** was formed through σ-π isomerization of **138** which is similar to other η^3-arylmethylenepalladium halides and could undergo intramolecular nucleophilic dearomatization interacting with hydroxypropyl group tethered at the 2-position of the furans. The success of this reaction is using the stable furfural hydrazones instead of 2-furylmethyl halides which is highly reactive and generally tends to aggregation (Scheme 9.44) [60].

Transition-metal-catalyzed difunctionalization reaction is an important topic in synthetic chemistry field. In 2016, Yin and coworkers developed a novel palladium-catalyzed 2,5-alkoxyarylation of furan rings, providing convenient, diastereospecific access to structurally novel spirooxindoles (Scheme 9.45). Firstly, they tested intramolecular reactions of N-2-bromo-phenyl-2-furamine with a nucleophilic hydroxyl group on the side chain. There are three possible pathways for this reaction as shown in Scheme 9.46. To gain more insight into the mechanism, two experiments were studied. When compound **145** reacted under the standard conditions, product **142a** was not achieved which suggested that the reaction could

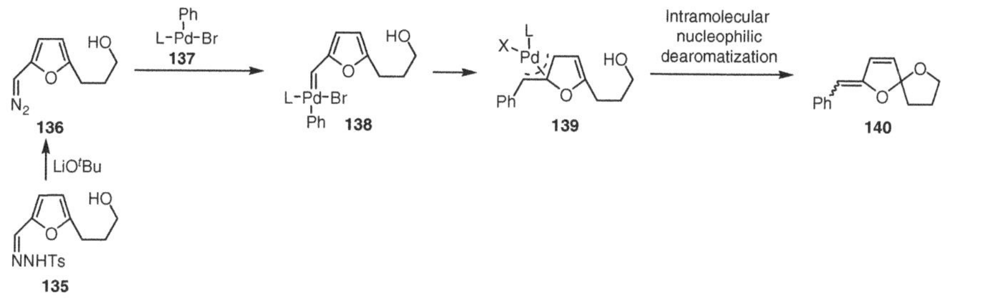

Scheme 9.44 Intramolecular nucleophilic dearomatization of furans.

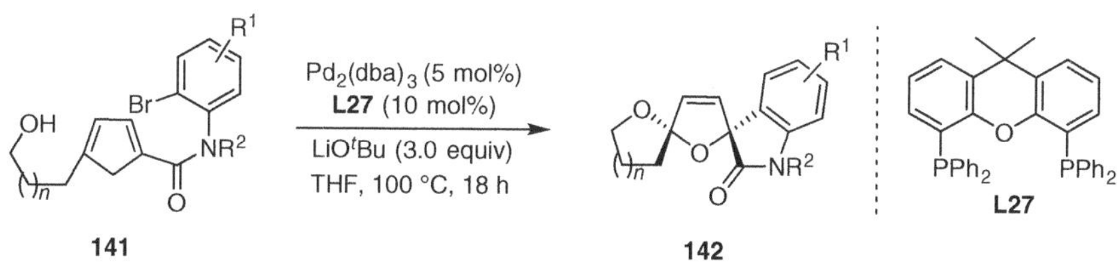

Scheme 9.45 Dearomatizing 2,5-alkoxyarylation of furan rings for spirooxindoles.

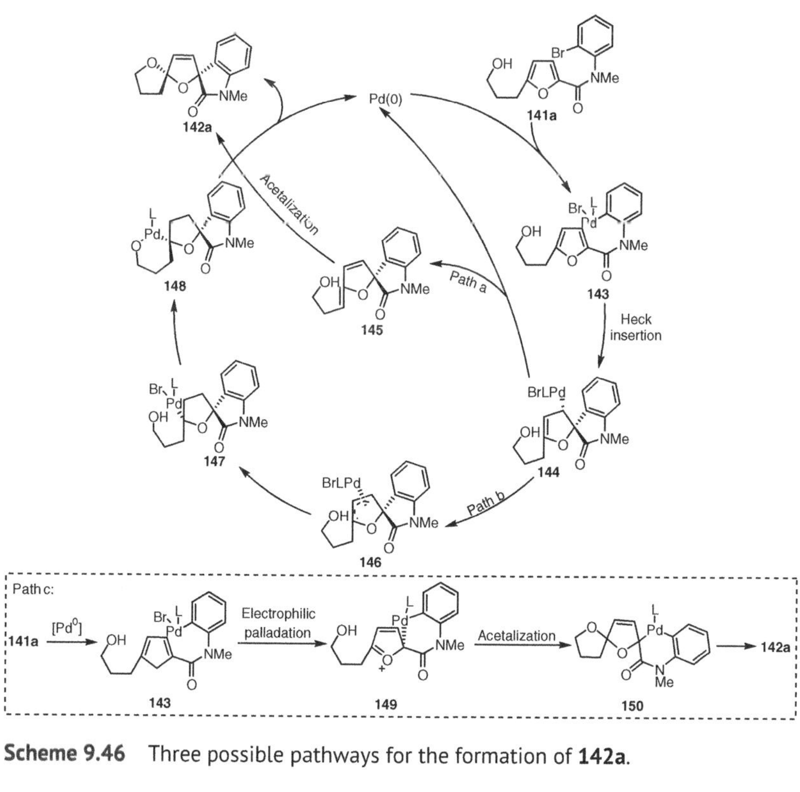

Scheme 9.46 Three possible pathways for the formation of **142a**.

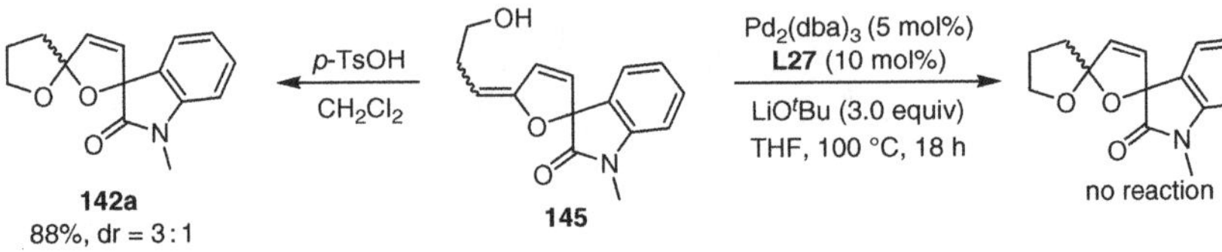

Scheme 9.47 Mechanistic experiments for dearomatizing 2,5-alkoxyarylation of furan rings.

not proceed in the path a. In addition, a 3 : 1 mixture of diastereomers was obtained when the compound was treated with *p*-TsOH (Scheme 9.47). This result implied that dearomative spirocyclization proceeds *via* Heck insertion instead of electrophilic palladation. Moreover, they also investigated the intermolecular version of the reaction. In the reaction, primary and secondary alcohols could be used as substrate and the reaction proceeds smoothly affording other spirooxindoles structures [61]. When the substituent on the furan was changed from chain alcohol to alkyl or benzyl group, the intermediate could be selectively transformed into structurally novel dispirooxindoles [62]. To gain more interesting biologically dispirooxindoles, they investigated furfurylcyclobutanols as the substrates catalyzed by palladium through the upper procedure giving the intermediate **151**. The hydroxyl group is coordinated with the palladium center on the same face of the molecule. Intermediate **151** subsequently undergoes a 1,2-alkyl shift on the face of the molecule opposite the palladium atom and then release of the palladium to complete the catalytic cycle (Scheme 9.48) [63].

Spiroacetals are present in numerous biologically active natural products and represent a privileged scaffold in drug discovery research. Yin and coworkers developed a protocol for the 2,5-oxyarylation of furan rings via Pd-catalyzed aerobic oxidative coupling of boronic acids with α-hydroxyalkyfurans [64]. Various furans were investigated and expected spiroacetals have been obtained in moderate or good yields. As shown in Scheme 9.49, the structure of the side chain clearly influenced the reaction outcome. The protocol opens a new avenue for accessing spiroacetals *via* dearomatization of furans.

9.6.3 Dearomative Spiroannulation of Thiophenes

Thiophenes are a class of readily available, cheap chemical feedstock and the functionalization of them has attracted great attention [65]. Generally, thiophene derivatives show great potential in the application of optoelectronic materials [66], especially in light-emitting diode, laser, field effect transistor, and so on. You and coworkers have been working on the incorporation of thiophene and other aromatic compounds. In 2018, they disclosed the first

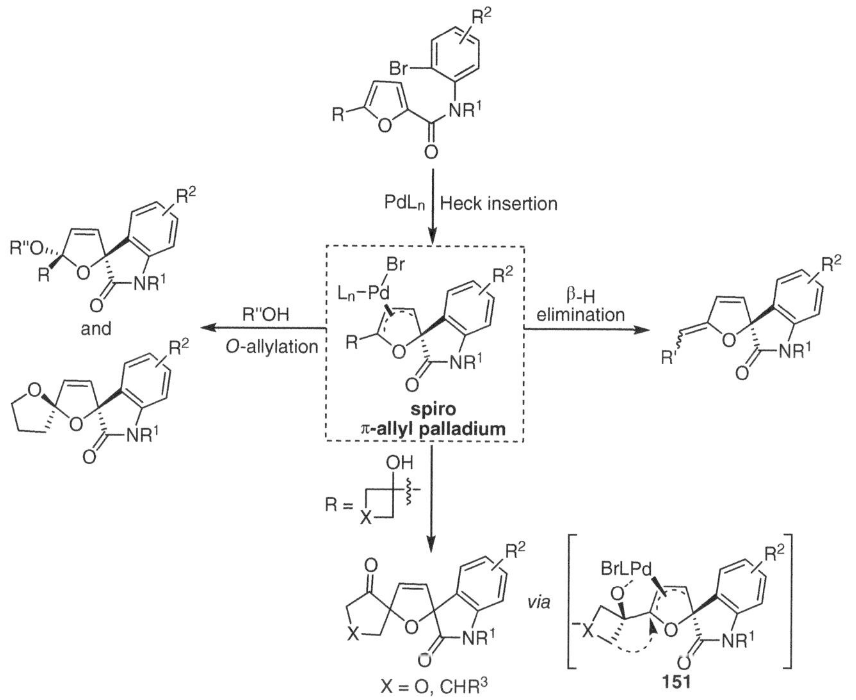

Scheme 9.48 Further transformations of spiro π-allyl palladium.

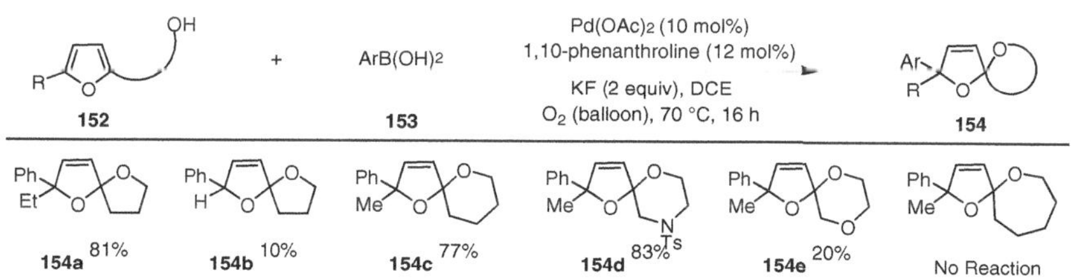

Scheme 9.49 Synthesis of spiroacetals from boronic acids and hydroxyalkylfurans.

Ir-catalyzed one-pot annulation reactions of thiophenes and α,β-unsaturated acids. When the reaction was carried out using 5 mol% of $[IrCp^*Cl_2]_2$, combined with 3 equiv of Ag_2O in HFIP at 120 °C for 12 hours, a series of thiophene-containing spirocyclic products have been acquired through a Heck-type pathway [67]. The selectivity of the spirocyclic products could only be defined by the configuration of the substrate. As they reported, when benzothiophenes were used as substrates, thiophene-fused coumarin-type frameworks were constructed (Scheme 9.50).

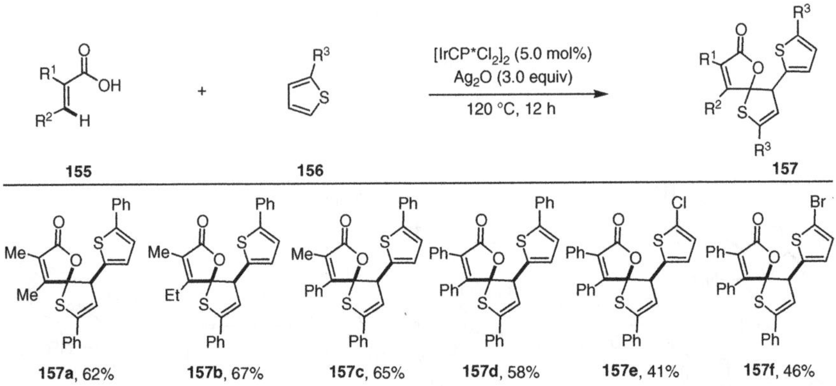

Scheme 9.50 Synthesis of thiophene-containing spirocyclic products via C–H activation of carboxylic acids.

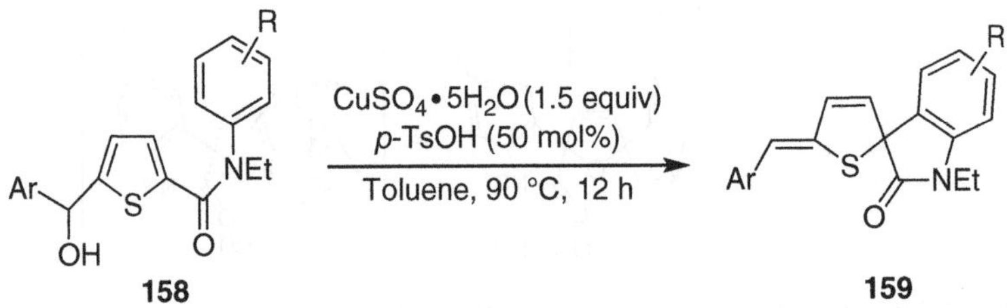

Scheme 9.51 Cu(II)-promoted transformations of α-thienylcarbinols into spirothienooxindoles.

In order to deliver a comprehensive library of heterocyclic spirooxindoles, Yin and coworkers aimed to employ α-thienylcarbinols as the substrates and explore transformation into unprecedented sulfether-containing spirooxindoles [68]. Various acids were tested, such as strong Lewis acid $SnCl_4$ and $BF_3{\cdot}Et_2O$, weaker Lewis acid $Dy(OTf)_3$, $CuSO_4{\cdot}5H_2O$, and $FeCl_3{\cdot}6H_2O$. Gratifyingly, the combination of $CuSO_4{\cdot}5H_2O$/*p*-TsOH behaved the best. Catalyzed by the system of acid, dearomatizing Friedel–Crafts reaction proceeded smoothly providing unprecedented spirothienooxindoles with thioether segment (Scheme 9.51). The selective halogenation of **159a** is extremely challenging because of the multiple halogenation cites, such as four olefinic carbons and two electron-rich phenyl rings. When compound **159a** was treated with 1 equiv $CuCl_2$ and 2 equiv LiCl, product **160** was formed as a mixture of Z/E isomers (Z/E = 1 : 2). Given the dearomative reaction is catalyzed by Cu(II), they imagined the halogenation could occurr directly from substrate **158a**. Fortunately, under the same reaction condition as the chlorination of **158a**, a series of **158** could be converted into the corresponding chlorinated spirothienooxindoles **160** in moderate Z/E ratios (Scheme 9.52).

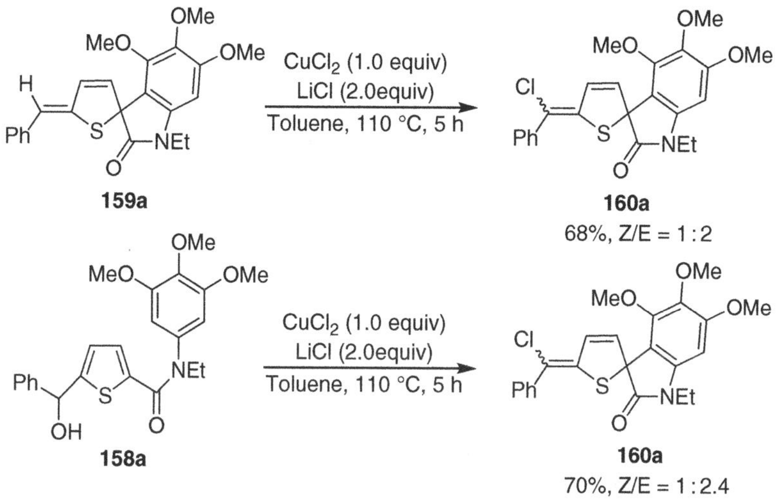

Scheme 9.52 Synthesis of halogenated spirothienooxindoles from **159a** and **158a**.

9.7 Dearomative of Other Aromatic Compounds

In recent years, dearomatization is regarded as one of the most efficient and reliable approaches to construct three-dimensional spirocycles from planar arenes. Despite significant progresses, the reported methods typically focus on using electronically activated aromatics, such as phenols and naphthols, and heteroarenes, such as indoles, pyrroles, thiophenes, and furans, as substrates owing to their favorable keto-enol tautomerism and/or lower resonance stabilization energy. Therefore, the dearomatization of unactivated halonaphthalenes is extremely challenging because of the higher activation energy. In 2016, Li and coworkers developed palladium-catalyzed reductive Heck-type vinylative dearomatization of inactive 1-halonaphthalenes. Firstly, migratory insertion of the two molecular of alkynes to the Pd–C bond occurred, then a dearomatizing Heck-type annulation occurred via olefin coordination/insertion giving the intermediate **163**, and finally the palladium vinyl intermediate could be trapped by the hydride source, providing novel spiro compounds **164** [69]. When the bromoarenes was substituted by alkyl group, the dearomative spirocyclization proceeded smoothly. Different from the former reaction, the final product was constructed via β-H elimination of intermediate **166**. Notably, the spiroannulation reaction using unsymmetrical alkynes may lead to three regioisomers. Fortunately, the attempts with unsymmetrical phenyl alkyl alkynes exclusively delivered the head-to-tail regioisomers (Scheme 9.53) [70].

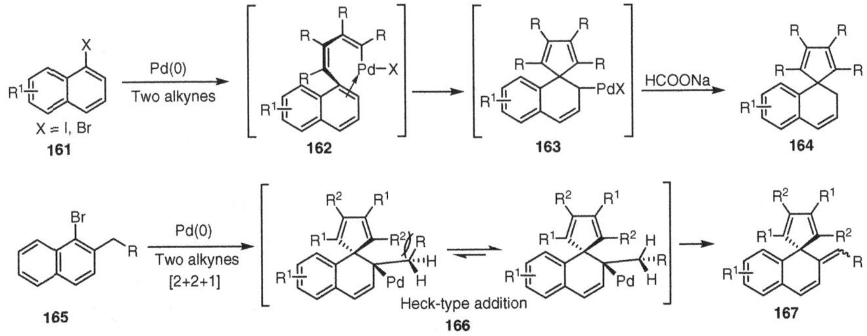

Scheme 9.53 Intermolecular Heck-type dearomative [2+2+1] spiroannulation of alkyl bromoarenes.

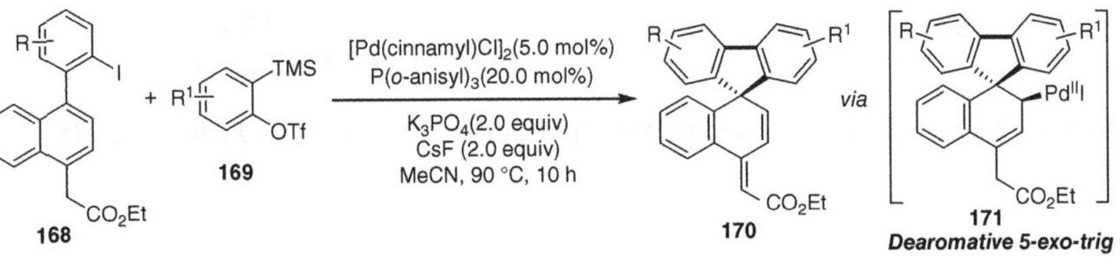

Scheme 9.54 Palladium-catalyzed aryne insertion/arene dearomatization to spirofluorenes.

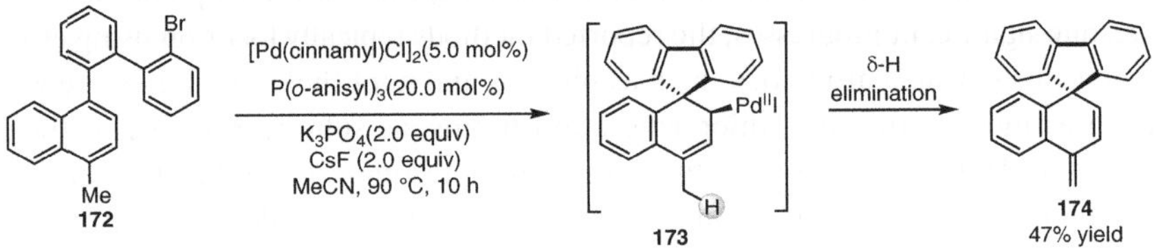

Scheme 9.55 Mechanism studies for the *5-exo-trig* spiroannulation.

In 2018, Luan and coworkers investigated reactions of naphthalene-based biaryls with arynes to construct spirofluorene architectures (Scheme 9.54). Different from above materials, the alkyl group on the naphthalene-based biaryl was replaced by electron-withdrawing ester groups which increased the acidity of the alkyl groups. So, the dearomatization was probably promoted by base. In order to reveal the reaction mechanism, they tested substrate **172** which was weakly acidic. Surprisingly, dearomative product **174** was obtained in 47% yield (Scheme 9.55). The result suggests that the arene dearomatization might take place through a 5-exo-trig spiroannulation and distal-hydride elimination

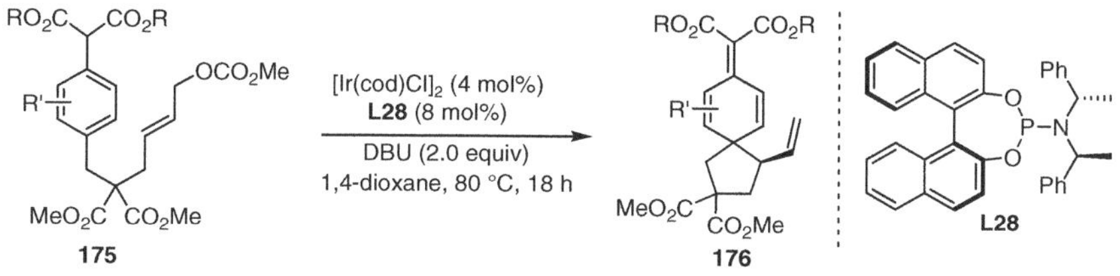

Scheme 9.56 Intramolecular asymmetric allylic dearomatization of benzene derivatives.

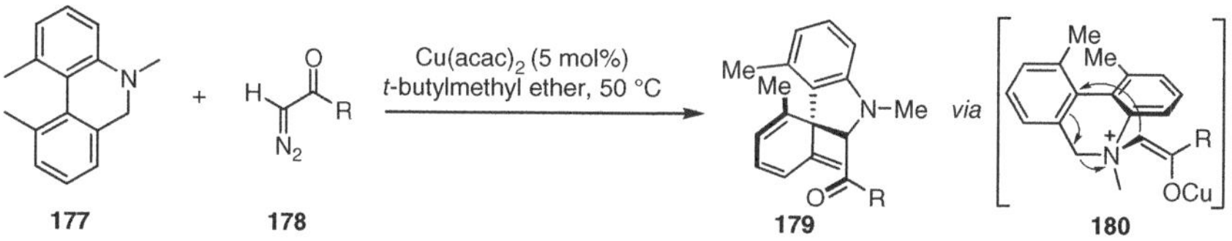

Scheme 9.57 A Cu-catalyzed Sommelet–Hauser dearomatization of dihydrophenanthridine and diazo.

Heck-type pathway [71]. In addition, You and coworkers also developed an Ir-catalyzed asymmetric intramolecular allylic dearomatization reaction of benzene derivatives. With an Ir complex generated from a commercially available iridium precursor and chiral phosphoramidite ligand, a series of spiro[4.5]cyclohexadiene compounds were delivered in reasonable yields and good enantioselectivity (Scheme 9.56) [72].

Sommelet–Hauser rearrangement is the reaction of ammonium ylide intermediate obtained from benzyl quaternary ammonium salt under alkali metal compound, and then [2,3]-Wittig rearrangement to *o*-substituted benzylamines. Sommelet–Hauser dearomatization can efficiently generate valuable three-dimensional compounds bearing quaternary carbon centers when the two ortho positions have non-hydrogen substituents. Though spiro-compounds could be produced by the Sommelet–Hauser rearrangement of benzocyclic ammonium salts, the efficiency of reported reactions was relatively low [73]. In 2018, a series of spiroindolines containing one quaternary center and one tertiary center were synthesized by Cu-catalyzed Sommelet–Hauser dearomatization of dihydrophenanthridine and diazo compounds in one step (Scheme 9.57) [74]. Moreover, the only one *ortho*-substituent has been tested in standard conditions and other optimized conditions, but the yield of dearomative products was reducing. The above experiment results suggest that the two *ortho*-substituents strongly influenced the efficiency of the Sommelet–Hauser reaction and selectivity for this type of rearrangement (Scheme 9.58).

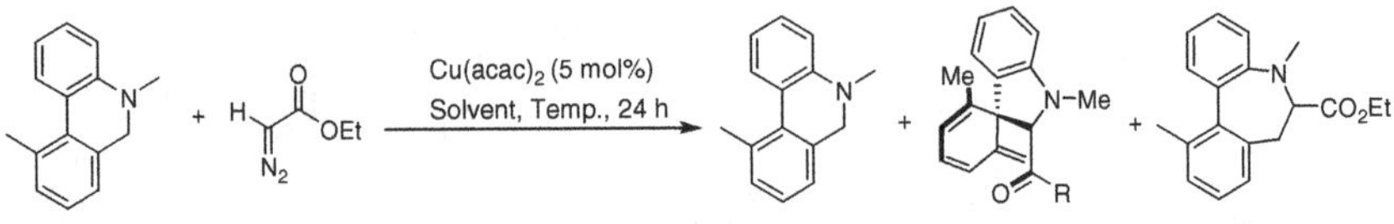

Solvent	Temp./°C	**182**	**183**	**184**
MTBE	50	64%	35%	–
MTBE	90	53%	34%	<5%
Toluene	90	43%	25%	<5%
Toluene	110	55%	<5%	33%

Scheme 9.58 Substituent effects.

9.8 Conclusion

In this chapter, the recent advances of spiroannulation of aromatic materials such as phenols, halonaphthalene, aryl diazo salt, arylamine derivatives and aromatic heterocyclic compounds including indoles, pyrroles, furans, or thiophens via transition-metal-catalyzed allylic alkylation and Pd-catalyzed arylation or alkylation were summarized. Contrary to traditional spiroannulation techniques, this strategy provides a versatile and efficient approach to highly diverse spirocyclic skeletons in one step from simple aromatic compounds. These enriched fused-polycyclic or spirocyclic skeleton databases based on the present methods are ubiquitous structural motifs in bioactive natural products, chiral ligands, and functional materials, indicating that the processes hold great promise for application in the field of total synthesis, ligand developments, as well as materials sciences.

References

1 (a) Tyman, J.H.P. (**1996**). *Synthetic and Natural Phenol.* New York: Elsevier. (b) Rappoport, Z. (**2003**). *The Chemistry of Phenols.* Chichester: Wiley. (c) Weber, M., Weber, M., and Kleine-Boymann, M. (**2004**). Phenol. In: *Ullmann's Encyclopedia of Industrial Chemistry.* Weinheim: Wiley.

2 For selected reviews, see:(a)Kündig, E.P. and Rape, A. (**2004**). *Top. Organomet. Chem.* 7: 95. (b) Quideau, S., Pouységu, L., and Deffieus, D. (**2008**). *Synlett* 2008: 467–495. (c) Roche, S.P. and Porco, J.A. Jr. (**2011**). *Angew. Chem. Int. Ed.* *50*: 4068–4093. (d) Zhuo, C.-X., Zhang, W., and You, S.-L. (**2012**). *Angew. Chem. Int. Ed.* *51*: 12662–12686.

3 Trost, B.M. and Van Vranken, D.L. (**1996**). *Chem. Rev. 96*: 395–422.

4 For selected reviews, see:(a)Zhuo, C.X., Zheng, C., and You, S. (**2014**). *Acc. Chem. Res.* 47: 2558–2573.(b) Wu, W., Zhang, L., and You, S.-L. (**2016**). *Chem. Soc. Rev. 45*: 1570–1580.(c) Sun, W., Li, G., Hong, L., and Wang, R. (**2016**). *Org. Biomol. Chem. 14*: 2164–2176.

5 Frlan, R. and Kikelj, D. (**2006**). Recent progress in diaryl ether synthesis. *Synthesis 14*: 2271–2285.

6 Nemoto, T., Ishige, Y., Yoshida, M. et al. (**2010**). *Org. Lett. 12*: 5020–5023.

7 Yoshida, M., Nemoto, T., Zhao, Z. et al. (**2012**). *Tetrahedron 23*: 859–866.

8 Donahue, M.G., Jentsch, N.G., and Realini, E.C. (**2017**). *Tetrahedron Lett. 58*: 3219–3222.

9 Wu, Q.-F., Liu, W.-B., Zhuo, C.-X. et al. (**2011**). *Angew. Chem. Int. Ed.* 50: 4455–4458.

10 Cheng, Q., Wang, Y., and You, S.-L. (**2016**). *Angew. Chem. Int. Ed. 55*: 3496–3499.

11 Nemoto, T., Zhao, Z., Yokosaka, T. et al. (**2013**). *Angew. Chem. Int. Ed. 52*: 2217–2220.

12 Nemoto, T., Nozaki, T., Yoshida, M., and Hamada, Y. (**2013**). *Adv. Synth. Catal. 355*: 2693–2700.

13 Rousseaux, S., García-Fortanet, J., Del Aguila Sanchez, M.A., and Buchwald, S.L. (**2011**). *J. Am. Chem. Soc. 133*: 9282–9285.

14 Xu, R.-Q., Yang, P., and You, S.-L. (**2017**). *Chem. Commun. 53*: 7553–7556.

15 Xu, R.-Q., Gu, Q., Wu, W.-T. et al. (**2014**). *J. Am. Chem. Soc. 136*: 15469–15472.

16 Zheng, H., Bai, L., Liu, J. et al. (**2015**). *Chem. Commun.* 51: 3061–3064.

17 Yang, L., Zheng, H., Luo, L. et al. (**2015**). *J. Am. Chem. Soc. 137*: 4876–4879.

18 Luo, L., Zheng, H., Liu, J. et al. (**2016**). *Org. Lett. 18*: 2082–2085.

19 Hu, W., Wang, H., Bai, L. et al. (**2018**). *Org. Lett. 20*: 880–883.

20 Bai, L., Yuan, Y., Liu, J. et al. (**2016**). *Angew. Chem. Int. Ed. 55*: 6946–6950.

21 (a) Lyons, T.W. and Sanford, M.S. (**2010**). *Chem. Rev. 110*: 1147–1169. (b) Wencel-Delord, J., Dröge, T., Liu, F., and Glorius, F. (**2011**). *Chem. Soc. Rev. 40*: 4740–4761. (c) Ackermann, L. (**2011**). *Chem. Rev. 111*: 1315–1345. (d) Yeung, C.S. and Dong, V.M. (**2011**). *Chem. Rev. 111*: 1215–1292. (e) Hartwig, J.F. (**2011**). *Chem. Soc. Rev.* 40: 1992–2002. (f) Engle, K.M., Mei, T.-S., Wasa, M., and Yu, J.-Q. (**2012**). *Acc. Chem. Res. 45*: 788–802. (g) Song, G., Wang, F., and Li, X. (**2012**). *Chem. Soc. Rev.* 41: 3651–3678. (h) Wencel-Delord, J. and Glorius, F. (**2013**). *Nat. Chem.* 5: 369–375.

22 Nan, J., Zuo, Z., Luo, L. et al. (**2013**). *J. Am. Chem. Soc. 135*: 17306–17309.

23 (a) Rudolph, A., Bos, P.H., Meetsma, A. et al. (**2011**). *Angew. Chem. Int. Ed. 50*: 5834–5838. (b) Oguma, T. and Katsuki, T. (**2012**). *J. Am. Chem. Soc. 134*: 20017–20020. (c) Zhuo, C. and You, S.-L. (**2013**). *Angew. Chem. Int. Ed. 52*: 10056–10059. (d) Zheng, C., Wang, L., Li, J. et al. (**2013**). *Org. Lett. 15*: 4046–4049. (e) Zhuo, C. and You, S.-L. (**2014**). *Adv. Synth. Catal. 356*: 2020–2028.

24 Zuo, Z., Yang, X., Liu, J. et al. (**2015**). *J. Org. Chem. 80*: 3349–3356.

25 Gu, S., Luo, L., Liu, J. et al. (**2014**). *Org. Lett. 16*: 6132–6135.

26 Jiang, P., Xu, Y., Sun, F. et al. (**2016**). *Org. Lett. 18*: 1426–1429.

27 Seoane, A., Casanova, N., Quinoñes, N. et al. (**2014**). *J. Am. Chem. Soc. 136*: 834–837.

28 Seoane, A., Casanova, N., Quinoñes, N. et al. (**2014**). *J. Am. Chem. Soc. 136*: 7607–7610.

29 Kujawa, S., Best, D., Burns, D.J., and Lam, H.W. (**2014**). *Chem. Eur. J. 20*: 8599–8602.

30 Khan, I., Chidipudi, S.R., and Lam, H.W. (**2015**). *Chem. Commun. 51*: 2613–2616.

31 Zheng, J., Wang, S.-B., Zheng, C., and You, S.-L. (**2015**). *J. Am. Chem. Soc. 137*: 4880–4883.

32 Catellani, M., Frignani, F., and Rangoni, A. (**1997**). *Angew. Chem. Int. Ed. 36*: 119–122.

33 Zuo, Z., Wang, H., Fan, L. et al. (**2017**). *Angew. Chem. Int. Ed. 56*: 2767–2771.

34 Nan, J., Yuan, Y., Bai, L. et al. (**2018**). *Org. Lett. 20*: 7731–7734.

35 (a) Jazzar, R., Hitce, J., Renaudat, A. et al. (**2010**). *Chem. Eur. J. 16*: 2654–2672. (b) Li, H., Li, B.-J., and Shi, Z.-J. (**2011**). *Catal. Sci. Technol. 1*: 191–206. (c) Baudoin, O. (**2011**). *Chem. Soc. Rev. 40*: 4902–4911. (d) Huang, Z., Lim, H.N., Mo, F. et al. (**2015**). *Chem. Soc. Rev. 44*: 7764–7786.(e) He, G., Wang, B., Nack, W.A., and Chen, G. (**2016**). *Acc. Chem. Res. 49*: 635–645. (f) He, J., Wasa, M., Chan, K.S.L. et al. (**2017**). *Chem. Rev. 117*: 8754–8786. (g) Baudoin, O. (**2017**). *Acc. Chem. Res. 50*: 1114–1123. (h) Chu, J.C.K. and Rovis, T. (**2018**). *Angew. Chem. Int. Ed. 57*: 62–101.

36 Tan, B., Bai, L., Ding, P. et al. (**2019**). *Angew. Chem. Int. Ed. 58*: 1474–1478.

37 Wu, Q.-F., He, H., Liu, W.-B., and You, S.-L. (**2010**). *J. Am. Chem. Soc. 132*: 11418–11419.

38 Zhang, X., Liu, W.-B., Wu, Q.-F., and You, S.-L. (**2013**). *Org. Lett. 15*: 3746–3749.

39 Wu, Q.-F., Zheng, C., and You, S.-L. (**2012**). *Angew. Chem. Int. Ed. 51*: 1680–1683.

40 Wang, Y., Zheng, C., and You, S.-L. (**2017**). *Angew. Chem. Int. Ed. 56*: 15093–15097.

41 Wu, K.-J., Dai, L.-X., and You, S.-L. (**2012**). *Org. Lett. 14*: 3772–3775.

42 Liu, R.-R., Xu, Y., Liang, R.-X. et al. (**2017**). *Org. Biomol. Chem. 15*: 2711–2715.

43 Li, X., Zhou, B., Yang, R.-Z. et al. (**2018**). *J. Am. Chem. Soc. 140*: 13945–13951.

44 Douki, K., Shimokawa, J., and Kitamura, M. (**2019**). *Org. Biomol. Chem. 17*: 1727–1730.

45 Peshkov, V.A., Pereshivko, O.P., and Van der Eycken, E.V. (**2012**). *Adv. Synth. Catal. 354*: 2841–2848.

46 (a)Ferrer, C. and Echavarren, A.M. (**2006**). *Angew. Chem. Int. Ed. 45*: 1105–1109. (b) Ferrer, C., Amijs, C.H.M., and Echavarren, A.M. (**2007**). *Chem. Eur. J. 13*: 1358–1373. (c) Ferrer, C., Escribano-Cuesta, A., and Echavarren, A.M. (**2009**). *Tetrahedron 5*: 9015–9020.

47 James, M.J., Cuthbertson, J.D., O'Brien, P. et al. (**2015**). *Angew. Chem. Int. Ed. 54*: 7640–7643.

48 Liddon, J.T.R., James, M.J., Clarke, A.K. et al. (**2016**). *Chem. Eur. J. 22*: 8777–8780.

49 Liddon, J.T.R., Clarke, A.K., Taylor, R.J.K., and Unsworth, W.P. (**2016**). *Org. Lett. 18*: 6328–6331.

50 Gao, R.-D., Liu, C., Dai, L.-X. et al. (**2014**). *Org. Lett. 16*: 3919–3921.

51 (a) Montgomery, T.D., Nibbs, A.E., Zhu, Y., and Rawal, V.H. (**2014**). *Org. Lett.* 16: 3480–3483. (b) Nibbs, A.E., Montgomery, T.D., Zhu, Y., and Rawal, V.H. (**2015**). *J. Org. Chem. 80*: 4928–4941.

52 Wittcoff, H.A., Reuben, B.G., and Plotkin, J.S. (eds.) (**2012**). *Industrial Organic Chemicals*. Wiley.

53 Dhankher, P., Benhamou, L., and Sheppard, T.D. (**2014**). *Chem. Eur. J. 20*: 13375–13381.

54 Wu, X.-X., Shen, Y., Chen, W.-L. et al. (**2015**). *Chem. Commun. 51*: 16798–16801.

55 Liu, Z.-S., Li, W.-K., Kang, T.-R. et al. (**2015**). *Org. Lett.* 17: 150–153.

56 Zhuo, C.-X., Liu, W.-B., Wu, Q.-F., and You, S.-L. (**2012**). *Chem. Sci. 3*: 205–208.

57 Wu, K.-J., Dai, L.-X., and You, S.-L. (**2013**). *Chem. Commun.* **49**: 8620–8622.

58 Zhuo, C.-X., Cheng, Q., Liu, W.-B. et al. (**2015**). *Angew. Chem. Int. Ed. 54*: 8475–8479.

59 Clarke, A.K., James, M.J., O'Brien, P. et al. (**2016**). *Angew. Chem. Int. Ed. 55*: 13798–13802.

60 Yin, B., Zhang, X., Liu, J. et al. (**2014**). *Chem. Commun. 50*: 8113–8116.

61 Liu, J., Xu, X., Li, J. et al. (**2016**). *Chem. Commun. 52*: 9550–9553.

62 Liu, J., Peng, H., Yang, Y. et al. (**2016**). *J. Org. Chem. 81*: 9695–9706.

63 Liu, J., Peng, H., Lu, L. et al. (**2016**). *Org. Lett. 18*: 6440–6443.

64 Li, J., Peng, H., Wang, F. et al. (**2016**). *Org. Lett. 18*: 3226–3229.

65 (a) Chen, C.-H., Hsieh, C.-H., Dubosc, M. et al. (**2010**). *Macromolecules 43*: 697–708. (b) Dou, L., Chen, C.-C., Yoshimura, K. et al. (**2013**). *Macromolecules 46*: 3384–3390. (c) Acharya, A., Gautam, V., and Ila, H. (**2017**). *J. Org. Chem. 82*: 7920.

66 (a) Mishra, A., Ma, C.-Q., and Bäuerle, P. (**2009**). *Chem. Rev. 109*: 1141–1276. (b) Chen, C.-P., Chan, S.-H., Chao, T.-C. et al. (**2008**). *J. Am. Chem. Soc. 130*: 12828–12833. (c) Xu, Y.-X., Chueh, C.-C., Yip, H.-L. et al. (**2012**). *Adv. Mater. 24*: 6356–6361.

67 Tan, G., You, Q., Lan, J., and You, J. (**2018**). *Angew. Chem. Int. Ed. 57*: 6309–6313.

68 Yin, B., Li, H., Zhang, X. et al. (**2012**). *J. Org. Chem. 77*: 6365–6370.

69 Dou, Y., Yang, L., Zhang, L. et al. (**2016**). *RSC Adv. 6*: 100632–100635.

70 Liao, X., Wang, D., Huang, Y. et al. (**2019**). *Org. Lett. 21*: 1152–1155.

71 Zuo, Z., Wang, H., Diao, Y. et al. (**2018**). *ACS Catal. 8*: 11029–11034.

72 Yang, Z.-P., Jiang, R., Wu, Q.-F. et al. (**2018**). *Angew. Chem. Int. Ed. 57*: 16190–16193.

73 (a) Lee, T.-J. and Holtz, W.J. (**1983**). *Tetrahedron Lett. 24*: 2071–2072. (b) Berger, R., Ziller, J.W., and Van Vranken, D.L. (**1998**). *J. Am. Chem. Soc. 120*: 841–842. (c) Li, Y., Shi, Y., Huang, Z. et al. (**2011**). *Org. Lett. 13*: 1210–1213. (d) Gassman, P.G., Van Bergen, T.J., Gilbert, D.P., and Cue, B.W. Jr. (**1974**). *J. Am. Chem. Soc. 96*: 5495–5517.

74 Pan, C., Guo, W., and Gu, Z. (**2018**). *Chem. Sci. 9*: 5850–5854.

10

Carbocyclic Spiro Compounds Occupying Higher Dimensions: Benzoannelated [5.5.5.5]Fenestranes and Beyond

Hak-Fun Chow[1] *and Dietmar Kuck*[2]

[1] *Department of Chemistry, Institute of Molecular Functional Materials and The Center of Novel Functional Molecules, The Chinese University of Hong Kong, Shatin, Hong Kong*
[2] *Department of Chemistry and Center for Molecular Materials (CM_2), Bielefeld University, Bielefeld, Germany*

10.1 Introduction

We learnt about "simple" spirocyclanes since Adolf von Baeyer's work published in 1900 [1]. We were aware of their two-dimensional, centrotetracyclic congeners, nowadays known as the fenestranes, since 1969, when Hass and Prelog published their first synthesis of the so-called vespirenes, which represent 9,9-spirobifluorenes bearing a second spiro axis [2]. Soon after, Georgian and Saltzman coined the name "fenestranes" and published the synthesis of two alicyclic fenestrane congeners [3]. And we even knew quite a number of the three-dimensional, centrohexacyclic versions of spiro compounds that first appeared in the form of the so-called Simmons-Paquette molecule in 1981 [4, 5]. This article describes a hopefully inspiring journey along selected stations of two- and three-dimensional carbocyclic spiro structures with fenestranes representing the key motif.

Spirocyclic organic structures have always had a special appeal to chemists because of the fact that two rings are fused by only one single, mostly carbon, atom [6]. On the one hand, the construction of quaternary centers has ever had a fascination on its own, with steric restrictions and the propensity for intervening rearrangements on the way to the final product [7]. On the other hand, the spiro center and the adjacent four α-atoms may enjoy a certain protection owing to the steric shielding about the neopentane core of the spiro structure. Furthermore, the conformational and configurational properties of spirane derivatives are remarkable: While spirocyclic molecules have a formal "spiro" axis crossing the central atom and two points at the remote edges of the rings, conformational forces give rise to two distinct half-axes connected at the spiro center. Therefore, if viewed as a nano-mechanical linking unit, a spirocyclic structure represents a molecular analog of a cardan (or U) joint in macroscopic dimensions. Hence, spiro structures are formally linear but intrinsically bent and particularly flexible.

Spiro Compounds: Synthesis and Applications, First Edition. Edited by Ramon Rios Torres.

In addition, the chirality of spiro compounds is an important subfield of stereochemistry [8] and gains continuous attention in enantioselective synthesis and catalysis [9, 10].

All these characteristic aspects and features of "simple" spirocyclic compounds apply also to fenestranes (Figure 10.1). Fusing two additional rings about the two pairs of α-carbon atoms of a given spirane structure generates an [*m.n.p.q*]fenestrane, with four rings of the ring sizes *m*, *n*, *p*, and *q* (Figure 10.1b). Starting from the original spirane of the ring sizes *m* and *p*, corresponding to an [(*m* − 1), (*p* − 1)]spiro structure, another [(*n* − 1), (*q* − 1)]spiro unit evolves with an orientation spatially orthogonal to the first one (Figure 10.2). Fusing only one such ring instead of two leads to what is called a "broken" fenestrane, sometimes termed [*m.n.p*]fenestrane (Figure 10.1d) [6]. Pushing this idea further, a spiro compound has been called an [*m.n*]- or a twofold broken fenestrane in early discussions on "structural pathologies" [11]. Pursuing the reverse direction of these logics, we can fuse one or even two more rings across the original neopentane core of a given fenestrane structure, such as **1** (e.g. **2**, Figure 10.2). This leads to so-called centropentacyclic or, respectively, even centrohexacyclic structures with truly three-dimensional stretch into space [12–16]. Clearly, the hexacyclic construction (**3**) represents the end point of this concept, with three spiro units

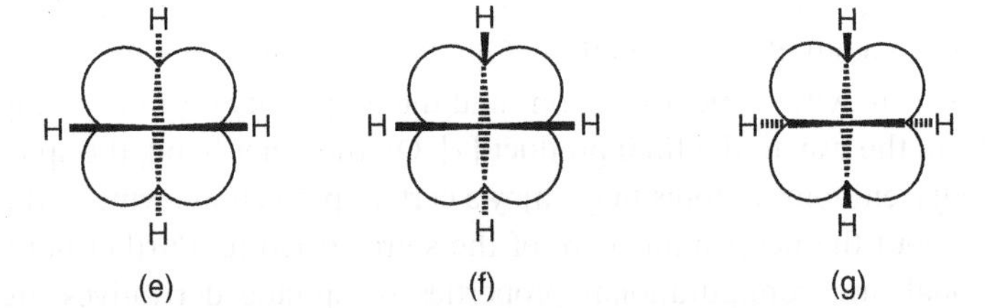

Figure 10.1 Fenestrane motifs and their multifariousness. Top: (a) A stylistic representation of the prototypical "molecular window," [4.4.4.4]fenestrane, (b) the general constitutional formula of any fenestrane, (c) the constitution of the smallest carbocyclic fenestrane core, known by experiment to date, a [4.4.4.5]fenestrane derivative, and (d) the general formula of "broken" fenestranes. Bottom: Three selected configurations of fenestranes, including (e) well-known all-*cis*-fenestranes, (f) rather well-known *cis,cis,cis,trans*-fenestranes, and (g) elusive all-*trans*-fenestranes. *Source:* Greenberg and Liebman [6].

merged to each other at a common central carbon atom and, in the same time, about a joint neopentane core. Again, all aspects and features of simple spiro structures apply to the centrohexacyclic congeners, occasionally termed "super" spiranes. Some additional ones concerning unusual topological properties of such centrohexacyclic (and likewise "super") fenestranes may be added [17, 18].

The name-giving [4.4.4.4]fenestrane hydrocarbon, C_9H_{12} (cf. Figure 10.1a), or a derivative thereof has never been synthesized, in spite of considerable efforts. Agosta and his group succeeded in preparing a derivative of all-*cis*-[4.4.4.5]-fenestrane (Figure 10.1c), the smallest fenestrane compound ever studied experimentally [19]. Molecular strain is an obvious feature of such polycyclic compounds and tolerable in the all-*cis*-[4.4.4.5]fenestrane framework. The skeleton of all-*cis*-[5.5.5.5]fenestrane, a "tetrafuso-centrotetraquinane" **2** [12] first reported by Keese in 1984 [20], is much less strained. It is accessible by several routes and many derivatives of this parent hydrocarbon have been synthesized, including the corresponding 2,5,8,11-tetraene by Cook [21] and the tetrabenzo analog by Kuck (see Section 10.2) [22]. However, molecular strain can be accumulated into [m.n.p.q]fenestranes if one or more of the four neopentane-type C–C bonds have *trans*- rather than *cis*-configuration (cf. Figure 10.1e) [23]. Quite a number of *cis,cis,cis,trans*-fenestranes (cf. Figure 10.1f), including small-ring congeners, have been prepared over the past decades; however, the diastereomeric form of the highest relative strain, an all-*trans*-isomer (cf. Figure 10.1g), has never been obtained by experiment.

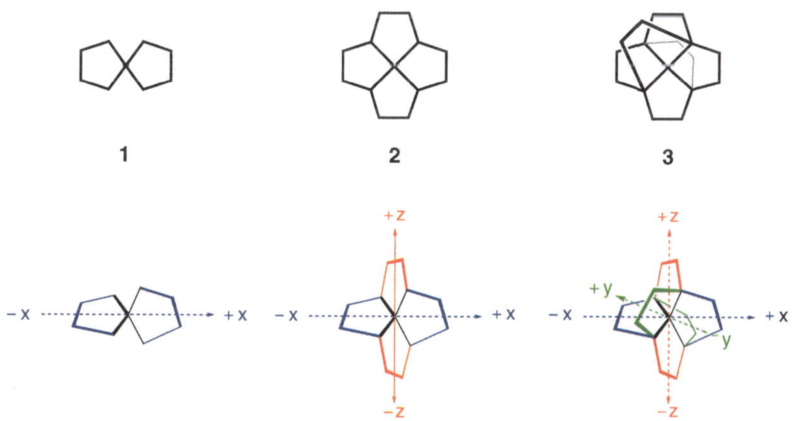

Figure 10.2 From spiranes to fenestranes and beyond. Top: Spiro[4.4]nonane (**1**), [5.5.5.5]fenestrane (**2**) and centrohexaquinane (**3**), bearing two or, respectively, even three such mutually fused spiro units. Bottom: Three-dimensional representations depicting the mutual orientation of the spiro axes in an all-*cis*-[5.5.5.5]fenestrane and in centrohexaquinane. Conformational distortions are ignored in all three cases.

Besides the strictly stepwise "aufbau" routes using a sequential construction of the four rings, there are two obvious synthesis strategies for [5.5.5.5]fenestranes and their lower and higher congeners. In both cases, it may be possible to avoid the isolation of the corresponding "broken" [5.5.5]fenestrane intermediate **4**. These two routes are depicted retrosynthetically in Scheme 10.1 for the case of [5.5.5.5]fenestrane (**2**). The first approach starts from a spiro compound derived from **1** and involves the introduction of a second spiro unit vertically to the first one, as already outlined conceptually above. The other strategy is less intuitive since it circumvents the spiro path. In this case, the starting point is a classical bicyclic structure, namely an [m.n.0]bicycloalkane derivative derived from **5**. In this case, two adjacent rings of the to-be-built fenestrane structure are already present in the starting material. Gund and Gund suggested to differentiate between three ways of merging two rings with each other, leading to spiro-, "fuso"-, and "ponte"-bicyclanes, depending on the number of atoms (one, two, or three, respectively) shared by these rings [12]. The respective structures of spiro[4.4]nonane (**1**), bicyclo[3.3.0]octane (**5**), and bicyclo[2.2.1]heptane (norbornane) are prototypical for these cases. The authors also introduced the term "centropolycyclanes," assigning polycyclic hydrocarbons in which all rings share one central carbon atom [12]. According to this nomenclature, we can design a "spiro-route" and a "fuso-route," both of which lead to the fenestrane framework. Interestingly, the latter approach has been used much more successfully for alicyclic fenestranes while the former route turned out to be more successful for the synthesis of benzoannelated fenestranes, including the vespirenes mentioned in the very beginning [2]. These

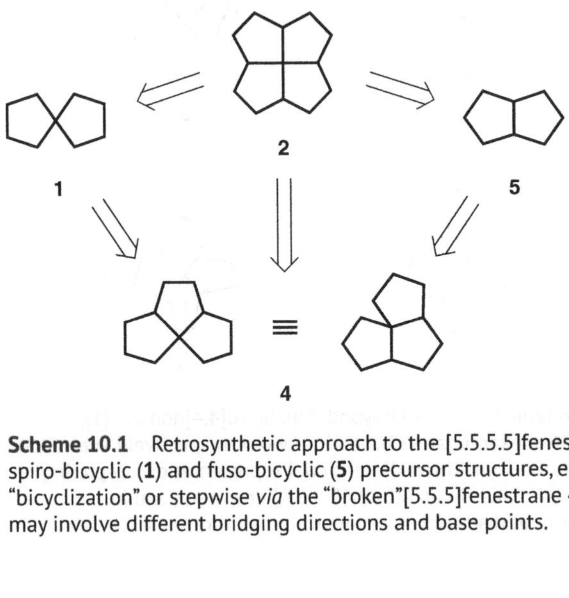

Scheme 10.1 Retrosynthetic approach to the [5.5.5.5]fenestrane skeleton from spiro-bicyclic (**1**) and fuso-bicyclic (**5**) precursor structures, either directly by "bicyclization" or stepwise *via* the "broken"[5.5.5]fenestrane **4**. The bicyclization routes may involve different bridging directions and base points.

two different synthesis strategies will be presented and discussed in detail in the following sections, and the reasons for the differing viability of the two routes may become obvious in the light of the introductory remarks made above.

Finally, it should be mentioned that several comprehensive reviews on fenestrane chemistry have been published over the past decades [24–32]. All of them discuss the major issues concerning the investigative work of the fenestrane structures. These motifs comprise natural compounds chemistry, the advanced art of organic synthesis, the rather complex stereochemistry of fenestranes, and, last but not least, the theoretical aspects leading back to the famous Van't Hoff–Le Bel hypothesis of the tetrahedrally tetracoordinated carbon atom and the quest for "planar" or "flattened" tetracoordinated carbon atoms embedded in molecular organic or metallo-organic structures [33–39]. The first review by Agosta et al. published in 1987 mainly dealt with the synthesis of small-ring [*m.n.p.q*]fenestranes, with $m = n = p = 4$ and only $q = 5$. Still to date, this represents the record for the most highly strained, experimentally accessed derivative [19]. As mentioned above, [4.4.4.4]fenestranes (and any lower congeners down to the formal limit, the [3.3.3.3]fenestranes) have never been synthesized. Starting even earlier, Keese published several reviews on mostly normal-ring fenestranes [24]. In 1998, we ourselves reported in a book chapter about benzoannelated fenestranes, a formerly almost unknown subfamily [2], first emerging systematically in 1986 [22, 27]. Keese's latest review on fenestranes populating the "carbon flatland" appeared in 2006 in a special issue comprising "Design of the Molecular World" [30], together with our own review on benzoannelated fenestranes and other "centropolyindanes" [31]. In 2013, Suffert and his team contributed another, and to date the most recent, review focusing on the syntheses of alicyclic fenestranes of various ring sizes and including the multifaceted field of heterocyclic fenestranes [32]. They also provided a compendium of virtually all fenestrane derivatives that had been synthesized thitherto.

In the present chapter, we will not dwell on those general treatments on fenestrane chemistry and the (still highly challenging!) planarization of tetracoordinate carbon. Rather, we will present selected examples for the successful synthesis of alicyclic and benzoannelated fenestranes that reflect the best viable strategies developed to date. By restricting ourselves to carbocyclic fenestranes that are built mainly from five-membered rings, we will display the development of fenestrane chemistry from the early syntheses of Keese's all-*cis*-[5.5.5.5]-fenestrane (**2**) and Cook's related all-*cis*-[5.5.5.5]fenestra-2,5,8,11-tetraene (**6**, Figure 10.3) to the parent all-*cis*-tetrabenzo[5.5.5.5] analog **7**, which we coined "fenestrindane," a particularly important polyaromatic structure. A considerable series of derivatives of the hydrocarbon **7** have been synthesized since. Beyond the fenestrindanes, the construction of larger three-dimensional "super spiro" analogs, such as centropentaindane (**8**) and centrohexaindane (**9**), and the

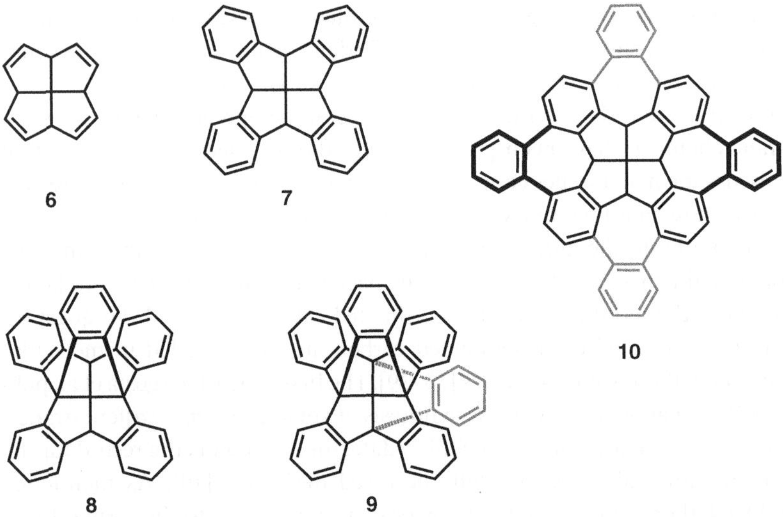

Figure 10.3 Systematic extensions of Keese's parent all-*cis*-[5.5.5.5]fenestrane (**2**) and Cook's all-*cis*-[5.5.5.5]fenestra-2,5,8,11-tetraene (**6**) by benzoannelation and 1,2-benzeno-bridging into higher dimensions: all-*cis*-tetrabenzo[5.5.5.5]fenestrane ("fenestrindane," **7**), the truly three-dimensional centropentaindane (**8**) and centrohexaindane (**9**), and the warped, quasi-three-dimensional polyaromatic derivative **10** bearing four bay-bridging 1,2-benzeno (or *ortho*-phenylene) units.

warped, quasi-three-dimensional polyaromatic derivatives, such as the large fourfold bay-bridged structure **10**, will be discussed. Some limitations to these approaches and efforts for alternative synthesis strategies will be touched upon as well.

We hope that these selected fruits grown on the still rather exotic field of fenestranes, which obviously extends slowly and reluctantly fertilizes larger, more general fields of chemistry, will find the readers' appreciation and inspire many chemists and material scientists to undertake their own research ventures in this exciting world of two- and threefold spiro compounds. We can never know where our ideas may fly to [40].

10.2 Selected Synthesis Strategies

The construction of carbocyclic fenestrane frameworks requires multistep syntheses. Here, we present only those of them which owe their elegance from the formation of more than one C–C bond in the key steps. As mentioned above, the bicyclo[3.3.0]octane skeleton has been favorably used several times as a basis for the annelation of two further rings. Keese and his group were the first to

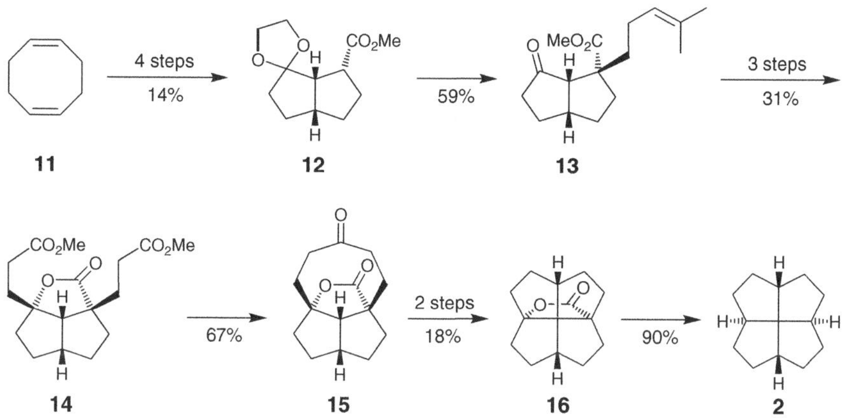

Scheme 10.2 Keese's first synthesis of all-*cis*-[5.5.5.5]fenestrane (**2**). *Sources:* Based on Luyten and Keese [20a]; Pfenninger et al. [20b].

demonstrate the success of this approach, as depicted in Scheme 10.2 [20]. The bicyclo[3.3.0]octane framework required as a groundwork had been constructed previously from cycloocta-1,5-diene (**11**) in two steps [41]. Subsequent refunctionalization in two further steps furnished the bicyclic *endo*-ester **12**. A first tentacle-like residue was then introduced with *exo*-orientation to give compound **13**, followed by a further sequence of steps affording two identical methyl propionate tentacles with fixed *exo*-stereochemistry in diester **14**. This allowed the authors to construct a cyclooctanone unit on top of the lactone-locked bicyclo[3.3.0]octane skeleton by Dieckmann condensation yielding tetracycle **15**. Conversion to the corresponding tosyl hydrazone and subsequent carbene insertion into the C–H bond of the central bridgehead position gave the pentacyclic lactone **16** – *a closed window but a carbon dioxide too many (!)*. Finally, reductive decarboxylation with an excess of palladium-on-charcoal at 310 °C furnished the desired all-*cis*-[5.5.5.5]fenestrane **2** for the very first time. In the following years, Keese and his group published modifications of this strategy to construct all-*cis*-[5.5.5.5]fenestrane (**2**), all of which were based on fuso-diquinane (i.e. bicyclo[3.3.0]octane) key intermediates decorated with two lateral aliphatic tentacles to annelate the third and fourth five-membered rings in the same time [20b, 42]. These efforts also included a futile attempt to construct the *cis,cis,cis,trans*-isomer of **2** (cf. Figure 10.1f) which resulted in another access to the all-*cis*-isomer [42b].

It should be mentioned that the pentacyclic compound **16** was already reported in 1979 and then represented the second known derivative of a carbocyclic [5.5.5.5]fenestrane [43]. The breakthrough to the parent hydrocarbon **2** was the rather harsh reductive decarboxylation step mentioned above. Interestingly, the synthesis of the fourfold unsaturated analog of **2**, all-*cis*-[5.5.5.5]-

fenestrane-2,5,8,1-tetrene (**6**) reported by Cook and his group in 1985 [21], suffered from a similar difficulty in the final step of the sequence. In fact, the very first carbocyclic [5.5.5.5]fenestrane derivative ever described was the 2,6,8,12-tetraketone **21** reported by Cook, Weiss, and their allies in as early as 1978 [44]. Remarkably, the name "staurane" (reminiscent for the Greek *stauros*, cross) suggested for the [5.5.5.5]fenestrane framework by these authors did not compete successfully with Saltzman's name coining.

Cook's synthesis of fenestratetrene **6** is illustrated in Scheme 10.3 [21]. Again, the construction of the [5.5.5.5]fenestrane (or tetrafuso-centrotetraquinane) skeleton is based on a diquinane derivative, bicyclo[3.3.0]octane-3,7-dione **19**, which bears a single residue at a bridgehead position of the bicyclo[3.3.0]octane unit in this case. This key intermediate was accessible from the monocyclic glyoxal hydrate derivative **18**, which in turn was made from cyclopentenecarboxylic acid **17**. Although the first four steps are quite efficient following a previously established work [45], overall eight preparative steps are required to synthesize the diquinanedione **19** [44]. Oxidative cleavage of the cyclopentene double bond in **19** produced the bicyclic diketo-diacid **20** bearing a singly based, doubly armed tentacle, which underwent efficient acid-catalyzed twofold condensation to give [5.5.5.5]fenestranetetrone **21**. In contrast to earlier attempts, reduction of the two 1,3-diketone groupings of **21** was successful when diborane was used as a reducing reagent. In fact, the fenestrane tetraalcohol **22** was obtained in very good yield as a mixture of stereoisomers without any (otherwise intervening) retroaldol-type cleavage of the strained fenestrane framework. In the final step, fourfold dehydration of **22** yielding all-*cis*-[5.5.5.5]fenestratetaene **6** was achieved in aprotic-dipolar medium (hexamethylphosphoramide, HMPA) at a high temperature to avoid Grob-type

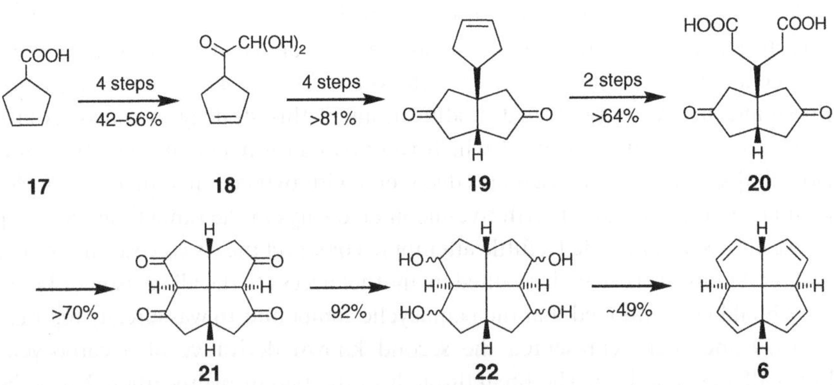

Scheme 10.3 Cook's first-reported synthesis of all-*cis*-[5.5.5.5]fenestra-2,5,8,11-tetraene (**6**). *Sources:* Based on Desphande et al. [21a]; Venkatachalam et al. [21b].

fragmentation processes. Later studies provided an alternative approach to the [5.5.5.5]fenestrane skeleton starting from the dialdehyde analog of diacid **20** [46].

Overall, both Keese's and Cook's syntheses of the parent all-*cis*-[5.5.5.5]fenestrane hydrocarbons **2** and **6**, respectively, represented important progresses in the field of non-natural fenestranes. While the syntheses of a remarkable number of other carbocyclic [*m.n.p.q*]fenestranes have been reported by these and other research groups over the past decades, most of them require long sequences of preparative steps. Relatively efficient constructions involve photolytic [2+2]-cycloaddition reactions generating strained cyclobutane rings. Other methods involve the Pauson–Khand reaction to achieve cyclopenta-annelation. The reader is referred to the comprehensive reviews by Keese [30] and Suffert and coworkers [32]. However, as mentioned above, it is highly remarkable that none of the successful syntheses of alicyclic (non-aromatic) fenestranes employ spiro compounds as the starting materials. Attempts may have been around, as can be inferred from extended work on modern spiroannelation using metathesis [47]. The main reason for this obvious lack in the field of fenestrane synthesis may be that the construction of fenestranes from spirane derivatives requires suitable – at least twofold – functionalization of the neopentane core of the spirocyclic framework. However, such 1,3-difunctionalized derivatives are prone to undergo fragmentation under various conditions, be it in the course of retro-aldol reactions or as Grob-type fragmentation on the level of the 1,3-diol derivatives [48]. Scheme 10.4 gives a simplified sketch on this situation. If the neopentane core of a spiro structure bears suitable functional groups at two of its α-carbons, either both in the same ring, as in (a), or one in each of the rings, as in (b), ring opening with concomitant formation of an exo- (e) or, respectively, endocyclic (f) C–C double bond can occur very easily. Because of this severe complication, tempting ideas to use the corresponding tri- (c) or even tetrafunctionalized (d) congeners as intermediates for the construction of fenestrane frameworks may not be viable to the best of our knowledge.

In fact, the same problem holds true for benzoannelated spiranes. Very telling early examples examined by Schönberg and coworkers were known since 1973 when they demonstrated that 1,3-indanediols (e.g. **24**) and [2.2]spirobiindane-1,3-diols (e.g. **28** and **30**) easily underwent Grob fragmentation in acidic media (Scheme 10.5) [49]. We ourselves also encountered such "destructive" interferences in our early studies toward the synthesis of benzoannelated fenestranes [22b, 50]. Ironically, in such cases, twofold cyclodehydration led to truly "flat" (i.e. planar) aromatic hydrocarbon structures, such as **26**, **29**, and **31**.

10.2.1 The Access to Benzoannelated Fenestranes

In contrast to the established syntheses of normal-ring alicyclic fenestranes, which are based on bicyclo[3.3.0]alkane derivatives as shown above, the

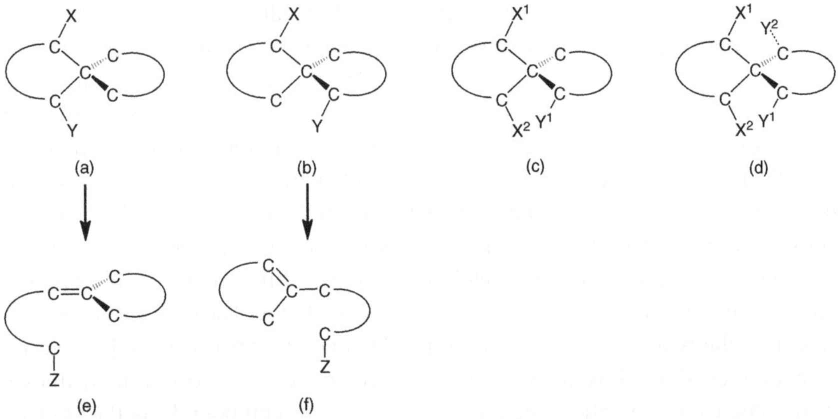

Scheme 10.4 Three hypothetical functionalized spiro structures as starting points for the construction of [*n.m.p.q*]fenestrane frameworks. Both (a) 1,3-difunctionalized and (b) 1,1′-difunctionalized precursors are prone to ring-opening reactions giving rise to the monocyclic products (e) and (f), respectively. The corresponding 1,1′,3-tri- and 1,1′,3,3′-tetrafunctionalized spiranes (c) and (d) should suffer such degradation processes even more rapidly. Note that concrete hybridization and stereochemistry at the functionalized carbon atoms are ignored.

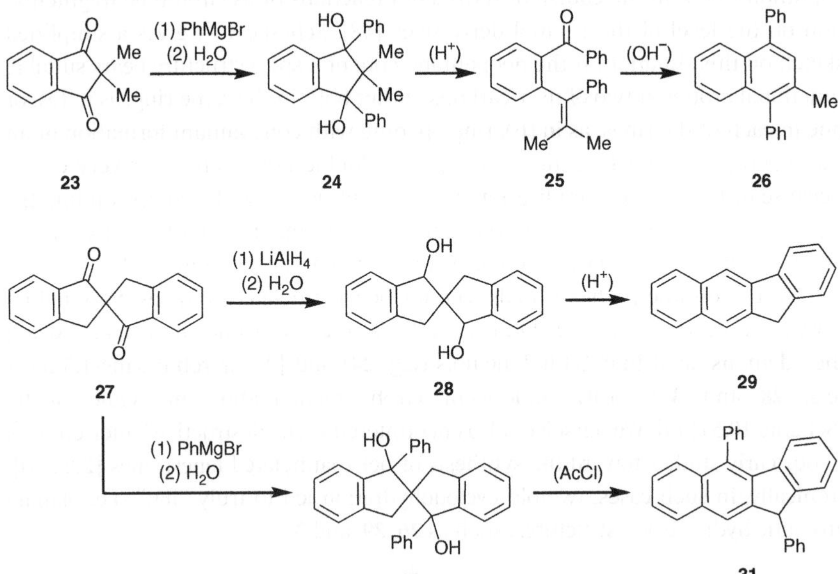

Scheme 10.5 Synthesis of 1,3-indanediol **24** and [2,2′]spirobiindane-1,1′-diols **28** and **30** from the corresponding diketone precursors **23** and **27** and their Grob fragmentation under acidic conditions. Cyclodehydration can be pushed forward by use of base for **25** or proceeds *in situ* under acid catalysis giving the respective fluorenes **29** and **31**. *Sources:* Based on Schönberg and Mamluk [49a]; Schönberg and Sidky [49b].

construction of the benzoannelated congeners turned out to be more successful if the spiro-route is pursued. First efforts in this line were published by Ten Hoeve andWynberg,whousedthespirocyclizationofcyclic1,3-diketoneswith1,5-diarylpenta-1,4-diene-3-ones to construct promising spirocyclic intermediates such as **32** (Scheme 10.6) [51]. In particular, two 5-methylfuryl residues were envisaged to provide the bridging elements to generate the second spiro axis in the hypothetical benzoannelated [5.6.6.6]fenestradienedione **34**. However, attempts to perform twofold aldol-type cyclocondensation of **33** after acid-catalyzed hydrolysis of the furyl units in compound **32** proved to be unsuccessful.

A few years later, we developed a very efficient access to the threefold benzoannelated broken [5.5.5]fenestrane **37**, which was prepared in two steps and high yield from 2,2-dibenzyl-1,3-indanedione **35** *via* the corresponding indanediol **36** [22b, 50]. The same twofold cyclodehydration (bicyclization) strategy was also the key step in the synthesis of tribenzotriquinacene (TBTQ) and its derivatives, another member of the cyclopolyindane family [31, 50, 52]. Therefore, we expected

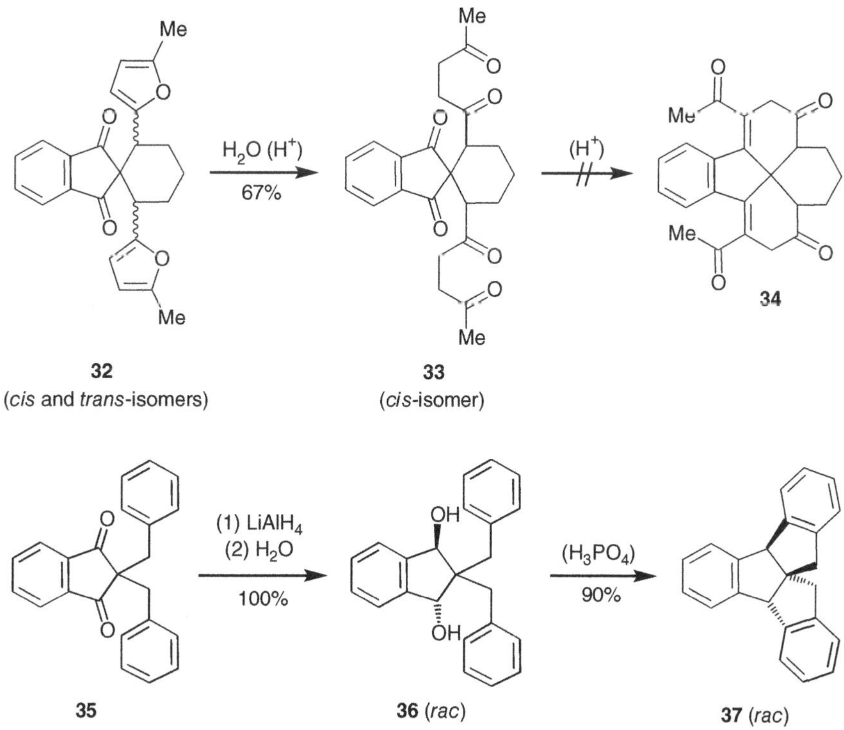

Scheme 10.6 Attempted construction of the benzoannelated [5.6.6.6]fenestrane derivative **34** [51] and successful synthesis of the threefold benzoannelated broken [5.5.5]fenestrane **37** by twofold cyclodehydration of the 2,2-bibenzyl-1,3-indanediol **35**. *Sources:* Based on Ten Hove and Wynberg [51a]; Kuck [50].

that spiro analogs of indanediol **36** bearing a cyclohexane moiety like in compound **32** would undergo a similar bicyclization reaction provided that the two aryl residues have the suitable *trans*-orientation. In fact, both the *cis*- and *trans*-diaryl-substituted spiro[cyclohexane-1,2′-indane]-1′,3′,4-trione isomers **38** and **42**, respectively, are accessible under suitable reaction conditions that enable either thermodynamic or, respectively, kinetic control of the double Michael spiroaddition process [53, 54]. This strategy proved to be highly successful not only with the *trans*-diphenylspirotriketone **42** as the precursor [22] but, unexpectedly, with the *cis*-diphenyl isomer **38** as well [55]. An overview of the selected synthesis routes that were investigated in great detail, including several mechanistic aspects, is given in Scheme 10.7.

Cis-2,6-diphenyl-substituted spiro[cyclohexane-1,2′-indane]-1′,3′,4-trione (**38**) and its *trans*-isomer **42** are the prototypical starting materials for the synthesis.

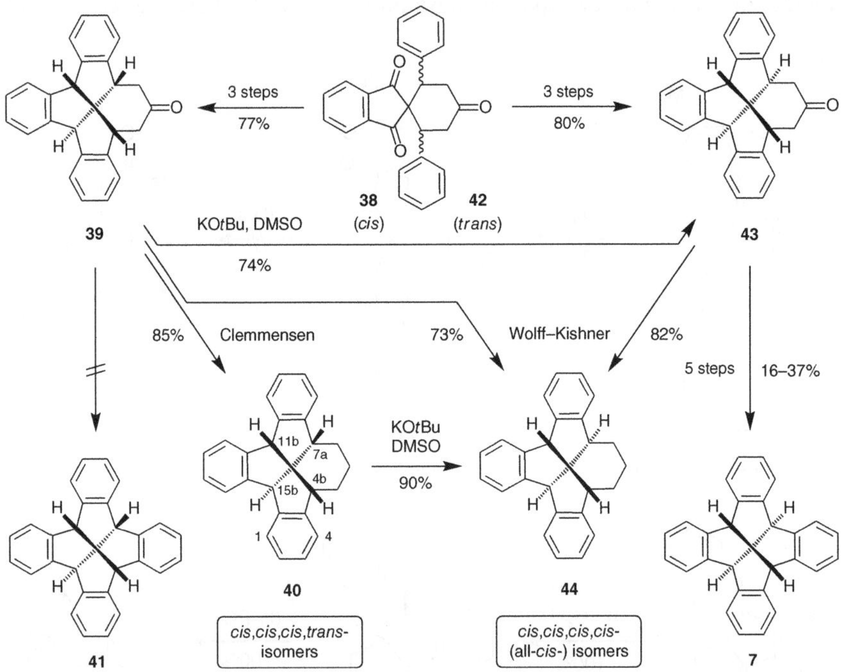

Scheme 10.7 Synthesis of the benzoannelated [5.5.5.6]fenestranes with *cis,cis,cis,trans*-(**39, 40**) and all-*cis*-configuration (**43, 44**). Owing to the based-induced epimerization at the "strained" bridgehead of the fenestranone **39**, giving **43**, the final target, fenestrindane (**7**) can be synthesized from both the *cis*- and the *trans*-diphenylspirotriketones **38** and **42**, whereas the *cis,cis,cis,trans*-isomer of **7**, "epi-fenestrindane" **41**, is not accessible by this strategy. *Sources:* Based on Kuck and Bögge [22a]; Kuck [22b].

Two pathways have been elaborated to prepare the stereoisomeric tribenzo-[5.5.5.6]fenestranones **39** and **43**, respectively, in three steps in both cases. Either the spirotriketones are first reduced to the corresponding spirotriols and then subjected to twofold cyclodehydration – in strict analogy to the conversion of indanediol **36** to the broken fenestrane **37** – with oxidation of the remaining cyclohexanol ring in the third step. Alternatively, the cyclohexanone unit in triketone **38** or **42** can be first selectively protected as an ethylene ketal, and the resulting diketone is then reduced to the corresponding spirodiol. In this case, the third step consists in an efficient twofold cyclodehydration with concomitant hydrolysis of the ketal grouping. In both ways, the *cis,cis,cis,trans*- and all-*cis*-[5.5.5.6]fenestranones **39** and **43** are obtained in good to excellent yields with perfect retention of the stereochemistry. Starting from suitably substituted 1,3-indanediones and 1,4-benzylideneacetones, only four steps are required overall. This spirane-based route to [5.5.5.6]fenestranes is highly versatile and has been used for the synthesis of a remarkably large number of benzoannelated fenestranes. Interestingly, the strained *cis,cis,cis,trans*-[5.5.5.6]fenestrane skeleton, being ca. 46 kJ/mol more strained than the all-*cis*-framework [55b], can be easily epimerized to the all-*cis*-framework under basic conditions. Thus, not only fenestranones, such as the parent compound **39**, can be stereochemically adjusted; epimerization can also be achieved with the corresponding tribenzo[5.5.5.6]fenestrane hydrocarbon **40**, which is nicely formed under reduction of the *cis,cis,cis,trans*-[5.5.5.6]fenestranone **39** in acidic medium, to give the all-*cis*-isomer **44** in good yield. Deuterium labeling experiments proved that the base-induced deprotonation–reprotonation processes employing potassium *tert*-butylate in dimethyl sulfoxide at 20 °C involves the two C–H bonds at the benzhydrylic bridgeheads (C-11b and C-15b) as well as the C–H bond at the strain-inducing bridgehead (C-7a) but not that of the other benzylic bridgehead (C-4b) [55]. Under less basic conditions (potassium hydroxide in diethylene glycol at 180 °C), however, only the epimerizing bridgehead suffers H/D exchange. This demonstrates that the C–H bond at the strained benzylic bridgehead of *cis,cis,cis,trans*-tribenzo[5.5.5.6]fenestrane **40** is more acidic than the C–H bonds at the two benzhydrylic bridgehead positions.

The classical synthesis of fenestrindane (**7**), the prototypical hydrocarbon for all higher benzoannelated fenestranes with all-*cis*-configuration, can be accomplished in five subsequent steps in moderate overall yield. The all-*cis*-stereochemistry of the fenestrane framework is not affected since it represents the most stable configuration throughout. As a consequence, the construction of the corresponding *cis,cis,cis,trans*-tribenzo[5.5.5.5]fenestrane **41** (dubbed "epi-fenestrindane") has remained futile in our hands. Not only is the contraction of the cyclohexanone ring in **39** (after α,α'-dibromination) a matter of base-assisted Favorskii reaction, which would give rise to epimerization, but also other ring contraction methods avoiding basic reaction media proved to be non-productive [55b, 56]. Therefore, epi-fenestrindane (**41**) which, according to semiempirical calculations, should be ca. 150 kJ/mol more highly strained than the *cis*-isomer **7**, may remain an elusive goal in fenestrane chemistry [55b]. It may be noted in this context that only one single, yet alicyclic, *cis,cis,cis,trans*-[5.5.5.5]fenestrane derivative is known. In this case, the strained configuration is stabilized by annelation of a further ring about the fenestrane core [57].

Turning back for a short while to the alternative spiro- and fuso-routes discussed in the beginning, it may be mentioned that the "long and winding road" to the world of fenestranes has sometimes appeared quite a challenge [51]. Our successful and well-established spiro-route to benzoannelated fenestranes, discussed above, is very efficient as far as the tribenzo[5.5.5.6]-congeners are concerned but anyway quite lengthy if we aim at fenestrindane (**7**) and its derivatives as the goal of synthesis. In fact, the synthesis becomes even more tedious when it comes to extended polyaromatic carbon frameworks, as will be shown below. Therefore, over the years, there have been attempts to switch back to the fuso-route and explore alternate approaches to fenestrindane. A series of such attempts, which turned out to be unsuccessful, though, was carried out some time ago already. An instructive example of these futile approaches, being reminiscent of Keese's first successful synthesis of the parent all-*cis*-[5.5.5.5]fenestrane **2** [20], is depicted in Scheme 10.8 [56]. The benzoannelated bicyclo[3.3.0]octanone **45**, bearing an *exo*-oriented phenyl residue, is a long-known, easily accessible, and valuable fuso-diindane building block, which undergoes attack by nucleophiles from the convex

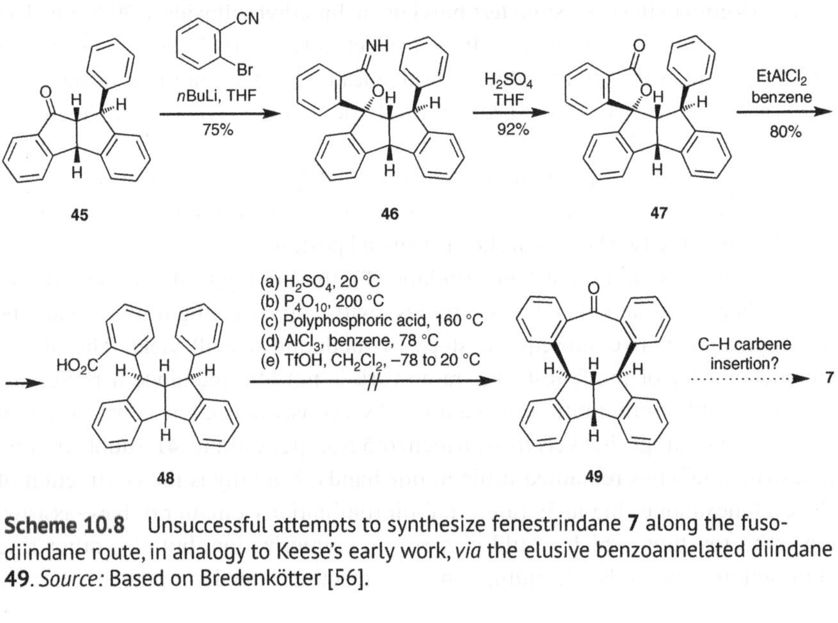

Scheme 10.8 Unsuccessful attempts to synthesize fenestrindane **7** along the fuso-diindane route, in analogy to Keese's early work, *via* the elusive benzoannelated diindane **49**. *Source:* Based on Bredenkötter [56].

molecular side [58, 59]. In the present case, 2-lithiobenzonitrile [60] was added to generate a spiro-annelated iminofuran ring in compound **46**, which was hydrolyzed to give the corresponding spirolactone **47**. Reductive cleavage of the lactone ring with ethylaluminum dichloride [61] in benzene gave the bis-*exo*-9,10-diphenyldiindane **48**, suitably equipped with a carboxyl functionality in one of the *ortho*-positions, in good overall yield. Unfortunately, all of the various methods examined to enforce cyclization of the carefully constructed benzoic acid **48** by electrophilic attack proved to be unsuccessful. Notably, no defined products could be identified either. Thus, formation of the cyclooctanone ring merged with a diindane moiety in the elusive ketone **49** failed and the envisaged carbene insertion technique toward fenestrindane **7** could not be tested. It appears that benzoannelation in intermediates like the cyclooctanone **49** is rather a handicap than an advantage because it may increase the conformational rigidity of the reactive intermediates as compared with the alicyclic analogs leading to all-*cis*-[5.5.5.5]-fenestrane **2**. Therefore, this and similar approaches along the fuso-route to fenestrindane **7** were abandoned [56].

10.2.2 Derivatives of [5.5.5.6]-Fenestranones and Fenestrindane

As mentioned above, the syntheses of a number of derivatives of benzoannelated [5.5.5.6]- and [5.5.5.5]fenestrane hydrocarbons were accomplished in our early studies. In particular, it appeared tempting to investigate the possibilities to construct larger three-dimensional polyaromatic carbon frameworks based on the fenestrindane nucleus. Owing to the saddle-like topography of the fenestrindane molecule, extended molecular constructions bearing two opposite concave surfaces in tangential fusion should be possible. An inspiring idea consisted of bridging all of the four molecular bays by C_2-units such as 1,2-ethylidene and *ortho*-phenylene (1,2-benzeno) units allows us to connect the otherwise isolated π-electron systems of the four benzene rings fused to the [5.5.5.5]fenestrane core [31]. To this end, we first undertook various efforts to use the spiro-route to the benzoannelated [5.5.5.6]- and [5.5.5.5]fenestranes by incorporating various aryl units from the very beginning [62]. A selection of such synthesis approaches and structural targets is collected in Schemes 10.9 and 10.10 and in Figure 10.4.

The three isomeric all-*cis*-benzodinaphtho[5.5.5.6]fenestranones **51**, **53,** and **54** were synthesized from 1,5-di(α-naphthyl)- or 1,5-di(β-naphthyl)penta-1,4-diene-3-one, respectively, and 1,3-indandione to give the corresponding spirotriketones **50** and **52** with the desired *trans*-dinaphthyl stereochemistry in 45 and 46% yield, respectively (Scheme 10.9) [62, 63a]. Reduction to the related spirotriols followed by twofold cyclodehydration gave the corresponding benzodinaphtho-all-*cis*-[5.5.5.6]fenestranols, which were subjected to oxidation in the last step. In the sequence involving the α-naphthyl derivative, the

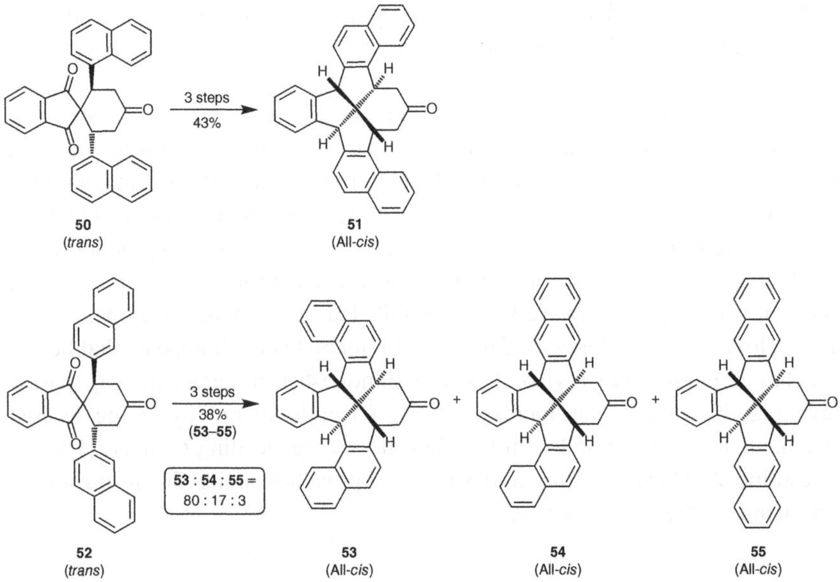

Scheme 10.9 Naphthoannelated all-*cis*-[5.5.5.6]fenestranones **51** and **53-55** accessible along the classical spirane route. The ratios given for the isomers **53-55** were deduced from the relative amounts of the corresponding fenestranol precursors obtained as a mixture. The isomers **53** and **54** were obtained in pure form, whereas isomer **55** was not isolated. *Sources:* Based on Seifert and Kuck [63a].

C_2-symmetrical benzodinaphtho-all-*cis*-[5.5.5.6]fenestranone **51** was isolated as the sole product of bicyclization in 43% overall yield in the last three steps. On the other hand, starting the sequence with the β-naphthyl derivative, a mixture of the three possible isomeric benzodinaphtho-all-*cis*-[5.5.5.6]fenestranones, including compounds **53** and **54** as the major components, was obtained in 38% yield in the last three steps. These isomers were also prepared by separate oxidation of the corresponding fenestranols. Notably, cyclodehydration with the di(α-naphthyl) fenestranetrioles does not involve peri-attack which would generate one or two new six-membered rings. In the case of the di(β-naphthyl)fenestranetrioles, cyclodehydration necessarily gives rise to two new five-membered rings. As expected, the attack at the α-position of the naphthyl residues is strongly dominant [62, 63a]. Further efforts aiming at the conversion of the fenestranones to the corresponding fenestrindane derivatives were not undertaken; however, it appears interesting to pursue these synthesis lines using more highly benzoannelated congeners.

The first phenyl-substituted [5.5.5.6]fenestrane derivatives **56–58** (Figure 10.4) were also synthesized with the perspective to eventually extend its the π-electron system by bridging the four bays of the fenestrindane skeleton with *ortho*-phenylene

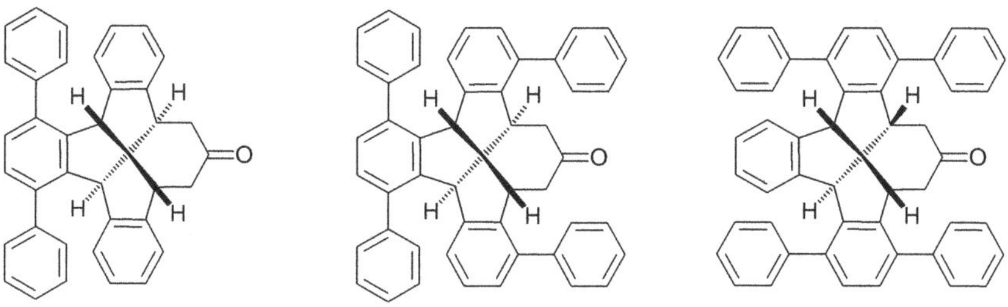

56 (All-*cis*) **57** (All-*cis*) **58** (*cis*, *cis*, *cis*, *trans*)

Figure 10.4 The first benzoannelated [5.5.5.6]fenestranones bearing two (**56**) or four (**57** and **58**) phenyl residues at the strategic *ortho*-positions, all synthesized along the classical spirane route. In contrast to the former two compounds, isomer **58** was obtained with the (unexpected) *cis,cis,cis,trans*-configuration. *Sources:* Based on Seifert et al. [63b].

units [62, 63b]. In these cases, *para*-terphenyl building blocks were employed in the starting materials and the classical spiro-route was pursued again. So far, this turned out to be successful. In the case of the diphenyl-substituted fenestranone **56**, 4,7-diphenyl-1,3-indanedione was used for the construction of the suitable spirotriketone and the reduction/cyclization/reoxidation sequence furnished compound **56** as the desired *trans*-isomer. This was one of the very first centropolyindanes that bear relatively bulky phenyl residues at the sterically shielded *ortho*-positions. This particular constitution was reflected by the characteristic magnetic shielding effect ($|\Delta\delta| \geq 1.0$ ppm) of these phenyl groups on the two protons at the *ortho* positions of the adjacent benzene nuclei, sited directly opposite and deeply within the magnetically anisotropic cone of the phenyl residues [62, 63b]. This spectral feature proved to be very useful in our later developments (see Section 10.2.3). In addition, the ^{1}H NMR spectrum confirmed the expected molecular C_2-symmetry of fenestranone **56**. The all-*cis*-[5.5.5.6]fenestranone **57** was synthesized in the same way by subjecting 4,7-diphenyl-1,3-indanedione and 1,5-bis(2-diphenylyl)penta-1,4-diene-3-one to the spiro-Michael addition in the first step. Again, the stereospecificity of the starting spirotriketone nicely translated to the all-*cis*-configuration of the fenestranone **57**. In the third case, in which we aimed at the construction of a likewise all-*cis*-configurated isomer of **58** that would bear two terphenyl moieties at opposite ends of a spiro axis, we were confronted with a partial failure. Here, we again started from parent 1,3-indanedione and the corresponding 1,5-bis(2′-*para*-terphenylyl)penta-1,4-diene-3-one. However, we were unable to generate the required *trans*-bis(terphenylyl)spirotriketone in satisfactory yield. Instead, the corresponding *cis*-spirotriketone was formed as the major isomer. Nevertheless, the subsequent three-step sequence did afford a tetraphenyl-substituted [5.5.5.6]fenestranone which obviously lacked the expected molecular C_2-symmetry, as reflected by its NMR spectrum. While we argued that the apparent C_1-symmetry may have resulted from conformational locking due to heavy intramolecular interactions between the aromatic entities, we later learned that we had obviously generated the tetraphenyl-substituted *cis,cis,cis,trans*-[5.5.5.6]fenestranone **58** along with the corresponding all-cis-isomer, which is also accessible, albeit in low yield [62, 63b]. At least in part because of this discouraging outcome, we then abandoned the otherwise promising goal to synthesize the fourfold *ortho*-phenylene-bridged fenestrindane **10** by starting from suitably phenyl-substituted building blocks. However, two decades later, an alternative aufbau strategy using the fourfold *ortho*-methoxy-substituted fenestrindane **86** proved to be highly successful (see Section 10.2.3).

Another project aiming at the construction of extended fenestrindanes that would bear completely closed π-electron systems about the central neopentane core started from building blocks that contained the phenanthrene unit instead of simple aryl rings (Scheme 10.10) [56, 64]. It was expected that the twofold-cyclodehydration methodology used that successfully for the construction

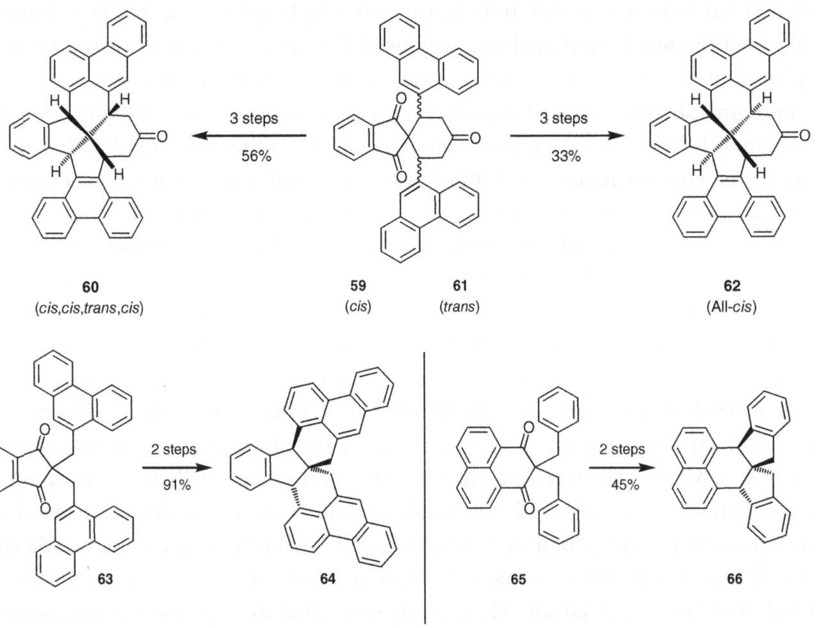

Scheme 10.10 Construction of extended benzoannelated fenestrane derivatives bearing *peri*-annelated phenanthrene units at stereoisomeric [5.5.6.6]-, [6.5.6]-, and [5.6.5]fenestrane cores. *Sources:* Based on Bredenkötter et al. [64]; Seifert and Kuck [63a].

of the [5.5.5.6]fenestrane framework and related centropolyindanes would be applicable to 9-phenanthyl-substituted spiro[cyclohexane-1,2′-indane]-1′,3′-diols as well. Therefore, we synthesized 1,5-bis[9-phenanthryl)-methyl]penta-1,4-diene-3-one as the suitable Michael component and subjected it to the double spiroaddition reaction with 1,3-indanedione. Again, both of the corresponding *cis*- and *trans*-diarylspirotriketones, **59** and **61**, respectively, were accessible in good yields. However, and much to our surprise, with the bulky end groups at the dienone acceptor, the conditions required for thermodynamic and kinetic control gave rise to the reversed stereochemical outcome: The less stable *trans*-isomer was obtained under basic conditions and the more stable *cis*-isomer was formed under in acidic medium. Nevertheless, the subsequent three steps involving protection of the cyclohexanone functionality, reduction to the corresponding 9-phenanthryl-substituted spiro[cyclohexane-1,2′-indane]-1′,3′-diols, and cyclodehydration–deprotection resulted in the stereospecific construction of two new polycyclic fenestrane ketones, namely the *cis,cis,trans,cis*-fenestranone **60** and the all-*cis*-fenestranone **62**. *Para*-toluenesulfonic acid was used as a co-catalyst here. However, instead of the expected formation of two new five-membered rings, both one five- and one six-membered ring were formed, giving rise to [5.5.6.6]fenestrane derivatives in both cases. Thus, one of the phenanthryl residues underwent the desired electrophilic attack at the 10-position of the quasi-olefinic phenanthrene C–C double bond, whereas the other phenanthryl residue suffered attack at the *peri*-position (C-8). Also of interest is the finding that the cyclohexanone ring and the newly formed six-membered ring share the *trans*-configurated central C–C bond in the strained *cis,cis,trans,cis*-isomer **60**. Thus, it appears that subtle individual contributions add to the overall strain of the novel, benzo-diphenanthro-annelated [5.5.6.6]fenestrane framework in the polycyclic ketones **60** and **62** [64].

For comparison, we also performed the twofold cyclodehydration sequence starting from di-[(9-phenanthryl)methyl]-1,3-indanedione (**63**) [64]. Much to our surprise once again, the absence of the cyclohexanone ring and thus of the original spiro moiety relieves the strain such that two new six-membered rings are formed, giving the broken [5.6.6]fenestranone **64** in excellent yield. Therefore, it appears that the twofold cyclodehydration with structurally flexible (9-phenanthryl)methyl residues like those provided in indanedione **63** occurs preferentially by electrophilic attack at the *peri*-positions [64]. A straightforward but reversed combination of ring sizes was achieved in the broken [5.5.6]fenestrane **66**. Here, the six-membered ring within a *peri*-annelated naphthalene unit was provided in the starting 2,3-dihydrophenalene-1,3-dione. The dibenzylated derivative **65** was converted to hydrocarbon **66** by reduction and cyclodehydration in two highly efficient steps [63a].

Another approach that should eventually lead to all-*cis*-[5.5.5.5]fenestranes with bridged or, at least, extended π-electron periphery required fenestrindane derivatives that bear methyl or methoxy substituents at the molecular rim [65]. Fenestrindanes equipped with one *ortho*-methyl group in that each four molecular bays, such as in hydrocarbon **69** (Scheme 10.11), have the potential to be expanded to fourfold bay-bridged derivatives. Therefore, we synthesized the corresponding stereoisomeric tetramethyl-substituted [5.5.5.6]fenestranones **67** and **68**. Using the established methodology, the *cis,cis,cis,trans*-isomer **67** was readily accessible from the corresponding bis(2,5-dimethylbenzylidene)acetone in 76% yield in three steps. However, we were unable to epimerize this unsuitable stereoisomer under basic conditions. In an extended series of efforts, and fortunately, a delicate control of the kinetics of the Michael spiroaddition reaction between 1,3-indanedione and the pentadienone component allowed us to prepare the

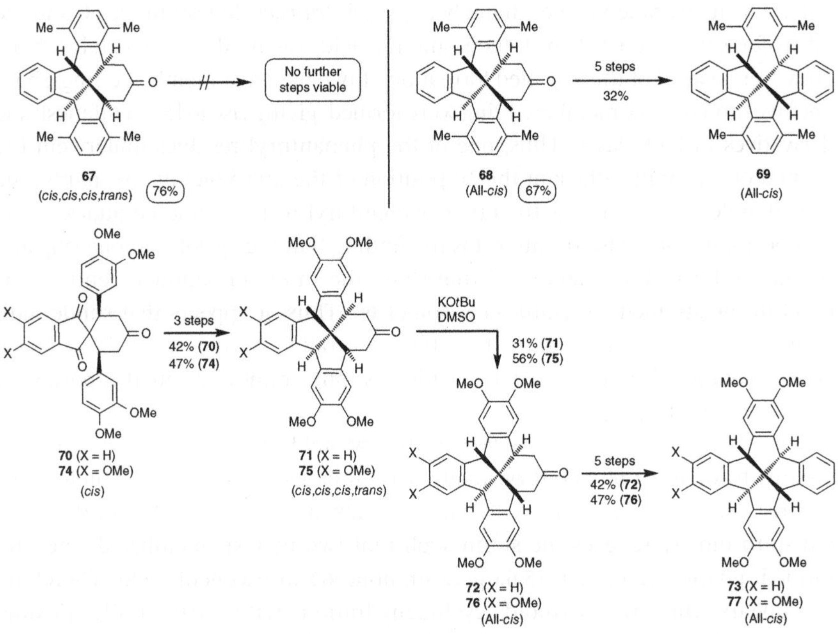

Scheme 10.11 Top: Both the stereoisomeric tetramethyl-substituted [5.5.5.6]fenestranones **67** and **68** are accessible in good yields; however, only the all-*cis*-isomer **68** is convertible into all-*cis*-1,4,9,12-tetramethylfenestrindane (**69**). Bottom: Only the *cis*-tetramethoxy- and *cis*-hexamethoxyspirotriketones **70** and **74** are accessible, whereas the corresponding *trans*-isomers are not. However, the *cis,cis,cis,trans*-configuration of the [5.5.5.6]fenestranone intermediates **71** and **75** is convertible to the all-*cis* form in the respective ketones **72** and **76** which allows the synthesis of the corresponding all-*cis*-1,4,9,12-tetramethoxy- and hexamethoxyfenestrindanes **73** and **77**. *Source:* Based on Bredenkötter et al. [65].

corresponding *trans*-dixylylspirotriketone in good yield. This compound was then converted stereospecifically and in 67% yield over three steps into the corresponding all-*cis*-[5.5.5.6]fenestranone **68**, and further five steps enabled the synthesis of 1,4,9,12-tetramethylfenestrindane (**69**) in 32% overall yield. Due to the quite elaborate and delicate sequence of steps, we did not further extend this approach to bay-substituted fenestrindanes and abandoned efforts to use the methyl groups in hydrocarbon **69** for potential bay bridging.

Multiply methoxy-substituted fenestrindanes were also synthesized in two different periods of our research and with different goals, one of which turned out to be extremely encouraging (see Section 10.2.3). The construction of the fenestrindanes **73** and **77** (Scheme 10.11) that bear four and, respectively, six methoxy groups at the outer arene positions of the molecular periphery was again accomplished by pursuing the spiro-route [65]. Notably, in these cases, the base-catalyzed adjustment of the stereochemistry on the stage of the threefold benzo-annelated [5.5.5.6]fenestranones was crucial. All attempts to prepare the corresponding *trans*-configurated diarylspirotriketones bearing four or six methoxy groups were not productive, whereas the respective *cis*-isomers **70** and **74** were readily synthesized in moderate (40%) or even excellent (90%) yields. The following three steps afforded the stereospecific construction of the corresponding *cis,cis,cis,trans*-[5.5.5.6]fenestranones **71** and **75** in 42 and 47% overall yield. In contrast to the tetramethyl congener **67**, which did not undergo based-catalyzed epimerization, the tetramethoxyfenestranones **71** and **75** were successfully converted into the all-*cis*-[5.5.5.6]fenestranones **72** and **76**, which were isolated in moderate yields (31 and 56%, respectively). Subsequent ring contraction and annealing of the fourth benzene ring furnished 2,3,10,11-tetramethoxyfenestrindanc (**73**) and 2,3,6,7,10,11-hexamethoxyfenestrindane (**77**) in 42 and 47% overall yield, respectively. This work again demonstrates that the synthesis of functionalized fenestrindanes has a certain stereochemical flexibility in cases where the stereoselectivity of the Michael spiroaddition in the very first step cannot be controlled in favor of the desired *trans*-diaryl spirotriketone. In turn, the synthesis of the *trans*-isomer can be achieved in suitable cases when the kinetic control of the Michael spiroaddition is optimized. Finally, it may be mentioned that, unfortunately, the synthesis of the 2,3,6,7,10,11,14,15-octamethoxy analog of the fenestrindanes **73** and **77** has not been pursued, although this most (formally D_{2d}-) symmetrical derivative would be of high interest for the extension of the four indane wings of the parent hydrocarbon **7**. However, the corresponding octabromo- and octaiodofenestrindanes are known [66]. Nevertheless, research aiming at the development of an efficient synthon that would allow for the annelation of a veratrole unit, instead of the simple benzene ring, to the given hexamethoxy-substituted tribenzo[5.5.5.5]fenenestrene intermediates readily obtained on the way to fenestrindane **77** would be very welcome.

10.2.3 Fenestrindanes as a Core Embedded in a Closed π-Electron Periphery

The challenging and long-lasting idea to construct *ortho*-phenylene or other unsaturated C_2-bridges about the molecular bays of the centropolyindanes, and of fenestrindane (**7**) in particular, became reality in the past decade [31, 66a, 67]. The first inspiring success was achieved with another member of the centropolyindane family, though. The bowl-shaped skeleton of a tribenzotriquinacene (TBTQ) was merged with the flat framework of hexa-*peri*-hexabenzocoronene (HBC), generating one single cycloheptatriene unit at the fusion region [68]. Parallel to our investigations of π-extended fenestrindanes, described below, we developed a synthesis route to threefold bay-bridged TBTQ derivatives [69]. Scholl-type cyclodehydrogenation techniques were used in all of these cases [70]. The early results in the TBTQ series encouraged us to investigate the possibility to merge HBC units to the fenestrindane core, as shown in Scheme 10.12 [71]. The easily accessible *trans*-di-(*para*-methoxyphenyl)-substituted spirotriketone **78** was the basis of the synthesis. The subsequent classical eight-step construction of 2,11-dimethoxyfenestrindane (**79**) was carried out with a respectable 11% overall yield. Subsequent ether cleavage using boron tribromide gave the corresponding dihydroxyfenestrindane, which was converted into the respective bis-trifluoromethanesulfonate **80** in 76% over these two steps. At this point, the decisive change of the synthesis strategy becomes obvious. Fenestrindane triflates such as **80** can be subjected to highly efficient C–C cross-coupling reactions to incorporate unsaturated hydrocarbon residues, such as acetylene and aryl groups, after the multistep construction of the centropolyindane framework – in contrast to our original strategy that involved phenyl-substituted building blocks from the very beginning of the synthesis sequence and intermediates such as fenestranones **56–58** (Figure 10.4).

In the present case, a twofold Sonogashira–Hagihara reaction of the fenestrindane-bis-triflate **80** with *para*-(*tert*-butylphenyl)acetylene gave the corresponding derivative **81** in 94% yield, a lower congener of a previously synthesized fenestrindane bearing a total of eight phenylethynyl groups at the molecular periphery [66a]. Subsequent cycloaddition of the two protruding phenylethynyl residues in **81** with tetrakis(*para*-*tert*-butylphenyl)cyclopentadienone afforded the corresponding bis(pentaphenylphenyl)fenestrindane **82** in 21% yield. The constitutional and stereochemical identity of this $C_{141}H_{148}$ hydrocarbon was unequivocally confirmed by mass spectrometry and NMR spectroscopy. Unfortunately, all attempts to perform an exhaustive, that is, fourteenfold Scholl cyclodehydrogenation in the last step of the synthesis failed. Whereas all of the twelve C–C bonds between the phenyl residues were nicely generated under the oxidative reaction conditions, resulting in two intact HBC entities, the envisioned bridging across the two bays of the fenestrindane core did not take place. As a result, the merged bis-HBC-fenestrindane **83** was

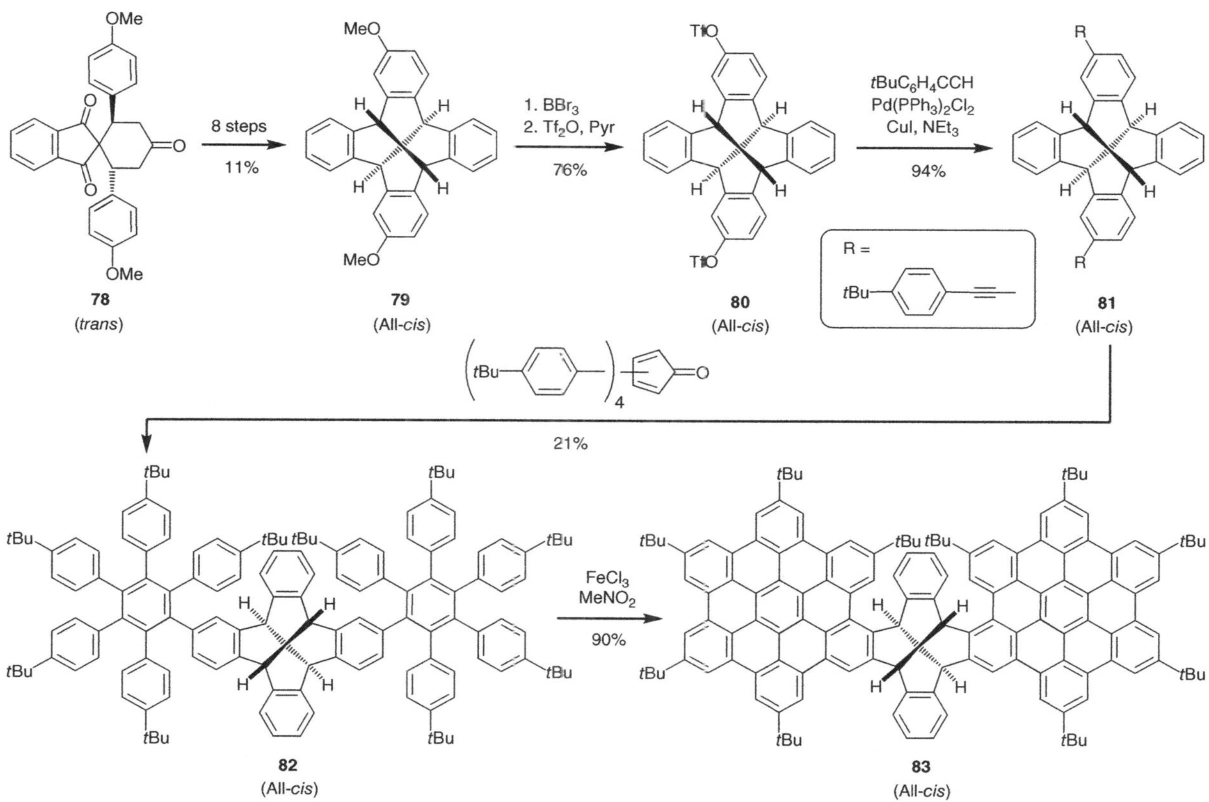

Scheme 10.12 Synthesis of the bis-(pentaarylphenyl)fenestrindane **82** from the *trans*-di(4-methoxyphenyl)spirotriketone **78** *via* the 2,11-dimethoxyfenestrindane **79** and the corresponding bis-triflate **80** (Tf = SO_2CF_3). The subsequent Sonogashira–Hagihara reaction gives the bis-phenylacetylene **81**, which is then condensed with tetra(*para-tert*-butyl)tetracyclone to give **82**. Multiple Scholl reaction leads to the product of twelvefold cyclodehydrogenation, **83**, in which the desired bridges across the two upper bays of the fenestrindane core remain unaccomplished. *Source:* Based on An et al. [71].

obtained and isolated in excellent yield (90%). It is obvious that structure **83** contains many *tert*-butyl groups that were introduced to increase the solubility. However, those two of them placed in close proximity to the pairs of *ortho*-positions that were to be coupled in the course of the Scholl reaction prevented the last two bridging C–C bond formation steps [71]. Nevertheless, to the best of our knowledge, the hydrocarbons **82** ($C_{141}H_{148}$) and **83** ($C_{141}H_{124}$) represent the largest derivatives of the parent [5.5.5.5]fenestranes **2** and **6** known so far.

After this nevertheless promising failure to construct the first large, graphene-type, bay-bridged fenestrindane derivative, we turned to a more conservative, step-by-step strategy [67]. In analogy to the 2,3,10,11-tetramethoxy-substituted fenestrindane **73** bearing the substituents at four opposite outer peripheral positions of the fenestrindane core, the corresponding 1,4,9,12-tetramethoxy isomer **86** was synthesized with an even better efficiency in part because the direct route starting from the *trans*-diarylspirotriketone **84** was viable in this case [67]. Thus, the classical ketalization–reduction–cyclodehydration/deprotection sequence led to the all-*cis*-[5.5.5.6]fenestranone **85** in only three steps and with 55% overall yield (Scheme 10.13). The subsequent ring contraction and benzoannelation sequence gave the key tetrafunctionalized fenestrindane **86** in 15% overall yield, which has recently been increased to even 24% [72]. Refunctionalization to the corresponding tetrakis-triflate **87** in 70% yield followed by fourfold Suzuki cross-coupling using the palladium acetate/SPhos catalyst system afforded the tetrakis(3,4-dimethoxyphenyl)fenestrindane **88** in excellent yield (95%), in spite of the considerable steric restrictions in the four bays of the fenestrindane skeleton. As expected, the ^{1}H NMR spectrum of this first fourfold *ortho*-aryl-substituted fenestrindane derivative again exhibits a characteristic magnetic shielding of the four protons at the remaining *ortho*-positions sited opposite to the four dimethoxyphenyl residues. In fact, the upfield shift observed for fenestrindane **88** (($|\Delta\delta| \geq 1.2$ ppm) slightly exceeded that of the *ortho*-phenyl-substituted tribenzo[5.5.5.6]fenestranones **56** and **57** [63b, 67].

In the final step, the tetraarylfenestrindane **88** was treated with 2,3-dichloro-5,6-dicyano-*para*-benzoquinone (DDQ) and trifluoromethylsulfonic acid in methylene chloride. Under these conditions, the product of fourfold cyclodehydrogenation, the fourfold bay-bridged saddle molecule **90** was formed in 45–50% yield, depending on the amount of DDQ used, along with the twofold cyclized congener **89** (10–5%) and the singly bridgehead hydroxylated derivative of **90** (4–7%, not shown in Scheme 10.13). Interestingly, the half-way cyclodehydrogenated compound **89** was the only observable intermediate even under sub-stoichiometric amounts of the oxidation reagent; neither the singly nor the triple cyclized congeners nor the other two possible twofold cyclized isomers of **89** were formed. This observation points to a delicate kinetic control of the possible cyclization steps, the origins of which were investigated in great detail recently [72].

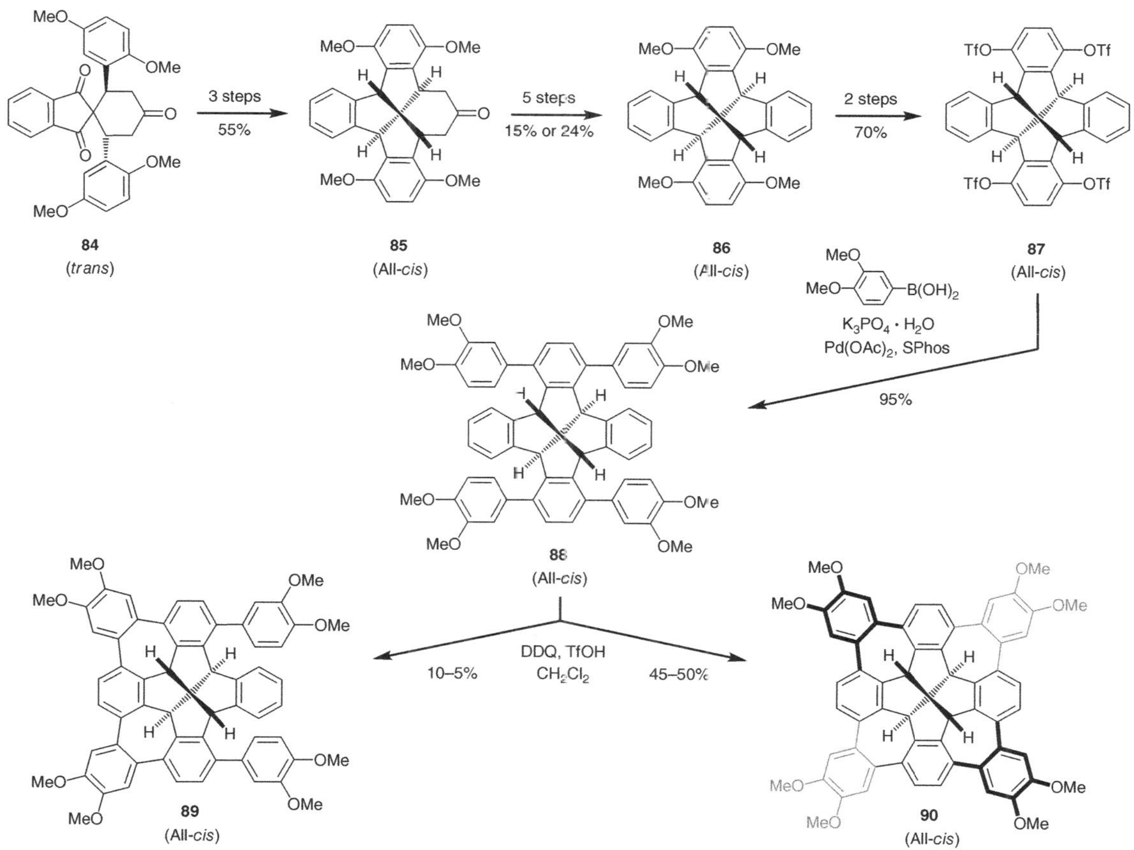

Scheme 10.13 Synthesis of saddle-shaped fourfold bay-bridged fenestrindane **90** from the *trans*-bis(2,5-dimethoxyphenyl)spirotriketone **84** *via* the [5.5.5.6]fenestranone **85** and the 1,4,9,12-substituted fenestrindanes **86–88**. The final step is a Scholl-type fourfold cyclodehydrogenation reaction of the 1,4,9,12-tetraarylfenestrindane **88** giving the saddle molecule **90** as the major product and the twofold cyclized fenestrane **89** as the only observable intermediate of the cyclodehydrogenation process. Minor amounts of the bridgehead-monoalcohol derived from **90** (not shown) are also formed. *Sources:* Based on Wong et al. [67]; Wong et al. [72].

Compound **90** represents the so far largest known bay-bridging extension of all-*cis*-[5.5.5.5]fenestrane **2** and the corresponding tetraene **6**. However, the chemistry of polyaromatic saddles based on the fenestrindane core should be broadened further. Recent studies have further confirmed that electron-rich aromatic residues facilitate the Scholl-type cyclodehydrogenation of the fourfold *ortho*-aryl-substituted fenestrindane substrates. For example, the dodecamethoxy analog **91** (Figure 10.5), bearing three adjacent methoxy groups in a symmetric pattern at the outer positions of each of the four benzosuberene units, was synthesized in good yield (52%) along with its bridgehead-monohydroxylated derivative (20%) [67]. Nevertheless, and very surprisingly, the parent hydrocarbon **10** is also accessible directly from the simple tetraphenyl precursor under the same reaction conditions [72]. Notably, both of the saddles **10** and **90** are achiral and could serve as the basis for multiple (re-)functionalization and further extension to likewise achiral, highly warped, and even larger fully conjugated π-electron systems. In contrast, the dodecamethoxy derivate **91** is chiral. The synthesis of the two enantiomers **91a** and **91b** (Figure 10.5) was achieved in two individual, stereospecific sequences. In particular, the enantiomerically pure tetramethoxy-substituted tribenzo[5.5.5.5]fenestrene intermediates were obtained from the corresponding racemic mixture by chiral-HPLC chromatography. With both of these enantiomers, the subsequent benzoannelation–refunctionalization sequence giving the enantiomers of tetrakis(triflate) **87** and the subsequent fourfold bridging procedures by Suzuki–Miyaura cross-coupling and Scholl-type cyclodehydrogenation were achieved throughout with the enantiomerically pure derivatives [73].

Molecular model calculations suggest that the fourfold bridging of the fenestrindane framework such as in **10** and **90** strongly rigidifies the polycyclic structure. As a consequence, the highly D_{2d}-symmetric conformation becomes the ground

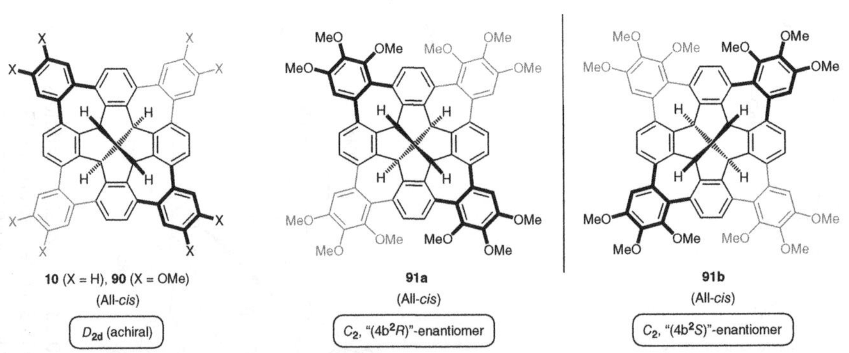

Figure 10.5 The parent fenestrindane saddle **10** and its octamethoxy derivative **90**, both being achiral, and the two enantiomers of the dodecamethoxy derivative **91**. *Sources:* Based on Wong et al. [72]; Wong et al. [73].

state, in contrast to parent fenestrindane **7** and its various bridgehead-substituted derivatives, which preferably exist as S_4-symmetric conformers in the solid state and in either dynamic or static equilibria in solution [27, 74]. The molecular D_{2d}-symmetry of the octamethoxy-substituted saddle **90** in the solid state was nicely confirmed by single-crystal X-ray diffraction analysis (Figure 10.6) [67].

The solid-state structure of the fourfold bridged fenestrindane **90** reveals several remarkable features: (i) the exocyclic C–C–C bond angles α and β at the central carbon atom are 116.4° and 117.1°, respectively (Figure 10.6a). These values are very close to those of the parent all-*cis*-fenestrindane **7** (116.4° and 116.6°) [22a]. Thus, the introduction of four *ortho*-phenylene units across the bays of the fenestrindane core of **90** does not give rise to flattening at the central quaternary carbon atom. In other words, the fitting of the four *ortho*-phenylene bridges is virtually perfect. (ii) The depth of the saddle at its upper and lower sides was

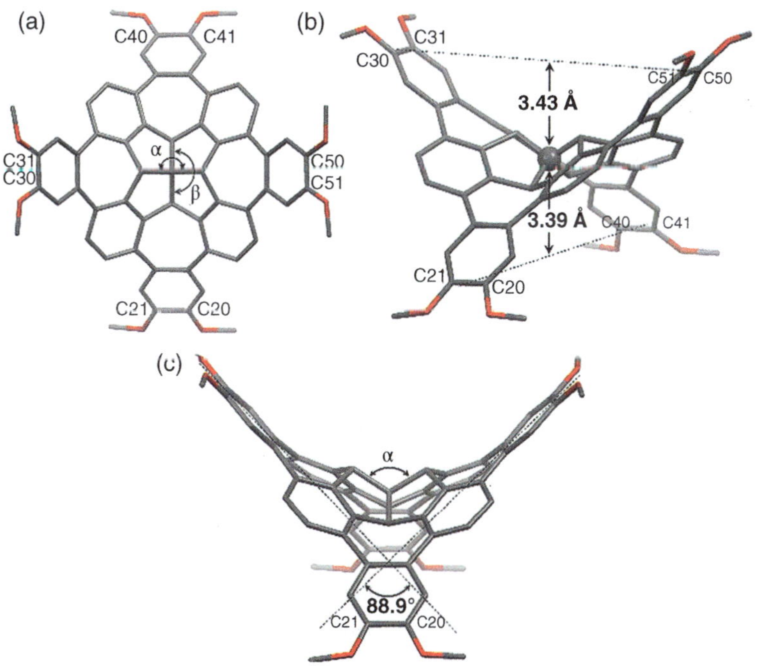

Figure 10.6 Solid-state structure of the fourfold bay-bridged fenestrindane **90**, as determined by X-ray diffraction. (a) Top view; (b) oblique view showing the depth of the saddle, which is defined as the distance between the central quaternary carbon atom and the centroid of the four carbon atoms defined by C20, C21, C40, and C41 or by C30, C31, C50, and C51; (c) front view showing the nearly orthogonal disposition (88.9°) of the two opposite veratrole moieties. Hydrogen atoms are omitted for clarity. *Source:* Reproduced with Permission from Wong et al. [67].

found to be 3.41 ± 0.02 Å (Figure 10.6b). Thus, the saddle is somewhat shallower than that in *o*,*p*,*o*,*p*,*o*,*p*,*o*,*p*-cyclooctaphenylene (3.74–3.97 Å), the corresponding aromatic hydrocarbon in which the central neopentane unit is lacking [75]. (iii) The spatial orientation of the two pairs of opposite veratrol units is also interesting: The two axes each of which bisects the central carbon atom and a pair of opposite dimethoxy-substituted rims of the saddle cross each other almost perfectly at right angles (Figure 10.6c). This translates the special geometry of the two mutually orthogonally fused spiro axes of the fenestrindane framework into a highly rigid and regular extension of the dimethoxy-functionalized arene rings into the 3D-space.

10.2.4 Three Spiranes Merged in a Centropolycyclic Framework: Centrohexaindane and Its Derivatives

As shown above, the saddle-shaped core of fenestrindane (**7**) provides a good basis for extensions into the 3D-space by annelating of up to four cycloheptatriene rings in the bays and further benzene rings at the extended molecular periphery thus formed. Another, and strictly three-dimensional extension consists of fusing a third spiro axis across the central neopentane unit, which is oriented again at right angles to the two spiro axes of the fenestrindane framework. As mentioned in the beginning of this treatise, the parent centrohexaquinane hydrocarbon **3** or any of its olefinic analogs, such as the corresponding T_d-symmetric hexaene, centrohexaquinacene, have never been synthesized. On the other hand, several benzoannelated congeners of **3** have already been reported. Centrohexaindane (**9**) was the very first "super spiro" compound within this series and three well-viable synthesis routes to **9** have been developed [16]. Owing to its three mutually orthogonal [2,2′]spirobiindane axes, this unusual polyaromatic hydrocarbon was dubbed a "Cartesian hexabenzene" [76]. A number of partially benzoannelated congeners were also synthesized, either by a suitable "aufbau" strategy [77] or by systematic oxidative dismantling ("abbau") of the benzene units [78]. The lowest congener, monobenzocentrohexaquinane, was obtained in minute amounts following the latter strategy [78, 79].

Two of the three well-viable synthesis routes to centrohexaindanes are illustrated in Scheme 10.14. The third and, notably, the most efficient one, is not depicted here because it involves propellane rather than fenestrane key intermediates [16b, 31, 77, 80]. The parent fully benzoannelated hydrocarbon **9** can be constructed in seven steps overall from 1,3-indanedione along the "broken fenestrane route," which involves the easily accessible centrotriindane **37**. Photoinduced bromination of **37** to the tetrabromo derivative **92**, formed as a mixture of stereoisomers, and subsequent twofold Friedel–Crafts reaction with benzene lead to centropentaindane (**8**) in 88% yield [81]. An apparently simple repetition of this

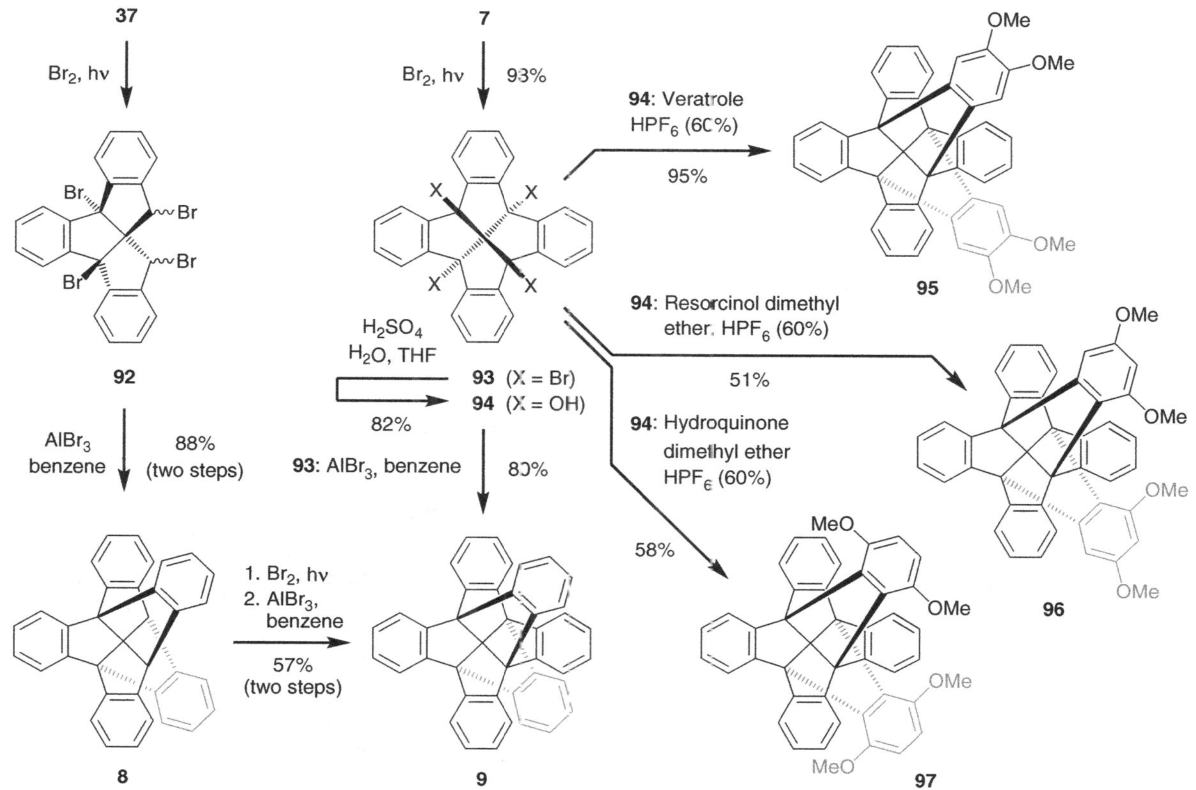

Scheme 10.14 Access to centrohexaindanes *via* the "broken fenestrane route" and the "fenestrane route." The former route involves the broken congener, centrotriindane **37**, and leads to the parent hydrocarbon **9** *via* centropentaindane (**8**). The latter route involves fenestrindane **7**, its tetrabromo derivative **93**, and, in the case of the tetramethoxycentrohexaindanes **95**–**97**, the tetrahydroxyfenestrindane **94**. *Sources:* Based on Kuck et al. [16, 81]; Tellenbröker et al. [82].

sequence incorporates the sixth benzene nucleus and affords centrohexaindane (**9**) in 57% yield [16b]. It should be noted that the formation of the tetrabromo derivatives **92** is difficult to control because of concomitant under- and over-bromination and that the dibromo intermediate generated from centropentaindane (**8**) is very unstable due to the steric repulsion between the two bromine substituents at the highly rigid polycylic framework. Nevertheless, several derivatives of centropentaindane (**8**) bearing functional groups at the two remaining bridgehead positions were synthesized *via* the dibromo intermediate [81b]. The second approach, the "fenestrane route," also starts from 1,3-indanedione and yields fenestrindane (**7**) after nine steps in total, as discussed above. However, in this case, the last two steps are well controllable: Photoinduced bromination gives tetrabromofenestrindane **93** as a stable intermediate in excellent yield (93%), and the subsequent Friedel–Crafts reaction of **93** with benzene gives the target hydrocarbon **9** in 80% yield [16]. Moreover, electron-rich arenes like anisole can be condensed with the fenestrindane skeleton using the easily accessible tetraalcohol **94** [74, 82, 83]. In this way, the three isomeric tetramethoxycentrohexaindanes **95–97** were synthesized in good to excellent yields [82]. The methoxy functionalities offer various possibilities to extend the polyaromatic framework of centrohexaindane into two or even four of the six Cartesian directions of space. The molecular structure of the rigid bis-veratrole derivative **95** obtained by X-ray diffraction is shown in Figure 10.7.

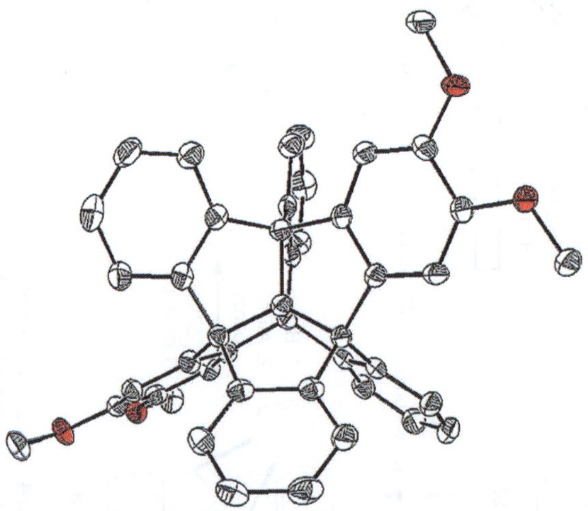

Figure 10.7 Solid-state molecular structure of tetramethoxycentrohexaindane **95**, as determined by X-ray diffraction (ORTEP plot). Oxygen atoms are colored in red, hydrogen atoms are omitted for clarity. *Source:* Adapted from Tellenbröker et al. [82].

The bis(hydroquinone dimethyl ether) derivative **97** may be of particular interest because it could allow us to incorporate again four electron-rich aryl residues by Suzuki–Miyaura coupling, in analogy to the three-step conversion of the tetramethoxyfenestrindane **86** presented above. In the case of the fourfold *ortho*-functionalized centrohexaindane skeleton of **97**, the four aryl groups introduced into the highly shielded cavities of the bulky hydrocarbon would suffer enormous steric strain. It would certainly be a great challenge to investigate the potentially fourfold Scholl-type cyclodehydrogenation of such a strained substrate. Beside derivatives of the saddle hydrocarbon **90** bearing two *ortho*-phenylene-bridges fused to the front- and back-side pairs of fenestrane bridgeheads, the formation of isomers can be envisioned. Again, using functionalized, electron-rich aryl groups, further three-dimensional extensions, all based on the unusual merging of three spiro axes in the centrohexaindane core, appear to be possible. Also, it may be mentioned that, in turn, fenestrindane-based saddle structures like **10** and **90** may possibly be converted to the same or similar extended centrohexaindane derivatives by pursuing the fenestrane route, i.e. though bridgehead functionalization followed by twofold condensation with suitable aromatic building blocks.

In our view, centrohexaindane (**9**) is a molecular sleeping giant. Besides C_{60}-fullerene, it represents the only known poly-carbocyclic molecule that can be functionalized with Cartesian orientation of the substituents in space. The resulting large and rigid, covalently bonded, and highly stable core bearing six octahedrally oriented functionalities should allow us to construct highly promising building blocks for nanoscopic three-dimensional scaffolds. Therefore, besides the tetramethoxy derivatives **95–97** presented above, other early studies were undertaken to synthesize derivatives of hydrocarbon **9** that contain substituents or functional groups at various positions. As mentioned above, the *ortho*-positions within the four three-dimensional bays of **9** are highly shielded; therefore, electrophilic aromatic substitution occurs exclusively at the outer positions of the molecular periphery. For example, Rieche formylation of **9** leads to the 2-carbaldehyde **98** in moderate yield (35%), the sole monosubstituted centrohexaindane derivative known so far [84]. As a consequence, the synthesis of *ortho*-substituted centrohexaindanes requires the *ab initio* use of suitably substituted building blocks. Thus, the 1,8-dimethyl- and the 1,4,8,28-tetramethyl-substituted congeners **99** and **100**, respectively, bearing all of the methyl groups at *ortho*-positions, were synthesized using the propellane route (Figure 10.8) [85]. These substituents could be a basis for the construction of larger functionalities within the bays.

Not surprisingly, no derivatives are known that would bear the full set of twelve substituents, such as methyl, methoxy, or hydroxyl groups, at these "inner" positions of the molecular periphery of centrohexaindane. In contrast, functionalization at the "outer" positions, including a few dodeca-substituted congeners, has reached a certain potential now. Nitration of hydrocarbon **9** under either classical

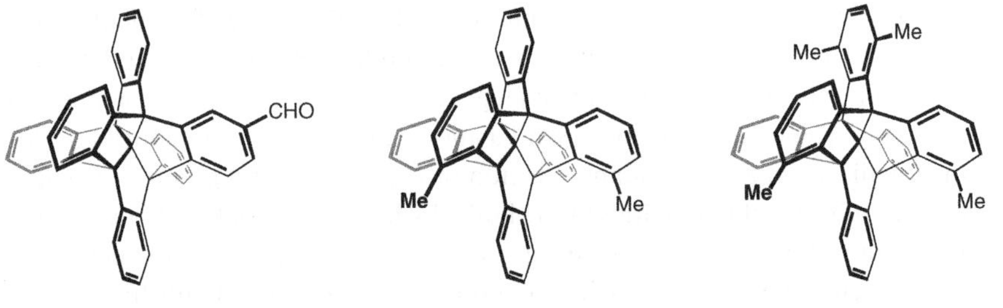

98 (C_s) **99** (C_s) **100** (C_s)

Figure 10.8 Selected derivatives of centrohexaindane (**9**) in a perspective presentation: The 2-carbaldehyde **98** and the di- and tetramethyl-substituted derivatives, **99** and **100**. In the latter hydrocarbon, each of the four methyl groups point into one of the four three-dimensional bays of the centrohexaindane skeleton. *Sources:* Based on Linke and Kuck [84]; Kuck et al. [85].

conditions (HNO_3/H_2SO_4, 20 °C, 20 h) [86] or, more readily, with sodium nitrate in trifluoroacetic acid (55 °C, 7 d) leads to a mixture of four hexanitrocentrohexaindanes, **101–104**, in excellent combined (crude) yield (96%) [87]. All of these constitutional isomers bear the nitro groups at the outer arene periphery and give rise to either C_3 or C_1 molecular symmetry; thus, they all represent racemates (Figure 10.9). Most remarkably, the isomers are generated (and isolated by chromatography) in a near-statistical ratio, that is, [**101** (C_3)] : [**102** (C_1)] : [**103** (C_3)] : [**104** (C_1)] = 1 : 3 : 1 : 3, in analogy to the threefold peripheral nitration of the lower congeners of the TBTQ series [88–90]. This finding indicates that the π-electron systems of the six benzene rings in centrohexaindane and its derivatives do not (or only negligibly) interact with each other [16a].

Both strategies, use of functionalized building blocks and post-functionalization of the complete polycyclic framework of **9**, have finally allowed us to synthesize twelvefold substituted centrohexaindane derivatives (Figure 10.10). Dodecamethoxycentrohexaindane **105** was synthesized from suitable veratrole-based synthons in low overall yield (6% in the last step) [91]. This derivative was the first centrohexaindane derivative in which all of the twelve positions of the outer molecular periphery were functionalized. Provided that the synthesis could be improved (which should be possible!), the access to the corresponding dodecahydroxy derivative appears to be within reach. This compound should be a highly interesting building block for the construction of covalent and non-covalent three-dimensional aggregates [92]. Also, while all attempts to perform a twelvefold nitration of hydrocarbon **9** turned out to be futile in our hands, the synthesis of the corresponding, so far elusive, dodecaaminocentrohexaindane should be feasible *via* reduction of the mixture of the isomeric hexanitro-derivatives **101–104** to the respective hexaamines and subsequent up-functionalization [31, 87, 93].

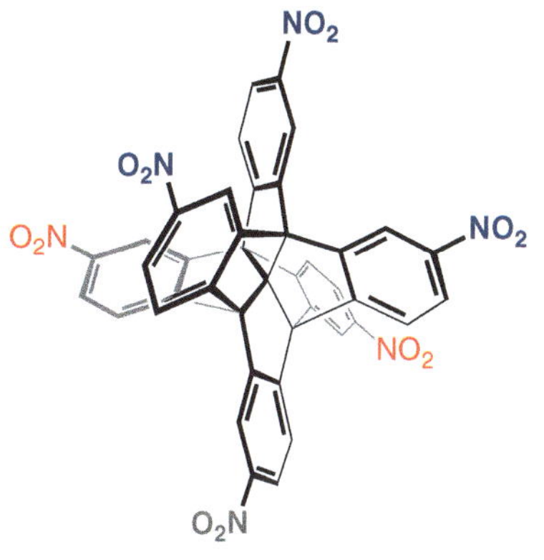

101 (C_3)

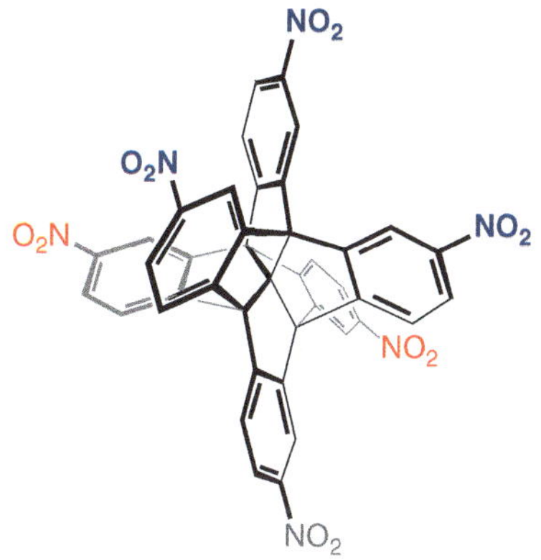

102 (C_1)

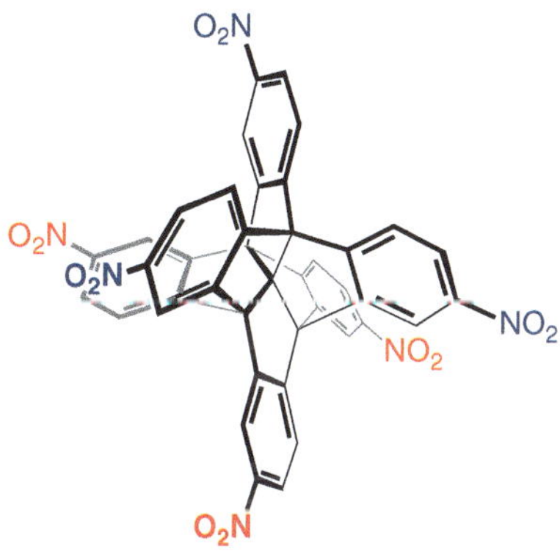

103 (C_3)

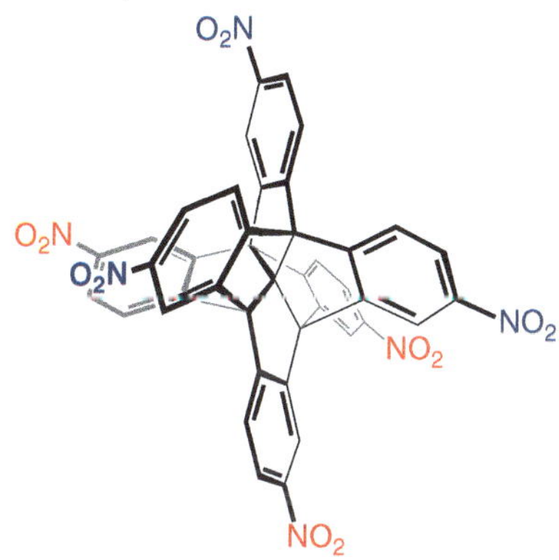

104 (C_1)

Figure 10.9 The four constitutional isomers of hexanitrocentrohexaindane, **101–104**. Each rim of the outer periphery bears one functional group [87]. The isomers are formed by nitration from the parent hydrocarbon **9** as four racemates present in the statistical ratio (1 : 3 : 1 : 3). Only one enantiomer is shown in each case; two sets of three nitro groups are displayed in different colors for easier distinction of the isomers. *Source:* Modified from Kuck et al. [87].

Most encouraging, we recently successfully accomplished our efforts to introduce twelve halogen functionalities at the outer positions of centrohexaindane (**9**) [94]. In contrast to classical bromination and iodination methods, all of which gave rise to mixtures of incompletely substituted derivatives [93], use of *N*-bromo- and *N*-iodosuccinimide, respectively, in highly acidic media (trifluoromethylsulfonic acid in 1,2-dichloroethane) under sonication or microwave irradiation of the reaction mixture afforded the corresponding dodecahalo derivatives **106** and **107** (Figure 10.10). The yields were excellent (> 90 and > 95%, respectively), reflecting

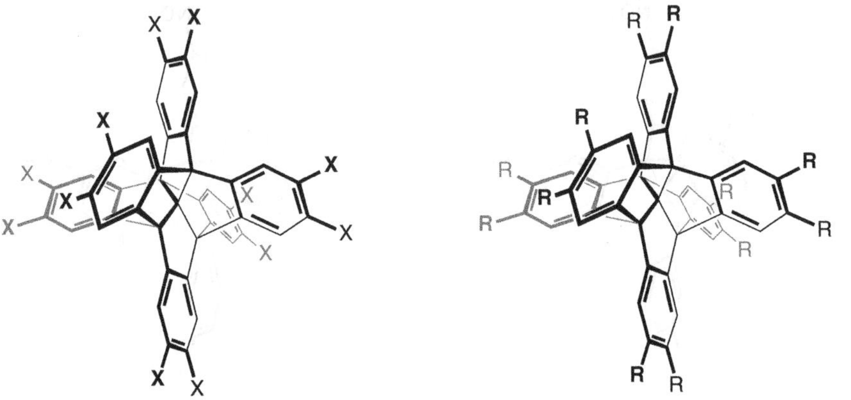

105 (X = MeO)
106 (X = Br)
107 (X = I)

(T_d)

108 (R = Me)
109 (R = CCPh)
110 (R = Ph)

Figure 10.10 Dodecafunctionalized centrohexaindanes **105–107**, synthesized by two different strategies, and the corresponding products of twelvefold C–C cross-coupling, hydrocarbons **108–110**. *Sources:* Based on Harig and Kuck [91]; Linke et al. [94].

the importance of the physical activation methods [94]. Both compounds exhibit extremely poor solubility in most organic solvents. Fortunately, pyridine was found to be a sufficiently good solvent to enable ^{1}H and ^{13}C NMR spectroscopic analysis of these compounds, which unequivocally confirmed their molecular T_d-symmetry. Electrospray ionization mass spectrometry in the negative ion mode (ESI-(−)) was the only means to determine the accurate molecular masses of compounds **106** and **107** from the corresponding $[M + Cl]^-$ adduct ions, and X-ray diffraction with single crystals obtained from pyridine solutions allowed us to elucidate the molecular structure in the solid state [94]. In spite of their extremely low solubility, we succeeded to subject both the dodecabromo- and the dodecaiodocentrohexaindane to twelvefold C–C cross-coupling reactions. Compound **106** was convertible to the corresponding dodecamethyl derivative **108** and, more interestingly, it underwent a twelvefold Sonogashira–Hagihara reaction with phenylacetylene to give the corresponding dodeca(phenylethynyl)centrohexaindane **109** in excellent yield (93%). It is obvious that the solubility of the intermediates formed in the course of the cross-coupling steps is drastically increased by symmetry-breaking and the introduction of nonpolar hydrocarbon residues. Similarly, the dodecaiodo derivative **107** undergoes twelvefold Suzuki–Miyaura reaction with phenylboronic acid to furnish the corresponding dodecaphenylcentrohexaindane **110** in good yield (63%) [94]. Again, both of the giant hydrocarbons **109** ($C_{137}H_{72}$) and **110** ($C_{113}H_{72}$) were fully characterized by mass spectrometry

(ESI-(+), generating the $[M+Ag]^+$ adduct ions this time) and NMR spectroscopy. Singly crystals of the dodecaphenyl derivative **110** were suitable for X-ray structure analysis, which further proved the identity of this truly three-dimensional super-spiro hydrocarbon. The solid-state molecular structure of hydrocarbon **110** is reproduced in Figure 10.11 [94]. It is obvious that further extensions by C–C and possibly hetero-atomic cross-coupling methodologies should open various routes to 3D-expanded centrohexaindanes and possibly to three-dimensional networks consisting of bispiro-junctions and super-spiro vertices.

Finally, we may add that further extremely exciting molecular design may arise from multiply functionalized super-spiro hydrocarbons such as centrohexaindane. We would like to suggest just two hypothetical examples here: First, considering the twelve inner positions of the arene periphery of **9**, we may imagine the incorporation of four cyclopropane rings, each of which would cap one of the four three-dimensional bays with nearly strainless fitting. This would give rise to a fourfold bullvalene derivative with expectably non-fluctuating π-electron systems (Figure 10.12a). Admittedly, the idea to construct a merged centrohexaindane-tetrakis-bullvalene was in our mind since the very outset of centropolyindane chemistry but has never been materialized [95]. Second, derivatives of **9** bearing suitable twelvefold functionalities at the outer periphery may eventually lead to

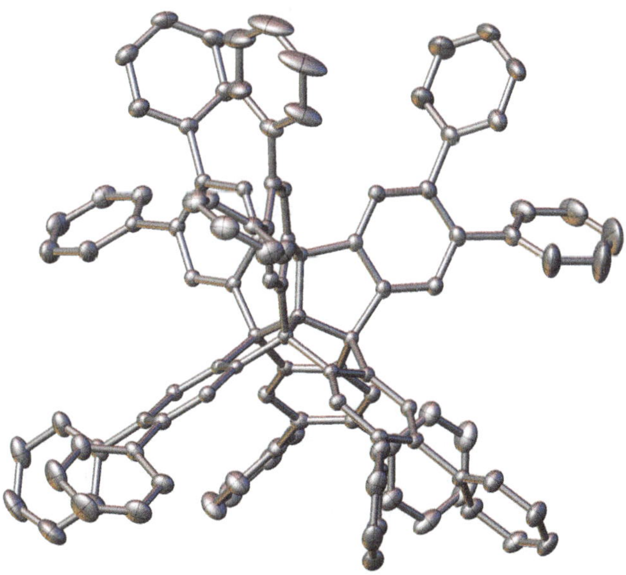

Figure 10.11 Solid-state molecular structure of dodecaphenylcentrohexaindane **110**, as determined by X-ray diffraction. Ellipsoids at 50% probability, hydrogen atoms, and solvent molecules are omitted for clarity. *Source:* Reproduced with Permission from Linke et al. [94].

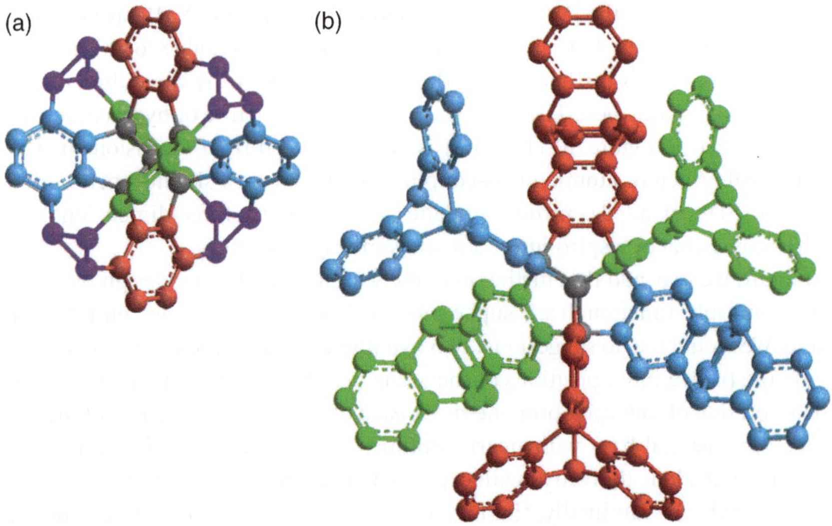

Figure 10.12 Two hypothetical extensions of the centrohexaindane core, modeled by MM2 calculations: (a) Capping of the four three-dimensional bays of **9** by cyclopropane units to generate four bullvalene entities. (b) Merging six triptycene units with 9 to give four deeply concave molecular surfaces with tetrahedral orientation in space. The top-down view shows one of the tribenzotriquinacene (TBTQ) units incorporated. In both cases, the three spiro axes crossing at the neopentane core are distinguished by coloring. Hydrogen atoms are omitted for clarity.

another giant hydrocarbon in which the centrohexaindane core is merged with a total of six triptycene units, generating four deep cavities in a tetrahedral orientation into space (Figure 10.12b). Very first and minute steps toward such a goal have been made recently [96].

10.3 Conclusions

The chemistry and structural variety of spirane derivatives have been carried to higher dimensions by fusing two or even three spiro units at a common neopentane core. Merging two spiro units with each other gives rise to the [*m.n.p.q*] fenestranes, representing tetrafuso-centrotetracyclic arrangements, and fusing even three spiro units leads to centrohexacyclic congeners. While spiro compounds can be regarded as quasi linear, one-dimensional molecular arrangements, fenestranes can be regarded as warped polycyclic structures stretched into two dimensions and centrohexacyclanes represent truly three-dimensional frameworks.

In this chapter, two major conceptually distinct synthesis strategies are described that, over the past decades, have enabled experimental access to either alicyclic and olefinic [5.5.5.5]fenestranes, on the one hand, or to benzoannelated [5.5.5.6]- and [5.5.5.5]fenestranes and their derivatives, on the other. The overall treatment is restricted to carbocyclic fenestranes but it demonstrates the stereochemical variability of the fenestranes and how it can be used to broaden the experimental access to the benzoannelated congeners. Quite an impressive set of derivatives of fenestrindane (**7**), the parent all-*cis*-tetrabenzo-[5.5.5.5]fenestrane hydrocarbon, has been synthesized along the spiro-route, that is, by first constructing one of the spiro axes by spiro-Michael addition and subsequently annealing the second spiro axis vertically to the first one by twofold cyclodehydration.

Various efforts, carried out over the years, to extend the fenestrindane core by circumferential benzoannelation and, in this way, to construct a highly warped but completely closed π-electron periphery are discussed. In particular, it is shown that functionalization of four strategically suitable *ortho*-positions, each of which at one of the bays of the fenestrindane framework, has eventually allowed us to bridge the four bays by 1,2-phenylene units and thus to close the π-electron system about the central neopentane. Even the parent hydrocarbon, **10**, of these novel polyaromatic compounds has become accessible. At the same time, four cycloheptatriene rings are fused with the fenestrindane core. Overall, a realistic perspective has arisen for the construction of saddle-shaped nanographene networks that bear a [5.5.5.5]fenestrane unit with five (or possible only one single!) carbon atoms in the very center.

Beyond the warped, saddle-shaped fenestrane-based nanographene motif, truly three-dimensional super-spiro frameworks can be generated by fusing the third spiro-axis about the given fenestrane skeleton. Centrohexaindane (**9**), a "Cartesian superspirane," comprises three mutually merged 2,2′-spirobindane unit and represents a conformationally rigid building block for the construction of various novel carbon-based networks in the three-dimensional space. Various first steps to functionalize the six benzene rings that point into the six directions of the Cartesian space are discussed, culminating in the synthesis of three twelvefold functionalized centrohexaindane derivatives (**105–107**) that can be used as a basis for covalent and non-covalent nano-scaled aggregates.

The structurally intriguing spirocyclic motif and the chemistry of spiro compounds have conquered higher dimensions within five decades of fenestrane chemistry. It appears that, eventually, fenestranes may enter even the field of materials sciences. Maybe fenestrindane and centrohexaindane will adopt hitherto unforeseeable roles as superior spiro compounds.

Acknowledgements

We thank all students and coworkers who contributed to the development of our fenestrane chemistry over the years. Most of their names appear in the list of references. Particular appreciation is expressed to Dr. Bernd Paisdor, Dr. Andreas Schuster, Dr. Monika Seifert, Dr. Detlef Gestmann, Dr. Jörg Tellenbröker, Dr. Björn Bredenkötter, Dr. Marco Harig, and Dr. Jens Linke (Bielefeld University) and to Dr. Peng An, Dr. Wai-Shing Wong, Mr. Wing-Wa Lau, Mr. Xiaoqing Sun, Mr. Ho-Wing Tse, and Mr. Enoch Cheung (Chinese University of Hong Kong) for their invaluable work.

References

1 von Baeyer, A. (**1900**). *Ber. Dtsch. Chem. Ges. 33*: 3771–3775.

2 (a) Haas, G. and Prelog, V. (**1969**). *Helv. Chim. Acta 52*: 1202–1218. (b) Haas, G., Hulbert, P.B., Klyne, W. et al. (**1971**). *Helv. Chim. Acta 54*: 491–509.

3 Georgian, V. and Saltzman, M. (**1981**). *Tetrahedron Lett.* 13: 4315–4317.

4 Simmons, H.E. III and Maggio, J.E. (**1981**). *Tetrahedron Lett. 22*: 287–290.

5 Paquette, L.A. and Vazeux, M. (**1981**). *Tetrahedron Lett. 22*: 291–294.

6 Greenberg, A. and Liebman, J.F. (**1978**, Chapt. 6). *Strained organic molecules*. In: *Vol. 38 of Organic Chemistry – A series of Monographs* (ed. H.H. Wasserman), 373–374. New York: Academic Press.

7 (a) Denissova, I. and Barriault, L. (**2003**). *Tetrahedron 59*: 10105–10146. (b) Martin, S.F. (**1980**). *Tetrahedron 36*: 419–460.

8 Eliel, E.L. and Wilen, S.H. (**1994**). Chirality in molecules devoid of chiral centers (chapter 14). In: *Stereochemistry of Organic Compounds*, 1e, 1119–1190. New York, NY: Wiley & Sons esp. 1119ff, 1138ff.

9 (a) Rios, R. (**2012**). *Chem. Soc. Rev. 41*: 1060–1074. (b) Ding, A., Meazza, M., Guo, H. et al. (**2018**). *Chem. Soc. Rev. 47*: 5946–5996.

10 Xie, X., Huang, W., Peng, C., and Han, B. (**2018**). *Adv. Synth. Catal. 360*: 194–228.

11 (a) Greenberg, J.F. and Greenberg, A. (**1976**). *Chem. Rev. 76*: 311–365. (b) Krohn, K. (**1991**). Fenestranes – a look at "structural pathologies". In: *Organic Synthesis Highlights* (eds. J. Mulzer, H.-J. Altenbach, M. Braun, et al.), 371–377. Weinheim: VCH. (c) Krohn, K. (**1987**). *Nachr. Chem. Techn. Lab. 35*: 264–266.

12 Gund, P. and Gund, T.M. (**1981**). *J. Am. Chem. Soc. 103*: 4458–4465.

13 (a) Maggio, J.E., Simmons, H.E. III, and Kouba, J.K. (**1981**). *J. Am. Chem. Soc. 103*: 1579–1581. (b) Benner, S.A., Maggio, J.E., and Simmons, H.E. III (**1981**). *J. Am. Chem. Soc. 103*: 1581–1582.

14 Paquette, L.A., Williams, R.V., Vazeux, M., and Browne, A.R. (**1984**). *J. Org. Chem. 49*: 2194–2197.

15 Ermer, O. (**1981**). *Aspekte von Kraftfeldrechnungen.* München: Wolfgang-Baur-Verlag Chapt. 4.6.3.

16 (a) Kuck, D. and Schuster, A. (**1988**). *Angew. Chem. Int. Ed. 27*: 1192–1194. (b) Kuck, D., Schuster, A., Paisdor, B., and Gestmann, D. (**1995**). *J. Chem. Soc. Perkin 1*: 721–732.

17 Kuck, D. (**1997**). *Liebigs Ann./Rec.*: 1043–1057.

18 (a) Kuck, D. (2009; Chapt. 9). Molecules with nonstandard topological properties: centrohexaindane, Kuratowski's cyclophane and other graph-theoretically nonplanar molecules. In: *Strained Hydrocarbons. Beyond the Van't Hoff and LeBel Hypothesis* (ed. H. Dodziuk), 425–447. Weinheim: Wiley-VCH. (b) Kuck, D. (2013, Chapt. 3). Experimental access to centropolycyclic carbon compounds containing the massive C_{17}-core: on the way to D_5 seeds. In: *Diamond and Related Nanostructures* (eds. M.V. Diudea and C.L. Nagy), Vol. 6 of *Carbon Materials: Chemistry and Physics*, F. Cataldo, P. Milani, Series Eds.), 49–73. Dordrecht: Springer.

19 Rao, V.B., George, C.F., Wolff, S., and Agosta, W.C. (**1985**). *J. Am. Chem. Soc. 107*: 5732–5739.

20 (a) Luyten, M. and Keese, R. (**1984**). *Angew. Chem. Int. Ed. 23*: 390–391. (b) Pfenninger, A., Roesle, A., and Keese, R. (**1985**). *Helv. Chim. Acta 68*: 493–507.

21 (a) Desphande, M.N., Jawdosiuk, M., Kubiak, G. et al. (**1985**). *J. Am. Chem. Soc. 107*: 4786–4788. (b) Venkatachalam, M., Desphande, M.N., Jawdosiuk, M. et al. (**1986**). *Tetrahedron 42*: 1597–1605.

22 (a) Kuck, D. and Bögge, H. (**1986**). *J. Am. Chem. Soc. 108*: 8107–8109. (b) Kuck, D. (**1994**). *Chem. Ber. 127*: 409–425.

23 (a) Luef, W. and Keese, R. (**1987**). *Helv. Chim. Acta 70*: 543–553. (b) Hirschi, D., Luef, W., Gerber, P., and Keese, R. (**1992**). *Helv. Chim. Acta 75*: 1897–1908.

24 (a) Keese, R. (**1982**). *Nachr. Chem. Tech. Lab. 30*: 844–849. (b) Keese, R. (**1986**). Planarization of tetracoordinate carbon. Synthesis and structure of [5.5.5.5]Fenestranes. In: *Organic Synthesis: Modern Trends* (ed. O.S. Chizhov), 43–52. Oxford: Blackwell. (c) W. Luef, R. Keese, Planarizing distortions in carbon compounds In: *Advances in Strain in Organic Chemistry, Vol. 3*, (ed. B. Halton), JAI Press, Greenwich, London, 1993, pp. 229–267.

25 (a) Venepalli, B.R. and Agosta, W.C. (**1987**). *Chem. Rev. 87*: 399–410. (b) W. C. Agosta, Inverted and planar carbon. In: *The Chemistry of the Alkanes and Cycloalkanes*, (eds. S. Patai, Z. Rappoport), Wiley-Interscience, Chichester, 1992, pp. 927–962.

26 Röttger, D. and Erker, G. (**1997**). *Angew. Chem. Int. Ed. 36*: 812–827.

27 D. Kuck, Benzoannelated fenestranes. In: *Advances in Theoretically Interesting Molecules*, (ed. R. P. Thummel), *Vol. 4*, JAI Press, Greenwich, London, 1998, pp. 81–155.

28 Siebert, W. and Gunale, A. (**1999**). *Chem. Soc. Rev. 28*: 367–371.

29 Hopf, H. (**2000**). *Classics in Hydrocarbon Chemistry*, 88–94. Weinhein: Wiley-VCH.
30 Keese, R. (**2006**). *Chem. Rev. 106*: 4787–4808.
31 Kuck, D. (**2006**). *Chem. Rev. 106*: 4885–4925.
32 Boudhar, A., Charpenay, M., Blond, G., and Suffert, J. (**2013**). *Angew. Chem. Int. Ed. 52*: 12786–12798.
33 Monkhorst, H.J. (**1968**). *Chem. Commun.*: 1111–1112.
34 (a) Hoffmann, R., Alder, R., and Wilcox, C.F. (**1970**). *J. Am. Chem. Soc. 92*: 4992–4993. (b) Hoffmann, R. (**1971**). *Pure Appl. Chem. 28*: 181–194.
35 Wiberg, K.B., Ellison, G.B., and Wendoloski, J.J. (**1976**). *J. Am. Chem. Soc. 98*: 1212–1218.
36 (a) Minkin, V.I., Miniyaev, R.M., Zakharov, I.I., and Avdeev, V.I. (**1978**). *Zh. Org. Khim. 14*: 3–15. (*J. Org. Chem. USSR* **1978**, *14*, 1-12). (b) Minyaev, R.M. and Minkin, V.I. (**1985**). *Zh. Org. Khim. 21*: 1361–1371.
37 Böhm, M.C., Gleiter, R., and Schang, P. (**1979**). *Tetrahedron Lett. 20*: 2575–2578.
38 (a) Chadrasekhar, J., Würthwein, E.U., and von Ragué Schleyer, P. (**1981**). *Tetrahedron 37*: 921–927. (b) Würthwein, E.U., Chadrasekhar, J., Jemmis, E.D., and von Ragué Schleyer, P. (**1981**). *Tetrahedron Lett. 22*: 843–846.
39 (a) McGrath, M.P. and Radom, L. (**1993**). *Chem. Soc. 115*: 3320–3321. (b) Rasmussen, D.R. and Radom, L. (**1999**). *Angew. Chem. Int. Ed. 38*: 2876–2878.
40 Kuck, D. (**2015**). *Chem. Rec. 15*: 1075–1109.
41 Fritz, H., Weis, C.D., and Winkler, T. (**1975**). *Helv. Chim. Acta 58*: 1345–1357.
42 (a) Luyten, M. and Keese, R. (**1984**). *Helv. Chim. Acta 67*: 2242–2245. (b) Luyten, M. and Keese, R. (**1986**). *Tetrahedron 42*: 1687–1691. (c) Brunvoll, J., Cuidetti-Crept, R., Hargittai, I., and Keese, R. (**1993**). *Helv. Chim. Acta 76*: 2838–2846.
43 Keese, R., Pfenninger, A., and Roesle, A. (**1979**). *Helv. Chim. Acta 62*: 326–334.
44 (a) Mitschka, R., Cook, J.M., and Weiss, U. (**1978**). *J. Am. Chem. Soc. 100*: 3973–3974. (b) Mitschka, R., Oehldrich, J., Takahashi, K. et al. (**1981**). *Tetrahedron 37*: 4521–4542.
45 Bestmann, H.J., Klein, O., Göthlich, L., and Buckschewski, H. (**1963**). *Chem. Ber. 96*: 2259–2265.
46 (a) Fu, X., Kubiak, G., Zhang, W. et al. (**1993**). *Tetrahedron 49*: 1511–1524. (b) Kubiak, G., Fu, X., Gupta, A.K., and Cook, J.M. (**1990**). *Tetrahedron Lett. 31*: 4285–4288.
47 (a) Kotha, S., Manivannan, E., Ganesh, T. et al. (**1999**). *Synlett* 10: 1618–1620. (b) Kotha, S. and Manivannan, E. (**2003**). *ARKIVOC*: 67–76.
48 Prantz, K. and Mulzer, J. (**2010**). *Chem. Rev. 110*: 3741–3766.
49 (a) Schönberg, A. and Mamluk, M. (**1973**). *Chem. Ber. 106*: 849–856. (b) Schönberg, A. and Sidky, M.M. (**1974**). *Chem. Ber. 107*: 2341–2344.
50 Kuck, D. (**1984**). *Angew. Chem. Int. Ed. 23*: 508–509.

51 (a) Ten Hoeve, W. and Wynberg, H. (**1980**). *J. Org. Chem. 45*: 2925–2930. (b) Ten Hoeve, W. (**1979**). Doctoral thesis. University of Groningen.

52 (a) Kuck, D. (**1996**). *Synlett* 7: 949–965. (b) Kuck, D. (**1998**). *Top. Curr. Chem. 196*: 167–220.

53 Ten Hoeve, W. and Wynberg, H. (**1979**). *J. Org. Chem. 44*: 1508–1514.

54 (a) Shternberga, I.Y. and Freimanis, Y.F. (**1968**). *J. Org. Chem. USSR 4*: 1044–1048. (b) Popelis, Y.Y., Pestunovich, V.A., Shternberga, I.Y., and Freimanis, Y.F. (**1972**). *J. Org. Chem. USSR 8*: 1907–1910. (c) Shternberga, I.Y. and Freimanis, Y.F. (**1972**). *Latv. PSR Zinat. Akad. Vestis, Kim. Ser.*: 207–214.

55 (a) Bredenkötter, B., Barth, D., and Kuck, D. (**1999**). *J. Chem. Soc. Chem. Commun.* 9: 847–848. (b) Bredenkötter, B., Flörke, U., and Kuck, D. (**2001**). *Chem. Eur. J. 7*: 3387–3400.

56 Bredenkötter, B. (**2000**). Doctoral thesis. Bielefeld University.

57 (a) Wender, P.A., Dore, T.M., and de Long, M.A. (**1996**). *Tetrahedron Lett. 37*: 7687–7690. (b) Wender, P.A., de Long, M.A., and Wireko, F.C. (**1997**). *Acta Cryst. C 53*: 954–956.

58 Baker, W., McOmie, J.E.W., Parfitt, S.D., and Watkins, D.A.M. (**1957**). *J. Chem. Soc.*: 4026–4037.

59 (a) Kuck, D., Neumann, E., and Schuster, A. (**1994**). *Chem. Ber. 127*: 151–164. (b) Kuck, D. and Seifert, M. (**1992**). *Chem. Ber. 125*: 1461–1469.

60 Parham, W.E. and Jones, L.D. (**1976**). *J. Org. Chem. 41*: 1187–1191.

61 Reinheckel, H., Sonnek, G., and Falk, F. (**1974**). *J. Prakt. Chem. 316*: 215–222.

62 Seifert, M. (**1991**). Doctoral thesis. Bielefeld University.

63 (a) Seifert, M. and Kuck, D. (**1996**). *Tetrahedron 52*: 13167–13180. (b) Seifert, M., Barth, D., and Kuck, D. (**2021**). *Eur. J. Org. Chem.*: 6435–6449.

64 Bredenkötter, B., Neumann, B., Stammler, H.-G., and Kuck, D. (**2014**). *Eur. J. Org. Chem.*: 53–65.

65 Bredenkötter, B., Linke, J., and Kuck, D. (**2017**). *Eur. J. Org. Chem.*: 4414–4428.

66 (a) Tellenbröker, J. and Kuck, D. (**2001**). *Eur. J. Org. Chem.*: 1483–1489. (b) Zhou, L., Cao, X.P., Neumann, G.B. et al. (**2005**). *Synlett* 16: 2771–2775.

67 Wong, W.-S., Ng, C.-F., Kuck, D., and Chow, H.-F. (**2017**). *Angew. Chem. Int. Ed. 56*: 12356–12360.

68 (a) Mughal, E.U. and Kuck, D. (**2012**). *Chem. Commun. 48*: 8880–8882. (b) Mughal, E.U., Neumann, B., Stammler, H.-G., and Kuck, D. (**2014**). *Eur. J. Org. Chem.*: 7469–7480.

69 (a) Ip, H.-W., Ng, C.-F., Chow, H.-F., and Kuck, D. (**2016**). *J. Am. Chem. Soc. 138*: 13778–13781. (b) Ip, H.-W., Chow, H.-F., and Kuck, D. (**2017**). *Org. Chem. Front. 4*: 817–822. (c) He, L., Ng, C.-F., Li, Y. et al. (**2018**). *Angew. Chem. Int. Ed. 57*: 13635–13639. (d) Ip, H.-W., Li, Y., Kuck, D., and Chow, H.-F. (**2021**). J. Org. Chem. 86, 5546–5551. (e) Ng, C.-F., Wong, W.S., Ip, H. W., Lau, W.-W., Sun, X.-Q., Tse, H.-W., He, L., Cheung, E., Kuck, D., Chow, H.-F. (**2020**) *Chin. J. Org. Chem. 40*, 3017–3025.

70 (a) Grzybowski, M., Skonieczny, K., Butenschön, H., and Gryko, D.T. (**2013**). *Angew. Chem. Int. Ed. 52*: 9900–9930. (b) Grzybowski, M., Sadowski, B., Butenschön, H., and Gryko, D. (**2019**). *Angew. Chem. Int. Ed.* https://doi.org/10.1002/anie.201904934. (c) Alsharif, M.A., Alam Raja, Q., Abdul Majeed, N.; Jassas, R.S.,Alsimaree, A.A., Sadiq, A., Naeem, N., Mughal, E.U., Alsantali, R.I., Moussa, Z., and Ahmed, S.A. (**2021**). *RSC Adv. 11*, 29826-29858.

71 An, P., Chow, H.-F., and Kuck, D. (**2016**). *Synlett 27*: 1255–1261.

72 Wong, W.-S., Lau, W.-W., Li, Y. et al. (**2020**). *Chem. Eur. J. 26*: 4310–4319.

73 Wong, W.-S., Tse, H.-W., Cheung, E. et al. (**2019**). *J. Org. Chem. 84*: 869–878.

74 Kuck, D., Schuster, A., and Krause, R.A. (**1991**). *J. Org. Chem. 56*: 3472–3475.

75 Sakamoto, Y. and Suzuki, T. (**2013**). *J. Am. Chem. Soc. 135*: 14074–14077.

76 Kuck, D. (**2006**). *Pure Appl. Chem. 78*: 749–775.

77 Kuck, D., Paisdor, B., and Gestmann, D. (**1994**). *Angew. Chem. Int. Ed. 33*: 1251–1253.

78 Gestmann, D., Pritzkow, H., and Kuck, D. (**1996**). *Liebigs Ann.*: 1349–1359.

79 Harig, M., Neumann, B., Stammler, H.-G., and Kuck, D. (**2017**). *ChemPlusChem 82*: 1078–1095.

80 (a) Kuck, D., Schuster, A., and Gestmann, D. (**1994**). *J. Chem. Soc. Chem. Commun.*: 609–610. (b) Kuck, D., Schuster, A., Gestmann, D. et al. (**1996**). *Chem. Eur. J. 2*: 58–67.

81 (a) Kuck, D., Schuster, A., Fusco, C. et al. (**1994**). *J. Am. Chem. Soc. 116*: 2375–2381. (b) Fusco, C., Fiorentino, M., Dinoi, A. et al. (**1996**). *J. Org. Chem. 61*: 8681–8684.

82 Tellenbröker, J., Barth, D., Neumann, B. et al. (**2005**). *Org. Biomol. Chem. 3*: 570–571.

83 Dilmac, A.M., Spuling, E., de Meijere, A., and Bräse, S. (**2017**). *Angew. Chem. Int. Ed. 56*: 5684–5718.

84 Linke, J. and Kuck, D. (**2016**). *Z. Naturforsch. B 71*: 897–904.

85 Kuck, D., Hackfort, T., Neumann, B., and Stammler, H.G. (**2007**). *Pol. J. Chem. 81*: 875–892.

86 Tellenbröker, J. and Kuck, D. (**1995**). Unpublished results: J. Tellenbröker, Diploma thesis. Bielefeld University.

87 Kuck, D., Linke, J., Teichmann, L.C. et al. (**2016**). *Phys. Chem. Chem. Phys. 18*: 11722–11737.

88 (a) Tellenbröker, J. and Kuck, D. (**2011**). *Beilstein J. Org. Chem. 7*: 329–337. (b) Strübe, J., Neumann, B., Stammler, H.-G., and Kuck, D. (**2009**). *Chem. Eur. J. 15*: 2256–2260.

89 Langhals, H., Rauscher, M., Strübe, J., and Kuck, D. (**2008**). *J. Org. Chem. 73*: 1113–1116.

90 (a) Mastalerz, M. (**2018**). *Acc. Chem. Res. 51*: 2411–2422. (b) Wagner, P., Rominger, F., and Mastalerz, M. (**2018**). *Angew. Chem. Int. Ed. 57*: 11321–11324.

91 Harig, M. and Kuck, D. (**2006**). *Eur. J. Org. Chem.*: 1647–1655.

92 (a) Vile, J., Carta, M., Bezzu, C.G., and McKeown, N.B. (**2011**). *Polym. Chem. 2*: 2257–2260. (b) Vile, J., Carta, M., Bezzu, C.G. et al. (**2014**). *Polymer 55*: 326–329.

93 Tellenbröker, J. (**1999**). Doctoral thesis. Bielefeld University.

94 Linke, J., Neumann, B., Stammler, H.-G., and Kuck, D. (**2018**). *Chem. Eur. J. 24*: 9316–9324.

95 (a) Kuck, D., Paisdor, B., and Grützmacher, H.-F. (**1987**). *Chem. Ber. 120*: 589–595. (b) Paisdor, B., Grützmacher, H.-F., and Kuck, D. (**1w988**). *Chem. Ber. 121*: 1307–1313.

96 Lucchesini, F., Grasse, M., Neumann, B. et al. (**2016**). *Eur. J. Org. Chem.*: 2828–2841.

11

The Synthesis of Natural Products Containing Spirocycles

Ramon Rios

University of Southampton, Southampton, England

Natural products are the most important source of active biological products. Organic chemists have attempted to recreate the biosynthetic pathways for their synthesis in laboratories, not only to better understand their structure but also to make small structural modifications to achieve improved biological effects. Examples of the power of organic synthesis applied to the synthesis of natural products include the recent synthesis of tryptorubin A by Baran [1].

One of the more difficult structures to synthesize is that of spirocycles because spiro compounds possess two or more rings that share only one common atom, and this spirocenter is frequently chiral. For these reasons, the challenges for overcoming their synthesis have been the focus of organic chemists for years.

A large number of spiro compounds have been isolated among natural products. Ranging from β-vetibone to griseusin, chemists in the field of synthetic organic materials have developed new synthetic routes for the production of such materials. This chapter presents some of the most relevant synthesis of natural products that contain spirocyclic motifs. The focus will be on (+)-calafianin (**1**), pseurotin A (**2**), (−)-ushikulide A (**3**), (−)-acutumine (**4**), and (−)-spirotryprostatin (**5**). The syntheses of these products and the different approaches to overcoming the complexity inherent in spirocenters will be the primary focus (Figure 11.1).

11.1 Porco's Synthesis of (+)-Calafianin [2]

First isolated from the sponge *Aplysina gerardogreeni n. sp* (Aplysinidae), (+)-calafianin (**1**) is a bromotyrosine-derived compound. The name *calafianin* was inspired by Amazonian mythology, wherein the Amazonian Queen, Calafia, was allegedly the ruler of the island of California. The structure was first elucidated by the research group of Christopherson in 2000 [3]. As shown in

Spiro Compounds: Synthesis and Applications, First Edition. Edited by Ramon Rios Torres.

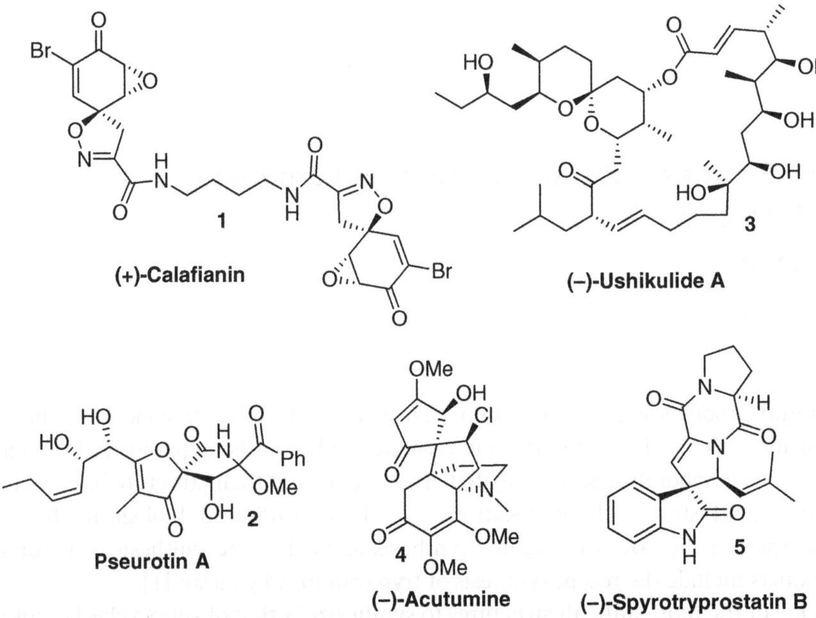

Figure 11.1 The different synthesized natural products discussed in this chapter.

Figure 11.2, the proposed structure comprises a dimeric spiroisoxazoline containing a stereogenic epoxide.

To confirm the synthesis and the stereochemical assignment of (+)-calafianin, Porco's research group designed a total synthesis process in which the key step was asymmetric nucleophilic epoxidation [4]. The retrosynthesis of the spirocycle is shown in Figure 11.3. The researchers proposed the preparation of the calafianin core **6** from a nitrile oxide cycloaddition of vinyl epoxide **7** to provide the spiroisoxazoline core, followed by acetal hydrolysis. Then, compound X will proceed from the stereoselective epoxidation of quinone monoketal **8**.

Starting from **8** (as per Figure 11.3), the initial reaction is the nucleophilic enantioselective epoxidation of the compound **8** using tartrate (*L*-DIPT) as a stereo source. Under typical epoxidation conditions (toluene, −50 °C), the final compound **9** was synthesized with good yields and excellent enantioselectivity (80% yield; 95% ee). Next, a Wittig reaction took place to transform epoxyketone **9** into an exocyclic vinyl epoxide **7**. This olefination was quite difficult due to the inherent lability of both the starting material and vinyl epoxide product. For example, the use of lithium strong bases such as BuLi of LiHMDS induced considerable decomposition of the compounds. Porco's group found an alternative in the use of KOtBu as a base and methyltriphenylphosphonium bromide at −40 °C to achieve the final compound **7** with 95% yield (see Scheme 11.1).

Figure 11.2 The structure of (+)-calafianin.

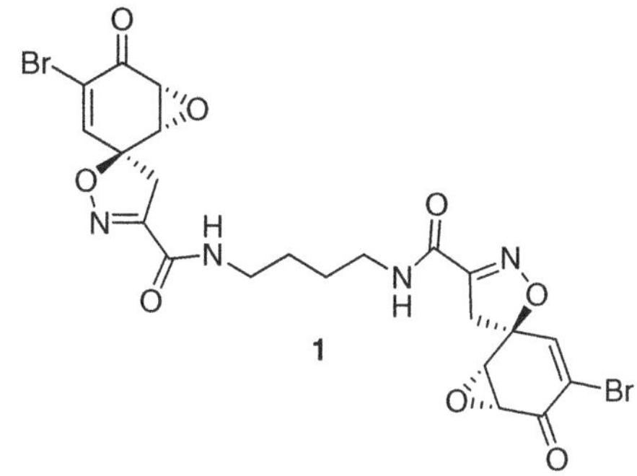

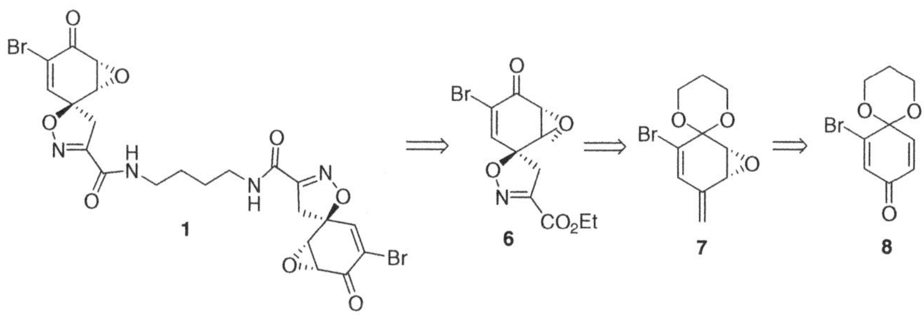

Figure 11.3 The proposed retrosynthesis for (+)-calafianin.

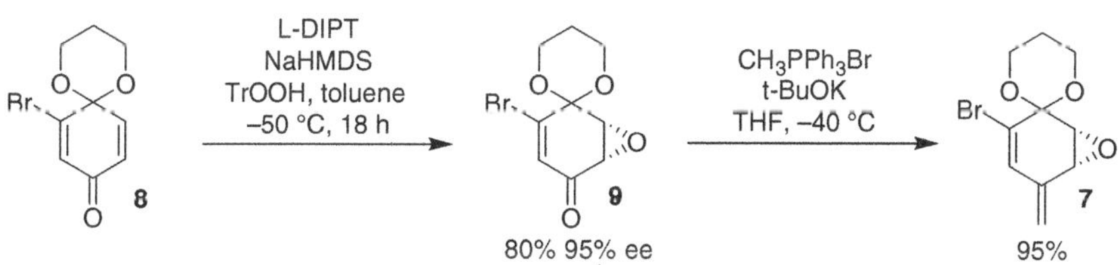

Scheme 11.1 The synthesis of exocyclic vinyl epoxide **7**.

Next, a 1,3-dipolar cycloaddition of the vinyl epoxide using an "*in situ*" nitrile oxide generated from ethyl chlorooximinoacetate and diisopropylamine took place to form the spiroisoxazoline core. Several reaction conditions were tested to achieve the final spiroisoxazoline, including the best use of Zr(IV) alkoxydes, previously reported in Kobayashi [5]. The use of Zr(OtBu)4 afforded high yields (93%) but moderate stereoselectivity, as the cycloaddition took place for the two faces of the alkene, leading to a mixture of diastereomers 1.8 (**6**):1 (**11**) (see Scheme 11.2).

Once they had synthesized the spiroisoxazoline ring, the authors proceeded to the diamidation step. The best conditions, in this case, were the use of 2-hydroxypyridine

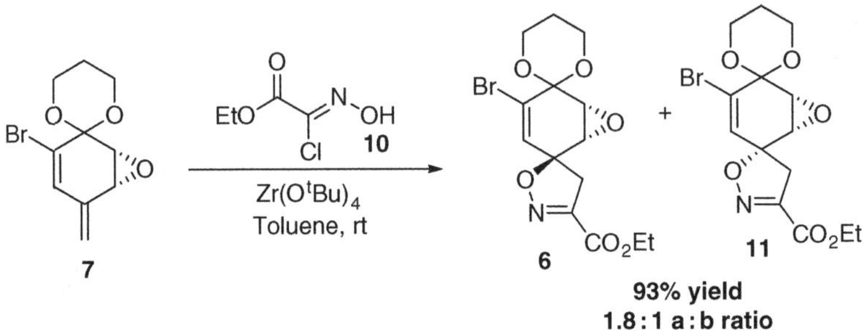

Scheme 11.2 The 1,3 dipolar cycloaddition leading to compounds **6** and **11**.

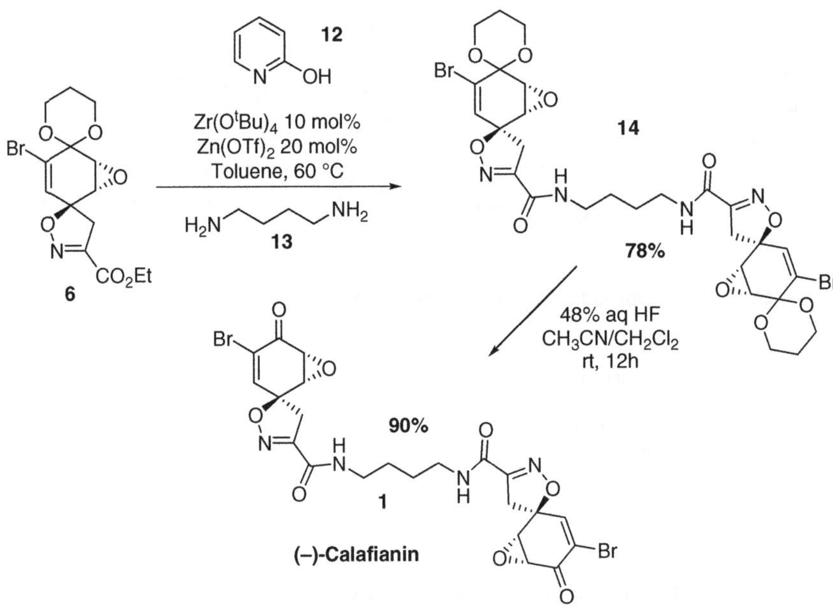

Scheme 11.3 The synthesis of **(+)-calafianin (1)**.

(**12**) as an activator in the presence of Zr(OtBu)4 and Zn(OTf)2. These reaction conditions lead to 78% yield in diamide **14**. Finally, the late-stage acetal deprotection phase took place using 48% HF in a binary solvent mixture of CH_3CN and CH_2Cl_2, affording clean hydrolysis of both cyclic acetals in high yield (90%) (Scheme 11.3).

In conclusion, Porco's research group synthesized (−)-calafianin in five steps with a global yield of 32% and moderate stereoselectivity. In this example, a key 1,3 dipolar cycloaddition was used for the synthesis of the spirocenter with excellent results.

11.2 Hayashi's Total Synthesis of Pseurotin A [6]

Pseurotin A is a microbial secondary metabolite that was isolated from the fermentation broth of Pseudeurotium ovalis (Ascomyceres) in 1976 [7]. Pseurotin A is an IgE production inhibitor that is employed in allergic illnesses [8], inhibit chitin synthase [9], and it is also found to induce cell differentiation of PC12 cells [10]. The structure of pseurotin A as it is shown in Figure 11.4 possesses a highly substituted 1-oxa-7-azaspiro[4,4]non-2-ene-4,6-dione skeleton. This "core" structure can be found in several other natural products like pseurotin B, C, D, E, F, and F2, and Azaspirene and Synerazol.

The retrosynthesis proposed by Hayashi is based on consecutive aldol and dehydration reactions between aldehyde **15** and ethyl ketone **16**, which contains γ-lactam moiety. This latter lactam can be synthesized from **17** by several functional group functionalizations, while the spiro will be synthesized though an amide addition to an alkyne as shown in Figure 11.5.

The synthesis of **16** starts with methyl pentenoate **19**, which undergoes Sharpless dihydroxylation using $(DHQ)_2PHAL$ as the chiral ligand. The corresponding diol **20** was afforded in 88% yield and was subsequently reacted with dimethoxypropane in the presence of catalytic amounts of TsOH to furnish acetal **21** in 79% yield and 95% ee. The following step consists in an aldol reaction, where treatment of **21**

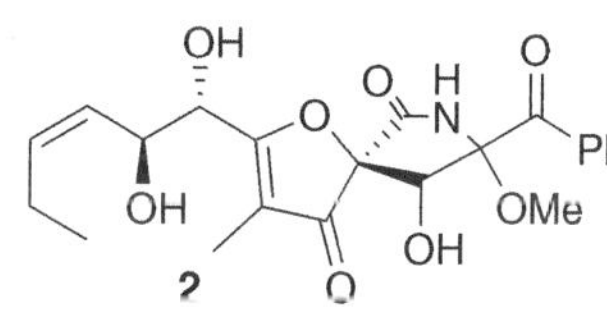

Figure 11.4 The structure of pseurotin A.

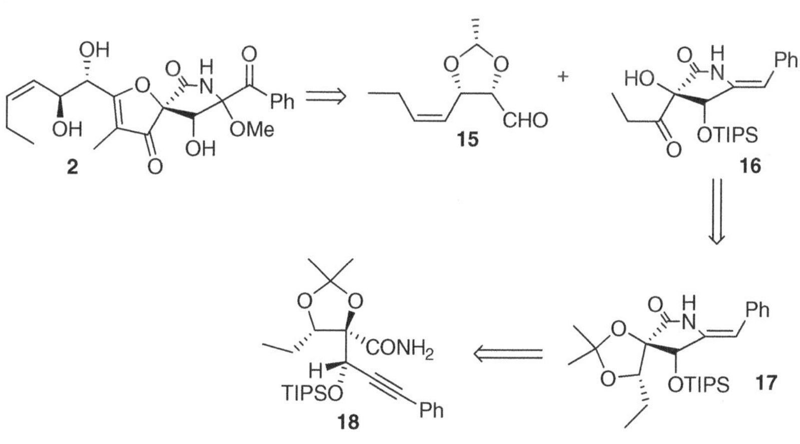

Figure 11.5 The retrosynthesis of pseurotin A.

with LDA generated the anion, which reacted with phenylpropargylaldehyde. This reaction was extremely challenging because it initially yielded a mixture of all possible diastereomers while obtaining the desired *syn* aldol (**22**) in 13% yield. An exhaustive screening of solvents, base, and additives were done; however, the results did not improve. A different approach was then selected based on the Mukaiyama aldol reaction. Following the conversion of **21** to its ketene silyl acetal, the reaction with phenylpropargylaldehyde was studied in the presence of various Lewis acids. The best conditions were observed when $MgBr_2$ was used as a Lewis acid, affording the desired product **22** in high diastereoselectivity and a good yield (69%) without sacrificing the enantioselectivity. Next, the recently formed alcohol was protected in the form of a TIPS silyl ether. Hydrolysis of the ester moiety followed by transformation of the corresponding acid to acid chloride and treatment with ammonia afforded the desired amide **18** in 45% yield. It should be noted that the direct amidation of ester **23** was not possible, likely due to the significant steric bulkiness of the ester moiety. At this stage, the relative stereochemistry of the compound was determined by single X-ray crystal analysis. The synthesis continued with the reaction of **18**, with NaH under sonication to afford the benzylidene γ-lactam **17** in 92% yield. Next, deacetalysation with TFA and oxidation of the secondary alcohol with Dess–Martin periodinane (DMP) in dichloromethane afforded chiral ketone **16** in good yields (see Scheme 11.4).

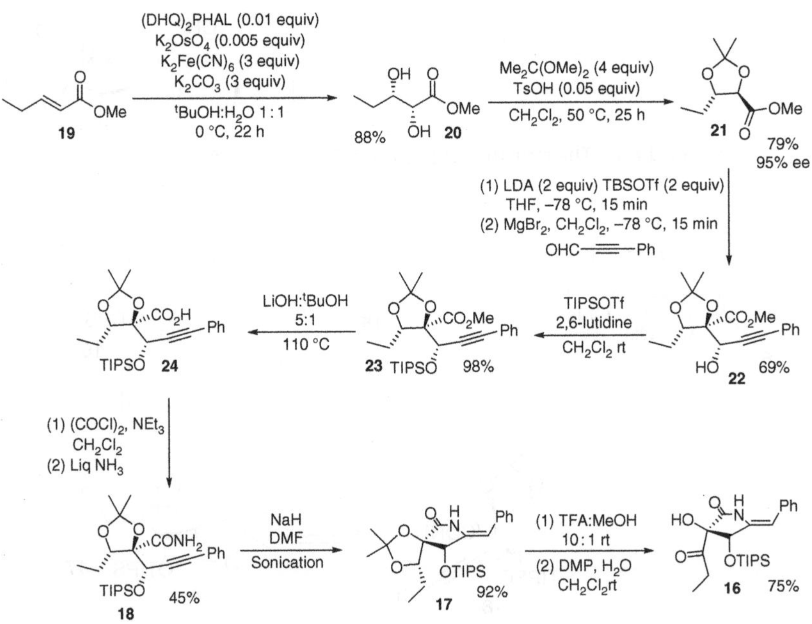

Scheme 11.4 The synthesis of **16**.

The treatment of **16** with LDA rendered the lithium enolate that reacted with chiral aldehyde **15** [11] to obtain the aldol product **25** in 54% yield and with a diastereomeric ratio of 2 : 1. Oxidation of the diastereomeric mixture with DMP in dichloromethane proceeded smoothly, yielding the 1,3-diketone **26**. Purification of **26** by thin-layer chromatography afforded compound **27** via a cyclization and dehydration reaction, which arose due to the mild acidity of the silica gel in 68% yield. Next, oxidation with dimethyldioxirane at a low temperature of −60 °C afforded the diol, which was selectively oxidized with DMP to obtain compound **28** with 42% yield for both steps. Deprotection of TIPS with NH_4F followed and treatment with AcCl in MeOH provided the final **pseurotin A** in 25% yield (see Scheme 11.5).

In conclusion, Hayashi et al. reported a total synthesis of pseurotin A using an acid-catalyzed hemiacetalization, followed by dehydration as key steps for the creation of the spirocenter.

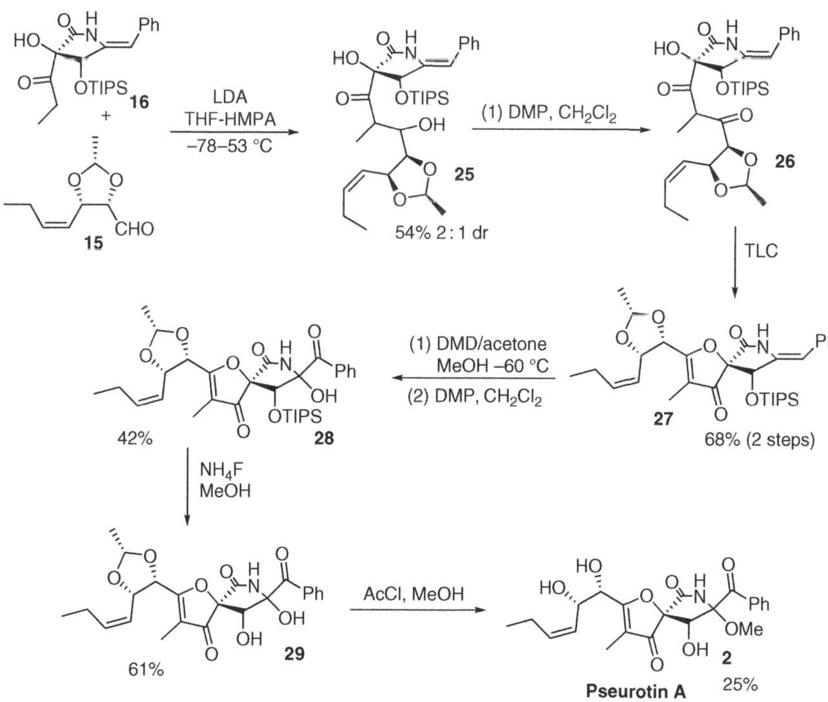

Scheme 11.5 The synthesis of pseurotin A.

11.3 Trost's Synthesis of (−)-Ushikulide A [12]

(−)-Ushikulide A (**3**) is an immunosuppressant found in the culture broth of *Streptomyces* sp. IUK-102. Its producing strain, Streptomyces sp. IUK-102, was isolated from a soil sample collected in Ibaraki prefecture, Japan, by Takahashi et al. in 2005 [13]. (−)-Ushikulide A exhibits potent activity against murine splenic lymphocyte proliferation, at a similar level as cyclosporine A and FK506 [14]. Trost reported the synthesis of (−)-Ushikulide A in 2009, in addition to confirm his stereochemical assignment (Figure 11.6).

The retrosynthesis proposed for (−)-ushikulide A was based first on a macrolactonization and a Suzuki coupling starting with fragments **30** (a spiroketal fragment) and **31** (a lineal fragment).

The lineal fragment **31** can be synthesized by an aldol reaction between **36** and **37** via a disconnection between the C_a and C_b bonds. The spiroketal fragment can be easily assembled by addition of alkyne anion **33** to aldehyde **32**. Alkyne **32** is expected to function as an electrophile in metal-catalyzed spiroketalization. The inherent polarity (by having an inductive electron-withdrawing group in the carbon of the alkyne) will give rise to the correct regiochemistry as a positive charge at the neighboring acetylenic carbon should be destabilized, directing the initial cyclization toward the other carbon of the triple bond. Compound **32** can be obtained analogously to **30** using a propargyl nucleophile **34** to react with aldehyde **35** (Figure 11.7).

The synthesis of spiroketal **30** starts with the reduction of methyl 3-oxopentanoate **38** using Noyori hydrogenation. This is followed by protection of the alcohol as a *p*-methoxybenzyl ether and its reduction to DIBAL-H to afford aldehyde **39** in good yield and almost total enantioselectivity (64 and 99% ee). Next, crotylation following the Brown method, followed by TBS protection of the resulting alcohol afforded alkene **40** in good yields (69% for both steps) and as a pure diastereomer. Hydroborylation/iodination and, subsequently, nucleophilic substitution with the lithium salt of TMS acetylene followed by TMS deprotection rendered alkyne **34** in 71% yield for both steps.

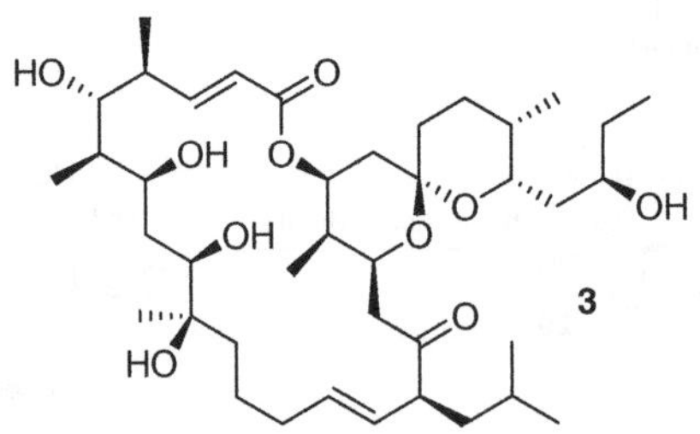

3

Figure 11.6 The structure of **(−)ushikulide A**.

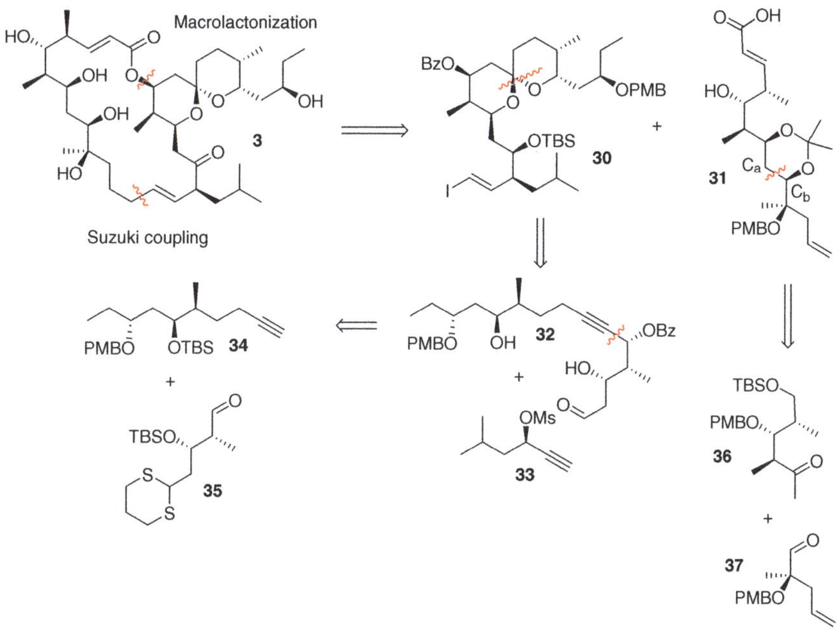

Figure 11.7 The retrosynthesis of **(−)ushikulide A.**

Contrastingly, the synthesis of aldehyde **35** began with an aldol reaction under Crimmins conditions, affording aldol **43** in excellent yields and as a single diastereomer. Next, protection of the alcohol as TMS ether and reduction with DIBAL-H provided aldehyde **35** in 74% yield over the three steps. To obtain propargyl mesylate **33**, a prophenol zinc alkynylation method was used to obtain adduct **46** from isovaleraldehyde (**44**) and methyl propiolate (**45**) in good yields (88%) and with excellent stereoselectivity (95% ee). Next, saponification of the methyl ester and decarboxylation by treatment with CuCl, followed by mesylation rendered compound **33** in good overall yield (71% for both steps) (Scheme 11.6).

The lithiation of alkyne **34** followed by the addition of aldehyde **35** rendered the alcohol in a diastereomeric ratio of 1.5 : 1 with good yield. The epimers were easily separated by SiO_2 chromatography and converted into the desired Benzoyl-protected compound **32** via direct benzoylation by BzCl, or via Mitsunobu conditions (which created the desired epimer following the inversion of the configuration to yield **32**). Spiroketal **47** was obtained after treatment with AuCl and PPTS. Hydrolysis of the 1,3-dithiane in the presence of methyl iodide afforded aldehyde **48**, which was consequently propargylated with **33** using Marshall's conditions [15], achieving a 3.8 : 1 mixture of epimers that could be separated by SiO_2 chromatography to obtain **49** in a diastereopure form. Protection of the alcohol

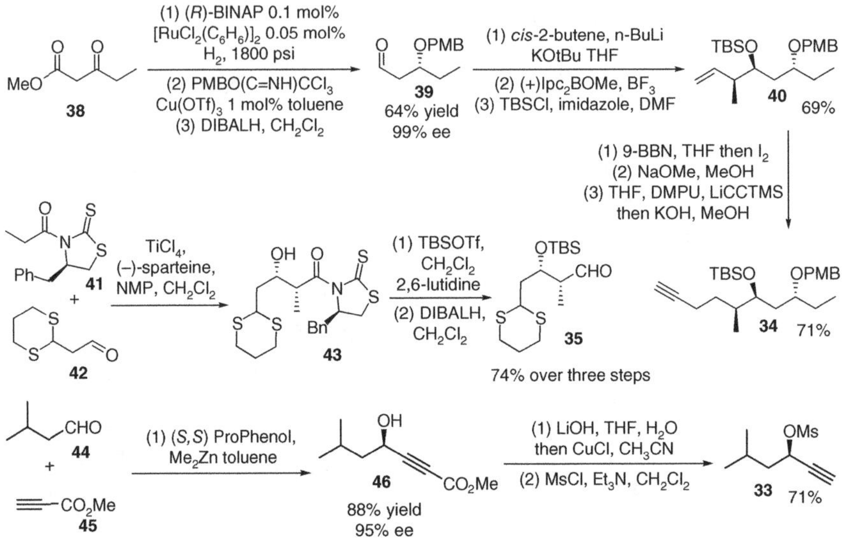

Scheme 11.6 The synthesis of compounds **33, 34,** and **35.**

with TBS and treatment with Bu_3SnH was followed by iodination furnished compound **30** in 90% yield for the three steps (Scheme 11.7).

Benzyl pyruvate **50** reacted with allyl bromide under tin (II) catalysis using Mukaiyama conditions to afford alcohol **51** [16]. The tertiary alcohol obtained was protected as PMB followed by reduction of the ester with DIBALH to obtain aldehyde **37** in 76% yield over two steps. Homoallylic alcohol **52** underwent Wacker oxidation [17] to generate ketone **36**. Next, **36** underwent an aldol reaction, with previously synthesized aldehyde **37** promoted by a dinuclear zinc-ProPhenol complex and achieved the desired diastereomer **53** with a 20 : 1 dr. *Syn* reduction of aldol **53** was followed by protection of diol as an acetonide to generate **54** in 71% yield. Next, deprotection of the terminal alcohol by TBAF followed by Dess–Martin oxidation and Wittig reaction, producing **31** in 92% yield across the three steps (Scheme 11.8).

The two main fragments were joined by a Suzuki coupling after treatment of **31** with 9-BBN. The addition of compound **30** under Pd catalysis created **55** at 67% yield. The macrolactonization took place using the Shiina methodology, creating compound **56** at 67% yield [18]. Next, deprotection of the TBS group using hydrogen fluoride pyridine, followed by oxidation using a Dess–Martin reagent, and final deprotection of the protected PMB alcohols by DDQ provided (−)-ushikulide A at 52% yield (the final three steps) as shown in Scheme 11.9.

In conclusion, Trost's group reported the synthesis of spiro-compound (−)-ushikulide A in 21 linear and 41 total steps at a global yield of 2.2%. The key

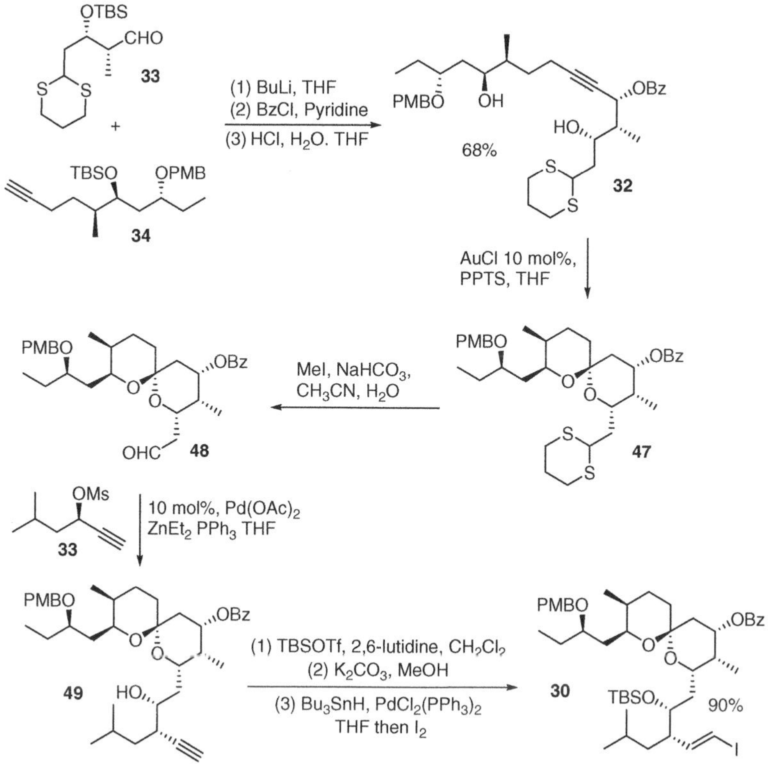

Scheme 11.7 The synthesis of compound **30**.

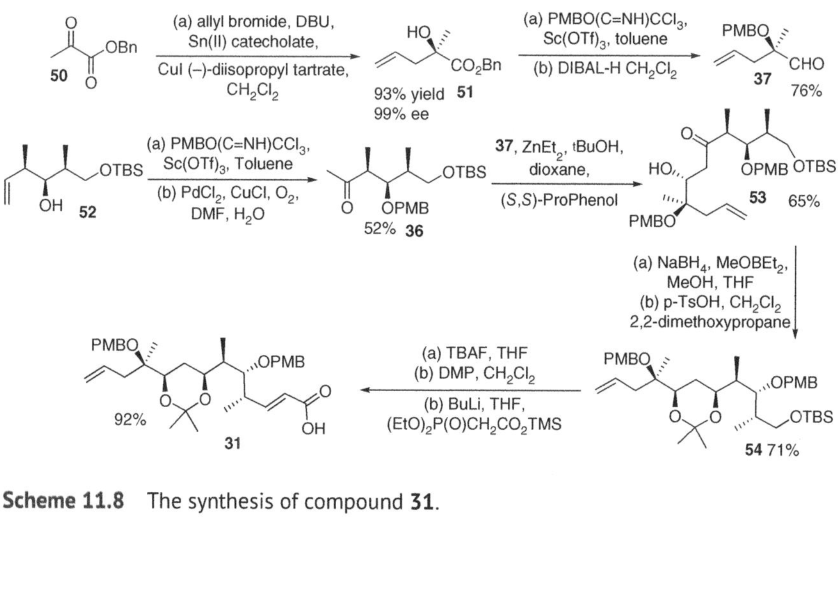

Scheme 11.8 The synthesis of compound **31**.

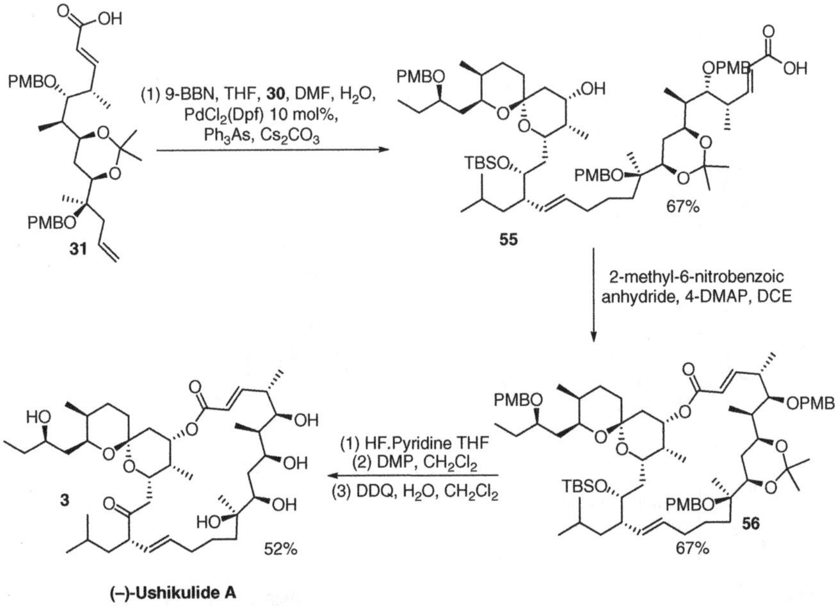

Scheme 11.9 The synthesis of **(−)ushikulide A.**

step for the preparation of the spirocenter was gold-catalyzed ketalization, which afforded the spiroketal in good yields and with regio- and stereoselectivity.

11.4 Castle's and Herzon's Syntheses of (−)Acutumine

Menispernum daericum DC. (Menispermaceae) occurs widely in China and its rhizome is a traditional Chinese medicine officially listed in the Chinese Pharmacopoeia as an analgesic and antipyretic. The roots of the plant contain a chlorine-containing alkaloid called acutumine, which is a tetracyclic alkaloid that was first isolated from the roots of Simonemium acutum by Goto and Sudzuki in 1929 [19]. Later, Tomita et al. determined its structure by X-ray crystallography after isolation from Menispernum daericum DC [20]. (−)-Acutumine possesses interesting bioactivity such as selective T-cell citotoxicity [21] and anti-amnesic properties [22].

(−)-Acutamine shares a degree of structural homology with the well-studied hasubanan alkaloids. The most important structural properties of (−)-acutamine include a propellane-type system and a spirocycle as shown in Figure 11.8.

Figure 11.8 The structure of **(−)-acutamine.**

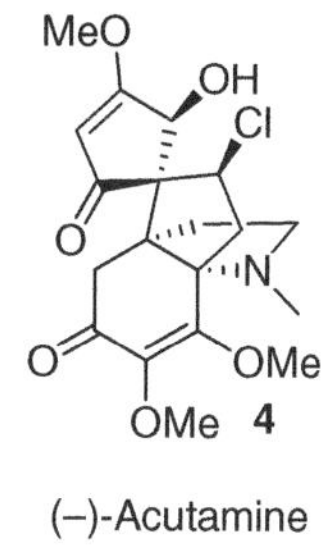

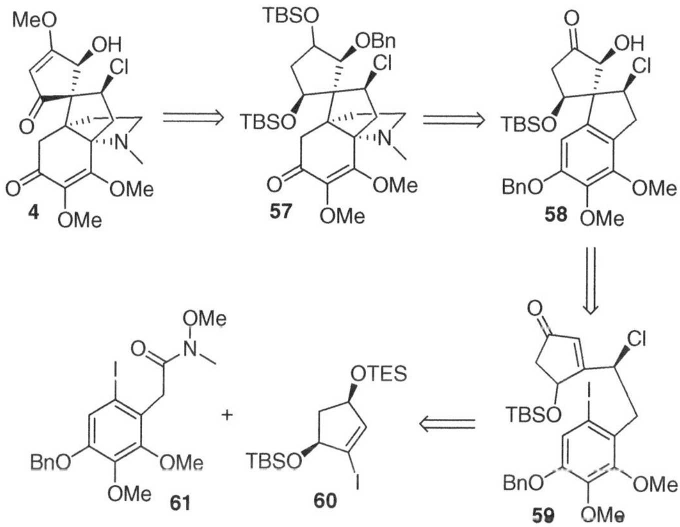

Figure 11.9 The proposed retrosynthesis procedure.

11.4.1 Castle's Synthesis of (−)-acutamine [23]

The first total synthesis of (−)-acutamine was reported by Castle in 2009. The retrosynthesis envisioned for (−)-acutamine is shown in Figure 11.9. First, manipulation of the oxygenated functional groups contained in the cyclopentene ring of **4** led to tetracyclic intermediate **57**. The tetracyclic intermediate **57** was derived from spirocyclic α-hydroxy ketone **58**. Next, disconnection of the aryl–alkyl C–C bond from the spirocyclic carbon led to the production of enone **59**. The α,β-unsaturated ketone was used as radical acceptor and caused the cyclization of the electron-rich aryl radical, leading to a 5-exo-trig cyclization. The transformation of **59** into **58** constituted a radical-polar crossover reaction that merged an intramolecular radical conjugate addition with an enolate hydroxylation. Finally, compound **59** could be synthesized via a Weinreb amide–Grignard

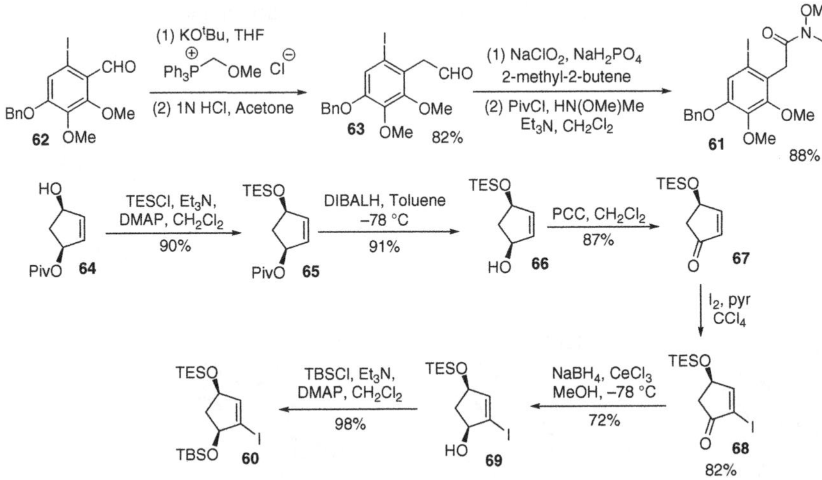

Scheme 11.10 The synthesis of compounds **60** and **61**.

reagent-coupling between **60** and **61**. The major difficulty in this regard was the use of two different protecting groups in **61**; the two orthogonal protecting groups were necessary to enable differentiating between them in the construction of enone moiety.

The synthesis of fragment **61** started with a Wittig homologation of pentasubstituted benzaldehyde **62**, providing aldehyde **63**. Next, **63** underwent oxidation and amidation to afford the Weinreb amide **61** with 72% yield over 4 steps. Contrastingly, cyclopentene 60 was obtained through the protection of enantiopure alcohol **64** as a TES ether, followed by Pivaloyl cleavage with DIBALH, affording alcohol **66**. Next, oxidation followed by iodination furnished iodoketone **68**, which was reduced using Luche conditions and followed by TBS protection of the resulting alcohol, thus affording **60** in excellent total yield (41% over 6 steps) (Scheme 11.10).

Next, the coupling of both fragments **60** and **61** took place. Treatment of **60** with isopropyl magnesium chloride generated the Grignard reagent, which reacted with Weinreb amide **61** to furnish **70** in a moderate yield (62%). The diastereoselective reduction of the ketone was accomplished using the Corey–Bakshi–Shibata catalyst and borane, this affording alcohol **71** in good yield and good diastereoselectivity (89% yield, 9 : 1 dr), which was transformed into the chloride following Corey's conditions. The corresponding chloride **72** was obtained via inversion of configuration of the alcohol (via the SN2 mechanism) in moderate yields (65%). Deprotection of the TES protected the alcohol; this was followed by oxidation, which delivered the α,β-unsaturated ketone **59** in good overall yields (59% over two steps) (Scheme 11.11).

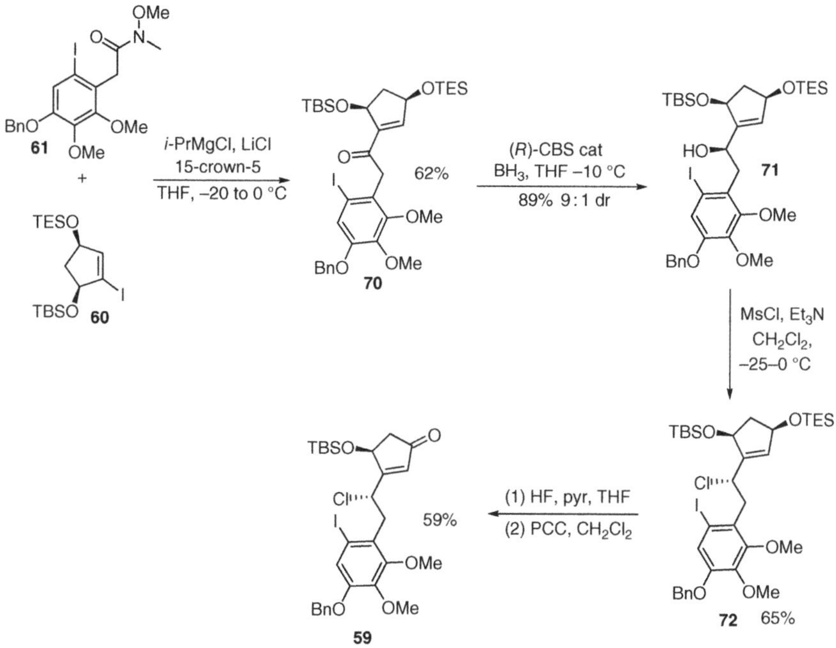

Scheme 11.11 The synthesis of compound **59**.

The next step is where the spirocycle is created. A radical-polar crossover reaction was employed as a strategy for doing so. A key element in the process was the α,β-unsaturated ketone, in which α-keto radical intermediate will further react following the 5-*exo*-trig cyclization to form the hydroxyl compound. The proposed cyclization hydroxylation of **59** will form two new stereocenters, leading to **58** and **73**. The authors hypothesized that the bulky silyloxy group would direct the aryl radical to the opposite face of the enone, while the aryl group would shield the *Re* face of the enolate intermediate; moreover, the pseudo-equatorial orientation of the silyl ether would minimize its impact on the steric environment, which would allow for hydroxylation to take place in the *Si* face. The best conditions for the radical-polar crossover reaction was using hexabutyl tin as a generator of the aryl radical, a sunlamp, Et_3Al as a promoter, and 3-(phenylsulfonyl)oxaziridine as an oxidant obtaining **58** in 62% yield, along with iodide **73** in 7% yield. Remarkably, the authors were able to convert **73** into the desired compound **58** using $ZnEt_2$, oxygen, and 3-(phenylsulfonyl)oxaziridine, in this manner increasing the total yield of this step (Scheme 11.12).

Once spirocycle **58** had been synthesized, the next step was the reduction of the ketone by *L*-selectride to obtain **74** in 88% yield and 9 : 1 dr. This step was

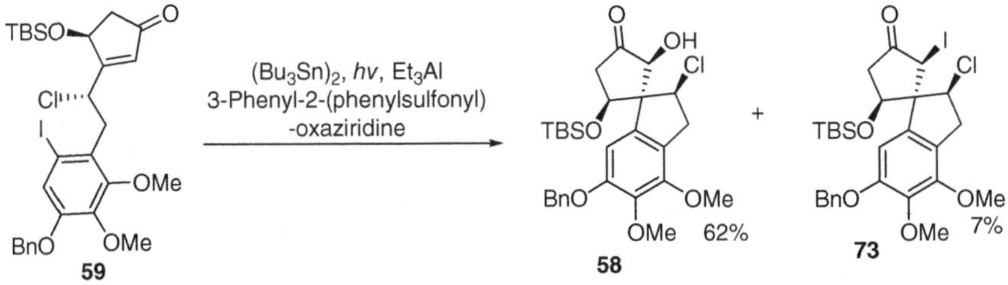

Scheme 11.12 The synthesis of **58**.

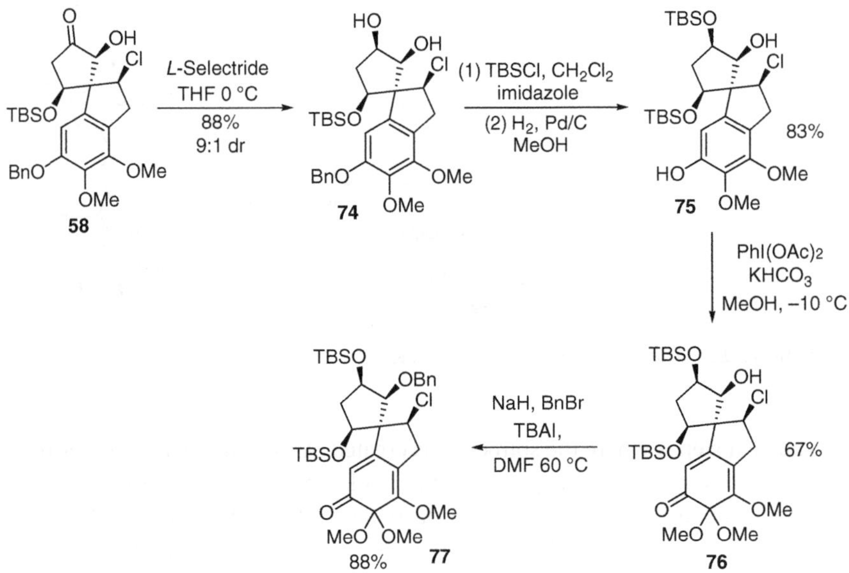

Scheme 11.13 The synthesis of **77**.

necessary due to the incompatibility of the ketone functional group with the future allylation. Selective silylation of the recently generated alcohol with TBSCl followed via treatment with H_2 and Pd/C to deprotect the benzyl alcohol and to generate **75** in good yield (83% over two steps). Phenolic oxidation of **75** by $PhI(OAc)_2$ in MeOH delivered masked *p*-benzoquinone **76** in 67% yield. Next, the neopentyl alcohol was benzylated to furnish **77** in good yield (88%) (Scheme 11.13).

Next, the crucial asymmetric ketone allylation took place. The authors used Nakamura's chiral allyl zinc reagent (**81**) to promote the diastereoselective allylation of **77**. In these conditions, the allylation proceeded in good yield and with excellent diastereoselectivity (79% and 93 : 7 dr) to afford allyl derivative **78**, which was then reacted with ^{t}BuOK and 18-crown-6, triggering an oxy-Cope

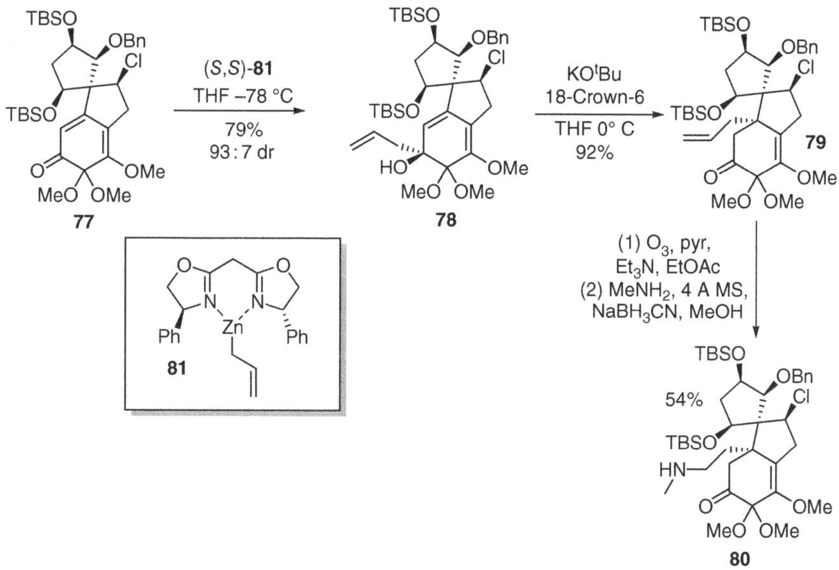

Scheme 11.14 The synthesis of compound **80**.

rearrangement that produced ketone **79** in excellent yield (92%). Next, ozonolysis of the double bond followed by reductive amination to produce amine compound **80** with good overall yield (see Scheme 11.14).

The next key cyclization was promoted by BCl_3 rendering compound **57** in moderate yield (45%), likely as a consequence of the congested nature of the formed C–N bond and the acid sensitivity of the substrate. Deprotection with TBAF of both silyl-protected alcohols followed by oxidation, generating diketone **82** in good yield (56% over two steps). Deprotection of the benzyl alcohol using H_2 and Pd/C followed by treatment with $TiCl_4$ in MeOH led to the final (−)-acutamine in excellent yield. Remarkably, in the final step, the opposite regioisomer (**84**) was obtained in a low yield (Scheme 11.15).

In summary, (−)-acutumine was synthesized using a 5-*exo* intramolecular conjugate addition of an aryl radical onto an enone acceptor, leading to the conversion of the resulting α-keto radical into an enolate and the hydroxylation of the enolate in a highly diastereoselective manner.

11.4.2 Herzon's Synthesis of (−)-Acutumine [24]

In 2013, Herzon et al. developed the second total synthesis of (−)-acutumine based on their previous synthesis of hasubanan alkaloids [25]. The retrosynthesis proposed by Herzon include several key intermediates. First, modification of the

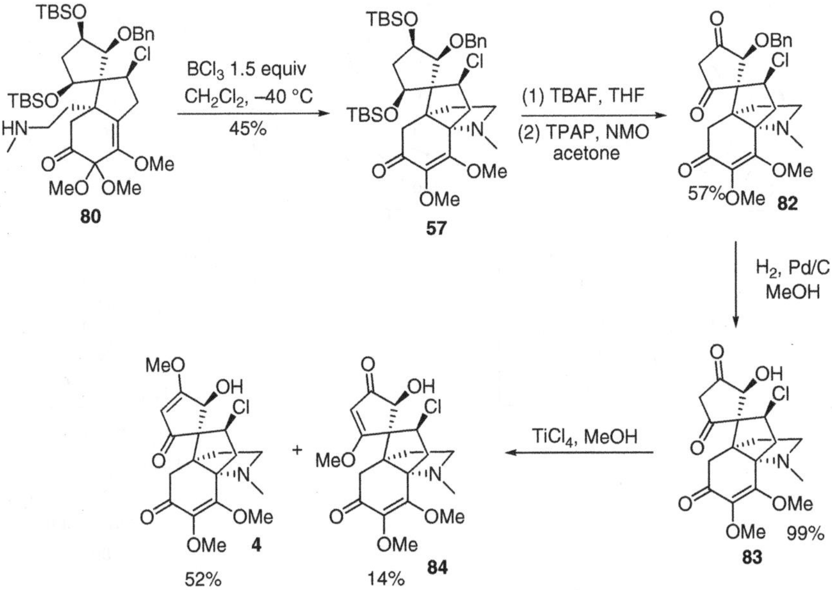

Scheme 11.15 The synthesis of (–)-acutumine.

cyclopentenone is effected, leading to compound **85**, which can be synthesized via a Hosomi–Sakurai allylation using **86**. The retrosynthesis of **86** is done via retro Diels–Alder reaction, followed by a propargyl addition between alkyne **88** and iminium **89**. Finally, iminium **89** can be synthesized from azide **90** in three steps (Figure 11.10).

The synthesis begins with the oxidation of azide **90** using hydrogen peroxide in formic acid, affording quinone **91** in a moderate yield (48%). Next, a regio- and stereoselective Diels–Alder reaction with 5-trimehylsilylcyclopentadiene, mediated by Corey–Bakshi–Shibata oxazaborolidine, took place and produced the *endo* adduct **92** in 78% yield and 93% ee. Compound **92** was reduced following a Staudinger procedure with trimethylphosphine, affording a quantitative yield of compound **89**, and intramolecular imine formation was followed by methylation of the imine by methyl triflate. Then, **89** was immediately treated with the lithium acetylide **88**, providing the 1,2 addition product **87** as single diastereomer in 85% yield (see Scheme 11.16).

The following step comprised a thermal retro Diels–Alder reaction that produced cyclohexadienone **86** in excellent yield (98%). Palladium-catalyzed regio- and stereoselective hydrostannylation of **86** afforded vinyl stannane **93**. Next, activation of the allylic TMS group by the addition of TBAF induced a Hosomi–Sakurai allylation to yield tetracycle **94** as a single diastereomer in modest yield (37%). A copper-mediated chloride stannylation of **94**, followed by the removal of

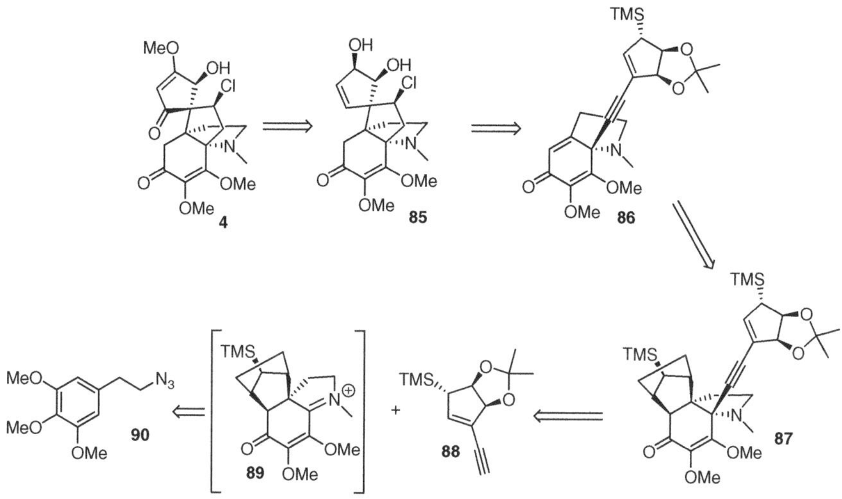

Figure 11.10 Herzon's retrosynthesis of (−)-acutumine.

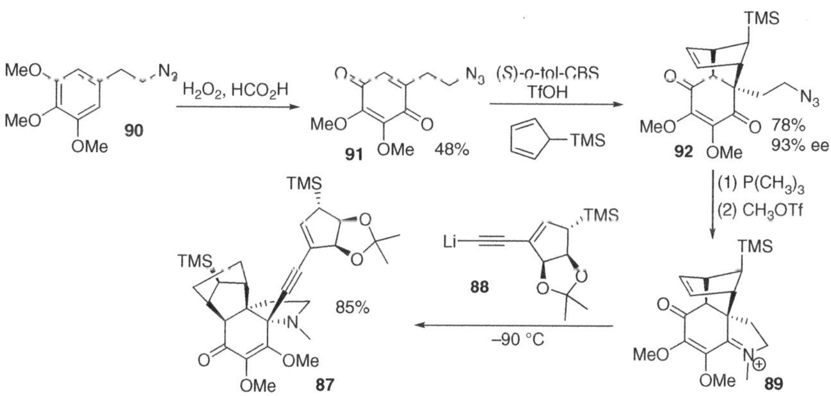

Scheme 11.16 The synthesis of compound **87**.

the acetonide using PTSA in water/methanol media afforded **85** in good yield (79%). Diol 85 was then oxidized by trifluoroacetic anhydride and DIPEA in DMSO and treated with sodium thiomethoxide to afford the Michael adduct **95** in excellent yield (97%) as it is depicted in Scheme 11.17.

Compound **95** was then treated with *N*-iodosuccinimide and formic acid, allowing for the oxidative elimination of the methanethiol, followed by 1,2 addition to a putative oxocarbenium ion to furnish compound **96**. Thermolysis of **96** using DIPEA at 100 °C promoted a [3,3]-sigmatropic rearrangement, which in turn afforded formiate **97**. Cleavage of the formyl group by ammonia in methanol followed by hemiketal formation produced compound **98**, which underwent

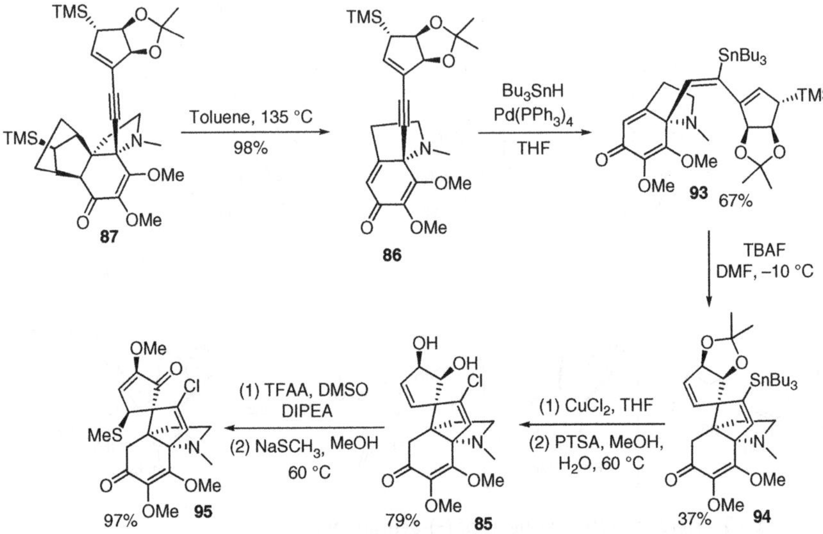

Scheme 11.17 The synthesis of compound **95**.

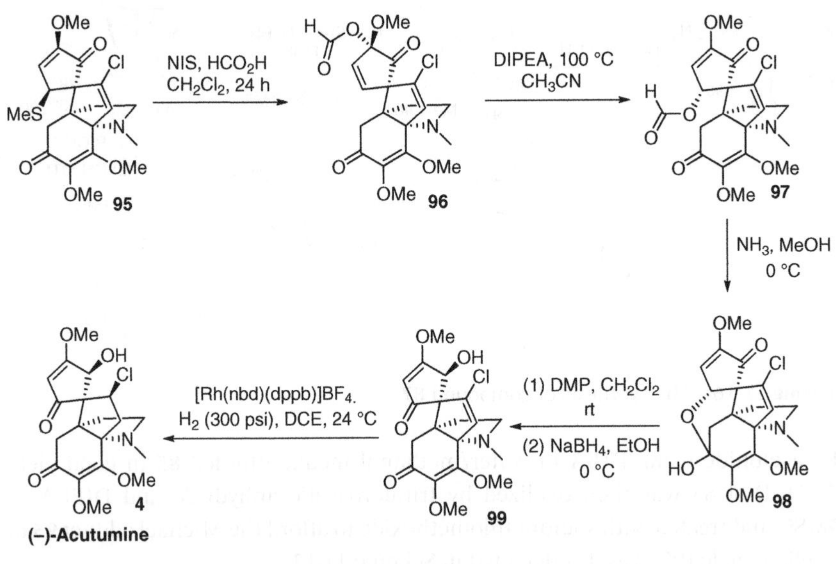

Scheme 11.18 The synthesis of (-)-acutumine.

Dess–Martin oxidation and formed a vinylogous diketone intermediate that was selectively reduced in situ by sodium borohydride to yield **99** in a moderate yield (28% over five steps). Next, compound **99** underwent hydrogenation promoted by ([Rh(nbd)(dppb)]BF_4 to produce **(−)-acutumine (4)** in 17% yield (Scheme 11.18).

In summary, Herzon et al. reported the synthesis of (−)-acutumine using Hosomi–Sakurai cyclization as a key step for the synthesis of the spirocycle feature to form two contiguous quaternary centers with excellent stereoselective control.

11.5 Overman's and Carreira's Synthesis of Spirotryprostatin B

Osada et al. obtained spirotryprostatin B from the fermentation broth of the fungus Aspergillus fumigatus in 1996 [26]. Spirotryprostatin B is a diketopiperazine alkaloid that inhibits the G2/M phase progression of the mammalian cell cycle at micromolar concentrations [27]. Spirotryprostatin B (Figure 11.11) is an extremely complex alkaloid that appears to arise biosynthetically by phenylation of a diketopiperazine derived from tryptophan and proline. Several approaches have been effected for their synthesis (see Danishefsky [28], Williams [29], Ganesan [30], Carreira [31], and Trost [32]).

11.5.1 Overman's Synthesis of Spirotryprostatin B [33]

Overman's strategy consists to correlate the relative configuration of the spirocenter and the consecutive tertiary center to the geometry of the internal double bond of a triene cyclization substrate **101** by capitalizing on the stereochemical selectivity of two palladium-catalyzed reactions: Heck insertions and the trapping of π3-allylpalladium intermediates by nitrogen nucleophiles (Figure 11.12). The retrosynthesis comprised a suprafacial intramolecular Heck reaction of **101** in a favored 5-exo sense, which generated the π-allyl palladium intermediate **100**, which was trapped by the nitrogen of the diketopiperazine and took place *anti* to the metal. The selection of the Palladium chiral ligand and the isomer of the triene **101** was crucial for the success of this approach. The triene compound **101** was easily synthesized from alcohol **103** or MBH adduct **104**, depending on which isomer of the double bond is required.

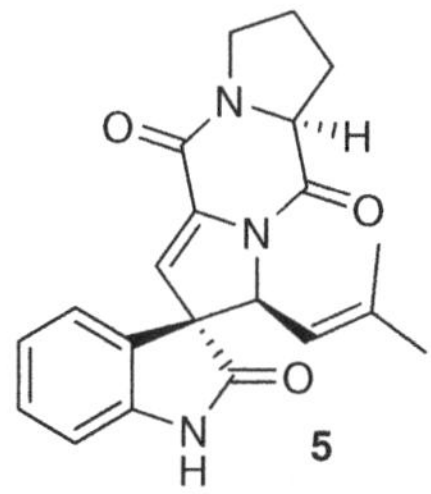

5

Spyrostryprostatin B

Figure 11.11 The structure of spirotryprostatin B.

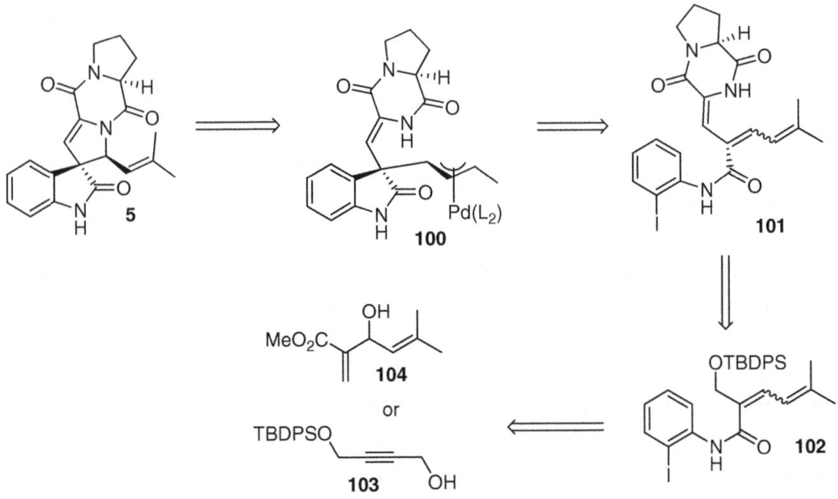

Figure 11.12 The proposed retrosynthesis.

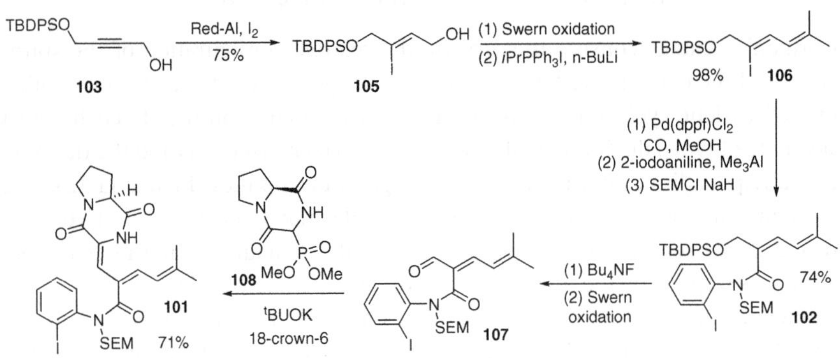

Scheme 11.19 The synthesis of compound **101**.

The first synthesis attempt began with alcohol **103**, which was treated with Red-Al and iodine to furnish vinyl iodide **105** in good yields (75%). Swern oxidation of the terminal alcohol, followed by Wittig reaction of the recently formed aldehyde rendered compound **106** in good yield. Next, palladium-catalyzed carbonylation in the presence of methanol, followed by aminolysis of the consequent ester with 2-iodoaniline and protection of the amide nitrogen as SEM afforded anilide **102** in 74% yield. Deprotection of the silylated alcohol by TBAF followed by Swern oxidation afforded **107** in good yield. Wittig reaction with compound **108** afforded compound **101** in good yield (71%) (Scheme 11.19).

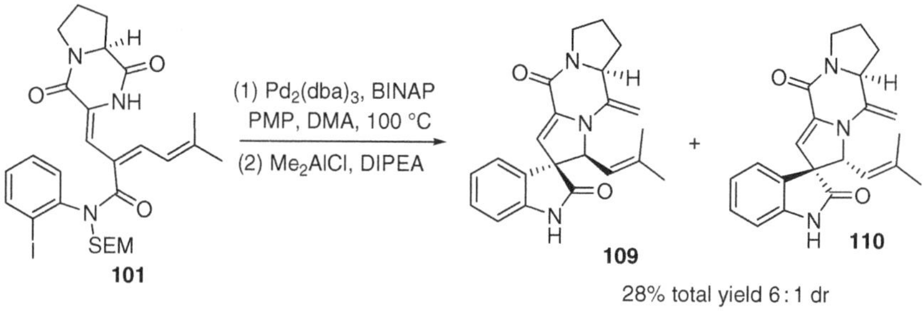

Scheme 11.20 The synthesis of compounds **109** and **110**.

With this isomer on hand, the cyclization of **101** with Pd/(*S*)-BINAP and an excess of PMP in DMA afforded a 6 : 1 mixture of **109** and **110** in 28% yield. Changing (*S*)-BINAP for (*R*)-BINAP proceeded similarly but with inversion of the ratio of the products. Unfortunately, both products had the wrong relative configuration compared with (−)-spirotryprostatin B (Scheme 11.20).

Based on these results, Overman and Rosen changed their approach using a *cis* triene rather than the *trans* triene **101** with the hope that the cyclization of this compound would lead to (−)-spirotryprostatin B. The new synthesis began with MBH adduct **104**, which reacted with acetic anhydride in pyridine followed by the treatment of $MgBr_2$ to give the primary bromide, which was displaced in situ with acetate to obtain (*E*)-dienoate **111**, in 94% yield. Hydrolysis of **111** followed by alcohol protection in the form of TBDPS and aminolysis with 2-iodoaniline furnished anilide **112** in excellent yield (78%). Protection of the nitrogen as SEM followed by deprotection of the silylated alcohol by TBAF, Swern oxidation to provide the aldehyde and Wittig reaction with compound **108** produced the desired (*E*) isomer **113** in 61% yield. The cyclization of **113** promoted by $Pd_2(dba)_3$ using tri(*ortho*)toluyl phosphine as a ligand with an excess of KOAc in THF afforded a mixture of pentacyles. The removal of the SEM group allowed for the formation of (−)-spirotryprostatin B in 21% yield and 3,18-bis-epi-spirotryprostatin B in 21% yield (Scheme 11.21).

In summary, Overman and Rosen reported the synthesis of (−)-spirotryprostatin B in 9% yield from methyl acrylate. Moreover, their different approaches allowed them to synthesize several diastereomers of (−)-spirotryprostatin B in similar yields.

11.5.2 Carreira's Synthesis of (−)-Spirotryprostatin B [31]

Carreira and Meyers reported in 2003 a new synthesis of (−)-spirotryprostatin B. The key step in this strategy for the synthesis of spiro[oxindole-3,3'-pyrrolidine] was a MgI_2-catalyzed ring expansion reaction between an oxindole vinyl

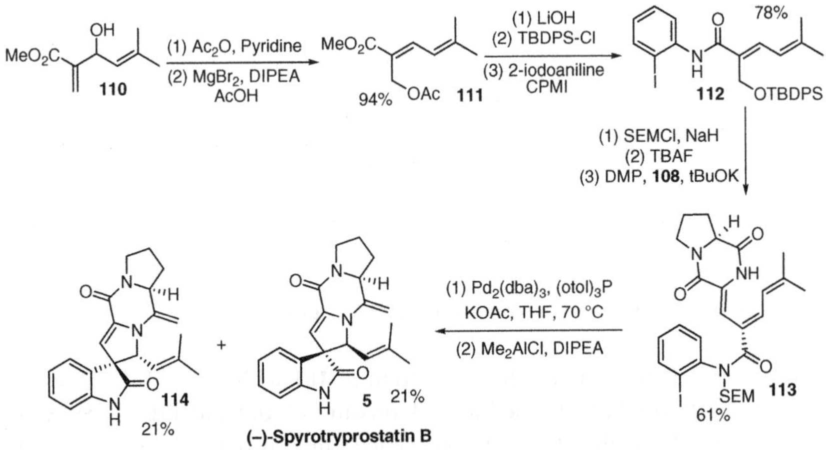

Scheme 11.21 The synthesis of (–)-spirotryprostatin B.

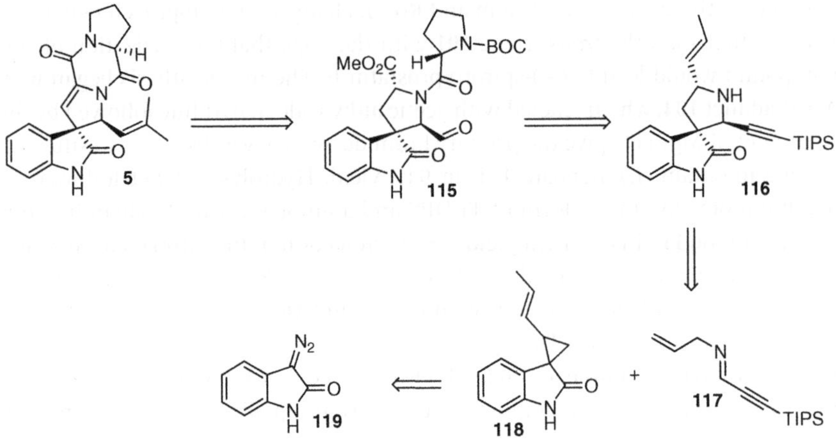

Figure 11.13 The retrosynthesis proposed by Carreira.

cyclopropane derivative and aldimines. The retrosynthesis proposed is shown in Figure 11.13. First, disconnection of the piperazine ring was effected, which led to compound **115**. Functional group interconversion and breaking the amide bound led to the creation of compound **114**. The pyrrolidine ring of **114** could be synthesized by a ring expansion starting from vinyl cyclopropane **118** and aldimine **117**. Finally, vinyl cyclopropane **118** was easily accessible from diazooxindole **119**.

The synthesis began with diazooxindole **119** undergoing cyclopropanation with piperylene catalyzed by Rh to afford **118** as a mixture of diastereomers. Next, compound **118** was treated with MgI_2 to furnish the intermediate **120**, which was reacted

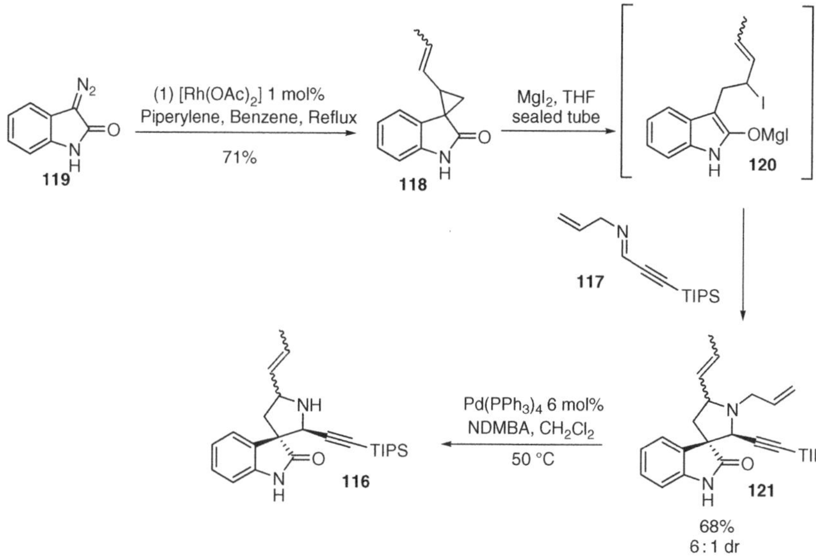

Scheme 11.22 The synthesis of compound **116**.

with aldimine **117** to produce compound **121** as a 6 : 1 mixture of diastereomers. Next, deprotection of the allyl amine promoted by Pd and 1,3-dimethylbarbituric acid produced **116** (Scheme 11.22).

At this stage, the mixture of diastereomers of **116** could be separated by column chromatography to obtain **116'**, which was the best diastereomer for finalizing the synthesis. Coupling of amine **116'** with the proline derivative **122** proceeded in 90% yield as a mixture of diastereomers that could be resolved by obtaining 45% of the enantiopure diastereomer **123**, which led to (−)-spirotryprostatin B. Next, oxidation of the double bond furnished aldehyde **124**, which was further oxidized to acid and methylated by CH_2N_2 to form **125** in 89% yield. Deprotection of the triple bond by TBAF followed by Lindlar reduction afforded **128** in 90% yield (for both steps). Oxidation of the alkene led to aldehyde **115** in 77% yield (Scheme 11.23).

The application of Kocienski modification of the Julia–Lythgoe olefination to aldehyde **115** rendered olefin **128** (78% yield), which was treated with PhSeCl using LHMDS as a base, followed by oxidation with DMDO to generate an insaturation in the pyrrolidine ring obtaining 129 in 74% yield. Deprotection of the proline ring with TFA and intramolecular amidation furnished (−)-spirotryprostatin B in 74% yield (Scheme 11.24).

In summary, Carreira and Meyers described the synthesis of (−)-spirotryprostatin B using a ring expansion reaction catalyzed by MgI_2 as a key step for the formation of the spirocenter.

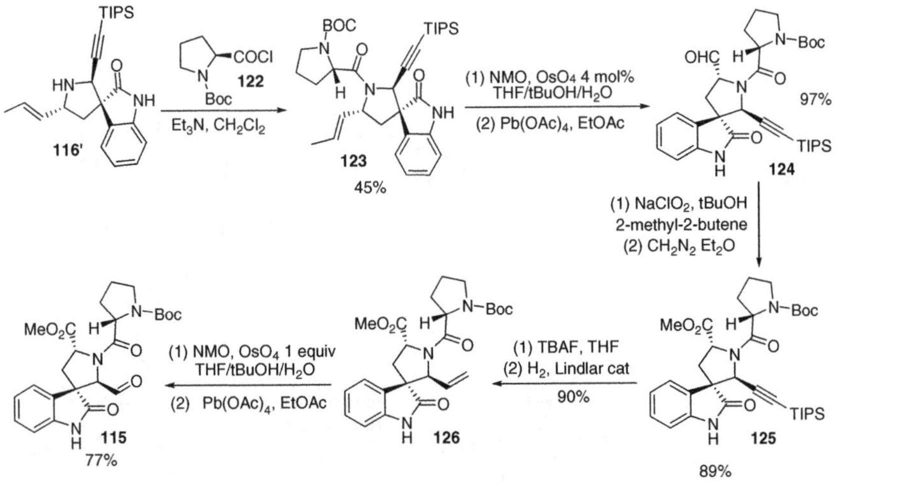

Scheme 11.23 The synthesis of compound **115**.

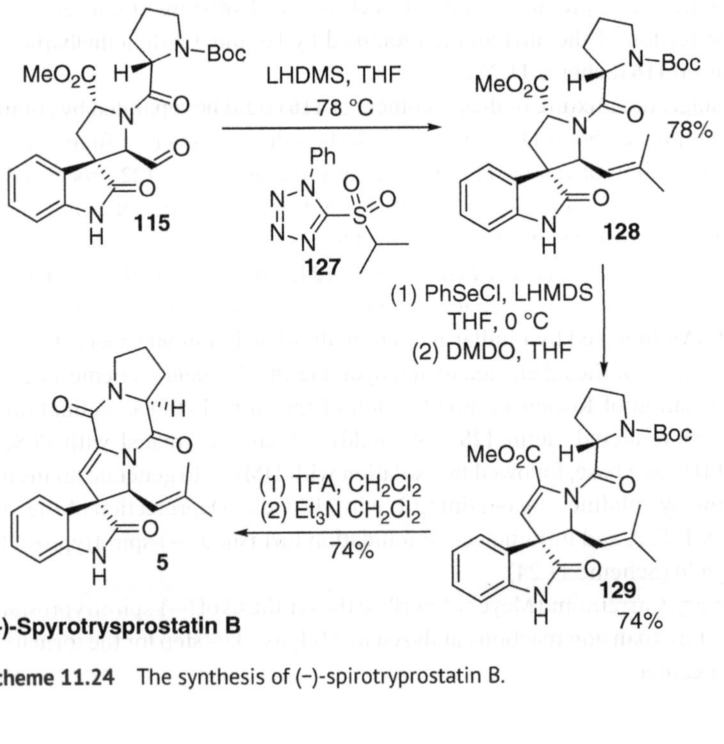

(−)-Spyrotrysprostatin B

Scheme 11.24 The synthesis of (−)-spirotryprostatin B.

11.6 Conclusion

In this chapter, we summarized the synthesis of five natural compounds that present in its structure a spirocycle. Seven syntheses were discussed in which different approaches were effected to provide the spirocenters. Including 1,3-dipolar cycloadditions, ring expansion reactions, and gold-catalyzed oxidation, these approaches overcame the inherent difficulties of spirocenters. These syntheses are important examples of how organic synthesis has evolved and improved to overcome the structural challenges presented by natural products. We hope that in the future, additional methodologies will be developed to allow organic chemists to compete with the natural environment in the synthesis of beautiful and complex structures such as spirocycles.

References

1 Reisberg, S.H., Gao, Y., Walker, A. et al. (**2020**). *Science 367*: 458–463.
2 Bardhan, S., Schmitt, D.C., and Porco, J.A. (**2006**). *Org. Lett. 8*: 927–930.
3 Encarnacion, R.D., Sandoval, E., Malmstrom, J., and Christopherson, C. (**2000**). *J. Nat. Prod. 83*: 874–875.
4 Li, C., Pace, E., Liang, M.C. et al. (**2001**). *J. Am. Chem. Soc. 123*: 11308.
5 Yamashita, Y. and Kobayashi, S. (**2004**). *J. Am. Chem. Soc. 126*: 11279–11282.
6 Hayashi, Y., Shoji, M., Yamaguchi, S. et al. (**2003**). *Org. Lett. 13*: 2287–2290.
7 Bloch, P., Tamm, C., Bollinger, P. et al. (**1976**). *Helv. Chim. Acta 59*: 133–137.
8 Ishikawa, M., Ninomiya, T., Akabane, H. et al. (**2009**). *Bioorg. Med. Chem. Lett. 19*: 1457–1460.
9 Wenke, J., Anke, H., and Sterner, O. (**1993**). *Biosci. Biotechnol. Biochem. 57*: 961–964.
10 Komagata, D., Fujita, S., Yamashita, N. et al. (**1996**). *J. Antibiot. 49*: 958–959.
11 Hayashi, Y., Shoji, M., Yamaguchi, J. et al. (**2002**). *J. Am. Chem. Soc. 124*: 12078–12079.
12 (a) Trost, B.M. and Boyle, B.M. (**2008**). *J. Am. Chem. Soc. 130*: 16190–16192. (b) Trost, B.M., O'Boyle, B.M., and Hund, D. (**2009**). *J. Am. Chem. Soc. 131*: 15061–15074.
13 Takahashi, K., Yoshihara, T., and Kurosawa, K. (**2005**). *J. Antiobiot. 58*: 420–424.
14 Kino, T., Hatanaka, H., Hashimoto, M. et al. (**1987**). *J. Antiobiot. 40*: 1249–1255.
15 Marshall, J.A. and Schaaf, G.M. (**2001**). *J. Org. Chem. 66*: 7825–7831.
16 Yamada, K., Tozawa, T., Nishida, M., and Mukaiyama, T. (**1997**). *Bull. Chem. Soc. Jpn. 70*: 2301–2308.
17 Kang, S.K., Jung, K.Y., Chung, J.U. et al. (**1995**). *J. Org. Chem. 60*: 4678–4679.

18 Shiina, I., Kubota, M., Oshiumi, H., and Hashizume, M. (**2004**). *J. Org. Chem. 69*: 1822–1830.
19 Goto, K. and Sudzuki, H. (**1929**). *Bull. Chem. Soc. Jpn. 4*: 220–224.
20 Tomita, M., Okamoto, Y., Kikuchi, T. et al. (**1971**). *Chem. Pharm. Bull. 19*: 770–791.
21 Yu, B.W., Chen, J.Y., Wang, Y.P. et al. (**2002**). *Phytochemistry 61*: 439–442.
22 Qin, G.W., Tang, X.C., Lestage, P. et al. (**2003**). *PCT Int. Appl.*: WO 2004000815.
23 Li, F., Tartakoff, S.S., and Castle, S.V. (**2009**). *J. Org. Chem. 74*: 9082–9093.
24 King, S.M., Calandra, N.A., and Herzon, S.B. (**2013**). *Angew. Chem. Int. Ed. 52*: 3642–3645.
25 Herzon, S.B., Calandra, N.A., and King, S.M. (**2011**). *Angew. Chem. Int. Ed. 50*: 8863–8866.
26 Cui, C.B., Kakeya, H., and Osada, H. (**1996**). *J. Antiobiot. 49*: 832–835.
27 Cui, C.B., Kakeya, H., and Osada, H. (**1996**). *Tetrahedron 52*: 12651–12666.
28 Von Nussbaum, F. and Danishefsky, S.J. (**2000**). *Angew. Chem. Int. Ed. 49*: 2175–2178.
29 Sebahar, P.R. and Williams, R.M. (**2000**). *J. Am. Chem. Soc. 122*: 5666–5667.
30 Wamg, H. and Ganesan, A. (**2000**). *J. Org. Chem. 65*: 4685–4693.
31 Meyers, C. and Carreira, E.M. (**2003**). *Angew. Chem. Int. Ed. 42*: 694–696.
32 Trost, B.M. and Stiles, D.T. (**2007**). *Org. Lett. 9*: 2763–2766.
33 Overman, L.E. and Rosen, M.D. (**2000**). *Angew. Chem. Int. Ed. 39*: 4596–4599.

Index

Spiro Compounds: Synthesis and Applications, First Edition. Edited by Ramon Rios Torres.

d